Current research on the origin and evolution of active galaxies is comprehensively surveyed in this collaborative volume. Both of the proposed types of central activity – active galactic nuclei and starbursts – are analysed with a particular emphasis on their relationship to the large-scale properties of the host galaxy. The crucial question is what triggers and fuels nuclear activity now and at earlier epochs? The topics covered here are gas flows near to massive black holes, the circumnuclear galactic regions, and the large-scale bars in disk galaxies. Aspects of nuclear bursts of star formation, and the relationship between central activity and the gas and stellar dynamics of the host galaxy are addressed as well. The contributors to this book for professionals and graduate students are world experts on galaxy evolution.

Mass-Transfer Induced Activity in Galaxies

Mass-Transfer Induced Activity in Galaxies

Edited by
ISAAC SHLOSMAN
University of Kentucky

PUBLISHED BY THE PRESS SYNDICATE OF THE UNIVERSITY OF CAMBRIDGE
The Pitt Building, Trumpington Street, Cambridge, United Kingdom

CAMBRIDGE UNIVERSITY PRESS
The Edinburgh Building, Cambridge CB2 2RU, UK
40 West 20th Street, New York NY 10011–4211, USA
477 Williamstown Road, Port Melbourne, VIC 3207, Australia
Ruiz de Alarcón 13, 28014 Madrid, Spain
Dock House, The Waterfront, Cape Town 8001, South Africa

http://www.cambridge.org

© Cambridge University Press 1994

First published 1994
First paperback edition 2003

A catalogue record for this book is available from the British Library

ISBN 0 521 47195 8 hardback
ISBN 0 521 54330 4 paperback

TABLE OF CONTENTS

III. THE CIRCUMNUCLEAR REGION

IV. GAS DYNAMICS AND STAR FORMATION IN BARRED AND NORMAL GALAXIES

V. NUCLEAR GAS AND LARGE-SCALE PROPERTIES OF AGN AND STARBURST HOSTS

VI. HOST GALAXY–AGN–NUCLEAR STARBURST CONNECTION

VII. GALAXY INTERACTIONS AND INDUCED ACTIVITY

VIII. GAS DYNAMICS IN ELLIPTICALS

IX. AGN AND STARBURST HOSTS AT LARGE REDSHIFTS

X. CONFERENCE SUMMARY

PREFACE

Where was that order and whence came this mocking illusion?
Stanislaw Lem, *Ananke*

This book includes contributions to the International Astrophysics Conference on *Mass–Tranfer Induced Activity in Galaxies,* held at the University of Kentucky, Lexington, on April 26–30, 1993. More than 140 participants from 17 countries attended, compared to 70 planned originally. We feel that such interest was *fueled* at least in part by the Conference being solely devoted to this fascinating subject.

This Conference was based on a number of review talks which prepared the audience for the follow-up discussion and contributed papers. We aimed at providing a balanced view of the field which by now has become well defined, reaching a certain degree of maturity. Some overlap between different review papers was planned and the careful reader will enjoy sometimes contradictory explanations of the same 'facts'.

The main question concerning the origin of active galaxies is tantalizingly simple: how does nature remove all but $\sim 10^{-7}$ of the angular momentum (in AGNs) initially residing in the gas a few kpc from the center, accomplishing this on the orbital timescale? And how do the nuclear starbursts fit within this picture — as a passive by-product or a major player forcing the gas inwards and/or outwards?

We proceed from 'inside out' (in the footsteps of the Creator?) by addressing first the angular momentum transfer within the central parsec. Next, the kinematics and molecular, atomic and ionized gas distributions in the circumnuclear region are reviewed followed by large-scale gas properties in barred and normal disk galaxies and in ellipticals. Galaxy interactions, the potential trigger of central activity were discussed at almost all sessions. And finally, the cosmological aspect of active galaxies which is closely related to the galaxy formation process was presented as well. Both observational, theoretical, and computational aspects of the problem have been extensively debated. The Conference, in fact, was summarized twice: during the opening remarks by Sterl Phinney and at the closing, by Richard Larson.

This Conference did not, of course, solve the problem of possible causal relationship between nuclear starbursts and Seyfert-like activity in galaxies, but it exposed this issue almost to the maximum possible. It highlighted some apparent misconceptions in our understanding of gas flows in barred galaxies and star formation in stellar bars. Probably more important than anything else, it resulted in more questions than answers provided. Although Proceedings should adequately represent the spirit of this meeting, one thing will be missing — the inspiring banquet speech given by Geoff Burbidge, which was definitely one of the highlights of the week.

There is a general feeling that real progress has been made during the last few years in understanding the whole problem. However, based primarily on his wishful thinking, this Editor would predict that a decade from now the issue of fueling the central activity in galaxies would still excite *not* just historians of astronomy ...

I am grateful to members of the International Scientific Committee for their help in soliciting the list of invited speakers: M. Begelman (USA), R. Blandford (USA), J. Frank (USA), T. Heckman (USA), R. Larson (USA), H. Netzer (Israel), M. Rees (Britain), I. Shlosman (USA), S. Tremaine (Canada), and J. van Gorkom (USA). I am indebted to Juhan Frank and Mitch Begelman for being ready to engage in hour-long telephone conversations at any time, and to Martin Rees for good advice and help all the way through.

This Conference would not take place without efficient help from members of the Local Organizing Committee and especially from Moshe Elitzur — who, it is my guess — should by now regret his policy of open doors. I also learned a lot from Tom Troland when going over the wine and beer list for the reception and the banquet, which made the whole process especially pleasant. Karen Heller helped to prepare the cover page of the *Abstracts* booklet and Clayton Heller designed the Conference poster and solved many other urgent problems.

This meeting was sponsored by the University of Kentucky, which together with NASA provided the financial support. We thank Dr. Lee Magid (UK Vice–President for Research and Graduate Studies) and Dr. John Connolly (Director of the Center for Computational Studies) for their encouragement and generous support. Special thanks go to my wife Cilia who arranged a trio and played for us during the *hors d'oeuvres* at the Conference banquet, and to my children Irma and Raphael for their patience and understanding during the preparation of this Conference and Proceedings.

The TeX macros used in these Proceedings were adopted from a package generously given to us by Christopher Mauche (LLNL).

October, 1993

Isaac Shlosman
Lexington

LIST OF PARTICIPANTS

Victor Andersen .... University of Alabama (USA)
Santiago Arribas .... Instituto de Astrofisica de Canarias (Spain)
Lia Athanassoula .... Observatoire de Marseille (France)
Dinshaw Balsara .... Johns Hopkins University (USA)
Stefi Baum .... Space Telescope Science Institute (USA)
Sara Beck .... Tel-Aviv University (Israel)
John Beckman .... Instituto de Astrofisica de Canarias (Spain)
Mitchell Begelman .... JILA, University of Colorado (USA)
Kenji Bekki .... Tohoku University (Japan)
Suketu Bhavsar .... University of Kentucky (USA)
Jonathan Bland-Hawthorn .... Rice University (USA)
Paul Bode .... Indiana University (USA)
Berto Boer .... Sterrewacht Leiden (Netherlands)
Carlotta Bonoli .... Osservatorio Astronomico di Padova (Italy)
Kirk Borne .... Space Telescope Science Institute (USA)
Albert Bosma .... Observatoire de Marseille (France)
Greg Bothun .... University of Oregon (USA)
Geoffrey Burbidge .... University of California at San Diego (USA)
Jordan Burkey .... University of Alabama (USA)
Gene Byrd .... University of Alabama (USA)
Eugene Capriotti .... Michigan State University (USA)
John Capriotti .... University of Minnesota (USA)
Timothy Carone .... SSL, UC Berkeley (USA)
Gerald Cecil .... University of North Carolina (USA)
Arthur Chernin .... Moscow State University (Russia)
Dimitris Christodoulou ...... University of Virginia (USA)
Luis Colina .... Universita Autonoma de Madrid (Spain)
Francoise Combes .... Meudon Observatoire (France)
Roger Coziol .... Universite de Montréal (Canada)
Anna Curir .... Osservatorio Astronomico di Torino (Italy)
Julianne Dalcanton .... Princeton University (USA)
Duilia de Mello Rabaca ...... University of Alabama (USA)
Nick Devereux .... New Mexico State University (USA)
Arjun Dey .... University of California at Berkeley (USA)
Tim de Zeeuw .... Sterrewacht Leiden (Netherlands)
Matthias Dietrich .... Universitatssternwarte Göttingen (Germany)
George Djorgovski .... California Institute of Technology (USA)
Rene Doyon .... Universite de Montréal (Canada)

Linda Dressel .... Applied Research Corporation (USA)

Deborah Dultzin-Hacyan ..... Instituto de Astronomia, UNAM (Mexico)

Wolfgang Duschl .... IfTA, Heidelberg (Germany)

Ifeanyi Ekejiuba .... Federal University of Technology (Nigeria)

Moshe Elitzur .... University of Kentucky (USA)

Brian Espey .... University of Pittsburgh (USA)

Heino Falcke .... MPI für Radioastronomie, Bonn (Germany)

Gary Ferland .... University of Kentucky (USA)

David Fisher .... University of California at Santa Cruz (USA)

Juhan Frank .... Louisiana State University (USA)

Daniel Friedli .... Geneva Observatory (Switzerland)

Jack Gallimore .... Space Telescope Science Institute (USA)

Prab Gondhalekar .... Rutherford Appleton Laboratory (Britain)

Ignacio Gonzalez-Serrano ..... Universidad de Cantabria (Spain)

Loretta Gregorini .... Inst. di Radioastronomia, Bologna (Italy)

Asao Habe .... Hokkaido University (Japan)

Fred Hamann .... Ohio State University (USA)

Timothy Heckman .... Johns Hopkins University (USA)

Clayton Heller .... University of Kentucky (USA)

Lars Hernquist .... UC Santa Cruz (USA)

Jarita Holbrook .... NASA/Goddard Space Flight Center (USA)

Jie H. Huang .... Nanking University (China)

Robert Hurt .... UC Los Angeles (USA)

Judith Irwin .... Queen's University (Canada)

Zeljko Ivezic .... University of Kentucky (USA)

Jelle Kaastra .... SRON Leiden (Netherlands)

Michele Kaufman .... Ohio State University (USA)

Demosthenes Kazanas .... NASA/Goddard Space Flight Center (USA)

William Keel .... University of Alabama (USA)

Jeffrey Kenney .... Yale University (USA)

Robert Kennicutt .... Steward Observatory (USA)

John Kielkopf .... University of Louisville (USA)

Mario Klaric .... University of Alabama (USA)

Johan Knapen .... Instituto de Astrofisica de Canarias (Spain)

Chris Kochanek .... Harvard University (USA)

Wolfram Kollatschny .... Universitatssternwarte Göttingen (Germany)

Julian Krolik .... Johns Hopkins University (USA)

Ari Laor .... IAS, Princeton (USA)

Richard Larson .... Yale University (USA)

Eija Laurikainen .... Turku University Observatory (Finland)

Joanna Lees .... University of Chicago (USA)

John MacKenty .... Space Telescope Science Institute (USA)

Roberto Maiolino .... Universita di Firenze (Italy)
Leonid Marochnik .... Computer Science Corporation (USA)
Andre Martel .... UC Santa Cruz (USA)
Pierre Martin .... Steward Observatory (USA)
Paolo Marziani .... University of Alabama (USA)
Smita Mathur .... Center for Astrophysics, Cambridge (USA)
Paolo Mazzei .... Osservatorio Astronomico di Padova (Italy)
Evencio Mediavilla .... Instituto de Astrofisica de Canarias (Spain)
Chris Mihos .... UC Santa Cruz (USA)
Richard Miller .... NASA Headquarters, Washington, DC (USA)
Felix Mirabel .... Service d'Astrophysique, Saclay (France)
Charles Nelson .... University of Virginia (USA)
Masafumi Noguchi .... Tohoku University (Japan)
Ray Norris .... Telescope National Facility (Australia)
Chris O'Dea .... Space Telescope Science Institute (USA)
Paolo Padovani .... II Universita di Roma (Italy)
Sterl Phinney .... California Institute of Technology (USA)
Shlomi Pistinner .... Technion (Israel)
Alice Quillen .... California Institute of Technology (USA)
Carlos Rabaca .... University of Alabama (USA)
Mario Radovich .... Universita di Padova (Italy)
Roberto Rampazzo .... OA di Brera, Milano (Italy)
Andrew Read .... University of Birmingham (Britain)
Luca Reduzzi .... OA di Brera, Milano (Italy)
Gail Reichert .... NASA/Goddard Space Flight Center (USA)
Nico Roos .... Sterrewacht Leiden (Netherlands)
Elaine Sadler .... Anglo–Australian Observatory (Australia)
Heikki Salo .... University of Oulu (Finland)
Hartmut Schulz .... Ruhr-Universität, Bochum (Germany)
Nick Scoville .... California Institute of Technology (USA)
Alla Shapovalova .... Special Astrophysical Observatory (Russia)
Martin Shaw .... University of Sheffield (Britain)
Giora Shaviv .... Technion (Israel)
Joseph Shields .... Ohio State University (USA)
Isaac Shlosman .... University of Kentucky (USA)
Susan Simkin .... Michigan State University (USA)
Michael Sitko .... University of Cincinnati (USA)
Eric Smith .... NASA/Goddard Space Flight Center (USA)
Steven Smith .... HAO/NCAR, Bouder (USA)
Ruggero Stanga .... Universita di Firenze (Italy)
Wayne Stein .... University of California at San Diego (USA)
Yoshiaki Taniguchi .... Tohoku University (Japan)

Jason Taylor ....NASA/Goddard Space Flight Center (USA)
Robert Thomson ....Institute of Astronomy, Cambridge (Britain)
Tom Troland ....University of Kentucky (USA)
Jean Turner ....UC Los Angeles (USA)
Wil van Breugel ....IGPP/Livermore National Lab. (USA)
Frank van den Bosch ....Sterrewacht Leiden (Netherlands)
Roeland van der Marel ......Sterrewacht Leiden (Netherlands)
Sylvain Veilleux ....NOAO/KPNO (USA)
Giampoalo Vettolani ....Instituto di Radioastronomia, Bologna (Italy)
Mark Voit ....California Institute of Technology (USA)
Susanne von Linden ....MPI für Radioastronomie, Bonn (Germany)
Frederick Vrba ....U. S. Naval Observatory, Flagstaff (USA)
Keiichi Wada ....Hokkaido University (Japan)
Zhong Wang ....California Institute of Technology (USA)
Rainer Wehrse ....IfTA, Heidelberg (Germany)
Melinda Weil ....UC Santa Cruz (USA)
Simon White ....Institute of Astronomy, Cambridge (Britain)
Mark Whittle ....University of Virginia (USA)
Paul Wiita ....Georgia State University (USA)
Xiao Lei Zhang ....Center for Astrophysics, Cambridge (USA)

Mass-Transfer Induced Activity in Galaxies: an Introduction

E. S. Phinney

Theoretical Astrophysics, 130-33 Caltech, Pasadena, CA 91125, U. S. A.

ABSTRACT

Galaxies with elevated metabolic rates get energy from their gaseous food by extracting its nuclear energy (in stars), and its gravitational energy (via accretion onto massive black holes). There is strong evidence that interactions with other galaxies trigger star formation activity, and weaker evidence that it triggers black hole accretion (nuclear activity). We review the processes by which interactions can remove angular momentum from gas, particularly gravitational torques and the $m = 2$, $m = 1$, Jeans and fission instabilities that give rise to them. There is ample evidence, both theoretical and observational, that these can remove enough angular momentum to move much of a galaxy's gas from ~ 3 kpc to ~ 300 pc. This is still many decades from the $\sim 10^{-5}$ pc scales of stars and black hole horizons. We discuss star formation, the interpretation of simulations, and cosmological implications. The evolution of binary supermassive black holes, and the problem of forming a dense ($\lesssim 1$ pc) nuclear star cluster are examined.

1 WHAT IS MASS-TRANSFER INDUCED ACTIVITY IN GALAXIES?

Before this conference, I wasn't sure. After this conference, I am sure I am not sure. Let me nevertheless attempt a definition, starting from the easy end, the back of the phrase. *Galaxies* are of course the island universes within which reside most of the stars, much of the gas, and a little of the mass in the cosmos. *Activity in Galaxies*, like that in animals, is defined by the metabolic rate. When this is well above the average 'resting' (aka. basal) level, a galaxy or animal is said to be active. For an animal of mass m kg, the mass-specific basal metabolic rate is about $m^{-1/4}$ cal s^{-1}kg^{-1} (Kleiber 1932). For galaxies, the mass-specific basal metabolic rate seems not to be strongly mass dependent, and is about $0.1 - 1$ L$_\odot$ M$_\odot^{-1}$ ($5 \times 10^{-6} - 5 \times 10^{-5}$ cal s^{-1}kg^{-1}; 1 L$_\odot$ M$_\odot^{-1}$ is the average rate of energy release per unit mass of cosmic baryons which fuse 10% of their hydrogen in a Hubble time). Animal athletes, such as horses running the Kentucky Derby, can reach metabolic rates 20 times their basal rate. Galaxy athletes like quasars and extreme star bursts are even more impressive, reaching $\gtrsim 10^3$ times their basal rates ($\gtrsim 10^5$ times in selected regions). Activity in animals is *induced* by many things: fright, anger, desire for food or sex ... . What induces activity in galaxies is less well understood, but collisions, tidal encounters, force-feeding and various instabilities are plausibly and commonly invoked.

We finally come to *mass-transfer*. Activity in animals is of course ultimately made possible by mass transfer, more commonly called eating. The work and heat generated during activity results from the metabolism of complex carbon molecules in food into CO_2 and water, releasing $\sim 5,000$ kcal per kg of food. The winds, jets and radiation characteristic of galaxy activity are believed to be powered mainly by two quite distinct metabolic pathways: fusion of hydrogen in stars ($\sim 2 \times 10^{10}$ kcal per kg of ZAMS stars), and accretion onto black holes and neutron stars ($\sim 2 \times 10^{12}$ kcal per kg accreted). The subtlety and major uncertainties are in the *mechanism* by which mass is transferred from the galactic scale on which it is distributed to the very much smaller scales of stars and black hole horizons on which it is metabolized.

The difficulty is in removing specific angular momentum. Material at the last stable orbit of a $10^8 M_8$ $M_\odot$ black hole has a specific angular momentum $\ell \sim 2 \times 10^{24} M_8$ cm^{2}s^{-1}, while material orbiting at the ~ 3 kpc scale of a typical galaxy has $\ell \sim 10^{29}$ cm^{2}s^{-1}, some five orders of magnitude larger. Similarly, material orbiting the surface of the sun has $\ell = 3 \times 10^{18}$ cm^{2}s^{-1}, while a 1 $M_\odot$ piece of even a perfectly smooth 10^9 $M_\odot$ disk of gas 3 kpc in radius has $\ell \sim 2 \times 10^{20}$ cm^{2}s^{-1}, and a 1 $M_\odot$ piece of a realistically clumpy and turbulent disk has ℓ one or two orders of magnitude larger. So if gas is to accrete onto a black hole, or form into stars, its angular momentum must be removed almost completely.

The feeding problem is illustrated in Figure 1. Animals have limbs, claws and teeth which they use in easily observed ways to funnel food from the wide world into their digestive tracts. The limbs, claws and teeth of galaxies unfortunately appear to be much less easily observed gravitational forces, magnetic stresses, and turbulent eddies. As you read the papers which follow, you will see that much progress has been made in predicting and observing how the outer one or two orders of magnitude of angular momentum are removed. The next few orders of magnitude are proving much more resistant.

2 INDUCED ACTIVITY IN GALAXIES

At the 'instant' defined by astronomers' past light cone, only a small fraction of galaxies have the extreme metabolic rate by which we define activity. Active galaxies do not appear to be special in their large-scale characteristics or environments (*e.g.* active galaxies are not all cluster-dominant cD galaxies at the centers of cooling flows), so it is natural to suppose that they are simply 'average' galaxies which were induced to become active by some recent accident.

2.1 Observations favoring induced activity

In Reader's Digest form, the observations adduced in support of the hypothesis of induced activity are as follows. Interacting galaxies have higher star formation rates than non-interacting galaxies (Lonsdale *et al.* 1984; Hummel *et al.* 1990; Keel and van Soest 1992). About 1/5 of elliptical galaxies with shells look as if they have just

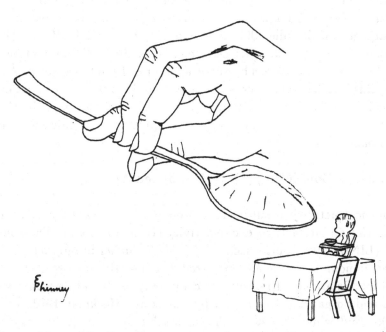

Fig. 1—The problem of feeding the monster: a large (angular momentum) spoon and a small (angular momentum) mouth. Hands and teeth (gravitational and magnetic forces, viscosity, . . .) are needed to guide and divide the food into morsels that can be metabolized during activity.

had a nuclear starburst (Carter *et al.* 1988). Barred galaxies also have a higher star formation rate than unbarred galaxies, as shown by enhanced Hα and far infrared flux (Kennicutt *et al.* 1987) and by enhanced radio emission from H II regions (Hummel *et al.* 1990). Indeed, most starburst galaxies (compared to only 30-50% of all disk galaxies) seem to be barred (Fricke and Kollatschny 1989; Devereux, these proceedings). Most ultraluminous *IRAS* galaxies appear to be recent mergers of galaxies — suggested by their tidal tails (Sanders *et al.* 1988; Melnick and Mirabel 1990) and double nuclei (Majewski *et al.* 1993). *IRAS* 19254–7245, the "superantennae", is the epitome of this class of objects — with one starburst nucleus, one Seyfert nucleus, and a pair of enormous 350 kpc tidal tails (Mirabel *et al.* 1991). A large fraction of classical double radio galaxies (Fanaroff-Riley type II radio galaxies) also show tidal tails and other evidence of interaction (Heckman *et al.* 1986; Smith and Heckman 1989; Colina and Perez-Fournon 1990; Baum *et al.* 1992). Since quasars outshine their galaxies, interactions would be harder to see, but a growing body of evidence suggests that quasars also have been involved in interactions (Hutchings and Neff 1992).

These observations suggest that an interaction with another galaxy, or the recent growth of a bar, may provide a trigger for activity in what would otherwise be a normal galaxy. There are striking concentrations of dense gas ($\gtrsim 10^9$ M$_\odot$, a Milky-Way's worth of gas, in $\lesssim 1$ kpc) in the central regions of interacting and starburst galaxies (Sargent and Scoville 1991; Solomon, Downes, and Radford 1992), and in barred galaxies (Kenney *et al.* 1992; Devereux *et al.* 1992). So interactions and bars both seem able to concentrate gas by about a factor of 10 in linear scale. As discussed in section 1, this is still several orders of magnitude shy of the concentration required to cause star formation directly, or to permit accretion onto a central black hole. But it is a respectable first step which might — or might not — unleash an inexorable cascade to smaller scales.

2.2 Observations Conflicting (?) with Induced Activity

The empirical picture just outlined is not obviously supported by all observational correlations. Not all interactions excite activity (Bushouse 1986). The abundance of nonthermal (LINER-type) nuclei seems only *slightly* higher in interacting and barred galaxies than in noninteracting and unbarred galaxies (Hummel *et al.* 1990). Seyfert nuclei and LINERs are found much more commonly in S0 and Sa galaxies, and only rarely in Sc and Sd galaxies (see review by Balick and Heckman 1982; also Hummel *et al.* 1990). And while almost all nuclear starbursts seem to be associated with bars (Devereux, these proceedings), Seyfert galaxies do not seem to be preferentially barred (Balick and Heckman 1982). Starburst galaxies with Seyfert nuclei have similar environments to starburst galaxies without Seyfert nuclei (MacKenty 1989). Finally, of the most isolated and non-interacting of galaxies, the giant low surface brightness galaxies, all (three out of three) seem to have Seyfert nuclei (Bothun, these proceedings)!

These results are not necessarily in conflict with the idea that activity is induced by bars and interactions. While a majority of interactions seem to be able to induce star formation, some speeds and orientations of interaction may be much more effective at inducing it than others. It is much less clear from the empirical evidence that bars and interactions affect non-stellar nuclear activity, at least at the level of LINERs and Seyferts. Not all galaxies may have suitable rotation curves or amounts or locations of gas for the trigger to have much effect. There may be a considerable delay between the triggering event and the onset of activity, by which time the companion may have passed on or been digested. Apologetics along these lines have been developed by Byrd *et al.* (1987), though the observational situation does not encourage the simplest models (Keel 1993, and see below). Finally there may be other triggers of activity besides bars and interactions (*e.g.* spontaneous self-propagating star formation for starbursts; dynamical evolution of a nuclear star cluster for AGN). To interpret the observed correlations, we need more quantitative models.

2.3 The Canonical Theoretical Model of Induced Activity

The minimal model for induced activity is perhaps the following sequence: a passing companion produces a tidal perturbation, predominantly quadrupole ($m = 2$), in a disk galaxy containing gas as well as stars. Even if the galaxy was stable against growth of infinitesimal $m = 2$ perturbations, the finite amplitude tidal perturbation can cause growth of a bar (Noguchi 1987; Sellwood 1989; Hernquist 1989; Gerin *et al.* 1990).

In the bar potential, the periodic orbits are crowded and may self-intersect, so that gas in the bar forms shocks on the leading sides of the stellar bar (*e.g.* Athanassoula 1992). The gas density is therefore larger on the leading side of the stellar bar. Consequently the gravitational torque from the stellar bar removes angular momentum from the gas, and the gas sinks in the potential (*e.g.* Wada and Habe 1992). As the gas accumulates, it may become self-gravitating, and fragment into one or more smaller lumps. If many lumps are formed, they can exchange angular momenta, leading to further concentration of a fraction of the gas (Christodoulou and Narayan 1992; Wada and Habe 1992, Christodoulou 1993).

This model is just about able to account for the observed concentrations of molecular gas, discussed at the end of section 2.1. The further orders of magnitude concentration required for activity are generally covered by prayer and handwaving — though Shlosman *et al.* (1989), following the maxim that one can never get too much of a good thing, proposed for AGN fuelling that the concentrated gas would be unstable to bar formation, and the process outlined above would repeat itself on successively smaller scales.

3 MODELS OF THE FIRST STAGE OF MASS TRANSFER (~ 3 KPC TO ~ 300 PC)

3.1 The Significance of Resonances

Suppose circular orbits in the axisymmetric potential of a galaxy have angular velocities $\Omega(r)$, where r is the distance from the axis of symmetry. Then not-quite-circular orbits can be described as the superposition of a circular motion about a guiding center r_c at $\Omega(r_c)$, and motion on a small epicyclic ellipse about the guiding center, at an angular frequency κ, where

$$\kappa^2 = \Omega^2 \left(4 + \frac{d \ln \Omega^2}{d \ln r} \right) \tag{1}$$

$$\kappa = \begin{cases} \Omega & \text{Keplerian point mass: } GM = \Omega^2 r^3 = \text{const;} \\ \sqrt{2}\Omega & \text{Flat rotation curve: } v = \Omega r = \text{const;} \\ 2\Omega & \text{Harmonic core potential: } \Omega^2 = (4\pi/3)G\rho = \text{const.} \end{cases} \tag{2}$$

The first and last results in (2) are obvious: the general orbit in a Keplerian potential is an off-center ellipse fixed in space, so the epicyclic frequency must be equal to the orbital frequency; and the general orbit in a harmonic potential is a centered ellipse fixed in space, so the epicyclic frequency must be twice the orbital frequency. For other potentials, the general orbit is a rosette (Bertrand's theorem).

As viewed from a reference frame rotating at angular frequency Ω_p, the guiding center lies at angle $\phi = (\Omega - \Omega_p)t + \text{const}$. So if $\Omega \equiv \Omega(r_c) = \Omega_p$, a particle on the orbit simply executes epicyclic oscillations about a fixed ϕ in that frame. If $\Omega_p = \Omega \pm \kappa/m$, then $\phi = \mp \kappa t/m + \text{const}$ and the radial displacement from the guiding center

$$\delta r \propto \text{Re } e^{i\kappa t} \propto \text{Re } e^{\mp im(\phi + \text{const})} \tag{3}$$

So for integer m, the rosette is transformed to an $m-$lobed closed orbit in the rotating frame. For $m = 2$, the orbits have the symmetry of a quadrupole tidal or bar perturbation, and the resonances are named

$$\Omega_p = \begin{cases} \Omega(r) + \kappa(r)/2 & \text{outer Lindblad resonance;} \\ \Omega(r) - \kappa(r)/2 & \text{inner Lindblad resonance.} \end{cases} \tag{4}$$

Besides the Lindblad resonances, there is another important resonance, the corotation resonance,

$$\Omega_p = \Omega(r) \qquad \text{corotation resonance.} \tag{5}$$

If a particle is displaced azimuthally rather than radially on its orbit, it will remain so displaced, since there is no restoring force. Thus in the guiding center frame, the natural frequency for azimuthal displacements is zero, while for radial displacements it is κ. This suggests that the orbits of stars and gas elements in their guiding center frame can be thought of as oscillators whose resonant frequencies for small amplitude oscillations are 0 and $\kappa(r_c)$.

Suppose now that the gravitational potential of the galaxy is not quite axisymmetric, perhaps because of a passing galaxy or a weak bar or spiral disturbance. Then the motion of the epicyclic oscillations will be driven by the perturbing, non-axisymmetric force. If in an inertial frame the perturbing potential has the form $\Phi_p \exp[im(\phi - \Omega_p t)]$,* then in the frame of the (unperturbed) guiding center of an orbit with guiding center radius r_c, the perturbing potential will vary as

$$\Phi_p \exp[im(\phi + [\Omega(r_c) - \Omega_p]t)]. \tag{6}$$

When the angular frequency of the perturbing force in this frame,

$$\omega_p = m[\Omega(r_c) - \Omega_p] \,, \tag{7}$$

is near in magnitude to either of the resonant frequencies κ or 0, the epicyclic motions will resonate, and will be driven to large amplitudes. Therefore we expect, as is indeed

---

* We take Ω_p positive when the perturbing potential rotates in the same sense as the guiding center orbit. If Ω_p varies with time, $\Omega_p t$ should be replaced throughout by $\int \Omega_p \, dt$.

the case, that the radial displacement will have three resonant denominators:

$$\Delta r \propto \frac{\Phi_p}{(\omega_p - 0)(\kappa - \omega_p)(\kappa + \omega_p)}.$$ (8)

As usual for driven oscillators, the response changes sign across resonance — *i.e.*, the maximum of Δr changes in phase by π/m relative to the perturbing potential's axis across a Lindblad resonance (if the perturbation is time-dependent rather than steady, the phase is not discontinuous, but changes smoothly across the narrow range of radii near resonance where the rate of rotation of the orbit angles in the pattern frame is slower than the rate of change of the perturbation).

For a typical rotation curve, the orientation of the perturbed direct orbits and the locations and sign changes across the resonances are shown in Figure 2.

What happens to the orbits of stars when we grow an $m = 2$ perturbation? When the $m = 2$ perturbation has pattern speed Ω_p, but zero strength, then in a frame rotating at Ω_p, stars with guiding centers at corotation execute little epicyclic ellipses about a fixed position, and stars with guiding centers at one of the Lindblad resonances execute elongated orbits with quadrupole symmetry about a fixed axis (see discussion above eq. [4]). In the same frame, stars with guiding centers near but not at corotation execute epicyclic ellipses whose centers slowly drift [at $\Omega(r_c) - \Omega_p$] about the origin, while stars with guiding centers near one of the Lindblad resonances execute elongated orbits whose axis of symmetry slowly rotates. When $\Phi_p = 0$, the angles describing these off-resonance orbits are like pendula rotating end-over-end in a zero gravitational field.* The pendula corresponding to orbits near resonance are rotating very slowly.

Now increase the strength of the $m = 2$ perturbation. This corresponds to increasing the gravitational field in which the pendula are rotating. At some critical gravitational field in this analogous system, the most slowly rotating pendula will find that they lack the energy to reach the vertical upside-down position, and they will start to oscillate instead of rotate. The same thing happens to the orbits of particles with guiding centers at some given distance from a resonance. If the resonance is corotation, at some critical value of Φ_p (which depends mostly on the distance from resonance, but also on the initial phase of the orbit), the centers of the epicyclic ellipses of a particle will cease to rotate slowly around the origin in the Ω_p frame, and will begin to oscillate back and forth about one of the Lagrange points. If the resonance is instead a Lindblad resonance, at some critical Φ_p, the orientation of the elongated elliptical orbit in the Ω_p frame will cease to rotate slowly, and will begin to librate about an axis either parallel or perpendicular to the axis of the perturbing potential. As the strength of the perturbing potential increases, orbits with guiding centers farther and farther from resonance will be trapped into libration. The range

---

* This analogy is mathematically quite exact for the corotation resonance, and captures the essential physics of the Lindblad resonances as well, though the pendulum analogy for them is not exact.

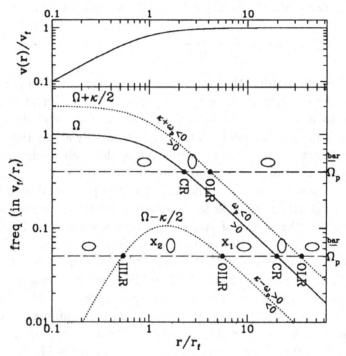

Fig. 2—Upper panel: rotation curve which asymptotes to a flat rotation curve with speed v_f and has rigid body rotation inside radius r_f. Lower panel: $m = 2$ pattern speeds which match the epicyclic (dotted) and corotation (solid) resonant frequencies, equations (4) and (5), as a function of guiding center radius. For two radius-independent pattern speeds, (horizontal dashed lines) the locations of the resonances are marked by dots and labelled OLR: outer Lindblad resonance, CR: corotation resonance, OILR: outer inner Lindblad resonance, IILR: inner inner Lindblad resonance. The ellipses indicate the orientations of orbits perturbed by a weak horizontal bar, rotating in the same sense as the orbits [see equations (8) and (7)]. Note how the orbit orientation changes by $\pi/2$ across each resonant curve, dividing the space into four regions. For strong bars, the two families of stable orbits inside corotation are denoted x_1 and x_2 as indicated.

of r_c over which orbits are trapped defines the 'width' of a resonance, and increases with the strength of the perturbing potential. When these and other higher order resonances overlap, particles are asked to librate in multiple incompatible ways, and their orbits often become chaotic. This may help concentrate gas in galaxies with very massive black holes (Friedli and Benz 1993).

3.2 Bars from Stars

The disks of galaxies containing both gas and stars seem to keep themseves on the verge of local instability (Kennicutt 1989), with Toomre's $Q = \sigma_r \kappa / (3.36 G \Sigma) \simeq 1$. Cooling of the gas reduces Q until instabilities begin to grow. These increase the stellar velocity dispersion and perhaps also lead to star formation and heating of the gas, increasing Q and suppressing the instabilities. Such a disk is, however, violently unstable to global non-axisymmetric modes, in particular to the $m = 2$ bar mode, unless it is stabilized by a massive spheroid (cf. review in Binney and Tremaine 1987).

The bar instability is ubiquitous in numerical simulations —indeed, makes it hard to maintain unbarred model galaxies, and to understand why two thirds of real disk galaxies have no very noticeable bar (see *e.g.*, Fig. 3 of Sellwood and Wilkinson's nice review [1993]). Analytic understanding of this instability, and therefore of the conditions which might suppress it, is still rather controversial. Different mechanisms may dominate in different circumstances.

What is now the textbook description of the bar instability is due to Toomre (1981), and is illustrated in the space-time diagram of Figure 3. The amplification cycle described in the caption of Figure 3 works in linear theory only if there is no inner Lindblad resonance. In other words, it works only if the pattern speed is sufficiently high (*cf.*, Fig. 2) and the galaxy has a constant density core. This is because the nonaxisymmetric perturbation will excite epicyclic motion of stars near the ILR, effectively absorbing the perturbation. However, Sellwood (1989) has shown that finite amplitude waves can saturate the resonance (by trapping orbits in libration) so that it is no longer an efficient absorber of the ingoing wave. Finite amplitude perturbations may thus grow even in the presence of an ILR.

The properties of the pre-perturbation galaxy may be hard to determine from the final, strongly barred state, however. The pattern speed of the bar may have been reduced by dynamical friction, and the bar often perturbs the potential so much that a 'rotation curve' and its naïvely computed resonance positions have little meaning. The presence or absence of an ILR generalizes to the presence or absence of stable orbits of the x_2 family (see Fig. 2). But even bars which form in galaxies which initially have no ILR (Sellwood 1981) might have x_2 orbits, though I am not aware that this has been demonstrated.

This picture of bar growth predicts, and numerical simulations of disks supported by rotation confirm, that the bar should end just inside corotation, and this will generally lie at 1–10 times the radius where the rotation curve flattens (see Fig. 2) — as observed in early-type spiral galaxies (Elmegreen and Elmegreen 1985).

A quite different type of bar growth was introduced by Lynden-Bell (1979). This is perhaps most easily visualized in a stellar disk supported by radial velocity dispersion rather than by rotation. Such a disk will be "Jeans unstable" in the azimuthal direction (a two-dimensional analogue of the radial orbit instability of spherical stellar systems formed by collisionless collapse — *cf.* Barnes *et al.* 1986). This mechanism leads to slowly rotating bars, which can end well inside corotation. A spirited defense of the relevance of such bars to real galaxies is given by Pasha and Polyachenko (1993). However this mechanism is likely to dominate only if disks formed by col-

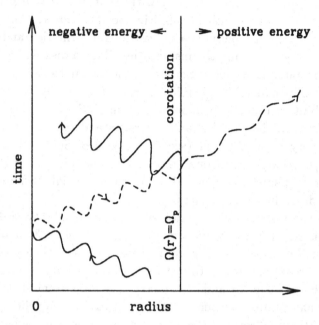

Fig. 3—Spiral density waves inside corotation have negative energy, and outside have positive energy. A weak nonaxisymmetric density perturbation moving inwards from corotation (lower solid curve) — for example, an $m = 2$ perturbation excited by a companion galaxy — will reflect at the center of the galaxy (if there is no ILR) as a leading wave (middle short dashed curve). Waves of very short wavelength cannot propagate through the corotation resonance, but waves of finite wavelength can, as in quantum mechanics, tunnel through the forbidden region. The tunneled wave (long dashed wave at right) has positive energy. Conservation of energy (more precisely, wave action) requires that the reflected part of the wave decrease its (negative) energy by an equal amount — *i.e.*, the reflected trailing wave (upper solid curve) has larger amplitude than the original wave. Repetition of the cycle leads to exponential growth of the nonaxisymmetric perturbation.

lisionless collapse of pancakes of stars — not popular in current cosmology, or in regions of solid-body rotation. The 'textbook' picture is also favored by the fact that the bar in the one galaxy whose bar pattern speed has been measured (Kent 1987, using the method introduced by Tremaine and Weinberg 1984) does seem to end near corotation. The offset dustlanes in barred galaxies also require the bars to end near corotation if the lanes are interpreted as the density maxima of simulations (Athanassoula 1992).

3.3 'Bars' from Black Holes

One way to guarantee a large $m = 2$ perturbation in the nucleus of a galaxy is to insert two supermassive black holes. Such a black hole binary will form in 3C 75 (a cD galaxy with a double nucleus, from each of which emerges a pair of radio jets — Owen *et al.* 1985), another in the similar 4-jet radio source 2149-158 (Parma *et al.* 1991), and others may form in the many ultraluminous *IRAS* galaxies with double nuclei (Majewski *et al.* 1993). That many of these have formed in the past is suggested by the ubiquitous evidence that a large fraction of elliptical galaxies, at least, have merged with other galaxies (Seitzer and Schweizer 1990), combined with evidence that a substantial fraction of all galaxies once contained active nuclei (Padovani *et al.* 1990).

A mass sinking by dynamical friction in a centrally concentrated spherical stellar system has its orbit circularized (Casertano, Phinney, and Villumsen 1987) because the drag is larger at pericenter than at apocenter. Thus Begelman *et al.* (1980) considered the evolution of a black hole binary, assuming that the second (captured) black hole would have had its orbit circularized before it became bound to the original black hole. They pointed out that the dynamical friction would become inefficient when the black holes' mutual binding energy exceeded the kinetic energy of the stars that approached them, and predicted that two 10^8 $M_\odot$ black holes would orbit each other at that radius for more than a Hubble time. Close binary black holes should then be common in galactic nuclei, with potentially spectacular effects on both gas and stars in their vicinity (Roos *et al.* 1993).

However, using the formalism of Casertano *et al.* (1987), it is easy to show that the orbit of a black hole in a constant density core keeps a constant shape as it shrinks under dynamical friction. If one adds a point mass to the potential, and continues to use the Chandrasekhar approximation, the same formalism predicts that the eccentricity of a slightly noncircular orbit will now *grow* as the orbit shrinks, $\epsilon \propto r^{-1.5} \propto t$. Direct integrations of the equations of motion, with dynamical friction included in the Chandrasekhar approximation (which assumes that stars passing near the black hole have the unperturbed density and velocity distribution, and have random and uncorrelated positions and velocities) show that the eccentricity continues to grow rapidly even when it is large. Fukushige *et al.* (1992) argued that the orbits of binary black holes in galactic nuclei would thus become so eccentric that gravitational radiation would become important and the black holes would merge in $\lesssim 10^7$ yr.

However, one might be suspicious of results obtained with the Chandrasekhar approximation for dynamical friction in these circumstances. As Begelman *et al.* (1980) already realized, as the black hole binary hardens, it loses so much energy and angular momentum as to significantly affect the orbits of the small fraction of the stars in the core which pass near enough to the binary to extract its energy and angular momentum. To study this, Jens Villumsen and I have performed several N–body simulations. We slowly grew one black hole in the center of a Plummer model star

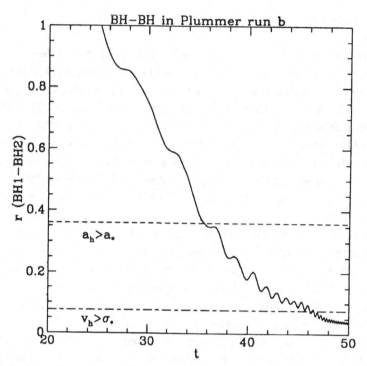

Fig. 4—Relative orbit of a black hole binary in an initially Plummer model distribution of 40,000 stars. The initial black hole, adiabatically grown at the center of the Plummer model, has a mass of 0.03, and the black hole infalling from $r = 2$ has a mass of 0.02 (in units of the total mass M_T in stars). The unit of length is the Plummer model scale length b. The time unit is $(b^3/GM_T)^{1/2}$. The gravitational forces have a softening length $\epsilon = 0.01$. Notice that the eccentricity of the orbit does not grow, in contrast to the claims of Fukushige *et al.* (1992). Below the upper dashed horizontal line, the acceleration of the black holes is dominated by their mutual attraction. Below the lower dashed line, the orbital velocity of the black holes exceeds the central velocity dispersion of the Plummer model stars. This is also approximately the radius below which the black hole binding energy exceeds the energy of all stars that pass near the binary, and loss-cone effects become important.

cluster (which thus developed a small cusp of stars around the black hole). We then dropped another black hole on an eccentric orbit from the outskirts of the Plummer model. The gravitational softening length was chosen to be much smaller than the final orbital radius of the black hole binary, so that 3-body encounters of stars with the binary were accurately integrated. One such simulation is illustrated in Figure 4. We find that the evolution is not at all well described by dynamical friction in the Chandrasekhar approximation (CA). In particular, beyond the point where the black hole becomes self-bound,

• The rate of orbit shrinkage is much slower than predicted by the CA. Dependence on 2-body relaxation sets in earlier than in the order-of-magnitude description of Begelman *et al.* (1980).

- The eccentricity of the orbit changes hardly at all. It does not grow rapidly, as predicted on the basis of CA by Fukushige *et al.* (1992). The original prediction that hard black hole binaries will be long-lived seems secure.
- The incoming black hole excites long-lived $m = 1$ oscillations of the core.

3.4 Instabilities with $m = 1$

Historically attention has focussed on $m = 2$ distortions (bar-like, or two-armed spirals). But other harmonics can be important too. Lopsided distortions, $m = 1$, have a displacement of the center from the center of mass. Distortions with $m = 1$ have a significance for point mass potentials similar to that of $m = 2$ distortions for harmonic (solid-body rotation) potentials. For the latter $\Omega - \kappa/2 \to 0$, while for the former, $\Omega - \kappa \to 0$ (but from negative frequency).*

The dominant global instability for self-gravitating disks around a central point mass tends to be an $m = 1$ distortion or fission instability (Shu *et al.* 1990; Christodoulou and Narayan 1992; Christodoulou 1993). Curiously, spherical non-rotating star clusters *without* central masses have a global $m = 1$ mode which is only weakly Landau-damped (as shown in the clever analytic analysis of Weinberg [1993]). Depending as one believes or not the dynamical evidence for a central black hole in M31 (Dressler and Richstone 1988; Kormendy 1988), either the growing mode or the weakly damped one could be responsible for the $m = 1$ mode in the nucleus of M31 discovered by Mould *et al.* (1989).

4 BEHAVIOR OF GAS

Unlike stars, gas particles can rapidly exchange momentum via pressure and viscosity, and can lose energy by radiative cooling. Unfortunately for simulators, gas is subject to thermal and gravitational instabilities which tend to drive much of the mass into dense molecular clouds. The gas also forms stars at poorly understood rates, which may depend on many things not usually simulated (magnetic fields, cosmic ray fluxes, ...). And the stars heat, clump, and give momentum to gas surrounding them. Simulations of these systems are thus at best schematic, and should perhaps be taken in the same spirit as computer simulations of national economies: general principles robust to the forms of the fudges adopted may emerge, but details should not be trusted — and even the general principles may ultimately prove less than robust.

---

* Realistic clothing of the point mass can prevent this, however. If the central black hole is surrounded by a star cluster with density $\rho \propto r^{-p}$, then $\kappa - \Omega \propto \Omega(r)r^{3-p} \propto r^{(3/2)-p}$. For a relaxed star cluster, $p = 7/4$, so $\kappa - \Omega \propto r^{-1/4}$, which diverges as $r \to 0$. For an unrelaxed cluster created by adiabatic growth of a black hole in a core, $p = 3/2$, and $\kappa - \Omega$ asymptotes to constant frequency.

Some precepts emerge from the considerations outlined in section 3.1.

- Equation (8) shows that the amplitude of the epicyclic 'pendula' will be large for strong perturbations, and rapidly changing near resonances. Pendula on neighboring guiding centers may thus collide — *i.e.*, orbits will intersect, and shocks will form in the cool, hence supersonic phases of the gas.

- Dissipation, whether by shocks or by viscosity means that the epicyclic pendula of gas particles, like any oscillator with damping, will in steady state have a phase lag with respect to the perturbation, and can thus exchange energy and angular momentum with the perturbation. The sign is generally such that gas is driven away from the corotation resonance.

- Trapping of orbits into libration depends on the rate of change of the perturbation potential, not just its amplitude. Past history can thus affect the final state of both stars and gas.

During the conference, we saw simulations of the effect of a passing galaxy and bar formation on disks containing gas and stars (Noguchi; Shlosman; White, these proceedings). Depending on the amount of gas, and the prescription adopted for gas cooling and star formation heating, these variously showed that

a) a stellar bar formed and drove the gas towards the ILR (if present) or nucleus.

b) the gas formed a ring, then fragmented into many compact lumps.

c) a massive binary molecular cloud formed by fission of a gas bar.

d) the original and gas-formed stars heated to make a thick disk or bulge.

Case *a* is a robust phenomenon in simulations of gas with negligible self-gravity and *imposed* non-axisymmetric potential (*cf.* Wada and Habe 1992 and references therein). Observations appear to be consistent with this phenomenon: if bar ends are identified with corotation, then rings of star formation and "twin-peaks" of molecular gas are commonly observed in the vicinity of the inferred OILR (Kenney *et al.* 1992; Devereux *et al.* 1992; Telesco *et al.* 1993). Gaseous (Zhang *et al.* 1993) and stellar (Friedli and Martinet 1993) bars are also observed inside the IILR. Simulations also show that gas outside corotation should be driven into a ring near the OLR (Schwarz 1981). Nature sometimes provides suggestive rings of star formation near the inferred OLR (*e.g.*, Thronson *et al.* 1989).

However the masses of gas deduced from CO and H I observations are such that self-gravity of the gas should be important. The Jeans or 'J' (case *b*) and fission or 'I' (case *c*) modes are both instabilities even of adiabatic single-phase gas tori (Christodoulou and Narayan 1992; Christodoulou 1993). Considering a sequence of gas tori of increasing density but fixed thickness in a centrally concentrated galaxy, the I mode would be the first gravitational instability to appear. However, instability in the J mode does not require much higher density, so it is perhaps not surprising that simulations with gas cooling (which by itself induces small-scale density fluctuations) and less idealized initial conditions than smooth tori sometimes exhibit the J-mode, case *b*, as well.

5 STAR FORMATION

Star formation and star-gas interactions are undoubtedly the place where simulations diverge most rapidly from reality. Divergences are likely to be important. Not simply because forming stars is one of more attention-grabbing things galaxies do, but because heating by stars, their winds, their protostellar jets and poststellar explosions can affect the motions, clumping and thermodynamic state of the gas (Krugel and Tutukov 1993). This in turn affects gas dynamics and gravitational instabilities (fission, Jeans or expulsion), and closing a feedback loop — rate of star formation.

Various prescriptions for star formation have been adopted by simulators: turning a gas cloud suddenly into stars when it collides with another cloud (Noguchi 1991), or when it reaches a critical density (Heller and Shlosman 1993), or turning it gradually into stars at a rate that increases with density (Mihos *et al.* 1992). Star-gas interactions are modelled by increasing the energy and/or momentum of gas near star-forming clouds. All of these are plausible and empirically motivated prescriptions. But reality is more complicated, perhaps in very significant ways. The interstellar medium contains not just the cold clouds usually modelled (sometimes with unrealistically high escape velocities and thus star-scattering efficiency), but hotter, volume-filling phases as well. These phases will also be heated: both by global shocks and by supernovae. The increased pressure of the hot phase may crush the cold clouds. This might aid star formation even if cloud-cloud collisions are unimportant (Jog and Solomon 1992; Jog and Das 1992).

However, in the only place where star formation is well studied, our local solar neighborhood, the rate of star formation is not limited by the rate of formation of dense molecular clouds. Instead it seems to be set by the rate at which magnetically supported clouds and protostars condense by ambipolar diffusion (McKee 1989), and by first massive stars to form, which disrupt clouds before more than a small fraction of their mass turns into stars. If the flux of cosmic rays were fixed, crushed clouds would be less ionized, have short ambipolar diffusion times, and could form stars rapidly. In regions of active star formation both the magnetic fields and the cosmic ray flux are likely to be quite different than in the solar neigborhood. Cosmic rays are accelerated at supernova shocks. They seem to contribute a significant pressure to the ISM, both locally and in starburst galaxies (Suchkov *et al.* 1993). A disk ISM with significant cosmic ray pressure is Rayleigh-Taylor unstable. Azimuthal magnetic fields are carried by the instability into the halo in vertically corrugated poloidal fields, which can undergo rapid reconnection at current sheets. As they are sheared by differential rotation into azimuthal fields, they close the loop for a fast magnetic dynamo (Parker 1992). This predicts that starburst galaxies should have very strong magnetic fields — as appears to be the case (Chi and Wolfendale 1993; note that the dynamo they rule out on the basis of magnetic field structure had fields orthogonal to those of a cosmic ray halo dynamo, whose vertical fields are in fact consistent with those seen in some edge-on starburst galaxies). Clouds may be crushed by magnetic pressure, and their formation assisted by the magnetic 'valleys' (Shibata and Matsumoto 1991). But if the flux of cosmic rays scales with total pressure (seemingly confirmed in M82; Suchkov *et al.* 1993), the ambipolar diffusion time would be constant, and crushed clouds might not in fact form stars rapidly.

To understand star formation in starbursts, it may prove helpful to test current theories of star formation by extrapolating them to the most extreme environments where stars are known to have formed: the inner parsecs of galaxies.

5.1 Nuclear ($\lesssim 10\,\mathrm{pc}$) Star Formation

As we saw in section 4, interactions and bars can drive gas to the inner hundreds of parsecs of a galaxy. Secondary bars may form and drive the gas to the central tens of parsecs (Shlosman *et al.* 1989; Friedli and Martinet 1993). Does this gas mainly accumulate on a central black hole, or does it form stars? On scales $\lesssim 0.1$ pc, the former is likely (Begelman and Rees 1978). But it is increasingly popular to explain various aspects of active galactic nuclei by the postulating the existence of a very dense ($\gtrsim 10^7$ M$_\odot$ pc^{-3}) star cluster around the central black hole: the broad-line clouds (Voit and Shull 1988), the heavy elements (Hamann and Ferland 1992), the fuelling and formation of the black hole (Murphy *et al.* 1991), or even replacing the black hole altogether (Terlevich and Melnick 1988; Stoeger *et al.* 1992; Morris [1993] discusses interesting possibilities for our Galaxy's nucleus). Relaxation and core collapse could create the required dense cluster cores, provided the stars initially form at $\lesssim 1$ pc.

Remarkably little work has been devoted to the actual formation of such nuclear star clusters. The conditions are dramatically different from those of star forming regions in the solar neighborhood, and so may teach us something about the problem of star formation.

Local molecular clouds have columns $N \sim 10^{22}$ cm^{-2}, and it has been argued (McKee 1989) that this is maintained by energy input from low mass star formation, regulated in turn by ambipolar diffusion, whose rate depends on the ionization state determined by the penetration of ambient interstellar far–UV photons — which depends on N, closing the loop. This mechanism for regulating the collapse of clouds cannot apply to star formation in galactic nuclei. If stars are to form there at all, the density of the gas from which they form must exceed the Roche density, $n_\mathrm{R} \sim 10^9\, r_\mathrm{pc}^{-3}$ cm^{-3} at distance r_pc pc from a $M_\bullet = 10^8$ M$_\odot$ black hole. Thus a gravitationally bound clump of gas which could form a 10 M$_\odot$ star must have a radius $R < (10\ \mathrm{M_\odot}/M_\bullet)^{1/3} r \sim 10^{16} r_\mathrm{pc}$ cm, and a column density $N > 10^{26} r_\mathrm{pc}^{-2}$ cm^{-2}. Such a clump is opaque except to > 100 MeV gamma rays, and to cosmic rays with $E > 1$ GeV (or higher if they do not follow straight line paths within the cloud). Photoionization is thus unlikely to regulate star formation in galactic nuclei.

The ionization fraction x, and therefore the ambipolar diffusion time $\sim 10^{14} x$ yrs, are likely to be determined by the cosmic ray flux, and thus by supernovae. The evolution of supernovae in such dense gas is quite different from that of familiar supernova remnants (Terlevich *et al.* 1992). Supernova blast waves become radiative after only a few months, while still travelling at $\gtrsim 5,000\,\mathrm{km\,s^{-1}}$, and having expanded only $\sim 10^{-3}\,\mathrm{pc}$. The cooled columns swept up are $\sim 10^{24}\,\mathrm{cm^{-2}}$. The supernova rate per unit volume in a $10^8\,\mathrm{M_\odot}$, 1 pc starburst is $\sim 10^{10}$ times that in the local solar neighborhood. One might therefore expect a huge cosmic ray flux. However, because the $\sim 10^{28}\,\mathrm{cm^{-2}}$ column density of gas across the central parsec greatly exceeds the stopping column for cosmic rays, the cosmic rays will be confined to the vicinity of the supernova remnant accelerating them. In contrast to the rather stable, non-local cosmic ray flux of the local solar neighborhood, the cosmic ray flux at a given location in a nuclear star formation region will be highly variable, on a timescale of a year. Cloud-cloud collisions at the $\gtrsim 10^3\,\mathrm{km\,s^{-1}}$ orbital velocities of the central parsec will also provide cosmic-ray accelerating shocks, important because of their proximity to the dense gas behind the radiative shock, which is most likely to be the gravitationally unstable site of star formation in galactic nuclei (see below).

Unless the mass of gas (in a disk of half-thickness h) exceeds h/r times the mass of the central black hole, the mean density of the gas will be less than the Roche density, and the gas will be stable against gravitational collapse (this is true even if the gas is in a thin disk; the Roche criterion is equivalent to the Safronov-Toomre $Q < 1$). Collisions of these low-density, stable clouds of gas will lead to radiative shocks. The gas cooling behind these will be of higher density, and, as pointed out in this context by Phinney (1989), can exceed the Roche density and become gravitationally unstable. Because of flux conservation in a disk of inflowing matter, or because of magnetic buoyancy limiting the Alfvén speed, we might scale the magnetic field $B \propto n^{1/2}$ from the local region of the Milky Way. Then we expect $B \sim 0.1(n/10^{10}\mathrm{cm^{-3}})^{1/2}\,\mathrm{G}$. If this is appropriate, the material swept up by supernovae will be magnetically cushioned against further compression when the post-shock temperature has fallen to $\sim 10^5\,\mathrm{K}$. We can then compute the density in the radiative shock at the point where magnetic cushioning sets in (Raymond 1979). This density is shown as a function of radius by the long dashed line in Figure 5. The Jeans mass in this region (for the parameters shown in the Figure) is $\sim 5 \times 10^4 r_{\mathrm{pc}}^{1.2}\,\mathrm{M_\odot}$. If ambipolar diffusion is effective at reducing the magnetic cushioning, even higher densities, and smaller Jeans masses may result (Phinney 1989).

A gas clump of mass M and density n threaded by a magnetic flux ϕ has a magnetic Jeans mass $M_\phi = 0.1\phi/G^{1/2}$ (Tomisaka *et al.* 1988), or

$$M_\phi = 2(M/10\,\mathrm{M_\odot})^{2/3} r_{\mathrm{pc}}^{1/2} (n/n_{\mathrm{R}})^{-1/6}\,\mathrm{M_\odot}\,. \qquad (9)$$

Thus just as for local molecular clouds, lumps containing the mass of a high mass star are magnetically supercritical, while those containing the mass of low mass stars may be magnetically subcritical, unable to collapse without ambipolar diffusion. The lumps, and the protostellar disks that might form from their collapse have one more hurdle. Because of the high density of protostellar lumps, and the short dynamical

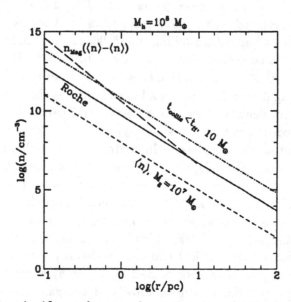

Fig. 5—Densities significant for star formation in galactic nuclei, as a function of the distance r from the nucleus (assumed to contain a central mass of 10^8 $M_\odot$ — black hole or stars and gas). Solid: Roche density. Gas clouds of lower density cannot collapse gravitationally. Dashed: mean density of $10^7(h/r)$ $M_\odot$ of gas in a disk of thickness h. This is gravitationally stable. Long dashed: post-shock density in the magnetically cushioned (see text) region of the radiative shock resulting from a collision at the virial velocity between two clouds of this mean density. Dot-dashed: minimum density required of 10 $M_\odot$ clumps of gas if they are not to collide in less than their own free-fall collapse time (assuming the entire $10^7(h/r)$ $M_\odot$ of gas is put in such clumps).

time, slowly contracting lumps are in danger of colliding with one another. Figure 5 shows the minimum density that 10 $M_\odot$ protostellar lumps must attain in order to avoid colliding with one another in less than their own free-fall collapse time, $n > 6 \times 10^{10} r_{\rm pc}^{-3}(10\,M_\odot/M)^{2/7}$. The lumps must smaller than $\sim$ 20 AU if they are not to collide in a 10^4 yr accretion time. The large collision rate suggests that multiple star systems or massive stars could be formed in large numbers (*cf.* Clarke and Pringle 1991; Artymowicz *et al.* 1993), if their protostars form disks.

However collapsing lumps may not form protostellar disks. If we cut $10M_{10}$ $M_\odot$ lumps out of a disk of density $10n_{\rm R}$, then if the rotation velocity at the edge of the lump is less than

$$v_{\rm c} = 0.1(r/1\,{\rm pc})^{-1}M_{10}^{2/3}\,{\rm km\ s^{-1}}, \tag{10}$$

the lump will be able to collapse directly to a main sequence star without forming a disk or losing angular momentum. The shear of a smooth gas disk gives velocities much lower than $v_{\rm c}$ for $r < 1\,{\rm kpc}$, but realistic lumpy disks with bar-induced motions will have much larger shears. Still, it is plausible that many condensations at $r \lesssim 1\,\mathrm{pc}$ may collapse directly without forming a protostellar disk.

6 CONCLUDING REMARKS

Our attention is naturally drawn to interactions between galaxies which result in large concentrations of gas and vigorous star formation. But to paraphrase Sherlock Holmes' discussion of the dog in the night time (Conan Doyle 1930),
'Is there any point to which you would wish to draw my attention?'
'To the curious incident of the galaxy in the interaction.'
'The galaxy did nothing in the interaction.'
'That was the curious incident.'
These galaxies may have much to teach us. Prograde encounters between galaxies excite much larger epicyclic motions than retrograde encounters, so simple cloud-collision models of star formation lead one to predict that the greatest activity should be seen in prograde encounters, and the least activity in retrograde ones (*e.g.* Byrd *et al.* 1987, Mihos *et al.* 1992). This does not appear to be borne out by observation (Keel 1993). The senses of encounters are not always unambiguous, and further observations will doubtless uncover subtleties. But if Keel's result is confirmed, current simulations are probably missing some significant element of star formation or structure in the interstellar gas.

Mass-transfer induced activity may have important lessons for cosmology. Galaxies in the past were certainly more gas rich, and probably had more frequent encounters with each other. Star formation induced by encounters could be responsible for the current appearance and past evolution of the stellar component of galaxies (Carlberg and Charlot 1992; Lacey *et al.* 1993), though the evidence for significant color evolution is controversial (Koo *et al.* [1993] finds that galaxy counts and redshift distributions can be fit with unevolving galaxies if account is taken of the dispersion of properties of local galaxies). Mass-transfer induced activity may thus be largely responsible for the morphology-density relation (Allington-Smith *et al.* 1993).

Elmegreen *et al.* (1990) present evidence that radial mass-transfer and star formation during an encounter can change an unbarred late-type galaxy into a barred early-type galaxy. This is an attractive idea, which might explain why galaxies in the Boötes void are still so huge and gas rich (Szomoru *et al.* 1993), and why Seyfert activity and nuclear star formation are so rare in late-type spiral galaxies (*cf.* the limits of Kormendy and McClure [1993] on the central mass in M33). But what then is one to make of the bulges and Seyfert nuclei of giant low surface brightness galaxies (Bothun, these proceedings)?

Finally, a young galaxy which becomes very active early may affect the birth of galaxies in a substantial region about it. Active nuclei through photoionization and jets, starburst galaxies by photoionization, superwinds and cosmic rays. These could modulate the threshold for galaxy formation (Rees 1993; Bower *et al.* 1993).

Such modulation, or slight differences in density or angular momentum of large-scale density fluctuations might cause 10% effects on galaxy interactions in large regions. If these translated into 10% effects on the sizes of bars or regions of star formation, or on the IMF and M/L, they could confuse Tully-Fisher estimates of peculiar velocities and mislead us about cosmology in serious ways (de Carvalho and Djorgovski 1992).

Like the star formation on which it depends, mass-transfer induced activity in galaxies is messy, subtle, and of ineluctable importance.

I thank Jens Villumsen for his collaboration in section 3.3, and Colin Norman for discussions of the topics in section 5.1. This research was supported in part by NASA grant NAGW-2144 and by the A. P. Sloan Foundation.

REFERENCES

Allington-Smith, J. R., Ellis, R. S., Zirbel, E. L., and Oemler, A. 1993, *Ap. J.*, **404**, 521.

Artymowicz, P., Lin, D. N. C., and Wampler, E. J. 1993, *Ap. J.*, **409**, 592.

Athanassoula, E. 1992, *Mon. Not. R. Astr. Soc.*, **259**, 345.

Balick, B. and Heckman, T. M. 1982, *Ann. Rev. Astr. Ap.*, **20**, 431.

Barnes, J., Goodman, J., and Hut, P. 1986, *Ap. J.*, **300**, 112.

Baum, S. A., Heckman, T. M., and van Breugel, W. 1992, *Ap. J.*, **389**, 208.

Begelman, M. C., Blandford, R. D., and Rees, M. J. 1980, *Nature*, **287**, 307.

Begelman, M. C., and Rees, M. J. 1978, *Mon. Not. R. Astr. Soc.*, **185**, 847.

Binney, J., and Tremaine, S. 1987, *Galactic Astronomy* (Princeton: Princeton Univ. Press).

Bower, R. G., Coles, P., Frenk, C. S., and White, S. D. M. 1993, *Ap. J.*, **405**, 403.

Bushouse, H. A. 1986, *A. J.*, **91**, 255.

Byrd, G. G., Sundelius, B., and Valtonen, M. 1987, *Astr. Ap.*, **171**, 16.

Carlberg, R. G., and Charlot, S. 1992, *Ap. J.*, **397**, 5.

Carter, D., Prieur, J. L., Wilkinson, A., Sparks, W. B., and Malin, D. 1988, *Mon. Not. R. Astr. Soc.*, **235**, 813.

Casertano, S., Phinney, E. S., and Villumsen, J. V. 1987, in IAU Symp. 127, on *Structure and Dynamics of Elliptical Galaxies*, ed. P. T. de Zeeuw (Dordrecht: Reidel), p. 475.

Chi, X. -Y., and Wolfendale, A. 1993, *Nature*, **362**, 610.

Christodoulou, D. M. 1993, *Ap. J.*, **412**, 696.

Christodoulou, D. M., and Narayan, R. 1992, *Ap. J.*, **388**, 451.

Clarke C. J., and Pringle, J. E. 1991, *Mon. Not. R. Astr. Soc.*, **249**, 588.

Colina, L., and Perez-Fournon, I. 1990, *Ap. J.*, **349**, 45.

Conan Doyle, A. 1930, *The Complete Sherlock Holmes* (Garden City: Doubleday), p. 347 ('Silver Blaze').

de Carvalho, R. R., and Djorgovski, S. 1992, *Ap. J. Lett.*, **389**, L49.

Devereux, N. A., Kenney, J. D. P., and Young, J. S. 1992, *A. J.*, **103**, 784.

Dressler, A., and Richstone, D. O. 1988, *Ap. J.*, **324**, 701.

Elmegreen, B. G., and Elmegreen, D. M. 1985, *Ap. J.*, **288**, 438.

Elmegreen, D. M., Elmegreen, B. G., and Bellin, A. D. 1990, *Ap. J.*, **364**, 415.

Fricke, K. J., and Kollatschny, W. 1989, in IAU Symp. 134 on *Active Galactic Nuclei*, eds. D. E. Osterbrock, and J. S. Miller (Dordrecht: Kluwer), p. 425.

Friedli, D., and Benz, W. 1993, *Astr. Ap.*, **268**, 65.

Friedli, D., and Martinet, L. 1993, *Astr. Ap.*, in press.

Fukushige, T., Ebisuzaki, T., and Makino, J. 1992, *P. A. S. J.*, **44**, 281.

Gerin, M., Combes, F., and Athanassoula, E. 1990, *Astr. Ap.*, **230**, 37.

Hamann, F., and Ferland, G. 1992, *Ap. J. Lett.*, **391**, L53.

Heckman, T. M. *et al.* 1986, *Ap. J.*, **311**, 526.

Heller, C. H., and Shlosman, I. 1993, *Ap. J.*, in press.

Hernquist, L. 1989, *Nature*, **340**, 687.

Hummel, E., van der Hulst, J. M., Kennicutt, R. C., and Keel, W. C. 1990, *Astr. Ap.*, **236**, 333.

Hutchings, J. B., and Neff, S. G. 1992, *A. J.*, **104**, 1.

Jog, C. J., and Solomon, P. M. 1992, *Ap. J.*, **387**, 152.

Jog, C. J., and Das, M. 1992, *Ap. J.*, **400**, 476.

Keel, W. C. 1993, *Rev. Mex. Astr. Af.*, in press.

Keel, W. C., and van Soest, E. T. M. 1992, *Astr. Ap. Suppl.*, **94**, 553.

Kenney, J. D. P. *et al.* 1992, *Ap. J. Lett.*, **395**, L79.

Kennicutt, R. C. *et al.* 1987, *A. J.*, **93**, 1011.

Kennicutt, R. C. 1989, *Ap. J.*, **344**, 685.

Kent, S. M. 1987, *A. J.*, **93**, 1062.

Kleiber, M. 1932, *Hilgardia*, **6**, 315.

Koo, D. C., Gronwall, C., and Bruzual A. G. 1993, *Ap. J. Lett.*, **415**, L21.

Kormendy, J. 1988, *Ap. J.*, **325**, 128.

Kormendy, J., and McClure, R. D. 1993, *A. J.*, **105**, 1793.

Krugel, E., and Tutukov, A. V. 1993, *Astr. Ap.*, **275**, 416.

Lacey, C., Guiderdoni, B., Rocca-Volmerange, B., and Silk, J. 1993, *Ap. J.*, **402**, 15.

Lonsdale, C. J., Persson, S. E., and Matthews, K. 1984, *Ap. J.*, **287**, 95.

Lynden-Bell, D. 1979, *Mon. Not. R. Astr. Soc.*, **187**, 101.

MacKenty, J. W. 1989, *Ap. J.*, **343**, 125.

Majewski, S. R. *et al.* 1993, *Ap. J.*, **402**, 125.

McKee, C. F. 1989, *Ap. J.*, **345**, 782.

Melnick, J., and Mirabel, I. F. 1990, *Astr. Ap.*, **231**, L19.

Mihos, J. C., Richstone, D. O., and Bothun, G. D. 1992, *Ap. J.*, **400**, 153.

Mirabel, I. F, Lutz, D., and Maza, J. 1991, *Astr. Ap.*, **243**, 367.

Morris, M. 1993, *Ap. J.*, **408**, 496.

Mould, J., Graham, J, Matthews, K., Soifer, B. T., and Phinney, E. S. 1989, *Ap. J. Lett.*, **339**, L21.

Murphy, B. W., Cohn, H. N., and Durisen, R. H. 1991, *Ap. J.*, **370**, 60.

Noguchi, M. 1987, *Mon. Not. R. Astr. Soc.*, **228**, 635.

Noguchi, M. 1991, *Mon. Not. R. Astr. Soc.*, **251**, 360.

Owen, F. N., O'Dea, C. P., Inoue, M., and Eilek, J. A. 1985, *Ap. J. Lett.*, **294**, L85.

Padovani, P., Burg, R., and Edelson, R. A. 1990, *Ap. J.*, **353**, 438.

Parker, E. N. 1992, *Ap. J.*, **401**, 137.

Parma, P., Cameron, R. A., and de Ruiter, H. R. 1991, *A. J.*, **102**, 1960.

Pasha, I. L., and Polyachenko, V. L. 1993, *Astr. Lett.*, **19**, 1.

Phinney, E. S. 1989, in IAU Symp. 136 on *The Center of the Galaxy*, ed. M. Morris (Dordrecht: Kluwer), p. 543.

Raymond, J. C. 1979, *Ap. J. Suppl.*, **39**, 1.

Rees, M. J. 1993, *Proc. Nat. Acad. Sci.*, **90**, 4840.

Roos, N., Kaastra, J. S., and Hummel, C. A. 1993, *Ap. J.*, **409** 130.

Sanders, D. B. *et al.* 1988, *Ap. J.*, **325**, 74.

Sargent, A., and Scoville, N. Z. 1991, *Ap. J. Lett.*, **366**, L1.

Schwarz, M. P. 1981, *Ap. J.*, **247**, 77.

Seitzer, P., and Schweizer, F. 1990, in *Dynamics and Interactions of Galaxies*, ed. R. Wielen (Berlin: Springer), p. 270.

Sellwood, J. A. 1981, *Astr. Ap.*, **99**, 362.

Sellwood, J. A. 1989, *Mon. Not. R. Astr. Soc.*, **238**, 115.

Sellwood, J. A., and Wilkinson, A. 1993, *Rep. Prog. Phys.*, **56**, 173.

Shibata, K., and Matsumoto, R. 1991, *Nature*, **353**, 633.

Shlosman, I., Frank, J. and Begelman, M. C. 1989, *Nature*, **338**, 45.

Shu, F. H., Tremaine, S., Adams, F. C., and Ruden, S. P. 1990, *Ap. J.*, **358**, 495.

Smith, E. P., and Heckman, T. M. 1989, *Ap. J.*, **341**, 658.

Solomon, P. M., Downes, D., and Radford, S. J. E. 1992, *Ap. J. Lett.*, **387**, L55.

Stoeger, W. R., Pacholczyk, A. G., and Stepinski, T. F. 1992, *Ap. J.*, **391**, 550.

Suchkov, A., Allen, R. J., and Heckman, T. M. 1993, *Ap. J.*, **413**, 542.

Szomoru, A., Van Gorkom, J. H., Gregg, M., and DeJong, R. S. 1993, *A. J.*, **105**, 464.

Telesco, C. M., Dressel, L. L., and Wolstencroft, R. D. 1993, *Ap. J.*, in press.

Terlevich, R., and Melnick, J. 1988, *Nature*, **333**, 239.

Terlevich, R., Tenorio-Tagle, G., Franco, J., and Melnick, J. 1992, *Mon. Not. R. Astr. Soc.*, **255**, 713.

Thronson, H. A. *et al.* 1989, *Ap. J.*, **343**, 158.

Tomisaka, K., Ikeuchi, S., and Nakamura, T. 1988, *Ap. J.*, **335**, 239.

Toomre, A. 1981, in *Structure and Evolution of Normal Galaxies*, eds. S. M. Fall and D. Lynden-Bell (Cambridge: Cambridge Univ. Press), p. 111.

Tremaine, S., and Weinberg, M. D. 1984, *Ap. J. Lett.*, **282**, L5.

Voit, G. M., and Shull, J. M. 1988, *Ap. J.*, **331**, 197.

Wada, K., and Habe, A. 1992, *Mon. Not. R. Astr. Soc.*, **258**, 82.

Weinberg, M. D. 1993, *Ap. J.*, submitted.

Zhang, X., Wright, M., and Alexander, P. 1993, *Ap. J.*, in press.

Angular Momentum Transfer in the Inner Parsec

Mitchell C. Begelman[1,2]

[1]Joint Institute for Laboratory Astrophysics, University of Colorado and
National Institute of Standards and Technology, Boulder, CO 80309-0440
[2]Also at Department of Astrophysical, Planetary, and Atmospheric Sciences,
University of Colorado, Boulder, CO 80309-0391

ABSTRACT

I discuss the various processes which may affect the transfer of angular momentum in gas within the inner parsec surrounding a supermassive black hole. Even after gas has been brought into the gravitational sphere of influence of the black hole, it still has 100–1000 times too much angular momentum to reach the event horizon. Angular momentum loss cannot be accommodated in a scaled-up standard thin accretion disk model, because local self-gravitational instabilities will lead to fragmentation of the disk, which will decrease the efficiency of angular momentum transport. Instead, some nonlocal mechanism or "external" trigger for angular momentum transfer is needed, such as a large-scale magnetized wind, stirring by winds from massive stars and supernovae, or global gravitational instabilities.

1 INTRODUCTION

Most participants in this conference seem to agree that large-scale, non-axisymmetric gravitational disturbances can be very effective in transferring angular momentum on scales of hundreds of parsecs or larger. These disturbances might be driven by tidal encounters or may be "self-starting", as in the "bars-in-bars" scenario. What is less certain is whether similar mechanisms can be effective all the way into the nucleus, say, into the inner parsec surrounding a supermassive black hole. Gravitational triggers for inflow usually involve at least a mild form of self-gravitational instability, which requires that the stars plus gas constitute a significant fraction of the total mass enclosed within the region under consideration (see, *e.g.*, Friedli, these proceedings). If the stars form a singular isothermal distribution with the Keplerian velocity given by $v_\phi = 100 v_{100}$ km s^{-1}, then a black hole of mass $M_{BH} = 10^8 M_8$ solar masses will dominate the potential at radii smaller than $R_H \sim 40 M_8 v_{100}^{-2}$ pc. Thus, unless the mass of the inflowing gas itself is comparable to that of the black hole, it is unlikely that a mechanism like "bars-in-bars" can operate all the way into the nucleus.

In principle, one cannot rule out the possibility that global self-gravity in the gaseous distribution is important down to a fraction of a parsec. Presumably, inflowing gas would collect near R_H until enough of it accumulates, then inflow could occur through the development of a bar. But this would require a *mean* gas density at R of order M_{BH}/R^3. Since the gas would cool readily, the local density would be much higher and the gas would certainly form very dense, *locally* self-gravitating clouds. It is hard to understand how to avoid rapid star formation in such a case. Therefore, before settling on gravitational disturbances as the universal mechanism of angular momentum transport in galactic inflows, it is worth asking whether some other means of angular momentum transport could drain the gas before a self-gravitating quantity of it is collected.

It turns out that such a mechanism is harder to find than one might guess. To power nuclear activity, even with a high efficiency of mass-to-energy conversion $\epsilon \sim 0.1$, fuel is required at a rate

$$\dot{M} \sim 0.2 \left(\frac{\epsilon}{0.1}\right) L_{45} \ \mathrm{M_\odot} \ \mathrm{yr}^{-1}, \tag{1}$$

where $L_{45} \equiv L/10^{45}$ ergs s^{-1} is the AGN luminosity. Gas in a Keplerian orbit, which has a specific angular momentum of $Rv_\phi = 3 \times 10^{25} v_{100}$ cm^2 s^{-1} at 1 pc, must lose between 99% and 99.9% of its angular momentum before it reaches the vicinity of the black hole, e.g., at 10 Schwarzschild radii. The mechanism which removes this angular momentum liberates about $10^{12}(\dot{M}/1 \ \mathrm{M_\odot} \ \mathrm{yr}^{-1}) \ \mathrm{L_\odot}$ of gravitational binding energy, $10^6 v_{100}^{-2}$ times more energy than binds the gas in the potential at 1 pc. Removing such a large amount of angular momentum and energy from such a small region is not easy.

2 SCALING UP THE STANDARD ACCRETION DISK

2.1 The α-Disk Paradigm

Whether or not we believe in a local α-type viscosity, a generalization of the standard model for thin accretion disks (Pringle 1981) provides a useful means of parametrizing the angular momentum transport problem. We retain the basic assumption that pressure forces are second-order, and that to first approximation the gas executes circular Keplerian orbits in the central potential. However, we generalize the "pressure" to include contributions from magnetic and turbulent stresses (radiation pressure is negligible on parsec scales), so that $P = P_{\mathrm{gas}} + P_{\mathrm{mag}} + P_{\mathrm{turb}}$. This pressure force thickens the disk normal to the orbital plane to a scale height H, where $H/R \sim (P/\rho v_\phi^2)^{1/2} \ll 1$ (ρ is the mean density inside the disk). The local viscous stress is then parametrized according to $\langle \tau_{\phi R} \rangle \sim \alpha P$, where $\alpha \lesssim 1$. The justification for the latter assumption is that $\alpha \sim 1$ corresponds to the largest stress that can be carried by any of the forms of energy associated with the generalized pressure (see, *e.g.*, Shakura and Sunyaev 1973; Lynden-Bell and Pringle 1974). However, the argument does not apply if the disk is acted on by "nonlocal" forces, such as large-scale gravitational disturbances or the stresses associated with an organized magnetic field (Blandford and Payne 1982).

According to this parametrization, the internal energy associated with the pressure P increases due to "viscous dissipation" on a time scale $t_{\mathrm{visc}} \sim \alpha^{-1} t_{\mathrm{Kep}}$, while angular momentum transport and inflow occurs on a much longer time scale, $t_{\text{in}} \sim \alpha^{-1}(R/H)^2 t_{\text{Kep}}$, where t_{Kep} is the Keplerian orbital time.

2.2 Application to Galactic Nuclei

To illustrate the difficulties encountered when one attempts to apply a simple α-type model to active galactic nuclei on pc scales, it is useful first to define a "vertical disk temperature",

$$T_{\text{vert}} \equiv 10^4 T_4 \text{ K} \equiv \left(\frac{H}{R}\right)^2 \frac{\mu v_\phi^2}{k} , \qquad (2)$$

where μ is the mean mass per particle and k is the Boltzmann constant. We then estimate the following accretion disk parameters, assuming that the disk is required to transfer $\dot{m}$ $M_\odot$ yr^{-1} over a distance R_{pc} pc. Inflow time scale:

$$t_{\text{in}} \sim 10^6 \alpha^{-1} T_4^{-1} v_{100} R_{\text{pc}} \text{ yr}; \qquad (3)$$

particle density:

$$n \sim 10^8 \dot{m} \alpha^{-1} T_4^{-3/2} v_{100}^2 R_{\text{pc}}^{-2} \text{ cm}^{-3}; \qquad (4)$$

column density:

$$N \sim 4 \times 10^{25} \dot{m} \alpha^{-1} T_4^{-1} v_{100} R_{\text{pc}}^{-1} \text{ cm}^{-2}; \qquad (5)$$

and pressure:

$$P \sim 10^{-4} \dot{m} \alpha^{-1} T_4^{-1/2} v_{100}^2 R_{\text{pc}}^{-2} \text{ dyn cm}^{-2}. \qquad (6)$$

Given the upper limit that we have imposed on α, it is clear that these conditions are going to be extreme compared to the gaseous media further out in the galaxy; whether they are self-consistent depends on the degree of self-gravitation in the gas, which in turn depends on the value of T_{vert}. What determines T_{vert}? If the disk is supported entirely by thermal gas pressure and the only source of heating is the viscous dissipation associated with angular momentum transport, then it will be highly optically thick, with a black-body surface temperature of

$$T_{\text{BB}} \sim 34 \dot{m}^{1/4} v_{100}^{1/2} R_{\text{pc}}^{-1/2} \text{ K}. \qquad (7)$$

The disk interior will be somewhat warmer than T_{BB} but still with $T_4 \lesssim 0.01$, implying very large values of n, N, and P. But such a low value of T_{vert} could never be reached in practice, because the disk would become gravitationally unstable first. Local (Jeans) instability sets in when the internal density of the disk exceeds M_{tot}/R^3, where M_{tot} is the total mass enclosed within R. Thus, the condition for local instability is

$$n \gtrsim n_{\text{J}} \sim \frac{v_\phi^2}{\mu G R^2} \sim 10^8 v_{100}^2 R_{\text{pc}}^{-2} \text{ cm}^{-3}, \qquad (8)$$

implying that a vertical temperature

$$T_{\text{vert}} \gtrsim 2 \times 10^4 \dot{m}^{2/3} \alpha^{-2/3} \text{ K} \qquad (9)$$

is required to suppress the instability. Note that this temperature exceeds $T_{\rm BB}$ by two orders of magnitude, implying that *a scaled-up version of a standard thin accretion disk cannot transport the required mass flux within the inner pc without becoming violently gravitationally unstable.* Lin and Pringle (1987) suggested that the turbulence generated by local gravitational instabilities could stir up the disk (increasing $P_{\rm turb}$ and $T_{\rm vert}$) and lead to an effective α greatly exceeding 1. However, Shlosman and Begelman (1987, 1989) found that such a disk was likely to fragment into self-gravitating clumps, which would cool rapidly and contract to such a degree that their mutual interactions would be ineffective at transferring angular momentum. Note that the thermal heating of the disk cannot even prevent it from becoming *globally* gravitationally unstable, which requires

$$T_{\rm vert} \gtrsim 2 \times 10^3 \dot{m} \alpha^{-1} v_{100}^{-1} \text{ K.} \qquad (10)$$

Indeed, the gas present in the inflow is likely to cool so rapidly that the thermal gas pressure is likely to be a small fraction of P. In order for angular momentum to be transferred by "local" mechanisms, one must appeal to processes which enhance the magnetic and/or turbulent components of the generalized pressure.

3 MAGNETIC FIELDS TO THE RESCUE?

If thermal pressure and the turbulence generated by local self-gravity are inadequate to cause angular momentum transfer at the required rate, might one have better luck appealing to hydromagnetic stresses? The viscous stress depends on $\langle B_{\rm R} B_\phi \rangle$. If there is any seed field present in the inflowing gas, a variety of effects could lead to dynamo action and amplification of a turbulent field. The most promising mechanism at present seems to be the robust instability re-discovered by Balbus and Hawley (Balbus and Hawley 1991, 1992; Hawley and Balbus 1991, 1992), which is capable of generating both radial ($B_{\rm R}$) and toroidal (B_ϕ) field components, given a poloidal seed field. Ordinary shear in the disk can further amplify B_ϕ, and dynamo effects can also be enhanced by the action of internal waves (Vishniac and Diamond 1992). Indeed, it seems likely that magnetic stresses are responsible for most of the angular momentum transport in disks.

However, it seems doubtful that $P_{\rm mag}$ could grow so large that it overwhelms the other forms of pressure. The modes studied by Balbus and Hawley apparently stop growing when $P_{\rm mag}$ becomes the dominant component of the pressure; moreover, a disk with $P_{\rm mag} \gtrsim P_{\rm gas} + P_{\rm turb}$ is expected to be subject to the Parker instability (Parker 1975), which rapidly removes toroidal magnetic field via buoyancy. At best, we might hope that the turbulent magnetic field is amplified to the point at which it transports angular momentum with an effective $\alpha \sim O(1)$, but the vertical temperature associated with the B-field is unlikely greatly to exceed that associated with the other forms of pressure. Therefore, in order to obtain efficient inflow, we would still need to find some mechanism which increases the hydrodynamic turbulent pressure.

An accretion disk in which angular momentum is transferred via magnetic stresses might have rather different observational properties than those envisaged by Shakura and Sunyaev (1973) or Pringle and Rees (1972). For one thing, there is no reason to assume that all or most of the binding energy released is dissipated deep within the disk. It is quite possible that buoyancy carries the twisted flux loops into the tenuous disk corona before reconnection releases the stored energy. While it is not known whether reconnection can accelerate relativistic electrons directly, the intense localized heating caused by reconnection could drive shocks into the surrounding coronal gas, accelerating particles efficiently via the Fermi mechanism. If most of the energy is dissipated in this way, it might not be too unrealistic to think of accretion disks as "thermally inert" slabs of gas heated externally by the activity around them (Begelman and de Kool 1991).

Finally, we note that angular momentum could be extracted hydromagnetically through a "nonlocal" mechanism if the disk is permeated by a sufficiently strong poloidal field (Blandford and Payne 1982). Such a field could be advected inward with the inflowing gas. If the "open" field lines piercing the disk make a sufficiently large angle with respect to the vertical, a fraction of the disk gas can be flung outward centrifugally (however, see Begelman 1993 for reservations about the simplest versions of this model). The tension in the field exerts a torque on the disk, extracting angular momentum and energy. Thus, it is possible that the gravitational binding energy could be extracted in the form of kinetic energy and Poynting flux, without much radiation being emitted at all!

4 ENERGY INPUT FROM STARS

It seems that, if we wish to make the local forms of angular momentum transfer work (*i.e.*, barring angular momentum extraction by large-scale magnetized winds or global self-gravitational instabilities), we need to find some external agent to stir up the disk, increasing T_{vert} through the turbulent energy density. Since we have already discussed the propensity of the disk to become locally self-gravitationally unstable, the energy input resulting from the formation of massive stars would seem to be a natural candidate. To have a hope of obtaining enough energy input, we need to use a portion of the accretion flow itself to form stars. Therefore, suppose that massive stars form at a rate $\sim 10^{-3}\zeta_{-3}\dot{M}$, and yield a kinetic energy output (through the combination of stellar winds and supernova explosions) of $\sim 10^{50}$ erg $M_\odot^{-1}$. Not all of this energy is usable for increasing the turbulence in the disk. Since the material comprising the disk is quite dense, the stellar energy input will drive radiative shocks into the disk gas, and only the residual buffeting due to the unevenness of the momentum input will contribute to the random kinetic energy. We optimistically take the efficiency of energy input to be $\eta \sim v_{\text{turb}}/v_{\text{w}} \sim 0.01 - 0.1$ for $0.1 < H/R < 1$, where $v_{\text{turb}} \sim (H/R)v_\phi$ is the turbulent velocity responsible for thickening the disk and $v_{\text{w}} \gtrsim 10^3$ km s^{-1} is the characteristic speed of the stellar winds or supernova ejecta. We therefore obtain an effective disk "heating" rate of

$$\dot{E}_{\rm W} \sim 3 \times 10^{37} \zeta_{-3} \dot{m} T_4^{1/2} \left(\frac{v_{\rm W}}{10^3 \ {\rm km \ s^{-1}}} \right)^{-1} \ {\rm erg \ s^{-1}}. \tag{11}$$

Now the energy required to "puff up" the disk is at least of order the rate at which gravitational binding energy is released:

$$\dot{E}_{\rm grav} \sim 3 \times 10^{39} \dot{m}_{\rm in} v_{100}^2 \ {\rm erg \ s^{-1}}, \tag{12}$$

where we now distinguish $\dot{m}_{\rm in}$, the rate at which material actually flows in towards the nucleus, from $\dot{m}$, the total rate at which material enters the star-forming region. Comparison of $\dot{E}_{\rm W}$ with $\dot{E}_{\rm grav}$ suggests two possible scenarios. 1) If the most of the mass entering the star-forming region continues in towards the nucleus, $\dot{M}_{\rm in} \sim \dot{M}$, then we would require very efficient massive star formation, $\zeta \gtrsim 10^{-2}$, as well as such violent stirring that the clouds comprising the accretion flow would be whipped into a quasi-spherical distribution, i.e., one with $H/R \sim 1$. I do not find this scenario very plausible. 2) Alternatively, one could have only a small fraction of the gas flowing in, $\dot{M}_{\rm in} \ll \dot{M}$. Most of $\dot{M}$ would either form stars, or be ejected in a wind. To drive 0.1–1 $M_\odot$ yr^{-1} of inflow towards the nucleus would then require mass to be supplied at the rate $(10 - 100)\zeta_{-3}^{-1} v_{100}^2 \ M_\odot$ yr^{-1}, and most of this would have to go into star formation.

Unfortunately, the latter picture is not really self-consistent, even if one is willing to accept all of the assumptions (efficient star formation, etc.) that go into it. To obtain the required $\dot{M}$, optimistically inflowing at a good fraction of v_ϕ, would require so much mass to be present that *global self-gravity would be significant*. Thus, we have come full circle, and have to consider...

5 SELF-GRAVITY AFTER ALL?

Thus, by the process of elimination we have been led back to the possibility that large-scale gravitational torques are responsible for transferring angular momentum, even within the inner parsec. The only viable alternative seems to be a magnetized wind, which is also a "nonlocal" mechanism. In any case, it is clear that angular momentum transport poses a problem even down to the smallest scales in galactic nuclei.

Acknowledgements: Many of the ideas discussed in this article were developed during a long collaboration with Isaac Shlosman and Juhan Frank. This work was partially supported by NSF grant AST91-20599 and NASA grant NAGW-766.

REFERENCES

Balbus, S. A., and Hawley, J. F. 1991, *Ap. J.*, **376**, 214.

Balbus, S. A., and Hawley, J. F. 1992, *Ap. J.*, **400**, 610.

Begelman, M. C. 1993, in *Astrophysical Jets*, ed. D. Burgarella, M. Livio, and C. O'Dea (Cambridge: Cambridge University Press), in press.

Begelman, M. C., and de Kool, M. 1991, in *Variability of Active Galactic Nuclei*, ed. H. R. Miller and P. J. Wiita (Cambridge: Cambridge University Press), p. 198.

Blandford, R. D., and Payne, D. G. 1982, *Mon. Not. R. Astr. Soc.*, **199**, 883.

Hawley, J. F., and Balbus, S. A. 1991, *Ap. J.*, **376**, 223.

Hawley, J. F., and Balbus, S. A. 1992, *Ap. J.*, **400**, 595.

Lin, D. N. C., and Pringle, J. E. 1987, *Mon. Not. R. Astr. Soc.*, **225**, 607.

Lynden-Bell, D., and Pringle, J. E. 1974, *Mon. Not. R. Astr. Soc.*, **108**, 603.

Parker, E. N. 1975, *Ap. J.*, **198**, 205.

Pringle, J. E. 1981, *Ann. Rev. Astr. Ap.*, **19**, 137.

Pringle, J. E., and Rees, M. J. 1972, *Astr. Ap.*, **21**, 1.

Shakura, N. I., and Sunyaev, R. A. 1973, *Astr. Ap.*, **24**, 337.

Shlosman, I., and Begelman, M. C. 1987, *Nature*, **329**, 810.

Shlosman, I., and Begelman, M. C. 1989, *Ap. J.*, **341**, 685.

Vishniac, E. T., and Diamond, P. 1992, *Ap. J.*, **398**, 561.

Can Supernovae and Accretion Disks Be Distinguished Spectroscopically?

Rainer Wehrse[1] and Giora Shaviv[2]

[1]Institut für Theoret. Astrophysik, Im Neuenheimer Feld 561, D69120
Heidelberg, Germany
[2]Department of Physics and Asher Space Research Institute, Technion, IL 32000
Haifa, Israel

ABSTRACT

By means of models for supernovae of type II during the photospheric phase and for accretion disks around (super-)massive black holes it is found that the pressures, temperatures and velocities in the spectrum-forming regions are quite comparable so that the energy distributions have to be similar. The observed similarity of moderate resolution optical spectra from supernovae of type II and from active galactic nuclei may therefore be just fortuitously. For the distinction of these classes of objects in particular line profiles can be used in addition to *gamma*-ray fluxes and polarization data.

1 INTRODUCTION

Optical spectra of type II supernovae (abbreviated subsequently "SNe II") during the photospheric phase look often very similar to those of active galactic nuclei (AGNs, see *e.g.* Filippenko 1992). This similarity was used by Terlevitch and Melnick (1985, see also subsequent papers by Terlevitch) to support the hypothesis that the activity of galaxies is caused by starbursts that produce lots of massive stars which subsequently explode as supernovae. On the other hand, there are many convincing, though indirect arguments that the optical/UV luminosity of AGNs is produced by disks around (super-)massive black holes (see *e.g.* Shields 1978; Rees 1984; Malkan 1991).

Since unique spectral signatures of accretion disks have still to be found this raises the following questions:

(i) is the optical light of AGNs indeed produced predominantly by SNe II?

(ii) what does the similarity tell us, if the AGN spectra are in fact produced by accretion disks?

(iii) are there ways to distinguish SNe II and accretion disk spectra from their ultraviolet to infrared spectra? Evidently, there are pronounced differences of AGNs and SNe II energy distributions in the γ-ray, X-ray and radio ranges as well as in the polarization properties of the emitted light; however, they are beyond the scope of this paper.

In this contribution we want to address aspects (ii) and (iii) on the basis of present model calculations for SNe and accretion disks. In the next section we briefly summarize the main features of these models, in section 3 we discuss the temperatures and pressures in the continuum and line forming regions. These physical conditions are the main factors that determine the level populations since they control the radiation fields together with the velocity field and the geometries. Continuous energy distributions and line profiles from both types of objects are compared in section 4. Effects of gas that might surround the objects and the relevance of the model calculations for the interpretation of AGN spectra are briefly examined in the final section 5.

2 SYNTHETIC SPECTRA FOR SNE II AND ACCRETION DISKS AROUND MASSIVE BLACK HOLES

2.1 Models of Supernova Photospheres

Present day models of SNe II photospheres during the coasting phase (*i.e.* approximately 1 to 4 weeks after outburst) are characterized by (see *e.g.* Eastman and Kirshner 1989; Best and Wehrse 1993):
- stationarity, *i.e.* the temporal evolution of an object is followed by a sequence of stationary models;
- spherical symmetry;
- no energy sources (as *e.g.* radioactive nuclei) in the atmosphere and energy transport by radiation only;
- increase of the expansion velocity proportional to the radius ("homologous expansion");
- a power law density distribution $\rho \propto r^{-n}$;
- no radiation incident on the supernova.

Usually, models involve some additional assumptions which – however– are not of importance in our context.

2.2 Models of Accretion Disks

In the construction of accretion disks discussed here the following assumptions are made (*cf.* Shaviv and Wehrse 1993):
- stationarity;
- axial symmetry;
- motion of matter around the center on Kepler orbits;
- vertical density distribution in hydrostatic equilibrium (which has to involve the vertical components of the central object's gravity, the self gravity, and the radiation pressure);
- local energy input into the radiation field proportional to the gas pressure;
- determination of the height of the disk by the condition that the integral of the local energy generation rate over height equals the prescribed total energy emitted per unit time and area;

- no illumination of the disk;
- Newtonian physics (the effects of the Kerr hole are in fact quite significant, *cf.* Abramowitz [1993], Viergutz [1993], but should not be of relevance for our discussion);

Again, as for supernovae there are additional properties that are however not important for this paper (see Störzer 1993; Shaviv and Wehrse 1993).

3 COMPARISON OF TEMPERATURES, PRESSURES AND VELOCITIES IN PHOTOSPHERIC LAYERS

From the data published for AGN accretion disks (Störzer 1991; Wehrse and Shaviv 1994) and for SNe II (Hauschildt, Shaviv, and Wehrse 1989) it is seen that in both types of objects one finds essentially the same temperatures and (very low) pressures in the spectrum forming regions. Therefore, similar ionization and excitation conditions are present and consequently the same lines should show up and the wavelength dependence of the continuum should also be similar.

For the temperature range 8,000 to 10,000 K (the range of maximal Balmer line emission), the surface layers of disks have orbital velocities of about 10^4 km s^{-1} for central masses of approximately 10^8 M$_\odot$. Such velocities are also typical for SNe II with fast expanding photospheres so that to first order the Doppler shifts seen in lines from both types of objects should be similar. As a consequence, differences between the spectra must be due to different temperature and pressure *gradients* and the different geometries (assuming that the chemical compositions are the same).

4 COMPARISON OF SPECTRA

4.1 Continua

The continuum energy distributions of accretion disks (*cf.* Shaviv and Wehrse 1993; Störzer 1993) show essentially power law dependencies from the UV to the IR. The Balmer jump and higher edges are very weak due to flat temperature gradients. As a consequence of the very high effective temperatures close to the black hole, the spectra extend very far to the EUV.

This is in contrast to the situation in supernovae (*cf.* Hauschildt, Shaviv, and Wehrse 1989; Hafner 1993) for which the surface temperatures do not go up so much. Although here, too, the hydrogen edges are weaker than for main sequence stars of the same temperatures due to the geometrical extension of the atmosphere, they are still more visible than for the accretion disks. Most important, however, also in this case the flux distributions follow closely power laws longwards of the Lyman jump but the exponent is somewhat different.

4.2 Line Profiles

The line profiles of accretion disks are symmetric to first order since we see an equal amount of matter that is moving towards us and that is moving away from us (without intervening gas, see Adam 1990). Asymmetries results from deviations from axial symmetry (as *e.g.* the dragging of space-time by a central Kerr hole) and from the fact that the transverse Doppler effect and Doppler boosting affect the red and blue line wings differently. For the hydrogen line forming regions these effects are usually small (Viergutz 1993; Papkalla 1993).

This is in contrast to the supernova situation where matter receding from us is only seen due to the geometrical extension of the atmosphere and it is only seen through absorbing matter (essentially at rest). Additional differences arise from the different velocity laws. Therefore, the line profiles from both types of objects are in fact quite distinct. Indeed, of the spectra presented by Filippenko (1992) the AGN line profiles show much more symmetric structure than the supernova ones although the spectrograph seems to have largely smeared out the effect.

5 DISCUSSION

The data presented above show that the optical spectra of accretion disks and type II supernovae are *fortuitously* rather similar. It implies that it is not allowed to infer from spectral similarities that the corresponding photons are indeed coming from the same type of object. We should note that this situation is like that for G77–61, the first carbon dwarf star identified (Dahn *et al.* 1977): it has a spectrum that is quite similar to that of a J–type giant. It was only possible to confirm the high gravity (and therefore the low luminosity) *spectroscopically* by detailed analysis of line profiles (Gass, Liebert, and Wehrse 1988). The same seems to hold in our case since the study of line profiles provides the easiest way to distinguish the two classes of objects if information from other wavelength regions (in particular the gamma regime) or polarization data are not available. Note that this behavior should not significantly be changed if surrounding matter is also taken into account since it should be of low density and optically thin in the continuum. Consequently, it should not influence the continuum and should modify the line profiles of major lines only by adding rather narrow (absorption or emission) spikes that can easily be identified.

Acknowledgment: R. W. thanks the Deutsche Forschungsgemeinschaft (SFB 328, project We 1148/2-2) and the DARA (project 50 OR 9106) for their financial support.

REFERENCES

Abramowitz, M. 1993, Proc. 2nd Haifa Conf. on *Cataclysmic Variables*, ed. J. Adler (Jerusalem: Israel Phys. Soc.), in press.

Adam, J. 1990, *Astr. Ap.*, **240**, 541.

Best, M., and Wehrse, R. 1993, *Astr. Ap.*, in press.

Dahn, C. C., Liebert, J., Kron, R. G., Spinrad, H., and Hintzen, P. M. 1977, *Ap. J.*, **216**, 757.

Eastman, R. G., Kirshner, R. P. 1989, *Ap. J.*, **347** 771.

Filippenko, A. V. 1992, in *Physics of Active Galactic Nuclei*, eds. W. Duschl and S. Wagner (Berlin: Springer), p. 345.

Gass, H. , Liebert, J., and Wehrse, R. 1988, *Astr. Ap.*, **189** , 194.

Hafner, M. 1993, in preparation.

Hauschildt, P. H., Shaviv, G., and Wehrse, R. 1989, *Astr. Ap. Suppl.*, **77**, 115.

Malkan, M. 1991, in *Structure and Emission Properties of Accretion Disks*, eds. C. Bertout, *et al.* (Paris: Frontieres), p. 165.

Papkalla, R. 1993, *Ph. D.* thesis, Heidelberg University.

Rees, M. J. 1984, *Ann. Rev. Astr. Ap.*, **22**, 471.

Shaviv, G., and Wehrse, R. 1993, in *Accretion Disks around Compact objects*, ed. C. Wheeler (World Scientific), in press.

Shields, G. A. 1978, *Nature*, **272**, 706.

Störzer, H. 1991, *Ph. D.* thesis, Heidelberg University.

Störzer, H. 1993, *Astr. Ap.*, **271**, 25.

Terlevitch, R., and Melnick, J. 1985, *Mon. Not. R. Astr. Soc.*, **224**, 193.

Viergutz, S. 1993, *Astr. Ap.*, **272**, 355.

Wehrse, R., and Shaviv, G. 1994, in preparation.

Low Ionization Broad Absorption Lines in Quasars

G. Mark Voit

Caltech, Mail Code 130-33, Pasadena, CA 91125

ABSTRACT

Low-ionization broad absorption line quasars, also known as Mg II BAL QSOs, are intriguing objects that might represent an evolutionary link between the infrared-luminous aftermaths of galaxy-galaxy collisions and the normal, unshrouded quasars thought by some to result from such collisions. This contribution briefly summarizes Mg II BAL QSO phenomenology, sketches the results of a more detailed analysis of Mg II BAL QSO spectra, and estimates the prodigious kinetic energy outputs of these objects. It appears likely that the kinetic luminosities of these quasars exceed 10% of their radiative luminosities.

1 INTRODUCTION TO MG II BAL QSOS

Broad absorption-line quasars (BAL QSOs), comprising $\sim$10% of all optically-selected radio-quiet quasars, have puzzled astronomers for over two decades. Broad, blueshifted absorption features in UV resonance lines like C IV, N V, Si IV, and Lyα show that these quasars are ejecting 10^4 K material at velocities of up to $0.1c$. In the spirit of this conference, we could call the BAL QSO phenomenon "Activity-Induced Mass Transfer in Galaxies".

We still do not understand how these quasars accelerate the absorbing material to such high velocities. Even though BAL outflows could be telling us something important about the innards of active galactic nuclei, BAL QSOs have been somewhat neglected in recent years, partly because we lack a basic understanding of the acceleration mechanism. The field has progressed more by ruling out possible acceleration mechanisms than by identifying schemes that might work. Weymann, Turnshek, and Christiansen (1985) and Begelman, de Kool, and Sikora (1991) provide particularly insightful analyses of the problems involved in accelerating BAL material.

Roughly 10% of optically-selected BAL QSOs with broad C IV and N V absorption profiles also show broad absorption features in lower-ionization species such as Mg II and Al III. These objects are known as Mg II BAL QSOs. Although rare in optically selected samples, these quasars might be more common than they appear. Sprayberry and Foltz (1992) have shown that the excessive reddenings of Mg II BAL QSOs, compared to other quasars, imply significant amounts of extinction, large enough to cause Mg II BAL QSOs to be undersampled by factors of up to 10 in optical surveys. Infrared samples of quasars, while plagued by small-number statistics, show much higher fractions of Mg II BAL QSOs than do optical surveys. In the latest IR census, that of Boroson and Meyers (1992), 4 of 19 infrared-selected quasars show broad low-ionization absorption.

The Boroson-Meyers survey contains another intriguing result: Mg II BAL QSOs tend to have little or no [O III] emission. This lack of [O III] emission indicates that much of the ionizing flux from these quasars is absorbed in regions too dense to emit forbidden lines efficiently. The four objects with broad low-ionization absorption have the four lowest [O III] equivalent widths in the survey. Since forbidden-line emission is thought to be a relatively isotropic property of quasars, this tight correlation suggests that the covering factors of low-ionization BAL regions are high; otherwise, we would expect to see a larger proportion of [O III]-dim objects without low-ionization BALs.

This brief summary of Mg II BAL QSO phenomenology points toward the possibility that Mg II BAL QSOs are dust-enshrouded quasars in the process of expelling surrounding dense material. In this respect, these objects might be close relatives of the ultraluminous *IRAS* galaxies that result from mergers of gas-rich spirals (Sanders *et al.* 1988). Perhaps they are even their direct descendants. They are certainly objects worthy of more detailed study than they have received thus far. This contribution outlines the results of a recent analysis of Mg II BAL QSOs by Voit, Weymann, and Korista (1993) and presents an additional assessment of the kinetic luminosities of these objects.

2 MG II BAL QSO IONIZATION AND VELOCITY STRUCTURE

Some systematic similarities in the spectra of Mg II BAL QSOs have enabled us to constrain the column densities and velocity fields of their BAL regions more tightly than has previously been possible. The analysis sketched here is presented in much greater detail in Voit, Weymann, and Korista (1993). The crucial features of Mg II BAL QSOs that allow us to perform this analysis are the following: (1) Al III resonance-line absorption at 1860Å is always stronger than Al II resonance-line absorption at 1670Å, (2) the Mg II absorption equivalent widths imply Mg II column densities of at least $4 \times 10^{14}\,\mathrm{cm^{-2}}$, and (3) the C IV absorption troughs are very deep and extend to much higher velocities, in the QSO rest frame, than the lower-ionization troughs.

The fact that Al III is always stronger than Al II implies that the ionization parameter U, defined to be the density of hydrogen-ionizing photons divided by the density of H nuclei, must be greater than $10^{-2.5}$, for typical QSO ionizing spectra. At this high an ionization parameter, Mg II is not very abundant. Absorbing regions with $U > 10^{-2.5}$ will contain a large enough column of Mg II only if they become optically thick in the He II ionizing continuum beginning at 54.4 eV. Beyond the He II ionization boundary, He II-ionizing photons no longer affect the ionization balance, Mg III becomes the highest attainable ionization state of magnesium, and Mg II becomes much more abundant. At these depths in a QSO-ionized cloud of solar abundances, the optical depths of Mg II and Al III are quite similar, in accord with observations. Al III stronger than Al II also implies that the BAL region is not thick enough to become significantly neutral in its most heavily shielded parts. If all the Lyman continuum photons were blocked, Al II would become the dominant ionization state, and the Al II absorption line would be much more prominent. These arguments constrain the column densities of low-ionization BAL regions to be about $(10^{23}\,\mathrm{cm}^{-2})\,U$, to within a factor of a few, depending on the shape of the ionizing spectrum.

Because the Mg II–Al III gas must sit behind a shielding layer of higher-ionization gas, the N V BAL gas and the Mg II BAL gas must be spatially distinct, with the higher-ionization material lying closer to the QSO. C IV should be abundant in both zones, since the C III ionization potential is lower than that of He II. If the BAL gas were accelerating monotonically outward along the line of sight from us to the active nucleus, the low-ionization absorption features would then be positioned at the high-velocity ends of the C IV troughs, but we see the opposite. Broad absorption-line material *cannot* be accelerating monotonically outward along our line of sight. If the BAL gas were *decelerating* monotonically, we would not expect to see N V absorption in the velocity range of the Mg II trough, but we *do* see N V absorption down to low velocities. Apparently the BAL outflow is radially non-monotonic in velocity along our line of sight. The most plausible arrangement for the BAL velocity field seems to be one in which a range of velocities is present in each coarse-grained radial interval. This could occur if a hot outflow, already moving at $\sim 0.1c$ when it reaches the BAL region, were ablating and entraining cool material from some dense source, accelerating it to high velocities while tearing it apart through hydrodynamic instabilities.

3 KINETIC LUMINOSITIES OF MG II BAL QSOS

The kinetic luminosities of high-ionization BAL QSOs have always been difficult to evaluate. Because the amount of saturation in BAL profiles cannot be measured easily, these features provide only lower limits to the total column density. Furthermore, the covering factors of high-ionization BAL regions could range from 0.1 to 1, although recent studies show that low covering factors (~ 0.1) are more likely (Hammann, Korista, and Morris 1993). In Mg II BAL QSOs we can constrain the column density to a linear function of the ionization parameter U, and the [O III] anticorrelation of Boroson and Myers suggests that the BAL covering factor is high. These findings allow us to estimate the kinetic luminosities of Mg II BAL QSOs with more confidence than has previously been possible.

To estimate the mass outflow rate from a Mg II BAL QSO, we multiply the surface area of the BAL region, equal to $4\pi R_{\mathrm{pc}}^2 f_c$ in square kiloparsecs, where f_c is the BAL covering factor, by the mass per unit area of the BAL region, approximately $(10^{23}\,\mathrm{cm}^{-2})\,U\,m_{\mathrm{H}}$, and divide this number by the flow time across the BAL region, which is of order the BAL region thickness, $\Delta R_{\rm pc}$ in parsecs, divided by the flow velocity, $0.1\,v_{0.1}$ in units of c. Thus the mass flow rate is

$$\dot{M}_{\rm BAL} \sim (300\,{\rm M_\odot\,yr^{-1}})\,R_{\rm pc}^2\,\Delta R_{\rm pc}^{-1}\,v_{0.1}\,f_c\,U\ .$$

To estimate the kinetic luminosity, we multiply by velocity squared and find

$$\dot{E}_{\rm BAL} \sim (2\times 10^{47}\,{\rm erg\,s^{-1}})R_{\rm pc}^2\,\Delta R_{\rm pc}^{-1}\,v_{0.1}^3\,f_c\,U\ .$$

Unless U is close to the lowest allowed values, this kinetic luminosity is enormous.

However, the lowest values of U do not necessarily lead to the lowest total energy fluxes. The physical thicknesses of BAL absorbing clouds are much smaller than the sizes of BAL regions, implying that some kind of confining medium exists. Regardless of whether the BAL clouds are ram-pressure confined or thermal-pressure confined, the energy flux in the confining outflow will be similar to the thermal pressure in the BAL clouds times the outflow velocity. Ionization equilibrium then gives

$$\dot{E}_{\rm conf} \sim 0.1\,v_{0.1}\,E_{\rm ph}^{-1}\,U^{-1}\,L_{\rm ion}\ ,$$

where $L_{\rm ion}$ is the ionizing luminosity of the active nucleus, $E_{\rm ph}$ is the average energy per ionizing photon in eV, and we have assumed that $kT \sim 1\,{\rm eV}$ in the photoionized clouds. Since $E_{\rm ph}$ is several tens of eV in quasars, the energy flux of the confining medium exceeds that of the BAL material for $U < 10^{-2}$. When $10^{-2} < U < 10^{-1}$ the energy fluxes in the BAL material and its confining flow are similar, and they sum to 10% or more of the total radiative luminosity from the central source.

I thank Tim Heckman for encouraging me to think more deeply about the kinetic energy outputs of Mg II BAL QSOs, NASA grant NAGW-2144 for funding this research, and Isaac Shlosman and the rest of the organizing committee for arranging an enjoyable and educational meeting.

REFERENCES

Begelman, M. C., de Kool, M., and Sikora, M. 1991, *Ap. J.*, **382**, 416.
Boroson, T. A., and Meyers, K. A. 1992, *Ap. J.*, **397**, 442.
Hammann, F., Korista, K. T., and Morris, S. L. 1993, *Ap.J.*, in press.
Sanders, D. B. *et al.* 1988, *Ap. J.*, **325**, 74.
Sprayberry, D., and Foltz, C. B. 1992, *Ap. J.*, **390**, 39.
Voit, G. M., Weymann, R. J., and Korista, K. T. 1993, *Ap. J.*, **413**, in press.
Weymann, R. J., Turnshek, D. A., and Christiansen, W. A. 1985, in *Astrophysics of Active Galaxies*, ed. J. Miller (Mill Valley: University Science Books), p. 333.

The Aftermath of Tidal Disruption

Christopher S. Kochanek

Harvard-Smithsonian Center for Astrophysics, Cambridge, MA 02138

ABSTRACT

The circularization and accretion of the debris created by tidally disrupting a star passing near a supermassive black hole depends on the transverse structure of the debris stream. The transverse structure is modified by crossing points in the stream where the orbits are focussed across the stream center or through the orbital plane, the velocity shear across the stream, self-gravity, recombination, shocks, and shear viscosity. Stream-stream collisions may have a weak effect on the orbits because of Lense-Thirring precession, mismatched geometric cross sections, and the kinematics of collisions.

1 INTRODUCTION

Tidal disruption of stars passing near supermassive black holes ($M \sim 10^6$ $M_\odot$) provides a mechanism for fueling low-luminosity active galactic nuclei (AGN). It is ineffective for more massive AGN ($M \gtrsim 10^8$ $M_\odot$) because the tidal gravity of the black hole is too small to destroy a star before it passes through the event horizon. Other processes in dense central star clusters such as star-star collisions may provide a steady accretion rate (Hills 1975, 1978; Frank 1978; Young, Shields, and Wheeler 1977), but the disruption of a star may lead to an observable flare in the luminosity of the AGN. Lacy, Townes, and Hollenbach (1982) first understood the kinematics of disruption, and the current picture of tidal disruption is reviewed in Rees (1988) and Phinney (1989). For a star of mass M_* and radius R_* passing at pericentric distance R_p from a black hole of mass M, the strength of the encounter can be parametrized by the square root of the ratio of the surface gravity to the tidal gravity $\eta = (M_* R_p^3 / M R_*^3)^{1/2}$ (Press and Teukolsky 1977). A star disrupts if its orbit has $\eta \lesssim 1$.

The hydrodynamics of stellar disruption were first treated using models based on the classical ellipsoids (Carter and Luminet 1983, 1985; Luminet and Carter 1986; Kochanek 1992; Kosovichev and Novikov 1992) which confirmed the disruption process and demonstrated that there was a qualitative difference between distant encounters ($\eta \sim 1$) and deep plunging encounters ($\eta \ll 1$). Smooth particle hydrodynamic (SPH) simulations by Nolthenius and Katz (1982) and Bicknell and Gingold (1983) confirmed the ellipsoid model. Evans and Kochanek (1989) examined the $\eta = 1$ encounter in great detail to determine the distribution of the debris in energy and angular momentum, calibrating the exact widths of the distributions in energy but otherwise confirming the analytic picture of the debris orbits. Recent SPH simulations by Laguna *et al.* (1993) and Khokhlov *et al.* (1993*a, b*) examined plunging orbits as well, and found that the shapes of debris orbits are strongly modified by relativistic precession, but the changes in the energy and angular momentum distributions are not important.

Thus we have a clear picture of how the tidal fields disrupt the star and how the debris is distributed afterwards from both a theoretical and a numerical perspective. There are also detailed studies of how an accretion disk evolves once it forms (Frank 1978; Gurzadyan and Ozernoy 1980; Rees 1988; Cannizzo *et al.* 1990). There is a gap, however, in our understanding of how the disk is formed from the debris. The assumption used to leap over this gap is to assume that it happens quickly, so that the rate at which material returns to pericenter is the same as the accretion rate into the disk. The physical picture used to justify this is that relativistic precession induces stream-stream collisions, and shocks dissipate the orbital kinetic energy in the collisions (Rees 1988). But the efficacy of shocks depends strongly the geometry of the collision; a strong interaction requires that the two streams have comparable cross sections when they collide, otherwise they can interpenetrate with only a weak interaction between them. Kochanek (1993) developed a technique to study the evolution of the debris stream in detail, and in this discussion we will review some of the results.

2 SOME PHYSICAL EFFECTS IN THIN GAS STREAMS

In this section we briefly discuss several processes that modify the transverse structure of the stream. We do not discuss other processes such as shocks and viscosity — a more complete discussion is given in Kochanek (1993). The structure of a stream in the orbital plane is shown in Figure 1.

2.1 Focal Points

The stream is dynamically "cold" in the sense that it is supported by the kinetic energy of the transverse motions rather than the pressure of the fluid. Most of the time the stream acts like a swarm of freely orbiting particles. The exception occurs when the free orbits are focussed either through the orbital plane or across the stream center and the thickness of the stream in one dimension approaches zero. As the thickness shrinks, the pressure rises until it is large large enough to prevent the crossing.

Crossings through the orbital plane are generic and occur during each orbit, while the crossings in the orbital plane depend on the transverse shear in the stream. If the centerline of the stream is an elliptical Keplerian orbit, then the crossings in the plane are generic for sufficiently elliptical orbits. The debris is not, however, following a

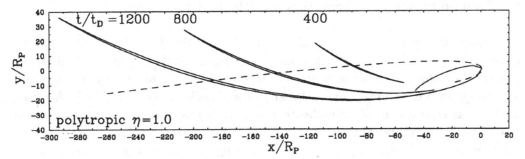

Fig. 1—The evolution of a polytropic debris stream. The figure shows the shape of the debris stream at various times in units of the dynamical time at pericenter $t_D = (R_p^3/GM)^{1/2} = 0.45 M_6^{-1/2} R_{p2}^{3/2}$ hours. The dashed line is the original parabolic orbit of the star, where the twisting of the orbit is caused by relativistic precession. For the sun passing a 10^6 $M_\odot$ black hole, the scale is $R_p = 100$ $R_\odot$. The black hole is located at the origin of the coordinates.

common orbit — each piece of the stream is following a different elliptical orbit, with a systematic gradient in energy and angular momentum along the stream. These gradients are large enough to prevent any stream crossings during the first orbit. On later orbits the gradients are weaker and crossings occur in the orbital plane.

2.2 Self-Gravity

After disruptions with $\eta \sim 1$ parts of the star remain self-gravitating in the transverse direction even though the debris is unbound in the direction along the stream. The ratio of the self-gravitational force to the tidal force in the vertical direction determines a critical density of

$$\frac{\rho_c}{\langle \rho_* \rangle} = \frac{2}{3} \frac{1}{\eta^2} \left(\frac{R_p}{r} \right)^3 \tag{1}$$

where $\langle \rho_* \rangle$ is the average density of the star, η is ratio of the self-gravity to tidal gravity at pericenter, and r is the current orbital radius. If the local stream density is greater than the critical density, then self-gravity is more important than tidal gravity. When the star leaves pericenter it is freely expanding, so the density scales as $\rho \propto r^{-3}$ which is the same as the radial variation in the critical density. Thus the central, high density regions of the star are dominated by self-gravity, and the outer, low density regions are dominated by tidal gravity. Deep plunging orbits for which $\eta \ll 1$ will be dominated purely by tidal gravity; the deepest orbit for which self-gravity matters is the orbit on which the critical density at pericenter is larger than the central density of the star ρ_{*0}, or $\eta \simeq (2\langle \rho_* \rangle/3\rho_{*0})^{1/2}$.

2.3 Recombination and Formation of Molecular Hydrogen

Adiabatic expansion rapidly cools the debris after the disruption, leading to the recombination of all ionized atoms. The recombination of Hydrogen and Helium can support the temperature of the debris for an extended period of time by storing internal energy until dynamical time scales and orbital radii are larger. If recombination begins near temperature T, then heating from recombination can maintain the gas near T for T_i/T dynamical times where $T_i \sim 10^5$ K is the recombination energy. The recombination of fraction T/T_i of the gas is enough to heat the gas to temperature T, and it takes one dynamical time for the expansion to remove the energy. The net effect is to keep the temperature about ten times higher than expected for a polytropic equation of state. The high densities make the three-body formation time scale for molecular Hydrogen (H_2) much shorter than the dynamical time scale, so the final state of the debris is H_2 not $H\,I$. There are probably a wide variety of chemical reactions occurring in the flow because of the high density, and the opacity of the material much later in the evolution of the system will change dramatically if the reactions lead to the formation of a large amount of dust.

2.4 The Ambient Medium and Radiative Heating

The stream interacts with both the accreting material needed to fuel the AGN and its radiation field. For the phase of evolution considered here, the ambient medium has no effect on the dynamics of the stream because of the extremely high density ratio between the stream and the medium. The ejected parts of the debris begin to interact strongly with the ambient medium near the broad line region of the AGN.

The stream is extremely optically thick, and the radiation from the AGN hitting the stream is absorbed to form a hot sheath of gas on the inner edge of the stream. The sheath is radiatively cooled, a process that is unimportant to the evolution of the stream in the absence of a strong heating mechanism like the AGN. The sheath shields the core of the stream from the effects of the AGN. The sheath is still kinematically cold, so it is dense and no larger than the stream core. This means that ionization levels in the sheath are near their equilibrium values and the sheath is not photoionized.

3 THE KINEMATICS OF STREAM-STREAM COLLISIONS

The streams remain thin after one orbit, so small amounts of Lense-Thirring precession can prevent the streams from intersecting. Typically the stream subtends about 0.2 degrees at the collision point as seen from the black hole, so a misalignment of the orbital and spin angular momenta of approximately 25 degrees may be enough to prevent a collision for a maximal Kerr hole. Even if the streams collide, they frequently have different geometric heights, in which case only parts of the larger stream participate in the collision.

The stream collisions are almost head on, with a typical angle between the streams of 130° to 140°. If the streams have velocities $\mathbf{v}_1$ and $\mathbf{v}_2$, and mass fluxes $\dot{M}_1$ and $\dot{M}_2$ then we can define a "center of mass" velocity $\mathbf{v}_c = (\dot{M}_1\mathbf{v}_1 + \dot{M}_2\mathbf{v}_2)/\dot{M}$ and a relative velocity $\Delta\mathbf{v} = (\dot{M}_1\dot{M}_2)^{1/2}(\mathbf{v}_1 - \mathbf{v}_2)/\dot{M}$, where $\dot{M} = \dot{M}_1 + \dot{M}_2$. Under most conditions $v_c \ll \Delta v$ and the gas "explodes" almost isotropically away from the collision point at velocity Δv with a small drift velocity v_c to conserve momentum. The average specific kinetic energy of the material is $(1/2)(v_c^2 + \Delta v^2)$ with a range from $(1/2)(v_c - \Delta v)^2$ to $(1/2)(v_c + \Delta v)^2$. This gives a range of binding energies comparable to the range between the binding energies of the streams before the collision. The collisions are never energetic enough to eject material from the system, and they are much more effective at changing the debris angular momentum than at changing the binding energy.

REFERENCES

Bicknell, G. V., and Gingold, R. A. 1983, *Ap. J.*, **273**, 749.

Cannizzo, J. K., Lee, H. M., and Goodman, J. 1990, *Ap. J.*, **351**, 38.

Carter, B., and Luminet, J. P. 1983, *Astr. Ap.*, **121**, 97.

Carter, B., and Luminet, J. P. 1985, *Mon. Not. Astr. Soc.*, **212**, 23.

Evans, C. R., and Kochanek, C. S. 1989, *Ap. J. Lett.*, **346**, L13.

Frank, J. 1978, *Mon. Not. R. Astr. Soc.*, **184**, 87.

Gurzadyan, V. G., and Ozernoy, L. M. 1980, *Astr. Ap.*, **86**, 315.

Hills, J. G. 1975, *Nature*, **254**, 295.

Hills, J. G. 1978, *Mon. Not. R. Astr. Soc.*, **182**, 517.

Khokhlov, A., Novikov, I. D., and Pethick, C. J. 1993a, *NORDITA* 92/74 preprint.

Khokhlov, A., Novikov, I. D., and Pethick, C. J. 1993b, *NORDITA* 92/69 preprint.

Kochanek, C. S. 1992, *Ap. J.*, **385**, 604.

Kochanek, C. S. 1993, *Ap. J.*, in press.

Kosovichev, A. G., and Novikov, I. D. 1992, *Mon. Not. R. Astr. Soc.*, **258**, 715.

Lacy, J. H., Townes, C. H. and Hollenbach, D. J. 1982, *Ap. J.*, **262**, 120.

Laguna, P., Miller, W. A., Zurek, W. H., and Davies, M. B. 1993, *LANL* 3716 preprint.

Luminet, J. P., and Carter, B. 1986, *Ap. J. Suppl.*, **61**, 219.

Nolthenius, R. A., and Katz, J. I., 1982, *Ap. J.*, **263**, 377.

Phinney, E. S. 1989, in Proc. IAU Symp. 136 on *The Center of the Galaxy*, ed. M. Morris (Dordrecht: Kluwer), p. 543.

Press, W. H., and Teukolsky, S. A. 1977, *Ap. J.*, **213**, 183.

Rees, M. J. 1988, *Nature*, **333**, 523.

Young, P. J., Shields, G. A., and Wheeler, J. C. 1977, *Ap. J.*, **212**, 367.

The Galactic Center — an AGN on a Starvation Diet

Heino Falcke and Peter L. Biermann

MPIfR, Auf dem Hügel 69, D-53010 Bonn, Germany

ABSTRACT

The Galactic Center shows evidence for the presence of three important AGN ingredients: a black hole ($M_{\bullet} \sim 10^6$ $M_{\odot}$), an accretion disk ($10^{-8.5} - 10^{-7}$ $M_{\odot}$ yr^{-1}) and a powerful jet (jet power $\geq$ 10% disk luminosity). However, the degree of activity is very low and can barely account for the energetics of the central region.

1 INTRODUCTION

The dynamical center of the Galaxy is the radio point source Sgr A*, which is also the center of the central star cluster (Eckart *et al.* 1993). Investigations of the enclosed mass in the central region show that there is evidence for a mass concentration of the order of 10^6 $M_{\odot}$ within the central arcsecond (Genzel and Townes 1987). There is good reason to assume that this "dark mass" indeed is the mass of a massive black hole (BH) powering Sgr A*. The total spectrum of this source from radio to NIR was compiled by Zylka *et al.* (1992). There is a flat radio spectrum up to 7 mm, and a steeply rising submm spectrum, which Zylka *et al.* interpret as thermal emission from a dust torus surrounding the BH. In the FIR one finds a spectral break at 30 μm indicated by upper limits and a third spectral component rising in the NIR, which has been interpreted as emission from an accretion disk around the BH.

2 HERTZSPRUNG-RUSSELL DIAGRAM FOR THE SGR A* DISK

Because of strong obscuration in the galactic plane we probably will never be able to measure exactly the optical and UV part of Sgr A*, which is needed to discriminate between different disk models. Nevertheless, there is a number of parameteres we can infer by indirect means to constrain possible models. For example, we know that in the outer part of the Galactic Center region almost 50% of the starlight is absorbed in the interstellar dust (Cox and Mezger 1989). IR measurements show that the dust concentration strongly peaks around Sgr A*, therefore we assume that all the luminosity of Sgr A* beyond the NIR is absorbed in the dust and reradiated at longer wavelength. Zylka *et al.* estimated the total luminosity in the central 30″ to be 1.5×10^6 $L_{\odot}$. Interestingly, the total luminosity of the central star cluster, as extrapolated from the outer parts (Falcke *et al.* 1993a) is of comparable order, thus one obtains an upper limit for the disk luminosity of $L_{\text{disk}} < 7 \times 10^5$ $L_{\odot}$. Further limits for L_{disk} and the effective temperature T_{eff} can be found by assuming that the NIR measurements represent the Rayleigh-Jeans tail of a black-body, leading to

Fig. 1—HR diagrams for BHs. The Figure on the left gives an overview for the whole parameter range of astrophysically relevant BHs. The small areas cover all possible values for effective temperature and luminosity of a BH of given mass and accretion rate but different inclination angles and angular momenta a. Areas at the same horizontal line have the same absolute accretion rate, areas at the same diagonal line from lower left to upper right have the same mass and areas at the same diagonal from the lower right to the upper left have the same accretion rate in terms of their Eddington rate. The boxes denote schematically the position of AGN, Sgr A* and stellar mass BHs. The Figure on the right is a zoomed version of the HR diagram for an edge-on disk and a fixed BH mass of 2×10^6 M$_\odot$.

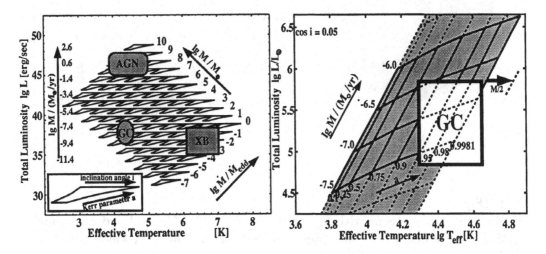

$L_{\text{disk}} > 7 \times 10^4$ L$_\odot$ and 20,000 K $< T_{\text{eff}} <$ 40,000 K.

On a HR diagram for BHs (Falcke *et al.* 1993a), we find that for a given $M_\bullet = 2 \times 10^6$ M$_\odot$ the data is consistent with a maximally rotating BH accreting $10^{-7} - 10^{-8.5}$ M$_\odot$ yr^{-1} in a disk seen edge-on (Fig. 1). A lighter BH with $M_\bullet = 10^3$ M$_\odot$ does not fit the current data. The accretion rate we find is more than 5 orders of magnitude lower than the Eddington limit of the BH and 7 orders of magnitude lower than in average AGN – the Galactic Center resembles an AGN on a starvation diet.

3 BLACK HOLE H II REGION

To learn more about the hidden UV spectrum of Sgr A*, one has to apply indirect methods, *i.e.* the heating and ionization of ambient gas by the disk spectrum. From the HR diagram one can see, that the disk is at the edge of producing an appreciable amount of ionizing UV photons, therefore the state of the ambient gas will be very sensitive towards changes of the parameters in the BH/accretion disk system. To test this we took the photoionization code HOTGAS developed by Schmutzler and Tscharnuter (1993) to calculate gas temperatures of a test cloud with constant density at different spatial positions around Sgr A* neglecting all dynamical and optical depth

Fig. 2—The temperatures of a test cloud of constant density ($n = 10^4$ cm^{-3}) exposed to a disk spectrum for an accretion rate of 10^{-7} M$_\odot$ yr^{-1} around maximally rotating BHs of mass 10^6 M$_\odot$ (left) and 10^3 M$_\odot$ (right). The disk lies horizontally in the center and is seen edge on.

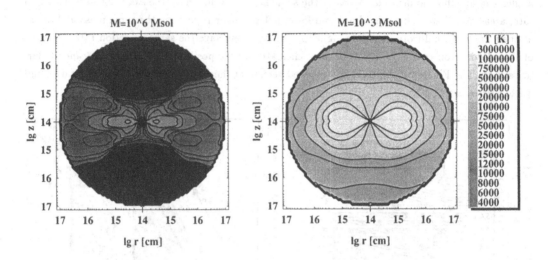

effects (Fig. 2).

For a set of parameters with $M_\bullet = 10^6$ M$_\odot$, $\dot{M} = 10^{-7}$ M$_\odot$ yr^{-1} and a maximally rotating BH, we find a very anisotropic temperature distribution, which reflects the relativistic beaming of the disk spectrum at high inclination angles, heating the gas to temperatures of $T_{\mathrm{gas}} \geq 10,000$ K. A moderate change of the angular momentum of the BH from $a = 0.9981$ to $a = 0.9$ would be enough to suppress the heating almost completely. On the other hand, a low mass BH (*e.g.* 10^3 M$_\odot$) inevitably leads to much higher temperatures ($10^5 - 10^6$ K) in the inner arcsec of the Galactic Center. No configuration is able to heat the total H II region Sgr A*. Nevertheless, once we are able to determine temperature and density distribution of the gas in the inner arcsecond of the GC we will have a powerful tool to discriminate between different models.

4 SGR A*: A JET?

After having shown that the IR–NIR spectrum is consistent with the presence of an accretion disk, it is straightforward to postulate the existence of a radio jet created at the inner edge of the disk producing the compact flat spectrum radio emission — a feature seemingly related to disks. This jet/disk link directly imposes an important constraint: the jet can not carry away more matter and more energy than is provided by the accretion process.

We adopt the Blandford and Königl (1979) jet model, which assumes a conically expanding, supersonic jet, with constant velocity and an internal gas pressure dominated by a turbulent magnetic field being in energy equipartition with relativistic electrons.

To account for the jet/disk link, we express the mass loss due to the jet $\dot{M}_{jet}$ and the total energy of the jet Q_{jet} in terms of the disk accretion rate $\dot{M}_{disk}$, such that

$$Q_{jet} = q_j \dot{M}_{disk} c^2 \quad \text{and} \quad \dot{M}_{jet} = q_m \dot{M}_{disk}. \tag{1}$$

Using the above parameterization, we find for the flux of Sgr A* a flat spectrum with $F_\nu = $ const and an absolute value of

$$F_\nu = 1 \text{ Jy} \times D(i, \gamma_j)^{13/6} \sin i^{1/6} \left(\frac{M}{3}\right)^{-11/6} \left(\frac{\Lambda}{9}\right)^{-5/6} \left(\gamma_j \beta_j \frac{q_m}{3\%} \frac{\dot{M}_{disk}}{10^{-7} \text{ M}_\odot \text{ yr}^{-1}}\right)^{17/12} \tag{2}$$

Here D is the Doppler factor, i the inclination of the jet axis, M the Mach number of the jet, β_j the velocity of the jet, γ_j the relativistic γ factor of the jet and Λ a logarithmic correction factor describing the relativistic electrons.

We see that even with the extremely low accretion rate in Sgr A* it is possible to feed a radio jet producing 1 Jy emission with a reasonable set of parameters. However, an appreciable fraction of the total mass accretion rate ($\geq 3\%$) has to be expelled by the jet. Using the energy equation of this system (Falcke *et al.* 1993b) this translates to a jet power of $Q_j \geq 0.04 \dot{M}_{disk} c^2$. Thus the ratio of the total jet power to the disk luminosity $L_{disk} \leq 0.3 \dot{M}_{disk} c^2$ has a lower limit of $Q_{jet}/L_{disk} \geq 0.12$.

Alas, there is a problem: for 20 years now, VLBI radio observations tell us that Sgr A* is a point source at all wavelengths. How can this be a jet? An answer is found by examining the structural predictions of the jet/disk model. The physical scale of the synchroton emission depends inversely on frequency yielding for the Sgr A* set of parameters

$$z \simeq 2 \cdot 10^{13} \text{ cm} \left(\frac{43 \text{ GHz}}{\nu}\right) \left(\frac{\gamma_j \beta_j}{\sqrt{1 + (\gamma_j \beta_j)^2}} \sqrt{\frac{9}{\Lambda} \frac{3}{M}} \frac{q_m}{3\%} \frac{\dot{M}_{disk}}{10^{-7} \text{ M}_\odot \text{ yr}^{-1}}\right)^{2/3}, \tag{3}$$

For the low accretion rate of Sgr A*, the scale of the jet then is smaller than the resolution of VLBI even at 43 GHz. Thus, one should see at best a marginally resolved central core. And indeed, this is confirmed by recent 43 GHz VLBI observations of Sgr A* (Krichbaum *et al.* 1993), where the emission ist still dominated by an unresolved central core which, however, is slightly elongated suggesting an underlying jet structure.

There is another interesting obervational consequence associated with equation (3). We can turn the argument around and ask: What is the shortest possible wavelength λ_{break} emitted by such a jet, namely the one emitted at the shortest length scale. For a given central mass, the smallest possible scale is the scale of a BH which is $R_g = 1.5 \times 10^{11}(M_\bullet/10^6 \text{ M}_\odot)$ cm. A reasonable and conservative guess for the smallest jet scale then also would be of the order of several R_g, say $z_{min} \geq 10R_g$. For a 10^6 $M_\odot$ BH we obtain $\lambda_{break} \leq 560$ μm and for a 10^3 $M_\odot$ BH we obtain $\lambda_{break} \leq 0.56$ μm. For wavelength shorter than λ_{break} we would expect to see a steepening of the spectral index from 0 to –0.5 or even steeper.

This explains very well the oberved lack of far infrared emission (shortwards 30 μm) from Sgr A* if there is indeed a 10^6 $M_\odot$ BH. On the other hand, there is no such argument for a 10^3 $M_\odot$ BH. As the break frequency should in this case be somewhere in the NIR we would rather expect a continuing flat spectrum visible also in the FIR, which is not observed.

4 CONCLUSIONS

Spectral and structural information of Sgr A* are consistent with the standard AGN triad black hole, jet and accretion disk, with converging evidence for a supermassive black hole (10^6 $M_\odot$), low accretion rate ($10^{-8.5} - 10^{-7}$ $M_\odot$ yr^{-1}) and a powerful radio-jet (as compared to the disk luminosity). This set of parameters is consistent with the NIR data, the dust luminosity, the radio spectrum, the size of the radio source and the lack of non-thermal FIR emission.

However, one question is still unanswered: Why is the central accretion rate so extremely low? Which mechanism prevents the large amount of gas in this region from being accreted onto the BH?

A plausible explanation would be to assume that the accretion process varies strongly radially and in time. Thus the central accretion rate could have been much higher in earlier epochs. A dust torus in the inner arcsecond, as suggested by sub-mm observations, could be just the outer part of a non-stationary accretion disk, serving as a reservoir where matter is temporarily stored until stronger accretion process in the inner parts sets in again.

Acknowledgement: We want to thank P. Mezger, W. Duschl, T. Krichbaum, K. Mannheim, M. G. Rieke and R. Zylka for extensive discussions on this topic. HF is supported by DFG grant (Bi 191/9). We thank the staff of the Vatican Observatory, where this article was written, for their kind hospitality.

REFERENCES

Blandford, R. D., and Königl, A. 1979, *Ap. J.*, **232**, 34.

Cox, P., and Meyger, P. G. 1989, *Astr. Ap. Rev.*, **1**, 49.

Eckart, A., Genzel, R., Hofmann, R., Sams, B. J., and Tacconi-Garman, L. E. 1993, *Ap. J. Lett.*, **407**, L77.

Falcke, H., Biermann, P. L., Duschl, W. J., Mezger, P. G. 1993a, *Astr. Ap.*, **270**, 102.

Falcke, H., Mannheim, K., and Biermann, P. L. 1993b, *Astr. Ap.*, submitted.

Schmutzler, T., and Tscharnuter, W. 1993, *Astr. Ap.*, **273**, 318.

Zylka, R., Mezger, P. G., and Lesch, H. 1992, *Astr. Ap.*, **261**, 119.

A Central Black Hole in M32?

Roeland P. van der Marel[1], Hans-Walter Rix[2], Dave Carter[3],
Marijn Franx[4], Simon D. M. White[5], and Tim de Zeeuw[1]

[1] Sterrewacht Leiden, Postbus 9513, 2300 RA Leiden, The Netherlands
[2] Institute for Advanced Study, Princeton, NJ 8540, U. S. A.
[3] Royal Greenwich Observatory, Madingley Road, Cambridge CB3 0EZ, England
[4] Kapteyn Instituut, Postbus 800, 9700 AV Groningen, The Netherlands
[5] Institute of Astronomy, Madingley Road, Cambridge, CB3 0HA, England

ABSTRACT

We have measured the line-of-sight velocity profiles of M32. The major axis velocity profiles are asymmetric, with opposite asymmetry on opposite sides of the nucleus. Existing models for M32 cannot account for these asymmetries. We present new models which assume the distribution function to be of the form $f = f(E, L_z)$. Such models require a central black hole of $\sim 1.8 \times 10^6 \, M_\odot$ to fit the observed rotation velocities and velocity dispersions. Without invoking any further free parameters, these models provide a good fit to the observed velocity profile asymmetries.

1 OBSERVED VELOCITY PROFILES

The presence of a massive black hole has been invoked to match the observed rotation velocities and velocity dispersions at the center of M32 (Tonry 1987; Richstone, Bower and Dressler 1990). Previous studies have assumed the line-of-sight velocity distributions of the stars, henceforth referred to as the *velocity profiles*, to be Gaussian. We have determined the velocity profile shapes of M32 from high S/N spectra taken with the William Herschel Telescope at La Palma (van der Marel *et al.* 1993), using the techniques of Rix and White (1992) and van der Marel and Franx (1993). The velocity profiles are asymmetric, with the asymmetry changing sign upon going from one side of the nucleus to the other (see Fig. 1). None of the existing models, in which the local (unprojected) velocity distributions of the stars are assumed to be Gaussian, can reproduce the observed asymmetries of the velocity profiles.

Fig. 1—Observed velocity profiles along the major axis of M32 at galactocentric distances of 2″, 1″, 0″, −1″, and −2″. Note the large velocity dispersion in the center and the rapid rotation and significant velocity profile asymmetry away from the center. The normalization is arbitrary.

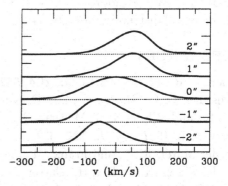

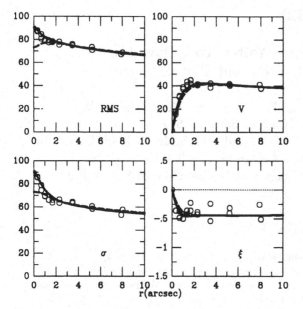

Fig. 2— $\sqrt{V^2 + \sigma^2}$, V, σ, and the skewness ξ for the major axis of M32. Circles are estimates for these quantities obtained from the observed velocity profiles. Results at negative radii were folded to positive radii, assuming reflection symmetry of the velocity profiles at opposite radii. The curves show the seeing convolved predictions of the two models described in the text (dashed curves: no black hole; solid curves $M_{BH} = 1.8 \times 10^6 \, M_\odot$).

2 MODELS

We infer the mass–density profile of M32 from the *HST* surface photometry (Lauer *et al.* 1992), assuming axisymmetry and constant M/L. The distribution function is assumed to be of the form $f(E, L_z)$ (*i.e.*, $\sigma_R = \sigma_z$). Given this, *all* projected velocity moments can be calculated from the (higher order) Jeans equations. The only freedom left is the separation of the second azimuthal velocity moment into streaming and dispersion. This freedom is used to fit the rotation V and dispersion σ separately, once the RMS projected line-of-sight velocity $\sqrt{V^2 + \sigma^2}$ has been calculated.

Figure 2 displays the results for the major axis of M32 (assumed edge-on). A model without a central black hole cannot fit the observed central peak in $\sqrt{V^2 + \sigma^2}$. However, a good fit is obtained when a central black hole of mass $1.8 \times 10^6 \, M_\odot$ is added. Both models provide an adequate fit to the observed skewness of the line profile, $\xi \equiv \overline{(v - V)^3}/\sigma^3$, without invoking any further free parameters! The black hole model also fits the data we have for four other slit position angles.

Our modelling does not yet imply that M32 *must* have a central black hole. To decide this issue, exploration of the full range of permissible anisotropic models with three integral distribution functions is required. Work along these lines is in progress.

REFERENCES

Lauer, T. R., *et al.* 1992, *A. J.*, **104**, 552.

Richstone, D. O., Bower, G., and Dressler, A. 1990, *Ap. J.*, **353**, 118.

Rix, H., and White, S. D. M. 1992, *Mon. Not. R. Soc.*, **254**, 389.

Tonry, J. L. 1987, *Ap. J.*, **322**, 632.

van der Marel, R. P., and Franx, M. 1993, *Ap. J.*, **407**, 525.

van der Marel, R. P., Rix, H., Carter, D., Franx, M., White, S. D. M., and de Zeeuw,
     P. T. 1993, *Mon. Not. R. Astr. Soc.*, submitted.

Stars and Disks around Massive Black Holes

Nico Roos[1] and Jelle S. Kaastra[2]

[1]Sterrewacht Leiden
[2]SRON Leiden

ABSTRACT

We calculated the orbital evolution of stars due to interaction with an accretion disk around a massive black hole in a galactic nucleus. After circularization the radius of a stellar orbit with initial inclination i to the disk shrinks by a factor $4/(1 + \cos i)^2$ before it settles in the plane of the accretion disk. Next, we calculate the rate at which stars from the star cluster around the hole are captured by a standard Shakura-Sunyaev disk. We find that the majority of captured stars are on retrograde orbits. These stars may reach a small separation from the hole before settling in the disk. AGN with $M \sim 10^6$ $M_\odot$ and $\dot{M} = M_{\rm Edd}$ are likely to have stars on inclined orbits with small separation from the hole, *i.e.* just outside the tidal disruption radius. Observational effects will be most conspicuous in low luminosity AGN.

1 EVOLUTION OF STAR ORBIT

The pre- and post-impact velocities $\mathbf{v}$ and $\mathbf{v_1}$ can be calculated from conservation of momentum: $m_* \mathbf{v} + \Delta m\ \mathbf{w} = (m_* + \Delta m)\mathbf{v_1}$, where Δm is the mass swept up by the star and $\mathbf{w}$ is the velocity of the accretion disk at the impact point. These three equations together with the requirement that the pre- and post-impact positions are the same, provide a set of four relations between the old and new orbital parameters. It is possible to find explicit expressions for the changes in the orbital parameters by linearizing these equations (Roos and Kaastra 1993, in preparation).

The evolution of the stellar orbit occurs in two stages: first, a *circularization stage* on a timescale $(2q\Delta m/pm_*)^{-1}$ times the orbital period (q and p are the apo- and pericenter distances, respectively), and second, a *grinding* stage in which both the radius and the inclination of the circular orbit decline on a timescale of order $(2\Delta m/m_*)^{-1}$ times the orbital period (see also Syer *et al.* 1991). An interesting result is that during the grinding stage the orbital radius shrinks by a factor $4/(1 + \cos i)^2$ before the star settles in the disk.

In Figure 1 some results are given for the orbital evolution time of a star interacting with an $\alpha = 1$ disk around a 10^6 solar mass black hole with $\dot{M} = 0.01$ $M_\odot$ yr^{-1}.

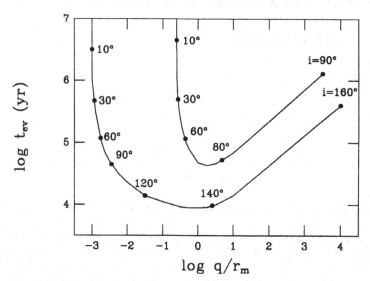

Fig. 1—Evolution time of the apocenter distance of stars with initial pericenter distance $p_0 = r_{\rm m}/2$ for an α-disk with $\alpha = 1$, $\dot{M} = 10^{-2}$ $M_\odot$ yr^{-1} and $M = 10^6$ $M_\odot$. The orbits have become nearly circular at $q/r_{\rm m} = 1$, where $r_{\rm m} = 1.5 \times 10^{14}$ cm. Solar-type stars are tidally disrupted at about 10^{13} cm.

2 CAPTURE RATE OF STARS.

Capture of stars by the disk occurs in the "meta-loss-cone" (Syer *et al.* 1991) where the two–body relaxation timescale for scattering of stars out of the cone is larger than the circularization time. We find that *most captured stars are on retrograde orbits* with initial pericenter distance around $r_{\rm m}$ and initial inclination $i \sim 130° - 150°$. These stars settle into the disk at typical radii of $\sim 0.1 - 0.01 r_{\rm m}$. The capture rate for an AGN like in Figure 1 is $\dot{N}_{\rm capt} \sim 10^{-4}$ yr^{-1}.

At such a small separation from the central hole the star will be destroyed by mass loss through Roche lobe overflow to the hole, by ablation or by a collision with another grinding star (on a more inclined orbit). Ablation occurs at a rate $\sim 3 \times 10^{-5} L_{\rm X,44}(r/10^{14}{\rm cm})^{-2}$ $M_\odot$ yr^{-1}. Roche lobe overflow is driven by angular momentum loss in star-disk collisions on a time scale $t_{\rm ev}$, which is of order 10^4 year for stars on inclined orbits. Note that in the case discussed above these time scales are comparable to the capture time $t_{\rm capt} = 1/\dot{N}_{\rm capt}$ so that the the expected number of stars orbiting at $r \sim 10^{13}$ cm is of order 1.

REFERENCES

Shakura, N. I., and Sunyaev, R. A., 1973, *Astr. Ap.*, **24**, 337.
Syer, D., Clarke, C. J., and Rees, M. J. 1991, *Mon. Not. R. Astr. Soc.*, **250**, 505.

Accretion onto Massive Binary Black Holes

Jelle S. Kaastra[1] and Nico Roos[2]

[1]SRON Leiden
[2]Sterrewacht Leiden

ABSTRACT

Massive binary black holes are expected to form in the nuclei of galaxies as a result of mergers between galaxies containing massive black holes. Evolutionary schemes (*e.g.* Roos 1988) where a galaxy merger leads to rapid evolution of a pre-existing wide binary towards a close binary predict that most binary black holes will be either wide (in a more or less undisturbed mode) or narrow (in a rather rapidly evolving stage). A nice example of a massive binary in such a rapidly evolving stage may have been found recently in the quasar 1928+738, where the observed wiggles in radio jets of this superluminal quasar could be interpreted as due to modulation of the jet velocity by the orbital motion of the binary (Roos, Kaastra, and Hummel 1993) We are doing numerical simulations of the (restricted) three body problem in order to study the evolution and accretion rate of massive binary black holes. Some results are presented.

1 SIMULATIONS

We are performing numerical simulations of the three body interactions between massive binaries with stars drawn randomly from a stellar cusp distribution around the holes. The equations of motion are solved using a pulsating-rotating coordinate system in which the binary is at rest (Szebehely 1967). The changes in orbital parameters of the binary are deduced from the change in energy and angular momentum of the star.

Most stars have a slingshot interaction with one of the holes, leading to (on average) energy loss of the binary. A small fraction of the stars approaching the binary is tidally disrupted or (for massive holes) swallowed whole by one of the black holes.

Figure 1 shows some results for runs with 10,000 interacting stars drawn from a cusp distribution containing 10^8 stars, with 1D velocity dispersion at the cusp radius of 200 km s^{-1}.

It is evident that the *stellar* accretion rate can be enhanced by a factor of 2–3 with respect to the single black hole case (limit of $\mu \rightarrow 0$). Also, most stars are accreted onto the primary hole; the fraction accreted to the secondary hole is minor unless the masses of the holes are comparable.

In Figure 2 we show an example of the evolution of a massive binary black hole neglecting loss-cone effects. Initially the evolution is rapid because many stars pass

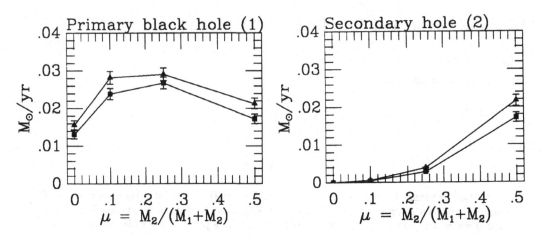

Fig. 1—Accretion of stars from a stellar cusp distribution onto the primary (left, $M_1 = 10^8 \, M_\odot$) and the secondary (right) of a massive binary black hole on a circular orbit. Results are given for an orbital radius of 3.5×10^{16} cm (triangles) and for 3.5×10^{17} cm (squares).

Fig. 2—Evolution of the orbital radius a_{bin} of a massive binary orbit ($M_1 = 10^8 \, M_\odot$, $M_2 = 10^7 \, M_\odot$) in a stellar cusp (outer radius 2×10^{19} cm, 10^8 stars). Loss-cone effects are neglected.

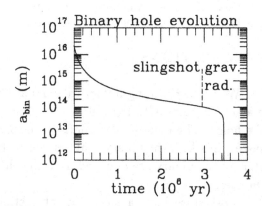

the wide binary orbit. As the orbit shrinks, fewer stars can have interaction and evolution slows down. The evolution time reaches a maximum of about 3×10^6 year when the energy loss via slingshot interactions becomes comparable to energy loss via emission of gravitational waves. Note that this happens at a separation of about 10^{16} cm, which is similar to that of the putative binary in 1928+738 (Roos *et al.* 1993).

REFERENCES

Kaastra, J. S., and Roos, N. 1992, *Astr. Ap.*, **254**, 96.
Roos, N. 1988, *Ap. J.*, **334**, 95.
Roos, N., Kaastra, J. S., and Hummel, C. A. 1993, *Ap. J.*, **409**, in press.
Szebehely, V. 1967, *Theory of Orbits* (New York: Academic Press).

Profile Variations of Broad-Line Radio Galaxies

Matthias Dietrich and Wolfram Kollatschny

Universitätssternwarte Göttingen, Geismarlandstr., D-37083 Göttingen, Germany

Broad-Line Radio Galaxies (BLRG) are the most extreme species of AGN regarding line width and structure of their optical emission line spectra. The FWZI amounts to $\Delta v \simeq 35,000$ km s^{-1} for some objects. Furthermore, the broad line profiles are more structured than these of Seyfert galaxies and show often double hump features (Chen, Halpern, and Filippenko 1989; Halpern 1990; Veilleux and Zheng 1991). The detailed analysis of the line profile variations of these AGN provides a powerful tool to get information about fundamental parameters of the Broad-Line Region (BLR) like size, geometry and especially the kinematics of the line emitting gas (Robinson, Pérez, and Binette 1990; Welsh and Horne 1991).

In 1989 we started a monitoring campaign of line profile variations of Broad-Line Radio Galaxies at Calar Alto Observatory/Spain on time scales of weeks to years. In the following we present some examples of emission line profile variations of BLRGs we are studying.

3C390.3 is a well known BLRG with prominent double-peaked Hα emission line profile. The spectra shown in Figure 1 were taken in October 89 and August 92. The line flux was scaled with respect to the narrow forbidden [O III]$\lambda\lambda$4959, 5007 and [O I]λ6300 emission lines. The corresponding difference spectrum is displayed at the bottom of Figure 1. The Hα and the Hβ line profiles show a double-peak structure in the difference spectrum. The humps are separated by $\Delta v \simeq 9,000$ km s^{-1}. A component at $v_{\rm rel} = 4600$ km s^{-1} described by Veilleux and Zheng (1991) is not visible for these epochs.

3C382 is a variable AGN with very broad emission lines, FWZI $\simeq 17,000$ km s^{-1} (Yee and Oke 1981). The spectra of 3C382 of two different epochs and the corresponding difference spectrum is presented in Figure 2. The spectra of the different epochs were scaled with respect to the [O III]$\lambda\lambda$4959, 5007 emission lines which are assumed to be constant. The difference spectrum show a characteristic double peak structure for Hα and Hβ (Fig. 2). The relative velocity shift of the blue and red hump is $v_{\rm rel} = -4,000$ km s^{-1} and $v_{\rm rel} = 1,000$ km s^{-1}, respectively.

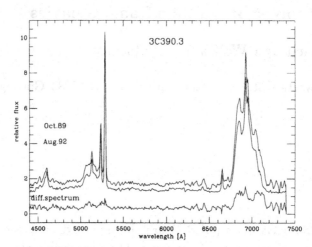

Fig. 1—Spectra of 3C390.3 taken at August 31, 1992 and October 28, 1989. The difference spectrum is shown at the bottom.

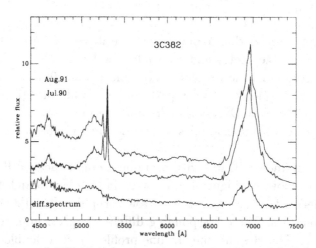

Fig. 2—Spectra of 3C382 taken at August 6, 1991 and July 16, 1990. The difference spectrum is shown at the bottom.

This work has been supported by DFG grant Ko857/13-1.

REFERENCES

Chen, K., Halpern, J. P., and Filippenko, A. V. 1989, *Ap. J.*, **339**, 742.
Halpern, J. P. 1990, *Ap. J. Lett.*, **365**, L51.
Robinson, A., Pérez, E., and Binette, L. 1990, *Mon. Not. R. Astr. Soc.*, **246**, 349.
Veilleux, S., and Zheng, W. 1991, *Ap. J.*, **377**, 89.
Welsh, W. F., and Horne, K. 1991, *Ap. J.*, **379**, 586.
Yee, H. K. C., and Oke, J. B. 1981, *Ap. J.*, **248**, 472.

The Budget of Energy in AGNs and Quasars

Prab M. Gondhalekar

Astrophysics Group, Rutherford Appleton Laboratory, Chilton, OXON, OX11 0QX, England

ABSTRACT

The balance of energy in AGNs and quasars is assessed. It is shown that there is no "energy budget" problem in AGNs and quasars and the covering factor of the broad line region in these objects is around 0.1-0.2 and this does not change with the luminosity.

1 INTRODUCTION

It has been claimed that there is an "energy budget problem" for some continuum energy distributions (CEDs) used to model the broad line region (BLR) in AGNs and quasars. In this paper the budget of energy in AGNs is assessed from UV and soft X-ray observations.

2 ENERGY BUDGET

Soft X-ray observations of AGNs were obtained from the *ROSAT/PSPC* archive (at the University of Leicester, England) and reduced in the standard way (see Gondhalekar *et al.* 1993 for details). A single power law was found to be adequate to represent the data and the power law index (Γ) is given in Table 1. The details of this analysis will be the subject of a separate publication. The soft X-ray spectra deviate significantly from the (best fit by eye) models of Laor and Netzer (1989) at energies higher than 0.5 keV.

The integrated ionizing radiation was obtained by integrating a least square interpolation from the UV to the X-ray data. The covering factor Ω was obtained from the relation $N_{\mathrm{Ly\alpha}} = f\Omega N_{\mathrm{ion}}$ where $N_{\mathrm{Ly\alpha}}$ is the integrated photon flux in the Lyα line, N_{ion} is the integrated flux of ionizing photons and f is the fraction of ionizing photons which eventually escape the BLR as Lyα photons ($f = 1.0$ is assumed here). The covering factor for six AGNs and quasars is given in Table 1. These covering factors are very similar to the covering factor of ~ 0.1 obtained for high redshift quasars (Smith *et al.* 1981) and suggests that the covering factor does not change with luminosity, this is in disagreement with the conclusions of Mushotzky and Ferland (1984).

Netzer (1985) has shown that observationally $E_{\mathrm{line}}/E_{\mathrm{Ly\alpha}} = 3 - 8$ where E_{line} is the total energy in the emission lines, $E_{\mathrm{line}} = \Omega E_{\mathrm{ion}}$ and E_{ion} is the total energy in the ionizing continuum. Some CEDs used to model the broad line spectrum do not have sufficient E_{ion}. Collin-Souffrin (1986) has claimed that there is an energy imbalance only if the high and low ionization lines are assumed to be emitted from the same region, she has postulated that these two groups of line may be emitted from different regions and the region from which the low ionization lines are emitted may be preferentially heated by either hard X-rays or a non-radiative process. The ratio $\Omega E_{ion}/E_{Ly\alpha}$ for six AGNs is shown in Table 1. Only a lower limit is obtained for NGC 5548 because the E_{ion} for this AGN was obtained by integrating the soft X-ray spectrum from 0.1 keV to $\sim$ 2.0 keV (Nandra *et al.* 1992). For these AGNs, which cover 6 dex in luminosity, there would appear to be no energy balance problem.

3 CONCLUSIONS

The ionizing continuum obtained from UV and soft X-ray observations has sufficient energy to account for the total energy in the emission lines and this is true over 6 dex in luminosity. The covering factor of the broad line region is 0.1–0.2 and this does not change with the luminosity.

TABLE 1
AGN CHARACTERISTICS

ID	m_v	z	M_{abs}	Ω	$\Omega E_{ion}/E_{Ly\alpha}$	Γ
NGC5548^a	13.73	0.017	-21.3	< 0.22	>1	
Mkn478^b	14.58	0.079	-23.8	0.15	4	-3.31
Q0414-06	15.94	0.773	-27.8	0.21	3	-2.74
Q1352+18	15.87	0.152	-23.9	0.14	3	-2.66
Q1411+14	14.99	0.089	-23.7	0.26	2	-3.11
Q1426+015	14.87	0.086	-23.7	0.22	4	-2.81

^{a}From Nandra *et al.* 1992.
^{b}From Gondhalekar *et al.* 1993.

REFERENCES

Collin-Souffrin, S. 1986, *Astr. Ap.*, **166**, 115.
Gondhalekar, P. M. *et al.* 1993, *Mon. Not. R. Astr. Soc.*, in press.
Laor, A., and Netzer, H. 1989, *Mon. Not. R. Astr. Soc.*, **238**, 897.
Mushotzky, R. F., and Ferland, G. J. 1984, *Ap. J.*, **278**,] 558.
Nandra, K., *et al.* 1992, *Mon. Not. R. Astr. Soc.*, **260**, 504.
Netzer, H. 1985, *Ap. J.*, **289**, 451.
Smith, M. G., *et al.* 1981, *Mon. Not. R. Astr. Soc.*, **195**, 437.

THE DETECTION OF MICROVARIATIONS FOR Akn 120

H. R. Miller[1] and J. C. Noble[2]

[1]NASA–JPL and Georgia State University
[2]Georgia State University

ABSTRACT

Rapid optical variations have been detected for the Seyfert 1 galaxy Akn 120 with time scales shorter than an hour. These variations are the most rapid which have been detected for any Seyfert galaxy, and, for the first time, clearly demonstrate that radio-quiet AGNs exhibit the phenomenon of microvariability. This result suggests that these variations are indepentent of the radio properties of these objects and are thus unlikely to be associated with any disturbance in a relativistic jet.

1 INTRODUCTION

Akn 120 was first identified as a Seyfert galaxy by Arakelian (1975) in a survey of high surface brightness galaxies. It has long been known to exhibit optical variations with time scales ranging from years to days (Miller 1979). In this paper, we report the first results of a program to determine if radio-quiet AGNs exhibit optical variations on time scales much shorter than a day.

2 OBSERVATIONS

The high time resolution optical observations of Akn 120 reported here were obtained with the 42-inch telescope at Lowell Observatory equipped with a direct CCD camera and an autoguider. The observations were made through a V filter with an RCA CCD. Repeated exposures of 60 seconds were obtained for the star field containing Akn 120 and several comparison stars. These standard stars, located on the same CCD frame as Akn 120 provide comparison stars for use in the data reduction process. The observations were reduced using the method described in Carini and Miller (1992). Each exposure is processed through an aperture photometry routine which reduces data as if it were produced by a multi-star photometer. Differential magnitudes can then be computed for any pair of stars on the frame. These simultaneous observations of Akn 120, several comparison stars, and the sky background allow one to remove variations which may be due to fluctuations in atmospheric transparency and extinction. The aperture photometry routine used for these observations is the apphot routine in *IRAF*.

3 RESULTS

The results of observations of Akn 120 which were obtained on 1993 January 1 are shown in the figure below. A well-defined outburst is detected near 6 hrs U.T. This outburst has a duration of approximately one hour and an amplitude of 0.05 mag. These observations are important because they demonstrate conclusively that microvariability is exhibited by (a) Seyfert galaxies and (b) radio-quiet AGNs. The latter point indicates that the origins of these variations are independent of the the radio properties of these objects and therefore of any relativistic jet which might be associated with the central engine. The authors thank Lowell Observatory for observing time and support from JPL and the CIF at GSU.

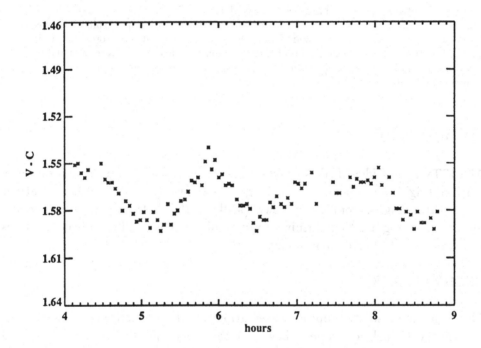

REFERENCES

Arakelian, M. A. 1975, *Soobshch. Byurak. Obs.*, **47**, 3.
Carini, M. T., and Miller, H. R. 1992, *Ap. J.*, **385**, 146.
Miller, H. R. 1979, *Astr. Ap. Suppl.*, **35**, 387.

The Contribution of Cool-Wind Reprocessing to AGN Spectra

Jason A. Taylor[1,2] and Demosthenes Kazanas[1]

[1]NASA/GSFC Code 655 LHEA, Greenbelt, MD 20771, U. S. A.
[2]University of Maryland, College Park MD 20742, U. S. A.

ABSTRACT

We have studied the effects on line emission due to the reprocessing of AGN continuum by cool ($T \lesssim 10^4$ K) stellar winds from stars, such as those from red giants, expected to be present in the region (Fabian 1979; Penston, 1988; Norman and Scoville 1988; Kazanas 1989). Using model stellar phase space distribution functions appropriate for stellar clusters with black holes, we have calculated the expected covering factors, line profiles, and two dimensional line transfer functions of these systems. Our stellar models contain between 10^7 and 10^8 stars within a parsec from the central source, of which we assume a fraction are able to reprocess continuum radiation. The line profiles of our models have FWHM up to 5,000 km sec^{-1} for black hole masses between 10^7 and 10^8 solar masses and covering factors up to approximately 10^{-1}. The line transfer functions we obtain peak at lags from 5 to 200 days for optically thick clouds. Forbidden line emission has also been studied for this effect. For some of our models, the wind line emission has been found to be quite significant.

1 INTRODUCTION

Active galactic nuclei (AGN) have very high stellar densities. That the effects of stars in the study of AGN has been largely ignored is somewhat surprising. In addition to the stellar densities, the X-ray continuum radiation is also quite high in AGN. We have calculated the magnitude and observational characteristics of line emission from reprocessing of the central continuum radiation of stellar winds of stars in AGN. For example, here we show covering factors, which give the magnitude of the line emission due to wind continuum reprocessing, line profiles, and transfer functions, which provide the time-dependent response in the lines due to a time-dependent continuum.

2 SUMMARY

Our approach is similar to that of Kazanas (1989), who suggested that the gas in the winds from the cool stars is heated from the continuum radiation until it becomes too hot to emit optical lines. The parameters of the winds for the model shown in Figs. 1 and 2 are essentially those used by Kazanas (1989).

From a dissertation to be submitted to the Graduate School, University of Maryland, by Jason Taylor in partial fulfillment of the requirements of the Ph. D. degree in Physics.

REFERENCES

Fabian, A. C. A. 1979, *Proc. R. Soc. Lond. A*, **366**, 449.

Kazanas, D. 1989, *Ap. J.*, **347**, 74.

Norman, C. A., and Scoville, N. Z. 1988, *Ap. J.*, **332**, 124.

Penston, M. V. 1988, *Mon. Not. R. Astr. Soc.*, **233**, 601.

Tout, C. A., Eggelton, P. P., Fabian, A. C., and Pringle, J. E. 1989, *Mon. Not. R. Astr. Soc.*, **238**, 427.

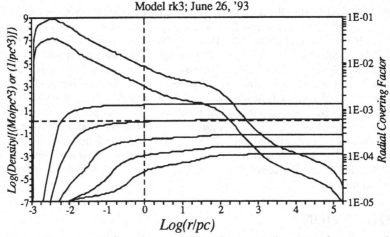

Fig. 1—From the upper left to lower right, the mass density of stars and number density of stars with reprocessing winds as a function of the distance from the continuum source for model rk3. Also shown is the radial covering factor for this model at luminosities of 10^{42} to 10^{46} erg sec^{-1}.

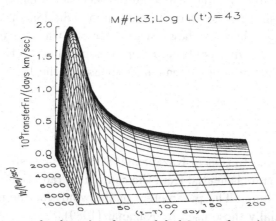

Fig. 2—The transfer function for model rk3 as a function of lag and velocity for an optically thick wind appropriate for luminosities close to 10^{43} erg sec^{-1}.

Ionized Gas Kinematics in Active and Related Galaxies

Mark Whittle

Department of Astronomy, University of Virginia

ABSTRACT

The kinematics of ionized gas in active (and normal) galaxies is reviewed. For clarity, discussion is divided first by emission region size, and then by galaxy type. Although a wide range of velocity fields are encountered on all scales, a number of recent developments are stressed : large scale outflows in Seyfert and Starburst galaxies (1 – 20 kpc); gravitationally dominated motion on intermediate scales in all galaxies (few $\times 10^2$ pc); radial flows of uncertain direction on small scales in Seyferts (3 – 100 pc); and possible continuity of velocity field in Seyferts down to very small scales ($\lesssim 0.1$ pc, BLR).

1 INTRODUCTION

Ionized gas can be found on many scales in both active and normal galaxies — on large scales $\sim 1 - 20$ kpc in the body and near environment of the host galaxy; on intermediate scales $\sim 0.1 - 1$ kpc in the bulge dominated regions; on small scales $\sim 3 - 100$ pc in the inner bulge cores; and on very small scales $\sim 0.01 - 1$ pc within the Broad Line Region (BLR). My intention is to review observations which shed light on the kinematics of this ionized gas, recognizing that the velocity fields on one scale may be quite unrelated to those on another.

Studies of ionized gas kinematics focus, of course, on emission lines and their Doppler profiles. First order information comes from the profile center and width, while higher order information comes from profile asymmetry, kurtosis, and substructure, as well as the comparison of these for different emission lines. Additional useful information comes from the host galaxy reference frame and/or gravitational field, which can be obtained from absorption line studies. Observational data is usually obtained in one of a few basic forms. Zero dimensional, single aperture spectra of the nuclear region ($\sim 2'' - 4'' \sim$ few $\times 10^2$ pc) yield a single set of emission line profiles. One dimensional, long-slit, spectra give velocity information along a line — usually the galaxy major or minor axis, or the radio axis. Two dimensional velocity information comes from multiple long-slit, multi-pupil, or Fabry-Perot observations. Not surprisingly, the number of objects studied in these various ways drops dramatically from zero- to one- to two- dimensional studies.

For clarity, our discussion will focus separately on the various different classes of (active) galaxy — Radio galaxies, Seyfert galaxies, LINERs, Starbursts, as well as brief comparisons with normal galaxies and our own galaxy. One aim, of course, is to look for systematic differences between the velocity fields in these different classes of galaxy. Such differences may highlight, for example, the role played by the active nucleus, the radio jet, or star formation.

Before describing the observational results, it is helpful to outline the basic velocity patterns — inflow, outflow, or some other form — and the physical contexts in which they might arise. Due to the extent of the subject, references are far from complete and I apologize in advance for the many omissions.

Inflows : The past decade has seen a large number of studies which aim to identify ways in which angular momentum can be drained from gas, allowing it to fall in towards the nucleus. These include studies of gas motion in spheroidal potentials (*e.g.,* Gunn 1977; Steiman-Cameron and Durisen 1984, 1988; Christodoulou *et al.* 1992; Habe and Ikeuchi 1988), bars (*e.g.,* Sanders and Huntley 1976; Shlosman *et al.* 1989, Athanassoula 1992), satellite passage (*e.g.,* Noguchi 1988), mergers (*e.g.,* Negroponte and White 1983; Barnes and Hernquist 1991; Noguchi 1991), self-gravitating disks (*e.g.,* Bailey 1980; Lin, Pringle, and Rees 1988; Shlosman and Begelman 1989), and magnetized disks (*e.g.,* Balbus and Hawley 1991; Christodoulou 1993). In a spheroidal potential, differential precession can lead to settling and infall on a timescale which depends on the orbital inclination and the form of the potential. While *net* inflow in bar simulations are undetectably low, streaming motions can include radial components which attain a significant fraction of circular velocity. In the case of satellite passage and mergers gravitational torques between the induced gas and stellar bars are sufficiently powerful to allow infall on dynamical timescales. A necessary condition for rapid infall in all these cases is that the gas cool efficiently. The infalling gas can settle into rings or disks which need not share the angular momentum of the host galaxy, leading to the possibility of a kinematically distinct gaseous subsystem or, if the gas forms stars, stellar subsystem. It is also worth noting that over a Hubble time, continued gas dissipation, settling, and subsequent star formation may be the principle factor which governs the shape and depth of the near nuclear potential, generating significantly deeper potentials than would otherwise result from the evolution of purely collisionless systems. Part of this dissipative evolution may, of course, include formation of a massive object.

Outflows : The most commonly considered form of outflow is a wind, driven either by stellar processes or an active nucleus. In the case of stellar input, the kinetic energy from supernova ejecta or winds from OB or Wolf Rayet stars is rapidly thermalized in shocks to produce a very hot bubble which expands, sweeping up and driving shocks into the ISM (*e.g.,* Chevalier and Clegg 1984; Norman and Ikeuchi 1989; MacLow *et al.* 1989). If the hot bubble attains a size comparable to the scale height of the disk gas, it expands rapidly along the minor axis in a "blowout" phase, entraining and carrying ISM well above the galactic plane. In the case of an AGN as the energy source, radiation pressure or radiative heating may drive an outflow (*e.g.,* Begelman *et al.* 1983, 1989), or a particle wind may flow directly from the AGN, perhaps as an MHD wind from an accretion disk (*e.g.,* Blandford and Payne 1982; Heyvaerts and Norman 1989). The final geometry of the wind depends in large part on the geometry of the confining medium and to a lesser extent on the geometry of the energy input, and can range from spherical, through conical, to bipolar. The most extreme bipolar flows are, of course, jets. Although classical radio jets do not themselves contain line emitting gas, they can interact with and accelerate ionized gas leaving emission line signatures. The form of the interaction is not yet clear, though possibilities include acceleration at the jet head through the bow shock (*e.g.,* Taylor *et al.* 1992), entrainment along the jet (*e.g.,* Blandford and Konigl 1979; Coleman and Bicknell 1985) or lateral acceleration around the jet driven by an expanding cocoon (*e.g.,* De Young 1986; Begelman and Cioffi 1989).

Other Forms : The majority of gas in galaxies is, of course, rotationally supported and the inflow/outflow velocity fields described above can be viewed as deviations from this overall graviational motion. Dispersion support may be dominant in contexts where gas has a low filling factor, perhaps as mass loss from bulge stars. Finally, there may be situations in which more general chaotic or turbulent motion is prevelant, either locally or globally, for example in cooling flows.

2 RESOLVED SCALES ($\sim 1 - 20$ KPC)

2.1 Radio Galaxies

The two most recent systematic studies of emission line gas in and around radio galaxies are those of Tadhunter *et al.* (1989) and Baum *et al.* (1992). They measured or compiled long-slit data on the extended emission line regions (EELRs) of ~ 40 low redshift radio galaxies, measuring both global velocity patterns and local velocity dispersions. Adopting the terminology of Baum *et al.*, the radio galaxies can be roughly divided into three kinematic groups. Calm Rotators show global rotation and low local linewidth. They tend to be found surrounding FR-II radio sources whose host galaxies show signs of distortion but are nevertheless isolated or in low density environments. The gas kinematic axis is aligned with the radio axis and the gas is of high ionization, probably photoionized by a hard nuclear spectrum. The interpretation of these sources is that they are recent post merger systems in which the EELR gas has been aquired externally, and is rapidly settling to fuel the nucleus. The overall lifetime is therefore quite short ($\sim 10^8$ yr). Calm Non-rotators have locally narrow lines but show no global rotation. They tend to be found surrounding FR-I radio sources whose hosts are giant elliptical or cD galaxies in a rich environment. A number are found in cooling flow clusters and the gas is generally of low ionization, possibly ionized and heated by shocks and/or the hot gaseous environment. The interpretation of these sources is that their EELRs originate either as cooling flows or as recycled ISM. The processes are therefore quasi-continuous leading to a long lifetime ($\sim 10^9$ yr). Violent Non-rotators share many of the characteristics of the

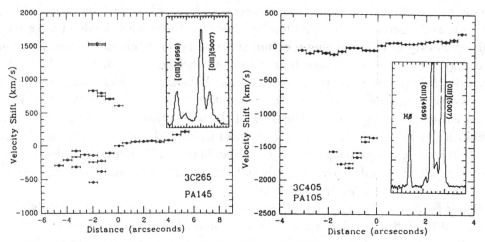

Fig. 1—Emission line velocities for 3C 265 (PA=145, left) and 3C 405 (PA=105, right). [O III]λ4959, 5007 profiles are inset. (From Tadhunter 1991).

calm rotators, while having higher local velocities, possibly related to radio source interactions.

Overall, the global and local velocity amplitudes are consistent with gravitational motion in the potential of the host galaxies. In a few cases ($\lesssim 20\%$) there is also evidence for interactions with the radio source. Two extreme examples of this are shown in Figure 1, taken from Tadhunter (1991). Along the radio axes of 3C 405 and 3C 265, a few arcseconds off nucleus, there are high velocity components red and blueshifted by $\sim 1,000 - 1,500$ km s^{-1}. In these cases it is interesting that the interactions with the ionized gas does not seem to have influenced the stability and propagation of the jets.

2.2 Seyfert Galaxies

Although the classical narrow line regions (NLR) of Seyferts are highly compact, most are slightly resolved in ground based spectra and there are a number of examples where emission is found extending over a few kpc. In some Seyferts there is a low surface brightness extended component (ENLR) thought to arise from normal disk ISM photoionized by the (possibly anisotropic) nuclear UV source (Unger *et al.* 1987).

Rotation. When resolved, the NLR of most Seyferts show at least some rotation, usually following the overall rotation curve of the host galaxy, as does the low surface brightness ENLR emission. There is evidence that the rotation curves of some Seyferts have large regions of solid body form (Keel 1993) and a local maximum near turnover (Afanasiev and Shapovalova 1993), though there are also cases of very steep, possibly unresolved, rotation (Whittle 1992*b*).

Biconical Outflows. Some Seyferts have well resolved NLR emission with clear signatures of minor axis outflow, probably in a biconical form (*e.g.*, NGC 7582,

Fig. 2—[O III] emission cone superimposed on continuum (dotted) contours
in NGC 3281. Double [O III] profiles from
cone center shown below. (From Storchi-
Bergmann *et al.* 1992).

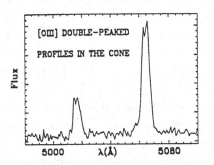

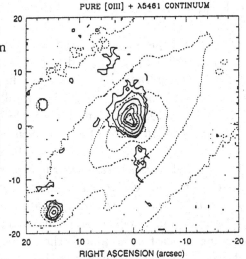

Morris *et al.* 1985; NGC 1365, Jorsater *et al.* 1984; Edmunds *et al.* 1988; NGC 7469,
Wilson *et al.* 1986; NGC 3281, Storchi-Bergmann *et al.* 1992). Figure 2 shows, for
NGC 3281, a classic [O III] cone of high excitation gas, with possible counter-cone
partially obscured behind the galaxy disk. Spectra show double peaks in the cone
center, becoming single peaks at the cone edge, suggesting conical outflow of a few
hundred km s^{-1}. It is not yet clear what drives the outflows in these cases. For the
Seyfert/Starburst hybrid systems NGC 7582 and NGC 1365, strong circumnuclear
star formation may drive the outflows. In NGC 3281, however, there are no obvious
signs of circumnuclear star formation and so its outflow may be driven by the AGN.

Jet Driven Outflows. Long slit studies of Seyferts with linear radio sources frequently show red and/or blue shifted components located close to the radio lobes
and superimposed on a more general rotation pattern (Whittle *et al.* 1988; Cecil and
Rose 1984). In the most clearcut cases the component velocities lie well outside the
rotation curve and can switch sign across the nucleus, strongly suggesting a bipolar
outflow. In more detailed Fabry-Perot studies of NGC 1068 and M51, high velocities
are found in the vicinity of the radio 'jet' although in these cases it is not clear whether
the flow is best described by radial acceleration, lateral cocoon driven expansion, or
even a fast wind (*e.g.*, Cecil *et al.* 1990; Cecil 1988).

Bar Driven Inflows. NGC 1068 is particularly suitable for studies of extended
kinematics because emission is found across the entire galaxy. Extensive Fabry-
Perot data in several emission lines has been discussed by Cecil *et al.* (1990). More
recent work considers the large scale velocity field (Bland-Hawthorn and Cecil 1993).
The iso-velocity map shows an approximately flat rotation curve with spiral (density
wave) perturbations beyond ~ 1.5 kpc. At intermediate radii, the contours twist in
a manner consistent with elliptic streaming in a barred potential with an orientation
consistant with the nuclear bar seen in the near IR. If a model of these components is
subtracted from the data, the residuals show 'bi-symmetric spiral streamers' spanning
$\sim 20'' - 5''$. This residual pattern seems to fit the classic pattern of gas slowing at a

bar shock, losing angular momentum, and flowing inwards with projected streaming velocities ~ 50 km s^{-1}.

Using multiple long slits and a multi-pupil fiber spectrograph, Afanasiev and Sil'chenko (1990) have mapped the velocity fields in several Seyferts, particularly those with linear radio sources. They find unusual velocity components spatially associated with the radio lobes, in overall agreement with the long-slit data from Whittle *et al.* (1988). Afanasiev and Sil'chenko, however, chose to interpret the velocity field not in terms of a jet driven outflow but in terms of inflow along a bar, where shocks at the leading edge of the bar give rise to the emission line components and linear radio structure. In order to explain the high infall velocities in some of the objects, they consider a counter-circulating vortex at the end of an inner bar. Possible objections to such an explanation center on the unusual nature of the Seyfert radio sources and velocity components compared to those of normal barred spirals, suggesting at least some link to the active nucleus. Their work nevertheless points to the need for more detailed two-dimensional studies and the possible roles played by nuclear bars and jets.

2.3 Starbursts

Rotation. To some extent, the kinematic properties of starbursts may depend on the way in which the sample has been selected. For a number of modest luminosity starbursts selected from the Markarian lists, DeRobertis and Shaw (1988) find approximately normal rotation.

Biconical Outflows. However, in samples of more luminous starbursts selected to have high far-infrared luminosities, clear evidence of minor axis outflows are found (*e.g.,* Heckman *et al.* 1990, 1992). In these objects a number of features suggest that an intense nuclear starburst has driven a wind out of the galaxy disk. Images in Hα show cones or bubbles along the minor axis. Recent *ROSAT* images show X-ray emission extended along the minor axis (*e.g.,* Petre 1993). Spectra show double emission lines and generally larger linewidths along the minor axis than along the major axis. Nuclear spectra show extended blue wings suggesting an outflow in which a dusty disk obscures the far side. In NGC 1808, spectra around NaD show blueshifted absorption and redshifted emission, confirming the outward direction of flow (Phillips 1993). In this object the NaD column suggests a neutral component of $\sim 10^8$ M$_\odot$ dominating the mass and at an outflow velocity ~ 400 km s^{-1} providing a kinetic energy $\sim 10^{56}$ ergs, comparable to the energy released in the observed nuclear starburst integrated over the lifetime of the superwind. The existance of these starburst driven superwinds has important implications for the chemical evolution of galaxies since evolved gas can be recycled back onto the disk via this nuclear 'fountain'. At higher redshifts, such enriched halo gas may provide the sites of metal line absorption systems seen in the spectra of background quasars.

Inflows. Inflow of gas to galactic nuclei is thought to predate the episode of rapid star formation which drives the outflowing winds described above. Evidence for such inflows is somewhat indirect, relying on morphological rather than kinematic arguments. For example, a number of studies show that interacting galaxies have enhanced *nuclear* emission relative to normal galaxies — including Hα, FIR, molecular, or radio (*e.g.,* Condon *et al.* 1982; Joseph *et al.* 1984; Keel *et al.* 1985). Presumably, as found in numerical simulations, gas moves down to the nuclear regions as the galaxy responds to tidal interactions. A nice demonstration of this effect comes from our recent Hα objective prism survey of ∼ 200 sprial galaxies in 8 nearby Abell clusters (Moss and Whittle 1993). We classify the Hα emission seen on the *prism* plates as compact (nuclear; median size ∼ 7″ ∼ 4 kpc) or diffuse (disk-wide; median size ∼ 18″ ∼ 10 kpc), while we independently classify the galaxy morphology on *direct* plates as disturbed or non-disturbed. Figure 3 shows not only that disturbed galaxies have a much greater likelyhood of being detected in Hα emission ($P_{\text{null}} \sim 10^{-4}$) but also that there is a strong association between galaxy disturbance and compact nuclear emission ($P_{\text{null}} \sim 10^{-10}$).

Fig. 3—Fraction of distorted and undistorted spiral galaxies detected with diffuse disk-wide emission (open shade) and compact nuclear emission (dense shade). Galaxy distortion leads not only to enhanced Hα emission, but specifically enhanced nuclear emission. (From objective prism survey of 8 Abell clusters by Moss and Whittle 1993).

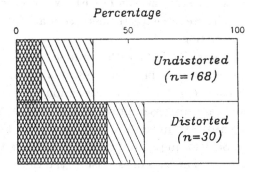

3 INTERMEDIATE SCALES (≲ FEW × 10² PC)

Data on ionized gas velocities on intermediate scales (∼ few × 10² pc) comes from the core of emission line profiles obtained with nuclear apertures. The basic parameter, FWHM, provides a rough estimate of the overall velocity amplitude. While inadequate for providing detailed velocity information it does allow us to identify the dominant acceleration mechanisms acting on the gas on these length scales.

3.1 Seyferts

The fact that early samples of Seyferts clearly had broader lines than starbursts, despite both samples drawing heavily on the Markarian lists, helped fuel the notion that Seyfert lines are broad because of the presence of an active nucleus. Before one can conclude this, however, it is important to test whether the line widths can be explained in terms of simple gravitational velocities in the host galaxy potential. Whittle (1992a, b, c) presents such an analysis for a sample of 140 Seyferts. The principle studies which provide line width information include Feldman *et al.* (1982), Heckman *et al.* (1984), DeRobertis and Osterbrock (1984), Whittle (1985a), Vrtilek and Carleton (1985), Busko and Steiner (1988), and the recent high resolution study by Veilleux (1991). Parameters which characterize the nuclear gravitational velocities are host galaxy rotation amplitude, ΔV_{rot}, and the absolute blue magnitude of the host galaxy bulge M_{bul}. A more direct gravitational parameter is the stellar velocity dispersion, σ_*, which has only recently been measured for a significant number of Seyferts (Terlevich *et al.* 1990; Nelson and Whittle 1993). Figure 4a is taken from Whittle 1992c and illustrates the basic result — for most Seyferts the dominant velocity field in the NLR is of gravitational origin and is <u>not</u> driven by the active nucleus. The linewidths correlate well with bulge luminosity, following closely the classic Faber-Jackson (1976) law for elliptical and spiral bulges (dashed line).

There are nevertheless secondary influences which contribue to the scatter on this correlation.

<u>Jets:</u> Seyferts with luminous linear radio sources (plotted as + symbols) have systematically broader lines at a given bulge luminosity, suggesting additional line broadening due to jet interactions. This is the same phenomenon identified in long slit data discussed previously in section 2.2. Digressing for a moment, if we want to identify examples of more powerful jet interactions we should look at objects similar to Seyferts in radio morphology and size but with much higher radio luminosity. Such a class of objects does exist — the Compact Steep Spectrum (CSS) radio galaxies and quasars. A high dispersion study of the CSS class indeed reveals broader forbidden lines than essentially all other classes of active galaxy, confirming the importance of jet interactions in at least some of these objects (Gelderman and Whittle 1993).

<u>Disturbance:</u> Seyferts which appear disturbed or have a nearby companion (plotted as Δ symbols in Figure 4a) also have systematically broader lines at a given bulge luminosity. Although such perturbations are identified on large scales, they evidently influence the velocity field in the inner regions.

<u>Bars:</u> The sample of Seyferts with sufficiently clear morphology to assign a reliable bar designation is significantly smaller than the sample plotted in Figure 4a. There is some indication, however, that unbarred Seyferts show a tighter correlation on the FWHM *vs* M_{bul} plot than the barred Seyferts, suggesting that bars may also influence the NLR velocity field.

<u>Inclination:</u> By analysing the inclination dependence of the scatter on the correlations between FWHM and both ΔV_{rot} and M_{bul} it is possible to identify to what extent the NLR velocity field is coplanar with the galaxy disk and therefore to what extent projection effects influence the observed linewidths. It seems that the observed FWHM contain both a projected and non-projected component (in approximately

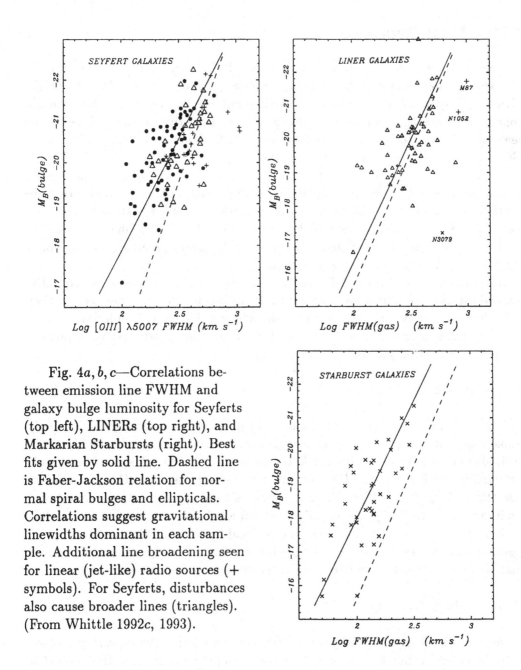

Fig. 4*a*, *b*, *c*—Correlations between emission line FWHM and galaxy bulge luminosity for Seyferts (top left), LINERs (top right), and Markarian Starbursts (right). Best fits given by solid line. Dashed line is Faber-Jackson relation for normal spiral bulges and ellipticals. Correlations suggest gravitational linewidths dominant in each sample. Additional line broadening seen for linear (jet-like) radio sources (+ symbols). For Seyferts, disturbances also cause broader lines (triangles). (From Whittle 1992*c*, 1993).

equal degree). While some of this is likely to result from including normal galaxy rotation inside the nuclear apertures, particularly for the more distant Seyferts, there nevertheless seems to be a genuine non-planar component to the NLR velocity field. Whether this reflects inner disk motion inclined to the galactic plane, or a true turbulent or dispersion supported component is not yet clear.

3.2 LINERs and Starbursts

Repeating much of the above analysis for LINERs and Markarian Starbursts yields very similar results (Whittle 1993). Figures 4*b* and 4*c* show FWHM ([O III], or Hα, or [N II]) *vs* M_{bul} for samples of LINER and Starburst galaxies. Statistically strong correlations again support nuclear gravity as the primary factor determining linewidth. Scatter to the high velocity side can be traced to strong winds (*e.g.* NGC 3079, see Veilleux *et al.* 1993) or probable jet interactions (*e.g.* NGC 1052, M87). Furthermore, overall offsets from the Faber-Jackson relation can be interpreted in terms of mass to light ratio. LINERs fit the F–J relation because their hosts are essentially normal spirals, the Seyferts are somewhat offset suggesting statistically lower M/L (which is also supported by offsets from the Tully-Fisher relation, see Whittle 1992*c*), while the Starbursts are significantly offset suggesting an even lower M/L ratio, as one expects from ongoing star formation.

We can now account for the difference in mean linewidth between Seyferts, LINERs, and (Markarian) Starbursts: Seyferts and LINERs have broad lines because they inhabit luminous early type spirals with high nuclear virial speeds, while (Markarian) Starbursts have narrow lines because they inhabit low luminosity late type galaxies with low nuclear virial speeds.

3.3 Normal Galaxy Nuclei

Since line emission is quite weak in normal galaxy nuclei, especially those of early Hubble type, obtaining reliable kinematic information is quite difficult. Phillips *et al.* (1986) give nuclear Hα or [N II] FWHM for a sample of E/S0 galaxies while Bertola *et al.* (1984) and Fillmore *et al.* (1986) present long-slit gas (Hα, [O II]) and stellar kinematics. Roughly speaking, the gas velocities have approximately gravitational amplitude. Inside the bulge, gas usually rotates faster than the stars, but slower than the circular velocity, and often has significant local linewidth. Fillmore *et al.* (1986) suggest the Hα comes from mass loss from bulge stars, reflecting rotation/dispersion dynamics for a time until it settles inwards or expands to become invisible.

3.4 Our Galaxy Nucleus

Despite superficial complexity arising from our proximity, the *overall* gas velocities in the inner few hundred parsecs have gravitational amplitude, in the sense that our galaxy would fit on the relations shown in Figure 4 (*e.g.* at $\lesssim$ 300 pc gas has $\Delta V \sim 200$ km s^{-1} and $M_{bul} \sim -18.0$). If we speculate that a Seyfert NLR would be somewhat smaller in a galaxy with M_{bul} of only -18, then it is appropriate to look at gas velocities on scales $\sim$ few pc in the galactic center for the purposes of comparison with Seyferts. The molecular and ionized gas on these scales, while having some non-circular and random components, still follows the rotation curve expected from the

2 μ stellar light distribution. At ~ 1.7 pc the circumnuclear disk is slightly inclined ($\lesssim 20°$) with moderate turbulent velocities. These characteristics are consistent with those found for Seyfert NLRs.

Given the similar origin for the near nuclear (few $\times 10^2$ pc) velocity field of ionized gas in Seyferts, LINERs, Starbursts, Normal galaxies, and even our own Galaxy, one is tempted to speculate that if a strong hard ionizing source were placed at the center of all the non-Seyfert galaxies, we would find kinematically normal NLRs. Thus, the ionization criteria used to define the separate galaxy classes distracts us from their kinematic similarities. It is interesting to further speculate that, apart from ionization degree, the *physical* properties (*e.g.* densities, pressures) of the near nuclear gas in all these classes also has a similar origin.

4 SMALL SCALES ($\sim 3 - 100$ PC)

Information on gas velocities in even more nuclear regions comes from profile bases and wings. Most of the information pertains to Seyfert galaxies. Typical Seyfert [O III]$\lambda 5007$ profiles have significantly broader wings relative to their cores than do Gaussians, suggesting velocities increase towards the nucleus. A more convincing demonstration of increasing velocities comes from the fact that the FWHM for lines of high critical density are frequently broader than those of low critical density. Because gas density almost certainly increases towards the nucleus, then so too must gas velocity.

In addition to being broader, the [O III] wings are frequently asymmetric, with the blue wing extending to higher velocities than the red wing. For Seyferts the statistical excess of blue asymmetries is very strong, with typical asymmetry parameters A20 $\sim 0.1 - 0.3$ (A20 characterizes the ratio of the difference between the blue and red extents in the base to the total base width). Since the lines are forbidden, the profile asymmetries indicate a significant radial velocity component with dust opacity suppressing the red side. Despite knowing about profile asymmetries for over a decade, the embarrassing fact is that the *direction* of the flow — in or out — is still not known, because the location of the dust is not known. If the dust is distributed throughout the NLR or in a nuclear disk then emission from the far side is suppressed, implying outflow. If the dust is inside the NLR clouds then we see less emission from clouds on the nearside (because they radiate preferentially on the side facing the central UV source), implying inflow.

Thus, we have two clear alternatives, (a) energy input by the AGN drives a fast outflow giving broad asymmetric wings, and (b) the nuclear gravitational potential is steep giving higher nuclear velocities and infall. Although there are a few observational results which have been interpreted to favour inflow (DeRobertis and Shaw 1990; Whittle 1985c) these are not conclusive and the question of flow direction is still unsettled. We can briefly weigh the relative merits of each. Outflow certainly seems plausible, even probable — given the proximity to the AGN and its possible energy input to the most nuclear regions. The fact that FIR luminous starburst galaxies also have nuclear blue asymmetries argues strongly, by analogy, for outflow since superwinds are unambiguously identified on larger scales in these galaxies. In this case one suspects a nuclear disk (perhaps the same disk which collimates the ionizing radiation in AGN) blocks radiation from the far side. In addition, there is a moderately good correlation between radio luminosity and wing width suggesting radio source expansion may be driving the outflow. The difficulty with AGN driven outflows is that the wing widths are largely oblivious of the overall level of activity, with similar velocities in modest luminosity Seyferts and luminous quasars.

Although gravity and infall may at first sight seem less appealing, it is still important to test this possibility (*e.g.* Whittle 1992*d*). The strongest general argument for the importance of gravity in the base and wing velocities is that the [O III] profiles have approximately similar shapes — a line with a broad core also has a broad base and wings, and a line with a narrow core also has a narrow base and wings. Thus, if the *core* velocities are gravitational then so are the base and wings. This is supported to some extent by correlations between bulge gravitational parameters and base and wing widths, although the strength of these correlations is less than for the core width.

If wing widths are gravitational, what constraints does this place on the form of the nuclear potential? We need to estimate the relative velocities and radii of the gas emitting the core and wings. Considering the core and wings as a two component Gaussian requires FWHM(core)/FWHM(wings) $\sim 0.4 - 0.6$ to give realistic values of profile kurtosis. To estimate radii, we assume the core gas has density $n_e \sim 10^2$ cm^{-3} (from [S II] line ratios), the wing gas has $n_e \sim 10^7$ (critical density), and the region as a whole has approximately constant radiation parameter, *i.e.* $n_e \propto r^{-2}$. Thus, $r(\text{core})/r(\text{wings}) \sim 10 - 100$. Combining these and characterizing the velocity field as a power law we obtain $V(r) \propto r^{-0.1} - r^{-0.3}$ which, for circular velocity, gives a mass density $\rho(r) \propto r^{-2.2} - r^{-2.7}$ and, for constant M/L, gives a surface brightness $I(r) \propto r^{-1.2} - r^{-1.7}$. Comparing these with data on Seyferts is thoroughly undermined by their nuclear light contribution. Even checking with normal galaxies is difficult since the length scales of interest are comparable to the seeing disk even for nearby galaxies. It does seem, however, that these relatively steep nuclear potentials are not excluded by the limits found from ground and space based imaging. Finally, we ask whether such potentials can readily explain the profile asymmetries in terms of infall. If we associate the difference between red and blue velocities as the radial component and the total velocity as the circular velocity, then observed asymmetry parameters imply $V(\text{radial})/V(\text{circular}) \sim 0.2 - 0.9$. Assuming rapid loss of angular momentum and infall on a dynamical time within the potentials defined above, we require free-fall through $\sim 10\% - 50\%$ of the initial radius to achieve the necessary radial velocity. Whether or not one regards this as reasonable depends on how efficiently angular momentum can be lost, and on whether simple elliptical streaming constitutes a viable velocity field incorporating a strong radial component.

6 VERY SMALL SCALES ($\lesssim 0.1$ PC, BLR)

Velocities of ionized gas in the broad line region (BLR) are still comparatively poorly understood, despite being easier to observe than velocities in the NLR. Line width *vs* luminosity relations have often been interpreted in terms of gravitational motion in the Keplerian potential of a central massive object, accreting at a significant fraction of the Eddington rate (*e.g.*, Padovani 1989; Wandel 1991). Comparison of line shapes (*e.g.*, Hα/Hβ, Hβ/He II, Hβ/He I, C IV/Lyα) generally indicate higher velocities at smaller radii (*e.g.*, Shuder 1982). This result is supported by the recent extensive variability studies which find shorter lag times for higher ionization lines (*e.g.*, Peterson 1993). Although line asymmetries are common, there are no clear systematic effects and the majority of the variability studies show little evidence for systematic radial flow.

Two interesting results point to a global link between NLR and BLR velocity fields. First, there is a loose but statistically significant correlation between FWHM of Hβ(broad) and [O III]λ5007 (Cohen 1983; Whittle 1985b). Second, in at least one object (PG 2251+113, Espey 1993), there is a continuous correlation between line width and critical density, ranging from the forbidden lines through the BLR semi-forbidden lines to the BLR permitted lines, where densities are estimated assuming the lines are emitting at thermal rates. A complete picture of the nuclear regions must account for these kinematic correlations which span 5 decades in radius. If the BLR velocities are indeed tied to a central object, then we require a globally coherent nuclear potential spanning the bulge, the inner nucleus, and the central object itself.

REFERENCES

Afanasiev, V. L., and Shapovalova, A. I. 1993, these proceedings.

Afanasiev, V. L., and Sil'chenko O. K. 1990, *Special Astrophysical Observatory*, Preprint Nos. 55, 57, 58, 59.

Athanassoula, E. 1992, *Mon. Not. R. Astr. Soc.*, **259**, 345.

Bailey, M. E. 1980, *Mon. Not. R. Astr. Soc.*, **191**, 195.

Balbus, S. A., and Hawley, J. F., 1991, *Ap. J.*, **376**, 214.

Barnes, J. E., and Hernquist, L. 1991, *Ap. J. Lett.*, **370**, L65.

Baum, S. A., Heckman, T. M., and van Breugel, W. J. M. 1992, *Ap. J.*, **389**, 208.

Begelman, M. C., and Cioffi, D. F. 1989, *Ap. J. Lett.*, **345**, L21.

Begelman, M. C., McKee, C. F., and Shields, G. A. 1983, *Ap. J.*, **271**, 70.

Begelman, M. C., de Kool, M., and Sikora, M. 1991, *Ap. J.*, **382**, 416.

Bertola, F., Bettone, D., Rusconi, L., and Sedmak, G. 1984, *A. J.*, **89**, 356.

Bland-Hawthorn, J., and Cecil, G. 1993, these proceedings.

Blandford, R. D., and Konigl, A. 1979, *Astrophys. Lett.*, **20**, 15.

Blandford, R. D., and Payne, D. G. 1982, *Mon. Not. R. Astr. Soc.*, **199**, 883.

Busko, I. C., and Steiner, J. E. 1988, *Mon. Not. R. Astr. Soc.*, **232**, 525.

Cecil, G. 1988, *Ap. J.*, **329**, 38.

Cecil, G., and Rose, J. A. 1984, *Ap. J.*, **287**, 131.

Cecil, G., Bland, J., and Tully, R. B. 1990, *Ap. J.*, **355**, 70.

Chevalier, R. A., and Clegg, A. W. 1985, *Nature*, **317**, 44.

Christodoulou, D. M. 1993, these proceedings.

Christodoulou, D. M., Katz, N., Rix, H. -W., and Habe, A. 1992, *Ap. J.*, **395**, 113.

Coleman, C. S., and Bicknell, G. V. 1985, *Mon. Not. R. Astr. Soc.*, **214**, 337.

Condon, J. J., Condon, M. A., Gisler, G., and Puschell, J. J. 1982, *Ap. J.*, **252**, 102.

De Robertis, M. M., and Osterbrock, D. E. 1984, *Ap. J.*, **286**, 171.

De Robertis, M. M., and Shaw, R. A. 1988, *Ap. J.*, **329**, 629.

De Robertis, M. M., and Shaw, R. A. 1990, *Ap. J.*, **348**, 421.

De Young. D. S. 1986, *Ap. J.*, **307**, 62.

Edmunds, M. M., Taylor, K., and Turtle, A. J. 1988, *Mon. Not. R. Astr. Soc.*, **234**, 155.

Espey, B. 1993, these proceedings.

Faber, S. M., and Jackson, R. E. 1976, *Ap. J.*, **204**, 668.

Feldman, F. R., Weedman, D. W., Balzano, V. A., and Ramsey, L. W. 1982, *Ap. J.*, **256**, 427.

Fillmore, J. A., Boroson, T. A., and Dressler, A. 1986, *Ap. J.*, **302**, 208.

Gelderman, R., and Whittle, M. 1993, *Ap. J. Suppl.*, in press.

Gunn, J. E. 1979, in *Active Galactic Nuclei*, ed. C. Hazard, and S. Mitton (New York: Cambridge University Press), p. 213.

Habe, A., and Ikeuchi, S. 1988, *Ap. J.*, **326**, 84.

Heckman, T. M., Miley, G. K., and Green, R. F. 1984, *Ap. J.*, **281**, 525.

Heckman, T. M., Armus, L., and Miley, G. K. 1990, *Ap. J.S.*, **74**, 833.

Heckman, T. M., Lehnert, M. D., and Armus, L. 1992, in *Evolution of Galaxies and Their Environments*, Grand Tetons (Dordrecht: Kluwer), in press.

Heyvaerts, J., and Norman, C. 1989, *Ap. J.*, **347**, 1005.

Jorsater, S., Lindblad, P.O., Boksenberg, A. 1984, *Astr. Ap.*, **140**, 288.

Joseph, R. D., Meikle, W. P. S., Robertson, N. A., and Wright, G. S. 1984, *Mon. Not. R. Astr. Soc.*, **209**, 111.

Keel, W. C. 1993, these proceedings.

Keel, W. C., Kennicutt, R. C., Hummel, E., and van der Hulst, J. M. 1985, *A. J.*, **90**, 708.

Lin, D. N. C., Pringle, J. E., and Rees, M. J. 1988, *Ap. J.*, **328**, 103.

MacLow, M. -M., McCray, R., and Norman, M. L. 1989, *Ap. J.*, **337**, 141.

Morris, S. L. Ward, M. J. Whittle, M., Wilson, A. S. and Taylor, K. 1985, *Mon. Not. R. Astr. Soc.*, **216**, 193.

Moss. C., and Whittle, M. 1993, *Ap. J. Lett.*, **407**, L17.

Negroponte, J., and White, S. D. M. 1983, *Mon. Not. R. Astr. Soc.*, **205**, 1009.

Nelson, C. H., and Whittle, M. 1993, these proceedings.

Noguchi, M. 1988, *Astr. Ap.*, **203**, 259.

Noguchi, M. 1991, *Mon. Not. R. Astr. Soc.*, **251**, 360.

Norman, C. A., and Ikeuchi, S. 1989, *Ap. J.*, **345**, 372.

Padovani, P. 1989, *Astr. Ap.*, **209**, 27.

Peterson, B. M. 1993, *P. A. S. P.*, **105**, 247.

Petre, R. 1993, in *The Nearest Active Galaxies*, ed. J. Beckman (Madrid: CSIC), in press.

Phillips, A. C. 1993, *A. J.* **105**, 486.

Phillips, M. M., Jenkins, C. R., Dopita, M. A., Sadler, E. M., and Binette, L. 1986, *A. J.*, **91**, 1062.

Sanders, R. H., and Huntly, J. M. 1976, *Ap. J.*, **209**, 53.

Shlosman, I., and Begelman, M. C. 1989, *Ap. J.*, **341**, 685.

Shlosman, I., Begelman, M. C., and Frank, J. 1989, *Nature*, **338**, 45.

Shuder, J. M. 1982, *Ap. J.*, **259**, 48.

Steiman-Cameron, T. Y., and Durisen, R. H. 1984, *Ap. J.*, **276**, 101.

Steiman-Cameron, T. Y., and Durisen, R. H. 1988, *Ap. J.*, **325**, 26.

Storchi-Bergmann, T., Wilson, A. S., and Baldwin, J. A. 1992, *Ap. J.*, **396**, 45.

Tadhunter, C. N. 1991, *Mon. Not. R. Astr. Soc.*, **251**, 46p.

Tadhunter, C. N., Fosbury, R. A. E., and Quinn, P. J. 1989, *Mon. Not. R. Astr. Soc.*, **240**, 225.

Taylor, D., Dyson, J. E., and Axon, D. J. 1992, *Mon. Not. R. Astr. Soc.*, **255**, 351.

Terlevich, E., Diaz, A.I., and Terlevich, R., 1990, *Mon. Not. R. Astr. Soc.*, **242**, 271.

Unger, S. W., Pedlar, A., Axon, D. J., Whittle, D. M., Meurs, E. J. A., and Ward, M. J. 1987, *Mon. Not. R. Astr. Soc.*, **228**, 671.

Veilleux, S. 1991, *Ap. J. Suppl.*, **75**, 383.

Veilleux, S., Cecil, G., Tully, R. B., Bland-Hawthorn, R., and Filippenko, A. V. 1993, these proceedings.

Vrtilek, J.M., and Carleton, N.P., 1985, *Ap. J.*, **294**, 106.

Wandel, A. 1991, *Astr. Ap.*, **241**, 5.

Whittle, M. 1985a, *Mon. Not. R. Astr. Soc.*, **213**, 1.

Whittle, M. 1985b, *Mon. Not. R. Astr. Soc.*, **213**, 33.

Whittle, M. 1985c, *Mon. Not. R. Astr. Soc.*, **216**, 817.

Whittle, M. 1992a, *Ap. J. Suppl.*, **79**, 49.

Whittle, M. 1992b, *Ap. J.*, **387**, 109.

Whittle, M. 1992c, *Ap. J.*, **387**, 121.

Whittle, M. 1992d, in *Testing the AGN Paradigm*, AIP Conference Proceedings No. 254, eds. S. S. Holt, S. G. Neff, and C. M. Urry, (New York: AIP), p. 607.

Whittle, M. 1993, in *The Nearest Active Galaxies*, ed. J. Beckman (Madrid: CSIC), in press.

Whittle, M., Pedlar, A., Meurs, E. J. A., Unger, S. W., Axon, D. J., and Ward, M. J. 1988, *Ap. J.*, **326**, 125.

Wilson, A. S., Baldwin, J. A., Sze-Dung Sun, and Wright, A. E. 1986, *Ap. J.*, **310**, 121.

Observations of Non-Circular Cold Gas Motions in the Circumnuclear Regions of Galaxies

Jeffrey D. P. Kenney

Astronomy Department, Yale University

ABSTRACT

Recent observations of non-circular cold gas motions in the central 100 pc — 1 kpc regions of galaxies are summarized, with an emphasis on relating the observed local motions to net inflow and outflow rates.

1 INTRODUCTION

Non-circular gas motions have been detected in many galaxies, providing empirical evidence that radial gas flows continue to affect the evolution of galaxies long after the initial epoch of galaxy formation. This paper reviews observations of non-circular motions of the cold (molecular and atomic) gas in the 100 pc – 1 kpc circumnuclear regions of disk galaxies. This is typically the region of galaxies where bulges begin to dominate the potential, where rotation curves change from steeply rising to flat, and where circumnuclear starbursts occur. Molecular gas is generally the dominant phase of the interstellar medium by mass in the central regions of gas-rich disk galaxies, and for this reason many of the observations discussed in this review are of CO or other tracers of molecular gas. The weakness of H I emission in circumnuclear regions, and the related difficulty of determining H I kinematics at high resolution, make H I results less common but valuable. Although on average inflowing gas is probably colder than outflowing gas, there are examples of both outflowing molecular gas and inflowing ionized gas.

The relationships between the observed local non-circular motions and the net radial flow rates vary considerably, and in several important cases the net flow rates are much smaller than the local non-circular motions. It is hard to directly detect net flows in galaxies for several reasons. 1.) Within disks, observations are sensitive to radial motions only near the minor axis, and one can't be sure whether radial motions are in the same or the opposite sense near the major axis. 2.) Gas may pile-up and be detectable only in regions where the radial motion is inward, and be less dense and undetectable on the part of the orbit where the radial motion is outward (Athanassoula 1992). 3.) Net radial flow rates are often small compared to circular speeds of $\sim$100 km s^{-1} and streaming and random motions $\sim$10 km s^{-1}, and in this case are virtually impossible to directly detect. Mass flow rates can be significant even if the net flow speeds are small: a flow speed of 1 km s^{-1} in a gas disk with a surface density of 300 M$_\odot$ pc^{-2} corresponds to a mass flow rate of 1 M$_\odot$ yr^{-1}. Outflows in galaxies can be easier to detect than inflows, since it is easier for the net flow speed to exceed the local circular speed for outflows than for inflows. In axisymmetric systems, gas must lose the z-component of its angular momentum to flow inward, and this usually limits the inflow speed to be less than the circular speed. No such constraint exists for outflowing gas, or for inflowing gas in strongly non-axisymmetric systems. 4.) Some inflows may happen rapidly (*e.g.,* in mergers or the collapse of gravitationally unstable gas lump) but the duration of the rapid flow is short, which means the chances of finding the system in this state is small.

2 INFLOWING GAS ALONG BARS

Gas responds strongly to bars, and bars are very common in galaxies. The torques on gas exerted by stellar bars are probably the most efficient mechanism for driving radial inflow of gas in non-interacting galaxies, at least on scales from a couple kpc down to a couple hundreds pc (Combes 1988). But a single bar can only move gas about an order of magnitude in radius, since bars themselves have inward as well as outward limits to their extent. For an extensive review of barred galaxy dynamics, including comparions with observations, see Sellwood and Wilkinson (1993).

The barred spiral galaxy M83 has relatively strong CO emission associated with the dust lanes along the leading edge of the optical bar (Handa *et al.* 1990), consistent with the theoretical prediction that the gas shocks here. The steep velocity gradients expected in a shock front have not been directly observed at the dust lane of any barred galaxy in H I or CO, perhaps due to insufficient spatial resolution or sensitivity. The best evidence that dust lanes trace shocks comes from optical emission lines in the barred spiral NGC 6221, which show a steep velocity gradient (150 km s^{-1} within 200 pc) across the dust lane of the bar (Pence and Blackman 1984). The streaming motion of the gas is outward before the dust lane, and abruptly shifts to inwards right after the dust lane, consistent with the idea that the dust lane traces a shock front. High resolution observations of the kinematics in the shock region, as well as the region near the bar minor axis where gas densities are low, are needed to relate local streaming motions to net inflow rates.

Some strongly barred galaxies have weak H I (Hunter *et al.* 1988) and CO (Kenney *et al.* 1992) emission at radii where the bar is strong, probably indicating that gas has been radially driven by the bars. Galaxies that do have H I (Sancisi, Allen, and Sullivan 1979) or CO (Handa *et al.* 1990) detected in the barred region generally show evidence for radial streaming motions, which in some cases are as high as 50-100 km s^{-1} . The isovelocity contours typically have a characteristic S-shape, similar to those which would be produced by elongated streamlines due to an oval distortion in the gravitational potential, and perhaps also similar to beam-smeared velocity discontinuities at shock fronts.

Many galaxies with weak stellar bars contain strongly barlike gas distributions. The center of nearby Scd galaxy IC 342 harbors a barlike molecular gas structure ∼500 pc in extent, and a modest nuclear starburst ∼70 pc in extent (Lo *et al.* 1984; Ishizuki *et al.* 1990*a*; Turner and Hurt 1992). The CO morphology and kinematics have been interpreted both as a gaseous response to a weak stellar bar although none has yet been detected (Lo *et al.* 1984; Ishizuki *et al.* 1990*a*), and as gaseous density wave spiral arms (Turner and Hurt 1992). Non-circular motions of ∼50 km s^{-1} in the molecular arms could plausibly be due to either bar or spiral arm streaming motions. Determining the true nature of the barlike gas structure is important for estimating the net radial flow rate from the observed local non-circular motions, since the inflow rate is expected to be different in the 2 cases. NGC 6946 (Ball *et al.* 1985; Ishizuki *et al.* 1990*b*) and M101 (Kenney, Scoville, and Wilson 1991) are other nearby Sc galaxies with mildly oval stellar potentials and strongly barlike molecular gas distributions.

Some people seem to think that the dust lanes along the leading edges of bars represent direct highways to the nucleus (or at least to the inner Lindblad resonance), but it's not that simple. Numerous numerical simulations of non-self-gravitating gas responding to quasi-steady state stellar bars show that gas moves inward after being shocked at the dust lane, but that the gas eventually leaves the dust lane and moves outward after passing the bar minor axis (*e.g.*, Sanders and Huntley 1976; Roberts, Huntley, and van Albada 1979). Thus gas loses a bit of its angular momentum at each dust lane crossing, but circulates about the bar and crosses a dust lane many times on its descent toward the circumnuclear region. Athanassoula (1992) recently studied the shape of the shock as a function of different parameters of an idealized analytical bar, and estimates that the net inflow rate is typically one or two orders of magnitude less than the maximum local inward radial velocity. This is important to keep in mind when interpreting velocity maps of barred galaxies, particularly since the largest inward motions are predicted to occur in the regions of highest gas density, which are often the only regions detected in the observations!

It has been known for many years that there are non-circular gas motions as large as 200 km s^{-1} in the central part of the Milky Way. Longitude-velocity maps of H I and CO show complex patterns which are clearly inconsistent with a simple model such as circular motions plus uniform expansion or inflow. Binney *et al.* (1991) and Blitz *et al.* (1993) have recently made a good case that most of the non-circular gas kinematics in the Milky Way can be explained if the gas follows orbits similar to the closed stellar orbits expected in a barred potential. The orbits elongated along the bar are called x_1 orbits, and a family of orbits nested deeper inside the bar, which are oriented perpendicular to the bar, are called x_2 orbits. If we viewing the bar edge-on, then these orbits form parallelogram-like patterns in the longitude-velocity plane, which agrees with many features of the observations. Their best-fitting bar has a viewing angle of 15 degrees and corotation at 2.4 kpc, roughly consistent with the bar inferred from near-infrared maps (Blitz and Spergal 1991) of our galaxy. An even better fit to the data should be obtained if deviations from the closed stellar orbits are considered. Gas probably piles up along the leading edge of the bar, as Bally *et al.* (1988) have proposed for the Sgr B and $l = 1.5°$ molecular gas complexes. Even the molecular gas at -190 km s^{-1} observed (in projection) near Sgr A* is likely to be foreground gas participating in bar streaming motions, rather than gas directly associated with events at the galactic nucleus (Liszt and Burton 1993). Even with this gain in understanding of kinematics, it is difficult to calculate a net inflow rate of gas, although Stark *et al.* (1991) effectively argue that the giant molecular clouds near the galactic center are doomed to spiral inwards.

The ends of many bars are sites of vigorous star formation. One such bar-spiral arm transition zone in M83 harbors a large molecular gas complex containing 2 spatially and kinematically distinct components (Kenney and Lord 1991). This complex may have formed by orbit crowding, where outward streaming gas from the bar region merges with inward streaming gas from the spiral arm. The origin of this gas complex is probably similar to the twin gas concentrations oriented perpendicular to large scale bars which are observed near the inner Lindblad resonances (ILRs) of some barred galaxies (Kenney *et al.* 1992).

3 FLOWS INSIDE INNER LINDBLAD RESONANCES

The central parts of many barred galaxies are rich in molecular gas, suggesting that radial inflow slows down at certain radii. In the barred galaxies NGC 4314 (Combes *et al.* 1992; Benedict, Smith and Kenney, in preparation) and NGC 1097 (Gerin, Nakai, and Combes 1988), there are CO rings near ILRs. The strongest CO emission is also found near the ILRs in NGC 3351 and NGC 6951, although in 2 peaks located symmetrically about the nucleus and oriented nearly perpendicular to the large-scale stellar bars (Kenney *et al.* 1992). These twin peaks are probably caused by orbit crowding, where gas flowing inward along the bar merges with gas on more circular orbits just inside the ILR. In NGC 3351 and NGC 6951 modest non-circular streaming motions in CO of $\sim$20 km s^{-1} exist in the rings. The CO kinematics in these galaxies may be similar to the Hα kinematics in NGC 1365. In the bar region of NGC 1365 the gas streams along highly elongated (x_1-like) orbits oriented along the bar, whereas inside the ILR, the gas streams along slightly elongated (x_2-like) orbits oriented perpendicular to the bar (Teuben *et al.* 1986).

The variety in circumnuclear gas distributions is probably due to different circumnuclear mass distributions, as well as evolutionary effects. In the barred circumnuclear starburst galaxy NGC 3504, the CO distribution is well fit by an azimuthally symmetric exponential disk which is centrally peaked (Kenney, Carlstrom, and Young 1993), while the CO velocity field exhibits only circular motions. The starburst and most of the central molecular gas disk in NGC 3504 seem to be located between the OILR and IILR, and this is consistent with the theoretical expectation that the radial inflow of gas from the bar region slows down at the OILR and creates a pileup of gas in between the OILR and IILR. The maximum value of $\Omega - \kappa/2$ is nearly twice as large as the pattern speed of the large-scale bar, so the OILR and IILR are well separated in NGC 3504. There is no strong enhancement in CO or Hα emission near the OILR, as there is in other galaxies with ILRs. Possible dynamical differences between NGC 3504 and non-starburst barred galaxies with enhanced emission near the OILR are that the OILR and IILR are well separated in NGC 3504 but may be close together in other galaxies, or that the circumnuclear mass distribution is azimuthally-symmetric in NGC 3504 but may be asymmetric in the others. The star formation rate per unit gas mass in NGC 3504 is 4 times higher inside the OILR than outside, suggesting that the starburst will eventually create a ringlike gas morphology near the OILR, similar to barred galaxies which are not presently undergoing starbursts. Thus radial variations in radial flow rates and star formation efficiency may both create ringlike gas morphologies.

Although radial inflow is apparently slowed at ILRs, recent numerical simulations and theoretical work have suggested ways to rapidly transport gas through the ILR barrier (Shlosman, Frank, and Begelman 1989; Wada and Habe 1992; Heller and Shlosman 1993; Combes, these proceedings). Some of this works suggests that the self-gravity of the gas can lead to instabilities and rapid inflow. Observations of gas in circumnuclear regions with sufficient resolution to compare with these theoretical predictions are not yet available. The predicted inflow rates are large enough to detect, although the duration of the rapid inflow is expected to be short-lived.

4 RADIAL FLOWS DUE TO SPIRAL ARMS

Radial and azimuthal streaming motions have been observed within the spiral arms of many galaxies in H I (*e.g.*, Visser 1980) or CO (*e.g.*, Vogel, Kulkarni, and Scoville 1988), with amplitudes as large as $\sim$100 km s^{-1}. The way in which the streaming motions change across a spiral arm are in broad accord with density wave theory, although the large arm-interarm mass density contrasts and large streaming velocities in some grand design spiral galaxies indicate that the density wave is often much stronger than a linear perturbation. There are at least 2 gravitational mechanisms by which density wave spiral arms and bars can drive radial flows of both stars and gas (see article by Larson in these proceedings). In addition, gas may be driven radially inwards with respect to the stars as a result of gas dissipation, which causes a spatial offset between the peak gas density and the peak stellar density. The gas and stars exert torques on each other due to the spatial offset, and are driven radially in opposite directions. Gas is driven inwards inside corotation, and outwards outside corotation. Since bars end at or before corotation, gas is driven inwards along bars. Spiral arms can exist on either side of corotation, so spiral arms drive gas inflows only inside corotation. Corotation in most galaxies occurs at a radius of several kpc (Elmegreen, Elmegreen, and Montenegro 1992) so within the central kpc, spiral arm driven flows are generally inward. Orbits are generally more circular and torques smaller for spiral arms than for bars, so spiral arm induced radial flows are typically smaller than bar induced flows. Orbit elongation, torques, and radial flows all increase as spiral arms become more open and more similar to bars. For both spiral arms and bars, the radial flow rate depends on the gas dissipation rate, which in turn depends on the magnitude of non-circular motions, although it is difficult to determine radial flow rates directly from the observed non-circular motions.

The nature of the non-circular motions in NGC 1068 are worthy of investigation since this galaxy harbors both Seyfert and starburst activity. NGC 1068 has a ringlike distribution of molecular gas at a radius of 1.5 kpc, which Planesas *et al.* (1991) interpret as a pair of tightly-wound spiral arms. If the line-of-nodes in the inner disk is the same as the outer disk, then there are local outward radial motions of $\sim$100 km s^{-1} along the minor axis in the molecular arms. These may be local streaming motions associated with the bar or spiral arms rather than a net expansion. Indeed, a detailed Fabry-Perot study of the ionized gas kinematics shows that the molecular arms are part of a more extensive spiral density wave pattern which exhibits strong streaming motions (Bland-Hawthorn and Cecil 1993).

5 INFLOWS IN MERGING AND INTERACTING GALAXIES

Net inflow rates of gas are probably large in tidally interacting and merging galaxies. Models of these rapidly evolving systems show gas driven to the central few hundred parsecs within a few rotation periods (Hernquist 1989; Hernquist and Barnes 1991). The interaction often causes the formation of a bar, which helps drive rapid gas inflow. In contrast to the quasi-steady state bar model of Athanassoula (1992), in which the net inflow rate of gas is a small fraction of the maximum local streaming motion, the net inflow rate in strongly interacting galaxies, and mergers is expected to be a large fraction of the maximum local streaming motion. A large fraction of elliptical (Kormendy and Djorgovski 1989) and S0 galaxies (Bertola, Buson, and Zeilinger 1992) contain a kinematically distinct gas or stellar core which rotates counter to the sense of the outer galaxy rotation, showing that accretion events and mergers are common. Significant counterrotating components in the S0 galaxy NGC 4550 (Rubin, Graham, and Kenney 1992; Rix *et al.* 1992) and the Sb galaxy NGC 4826 (Braun, Walterbos, and Kennicutt 1992) demonstrate that disks can survive significant accretion events, so even most spirals probably accrete gas after their "formation". Some of the high-velocity H I gas in M101 is probably due to infalling gas (van der Hulst and Sancisi 1988; Kamphuis 1993). Ionized gas has been studied in most systems, and here we briefly mention a few results on atomic or molecular gas.

H I absorption studies have been used to measure the radial motions of the H I gas which happens to lie along the line of sight to strong nuclear radio continuum sources. Among systems studied so far, H I inflow is more common than H I outflow. In a sample of 19 galaxies containing interacting systems, Seyferts, and starbursts,

Dickey (1986) found H I absorption radial velocities ranging from 100 km s^{-1} outward to 500 km s^{-1} inward, with typical uncertainties of 100 km s^{-1}. The linewidths can help determine the nature of the absorbing systems. Some systems studied by Dickey have linewidths as large as 100 km s^{-1}, consistent with a disturbed interacting galaxy, and inconsistent with single narrow line absorbing clouds like those within the disk of the Milky Way. While an ensemble of narrow line clouds experiencing bar or spiral arm streaming motions might produce a large linewidth, it is unlikely that many systems will be viewed sufficiently edge-on to detect these non-circular disk motions. An H I cloud moving at 400 km s^{-1} toward the powerful radio galaxy 4C 31.04 has been detected in absorption by Mirabel (1990). The narrow linewidth of ~10 km s^{-1} in this cloud is similar to both Milky Way high velocity clouds, which have infall speeds of 200–400 km s^{-1}, and normal Milky Way disk clouds. This high velocity cloud likely originates from accreting gas, or possibly from internal gas somehow driven inward. While the redshift or blueshift of H I gas with respect to the nucleus can show whether the H I is moving toward or away from the nucleus, it is difficult to determine the proximity of the gas to the nucleus, or the 3-dimensional flow pattern from the line-of-sight absorption measurement.

The "Atoms-for-Peace" galaxy, NGC 7252, is a merger in an advanced state, probably evolving into an elliptical. There is a large amount of molecular and ionized gas in central few kpc which is counterrotating with respect to H I gas in the outer galaxy (Wang, Schweizer, and Scoville 1992). In the merger simulations of Hernquist and Barnes (1991), stars and gas can become kinematically decoupled, so the lump of gas which sinks to the center need not retain its initial angular momentum. The CO kinematics of the irregularly shaped molecular complex in the central kpc appear consistent with circular motions and solid body rotation. Circular motions show that gas in the circumnuclear region is no longer infalling rapidly, and this may not be surprising since dynamical timescales are short near the nucleus, and the merger is in an advanced state.

The nearby "Evil Eye" Sb galaxy NGC 4826 is a dramatic example of a disk galaxy containing significant counterrotating gas disks (Braun, Walterbos, and Kennicutt 1992). H I in the outer disk is counterrotating with respect to an H I disk in the inner 1 kpc and to stars throughout the galaxy, indicating an accretion event in the outer galaxy (R. Kennicutt, private communication; V. Rubin, private communication). While most E's and S0's with kinematically distinct components contain an inner gas or stellar core which rotates opposite to the sense of the outer rotation, in NGC 4826 the outer gas rotates opposite to the single sense of rotation of stars within the entire disk. Circular motions dominate in both the inner and outer H I disks, but in the gas-poor transition zone at 2–3 kpc, the velocity field is irregular and there are large non-circular motions. Further observations and modelling can help determine whether gas in the circumnuclear region originated in the main galaxy, and how far it was driven radially inward by the merger.

6 OUTFLOWING MOLECULAR GAS FROM ACTIVE NUCLEI

The energetic monsters which inhabit 2 nearby galaxy nuclei have recently been shown to influence the surrounding molecular medium.

A high resolution circumnuclear CO map of the radio lobe spiral galaxy NGC 3079 (Irwin and Sofue 1992; Sofue and Irwin 1992) shows several components, one of which is interpreted by Irwin and Sofue as molecular gas accelerated out of the plane by a nuclear jet. This component is tilted by 40° from the major axis and is aligned with the direction of a *VLBI* jet. It shows no evidence for rotational or radial motions, and is clearly kinematically distinct from a relatively normal, rapidly rotating CO disk. An association of this unusual component with the active nucleus seems reasonable, although outflowing motion is not directly detected. The morphology and kinematics are consistent with a collimated feature which may have an unobserved component of velocity in the plane of the sky, and are inconsistent with an expanding or contracting ring or sphere. While there are many uncertainties, it seems energetically plausible for the jet to accelerate the $\sim 10^8$ $M_\odot$ of gas observed in this component.

An interaction between a nuclear jet and a molecular gas complex in NGC 4258 has been suggested from CO observations by Plante *et al.* (1991). NGC 4258 has a pair of peculiar, "extra" spiral arms which emit strongly in optical lines and the radio continuum, but are not arms in the dense gas or stars, and have been generally interpreted as material ejected from an active nucleus. Plante *et al.* demonstrate a spatial anti-correlation of circumnuclear CO peaks with one peculiar arm, and suggest that the anomalous location of one CO complex in the spatial-velocity plane is evidence for a jet-molecular cloud interaction which deflects the jet and causes the molecular cloud to recoil. This interpretation is consistent with but not highly constrained by the data, and this galaxy clearly deserves further attention. The central 2 pc of NGC 4258 hosts H_2O maser emission lines with velocities which differ from the systemic velocity by ± 900 km s^{-1} (Nakai, Inoue, and Miyoshi 1993), which is the highest velocity molecular gas yet observed in any galaxy. Such high velocity gas would have been missed in most millimeter wave observations of galaxies due to the limited bandwidths.

7 OUTFLOWING MOLECULAR GAS FROM STARBURSTS

Outflowing warm or hot gas from starbursts is well-known (Heckman, Armus, and Miley 1990) but outflowing cold gas is also known or suspected in a few galaxies. Molecular gas is generally the dominant form of the ISM by mass in the central regions of starbursts, so starburst evolution depends critically on what happens to the reservoir of dense star-forming material. Observations are beginning to reveal the energies and mass flow rates from the circumnuclear regions.

The nearby and highly inclined starburst galaxy M82 has an unusually large extent of CO emission along its minor axis of ~ 0.5–2 kpc (Young and Scoville 1984;

Nakai *et al.* 1987; Yun 1992), and some of this appears to be due to circumnuclear outflow. Nakai *et al.* (1987) detected spurlike structures in CO along the minor axis associated with the optical filaments which trace outflowing warm gas, and modelled the system as a chimney of molecular gas filled with hotter outflowing gas. Yun (1992) found non-rotational motions as high as $\sim$100 km s^{-1} in CO above the plane, and arc-shaped features in spatial-velocity plots consistent with radial expansion. The H I in the central part of M82 is also extended along the minor axis, and shows non-circular motions as high as $\sim$200 km s^{-1} which may be related to the outflow (Yun 1992). The large CO scale height extends far beyond the circumnuclear starburst region, and it is unclear whether the outer galaxy high-z gas is previously ejected gas in the process of resettling, a puffed-up disk caused by tidal heating from a galaxy-galaxy collision, or something else.

NGC 253, the other nearby and highly inclined starburst galaxy, has a plume of OH-emitting molecular gas extending from the nuclear region to a height of 1.5 kpc above the plane (Turner 1985). All parts of the plume have a velocity 35 km s^{-1} higher than the systemic velocity of the galaxy, suggesting that the true radial motions are $\sim$100–200 km s^{-1}. H I absorption lines detected toward the nucleus are blueshifted, indicating outflow rather than inflow (Gottesman *et al.* 1976). The plume-like OH morphology of NGC 253 appears different than the chimney-like CO morphology of M82, and it is presently unclear whether this is a difference in molecular tracers or in the outflows of the 2 galaxies. The plume is estimated to have a gas mass of a few times 10^7 M$_\odot$, and an associated outflow rate of a few M$_\odot$ yr^{-1}. The extent and kinematics of the plume suggest a steady outflow lasting $\sim$$10^7$ yrs. Such a high mass loss rate over such an extended period is enough to remove several percent of the dynamical mass in the nuclear region.

Maffei 2 is a heavily obscured, nearby, late type spiral undergoing a modest starburst. In the central kiloparsec, CO is distributed in a narrow ridge, consistent with gas piled up along leading edge of a bar. A CO velocity map with 5″ ($\sim$135 pc) resolution of the central region by Ishiguro *et al.* (1989) showed a peculiar asymmetric pattern, with a feature close to the nucleus which was interpreted by the authors as an expanding ring perhaps caused by an explosive event. However, a ^{13}CO map with similar resolution and sensitivity by Hurt and Turner (1991) shows a symmetric velocity pattern consistent with rotation plus bar streaming motions in the same region, so the reality of the expanding ring feature is questionable.

The presence of neutral gas in a starburst outflow is particularly evident in the spiral galaxy NGC 1808, known for its dusty optical filaments which extend from the circumnuclear region to $\sim$3 kpc above the disk. Optical spectroscopy by Phillips (1993) shows the Na I D line blueshifted in absorption and redshifted in emission, clearly demonstrating an outflow with velocities of 400–700 km s^{-1} containing large amounts of neutral gas. A high-resolution *VLA* study shows H I in absorption arising from a circumnuclear torus of star-forming dense gas with radius 500 pc and rotational velocity of 250 km s^{-1} (Koribalski, Dickey, and Mebold 1993*a*). While the kinematics of this circumnuclear H I are consistent with pure rotation, the kinematics of H I in emission suggests a possible outflow of H I beyond the circumnuclear region on scales of $\sim$1–3 kpc. The large-scale H I velocity field in NGC 1808 shows non-circular motions due to a bar and perhaps a warp. After modelling these features, (Koribalski *et al.* 1993*b*) find a region of velocity residuals spatially coincident with the dusty optical filaments, and suggest that this indicates H I participating in the starburst outflow.

Smaller versions of gaseous outflows from disks occur far beyond the nuclear regions of some galaxies. The most remarkable known case is the H I superbubble in M101 (Kamphuis, Sancisi, and van der Hulst 1991), which is an expanding H I shell associated with a "hole" in the H I disk located $\sim$10 kpc from the nucleus. The existence of both redshifted and blueshifted H I in this superbubble and the spatial correlation with an extremely luminous complex of H II regions provides good evidence that star formation activity from within the disk generated the outflow. The characteristic size (1.5 kpc), mass (3×10^7 $M_\odot$), and velocity (50 km s^{-1}) in this superbubble are significantly smaller than circumnuclear starburst superwinds, but energetically still require the equivalent of $\sim$1,000 supernovae.

Although it cannot be stated with certainty that the cold gas detected in starburst outflows has remained in the same atomic or molecular phase during the outflow period, the large masses of outflowing cold gas suggest that much of it was cold prior to being driven from the disk. The ejection of the cold star-forming medium from circumnuclear regions surely helps to curtail bursts of star formation. Yet it is still unclear what fraction of the initial gas supply in a starburst gets consumed by star formation, and what fraction is ejected in an outflow, or what fraction of the outflowing gas returns to the disk of the galaxy to fuel future star formation, and what fraction escapes to intergalactic space. Starbursts and their outflows dramatically affect galaxy evolution, so it is worth the effort to learn the answers to these questions.

REFERENCES

Athanassoula, E. 1992, *Mon. Not. R. Astr. Soc.*, **259** 345.

Ball, R., Sargent, A. I., Scoville, N. Z., Lo, K. Y., and Scott, S. L. 1985, *Ap. J. Lett.*, **298**, L21.

Bally, J., Stark, A. A., Wilson, R. W., and Henkel, C. 1988, *Ap. J.*, **324**, 223.

Bertola, F., Buson, L. M., and Zeilinger, W. W. 1992, *Ap. J. Lett.*, **401**, L79.

Binney, J., Gerhard, O. E., Stark, A. A., Bally, J., and Uchida, K. I. 1991, *Mon. Not. R. Astr. Soc.*, **252**, 210.

Bland-Hawthorn, J., and Cecil, G. N. 1993, these proceedings.

Blitz, L., Binney, J., Lo, K. Y., Bally, J., and Ho, P. T. P. 1993, *Nature*, **361**, 417.

Blitz, L., and Spergal, D. N. 1991, *Ap. J.*, **379**, 631.

Braun, R., Walterbos, R. A. M., and Kennicutt, R. C., Jr. 1992, *Nature*, **360**, 442.

Combes, F. 1988, in *Galactic and Extragalactic Star Formation*, eds. R. E. Pudritz and M. Fich (Dordrecht: Kluwer), p. 475.

Combes, F., Gerin, M., Nakai, N., Kawabe, R., and Shaw, M. A. 1992, *Astr. Ap.*, **259**, L27.

Dickey, J. M. 1986, *Ap. J.*, **300**, 190.

Elmegreen, B. G., Elmegreen, D. M., and Montenegro, L. 1992, *Ap. J. Suppl.*, **79**, 37.

Gerin, M., Nakai, N., and Combes, F. 1988, *Astr. Ap.*, **203**, 44.

Gottesman, S. T., Lucas, R., Weliachew, L., and Wright, M. C. H. 1976, *Ap. J.*, **204**, 699.

Handa, T., Nakai, N., Sofue, Y., Hayashi, M., and Fujimoto, M. 1990, *P. A. S. J.*, **42**, 1.

Heckman, T. M., Armus, L., and Miley, G. K. 1990, *Ap. J. Suppl.*, **74**, 833.

Heller, C. H., and Shlosman, I. 1993, *Ap. J.*, in press.

Hernquist, L. 1989, *Nature*, **340**, 687.

Hernquist, L., and Barnes, J. 1991, *Nature*, **354**, 210.

Hunter, J. H., Jr., Ball, R., Huntley, M. N., England, M. N., and Gottesman, S. T. 1988, *Ap. J.*, **324**, 721.

Hurt, R. L. and Turner, J. L. 1991, *Ap. J.*, **377**, 434.

Irwin, J. A., and Sofue, Y. 1992, *Ap. J. Lett.*, **396**, L75.

Ishiguro, M. *et al.* 1989, *Ap. J.*, **344**, 763.

Ishizuki, S., Kawabe, R., Ishiguro, M., Okumura, S. K., Morita, K. -I., Chikada, Y., and Kasuga, T. 1990a, *Nature*, **344**, 224.

Ishizuki, S., Kawabe, R., Ishiguro, M., Okumura, S. K., Morita, K. -I., Chikada, Y., Kasuga, T., and Doi, M. 1990b, *Ap. J.*, 355 436.

Kamphuis, J. 1993, *Ph. D.* thesis, University of Groningen.

Kamphuis, J., Sancisi, R., and van der Hulst, T. 1991, *Astr. Ap.*, **244**, L29.

Kenney, J. D. P., Carlstrom, J. E., and Young, J. S. 1993, *Ap. J.*, **417**, in press.

Kenney, J. D. P., and Lord, S. D. 1991, *Ap. J.*, **381**, 118.

Kenney, J. D. P., Scoville, N. Z., and Wilson, C. D. 1991, *Ap. J.*, **366**, 432.

Kenney, J. D. P., Wilson, C. D. Scoville, N. Z., Devereux, N., and Young, J. S. 1992, *Ap. J. Lett.*, **395**, L79.

Koribalski, B., Dahlem, M., Mebold, U., and Brinks, E. 1993b, *Astr. Ap.*, **268**, 14.

Koribalski, B., Dickey, J. M., and Mebold, U. 1993a, *Ap. J. Lett.*, **402**, L41.

Kormendy, J., and Djorgovski, S. G. 1989, *Ann. Rev. Astr. Ap.*, **27**, 235.

Liszt, H. S., and Burton, W. B. 1993, *Ap. J. Lett.*, **407**, L25.

Lo, K. Y. *et al.* 1984, *Ap. J. Lett.*, **282**, L59.

Mirabel, I. F. 1990, *Ap. J. Lett.*, **352**, L37.

Nakai, N., Hayashi, M., Handa, T., Sofue, Y., Hasegawa, T. and Sasaki, M. 1987, *P. A. S. J.*, **39**, 685.

Nakai, N., Inoue, M., and Miyoshi, M. 1993, *Nature*, **361**, 45.

Pence, W. D., and Blackman, C. P. 1984, *Mon. Not. R. Astr. Soc.*, **207**, 9.

Phillips, A. C. 1993, *A. J.*, **105**, 486.

Plante, R. L., Lo, K. Y., Roy, J. -R., Martin, P., and Noreau, L. 1991, *Ap. J.*, **381**, 110.

Planesas, P., Scoville, N. Z., and Myers, S. T. 1991, *Ap. J.*, **369**, 364.

Rix, H. -W., Franx, M., Fisher, D., and Illingworth, G. 1992, *Ap. J. Lett.*, **400**, L5.

Roberts, W. W., Huntley, J. M., and van Albada, G. D. 1979, *Ap. J.*, **233**, 67.

Rubin, V., Graham, J. A., and Kenney, J. D. P. 1992, *Ap. J. Lett.*, **394**, L9.

Sancisi, R., Allen, R. J., and Sullivan, W. T., III 1979, *Astr. Ap.*, **78**, 217.

Sanders, R. H., and Huntley, J. M. 1976, *Ap. J.*, **209**, 53.

Sellwood, J. A., and Wilkinson, A. 1993, *Rep. Prog. Phys.*, **56**, 173.

Shlosman, I., Frank, J., and Begelman, M. C. 1989, *Nature*, **338**, 45.

Sofue, Y., and Irwin, J. A. 1992, *P. A. S. J.*, **44**, 353.

Stark, A. A., Gerhard, O. E., Binney, J., and Bally, J. 1991, *Mon. Not. R. Astr. Soc.*, **248**, 14p.

Teuben, P. J., Sanders, R. H., Atherton, P. D., and van Albada, G. D. 1986, *Mon. Not. R. Astr. Soc.*, **221**, 1.

Turner, B. E. 1985, *Ap. J.*, **299**, 312.

Turner, J. L., and Hurt, R. L. 1992, *Ap. J.*, **384**, 72.

Visser, H. C. D. 1980, *Astr. Ap.*, **88**, 159.

van der Hulst, J. M., and Sancisi, R. 1988, *Astr. Ap.*, **95**, 1354.

Vogel, S. N., Kulkarni, S. R., and Scoville, N. Z. 1988, *Nature*, **334**, 402.

Wada, K., and Habe, A. 1992, *Mon. Not. R. Astr. Soc.*, **259**, 82.

Wang, Z., Schweizer, F., and Scoville, N. Z. 1992, *Ap. J.*, **396**, 510.

Young, J. S., and Scoville, N. Z. 1984, *Ap. J.*, **287**, 153.

Yun, M. S. 1992, *Ph. D.* thesis, Harvard University.

Are the High Molecular Mass Fractions in Nearby Spiral Nuclei Real?

Jean L. Turner

Department of Astronomy, UCLA

ABSTRACT

We examine the question of whether the molecular mass fractions of >50% in the inner 100–200 pc of nearby starburst galaxies are real. If so, this result would imply that molecular gas has a significant impact on the dynamics of the nuclear regions of these galaxies.

1 INTRODUCTION

It has been known for some time that the centers of spiral galaxies are often rich repositories of molecular gas (Young and Scoville 1991), often much richer than the spiral disks. Molecular gas masses are sufficient to fuel the vigorous star formation seen in galactic nuclei, even in "starbursts". Recent interferometric maps have revealed that nuclear gas is distributed in coherent, nonaxisymmetric structures which are often described as "bar-like", structures which are generally seen reflected in the starburst as well.

An example of a gas-rich nucleus with a high star formation (SF) rate is the center of the nearby spiral galaxy, IC 342. IC 342 has a barlike CO distribution in the nucleus (Lo *et al.* 1984), which at high resolution resolves into two very open arms of gas that continue to within 50 pc of the nucleus (Ishizuki *et al.* 1990; Turner and Hurt 1992). There are also spiral arms observed in Hα (J. S. Young, private communication). The CO and Hα arms in IC 342 are offset by about 5″, or 50–100 pc, with the CO arms lying to the inner, concave, portion of the Hα arms. Turner and Hurt (1992) suggest that the offset of the CO and Hα arms and the noncircular motions of the gas can be explained as the response of the gas to a spiral density wave, and that the density wave may be gaseous and not stellar. Others have suggested that the bar-like structures may be due to gas streaming in the oval potential of a stellar bar (Kenney *et al.* 1992). In either case, the dissipative and strongly shearing flow seen here is likely to drive a slow radial inflow of gas into the nucleus (Athanassoula 1992) to feed the SF there.

Since the molecular gas drives the SF activity in the nuclear regions of these galaxies, what determines the molecular structure is of crucial importance to our understanding of the causes and evolution of SF in the centers of galaxies. What

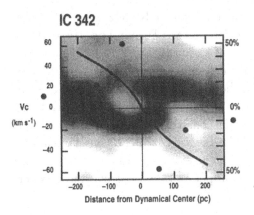

Fig. 1—The CO rotation curve in the central 300 pc of IC 342 (Turner and Hurt 1992), overlaid on the CO intensity map (Levine *et al.* 1993). Dots represent molecular mass fraction.

drives the dynamics of the nuclear gas? Is it the stars or the gas or both? A critical datum is the fraction of the total mass that is gaseous.

2 MOLECULAR MASS FRACTIONS

To estimate what fraction of the total mass is molecular gas we need to estimate both the molecular and dynamical masses. Molecular masses are determined from the CO integrated intensities either by assuming the usual Galactic conversion factor for I_{CO}/N_{H_2} (*e.g.*, Young and Scoville 1991) or by assuming the gas is optically thin, and adopting a relative CO abundance, in the case of ^{13}CO. Dynamical masses are determined by calculating a rotation curve and assuming a mass distribution, usually spherical. Getting rotation curves is not straightforward in galactic nuclei, where there are often substantial radial motions. Furthermore, CO seems to be enhanced in regions of radial streaming (Turner and Hurt 1992). However an estimate of the radial and circular components of flow can be obtained from velocity patterns, which are in agreement with predicted gas density distribution and flow in oval potentials (Roberts *et al.* 1979; Athanassoula 1992). We estimate that there is uncertainty of $< 50\%$ in calculated dynamical masses based on this comparison (Turner and Hurt).

Molecular mass fractions obtained in this way and from the literature are presented in Table 1. Also listed is the sizescale corresponding to each measurement. Listed first are the two closest spiral galaxies: the nuclei of both M31 and the Milky Way are relatively starved of molecular gas, with molecular mass fractions of less than 1–2%. This can be compared to Galactic disk molecular mass fractions of ~ 10–15% (Sanders, Solomon, and Scoville 1984)

The next four galaxies in the table, IC 342, Maffei 2, M83, and NGC 253, all have nuclear starbursts. The starbursts range in luminosity from $6 \times 10^8\,L_\odot$ (IC 342) to $2 \times 10^{10}\,L_\odot$ for NGC 253, corresponding to star formation rates of ~ 0.01–2 $M_\odot\,yr^{-1}$. All of these galaxies have higher molecular mass fractions of more than 40–50% in the inner 100–200 pc, while having "typical" disk values farther out. The increase in molecular mass fraction in the nucleus is clearly seen in IC 342 (Fig. 1); both single dish (Sage and Solomon 1992) and interferometer (Turner and Hurt 1992) measurements indicate a molecular mass fraction of about 10% at a radius of 500 pc. This fraction increases steadily in toward the center, peaking at 50%.

Continuing in order of infrared luminosity, the next galaxies in the table are the more active Seyferts and luminous *IRAS* galaxies (LIRAS). Although there are large uncertainties in these mass fractions, these galaxies consistently appear to have large amounts of molecular gas. In Arp 220, 90% of the total dynamical mass appears to be in molecular gas (Scoville *et al.* 1991). Although the evidence is sketchier for these galaxies, since they are more distant, there does appear to be a correlation between molecular mass fraction and infrared activity.

TABLE 1

MOLECULAR MASS FRACTIONS

Galaxy	Sizescale	Molecular Mass Fraction	Reference
Milky Way	500 pc	1–2%	Sanders *et al.* 1984; Gusten 1987
M31	500 pc	<0.1%	Barnbaum and Turner 1993
IC 342	500 pc	10%	Sage and Solomon 1992
IC 342	100 pc	>50%	Turner and Hurt 1992
Maffei 2	200 pc	>50%	Hurt and Turner 1991
M83	500 pc	15%	Handa *et al.* 1990
M83	200 pc	>40%	Turner, Kenney, and Hurt 1993
NGC 253	500 pc	>400%	Canzian, Mundy, and Scoville 1988
Seyferts	500 pc	10–100%	Meixner *et al.* 1990, Sanders *et al.* 1988
LIRAS	500 pc	20–90%	Scoville *et al.* 1991

3 UNCERTAINTIES

The worrisome thing about these high molecular mass fractions is that they are calculated for nuclear regions of galaxies, which may be very different in their molecular properties from the disk gas with which we are familiar. Although the increase in molecular mass fraction observed in nearby galaxies suggests that this is a real effect, there are still possible systematic sources of uncertainty. Is it likely that the nuclear molecular mass fractions are not really high, but are an artifact of different gas properties, such as metallicity or gas temperature, in the nucleus?

A first source of uncertainty in molecular mass fractions is distance. Since molecular masses are measured from CO luminosities, the mass will scale as the distance squared. Dynamical masses depend on $1/r$, which goes as the distance. Therefore, molecular mass fractions will be proportional to the distance to the galaxy. Not much can be done about this. With a large enough sample, this effect should wash out, giving scatter but not bias in the mass fractions.

A second source of uncertainty is in the derivation of H_2 masses from CO luminosities. Masses have been calculated both using the normal, optically thick $^{12}C^{16}O$ (CO) using the standard conversion factor, and with the optically thinner transition of $^{13}C^{16}O$. The uncertainties are different for these two methods. The optically thin ^{13}CO might at first appear more reliable but this is probably not true. ^{13}CO suffers fractionation effects at modest ($T_K \sim 35$ K) temperatures, causing the abundance of ^{13}CO relative to CO to vary by factors of 2–3, and leading to an apparent overabundance of ^{13}CO in the Galactic Center (Langer and Penzias 1990). In addition to abundance uncertainties, ^{13}CO emissitivity is proportional to $1/T$. Since galactic center clouds may have higher temperatures than typical disk clouds, ^{13}CO could be suppressed. It is therefore conceivable that masses derived from ^{13}CO fluxes could be off by factors of 2–3. In addition to variation in the $^{13}CO/CO$ ratio, there is also uncertainty in the abundance of CO relative to H_2. The effects of metallicity gradients on this quantity are unknown and could be substantial in the absence of radial gas mixing. At any rate, ^{13}CO does give a useful comparison mass.

Masses are obtained from the optically thick CO transition by using the standard conversion factor, N_{H_2}/I_{CO}, about 3×10^{20} cm^{-2} (K km s^{-1})$^{-1}$. This conversion factor works because the optically thick line is a product of the cloud temperature and the linewidth; and the linewidths of molecular clouds are proportional to their masses (Scoville and Sanders 1987; Maloney and Black 1988). Although the conversion factor is uncertain, typically quoted at a factor of two, it is actually a more robust measure of cloud masses in galactic centers than abundance-sensitive thin transitions since it is dominated by cloud linewidth. CO masses in galactic nuclei are probably no more uncertain than they are in our Galactic disk.

Masses have been obtained using both transitions, ^{13}CO and CO, in two galaxies, Maffei 2 (Hurt and Turner 1991) and IC 342 (Turner and Hurt 1992). The ^{13}CO and CO masses agree to within 30%. Since these methods are relatively independent, this suggests that the properties of nuclear molecular clouds are not so different from Galactic molecular clouds, and that the nuclear molecular mass fractions therefore are significantly higher than in our Galactic disk. We also note that in the galaxies for which nuclear gas masses have been obtained, the masses are consistent with the inferred star formation rate, *i.e.*, the SF efficiencies are similar to what is observed in the Galaxy.

4 CONCLUSIONS AND RAMIFICATIONS

We conclude that the high molecular mass fractions in the nuclei of active infrared galaxies are likely to be significantly higher, greater than 40–50%, than is observed in the disk of our Galaxy. This is based on mass measurements from CO. There is an absolute uncertainty of about a factor of two in these mass fractions. However, two things are relatively more certain: first, the molecular mass fraction does increase within galaxies towards the nucleus; and second, the molecular mass fractions are significantly higher than in our Galactic disk. There is a tantalizing suggestion that the more infrared-luminous the galaxy, the higher the molecular mass fraction. Clearly gas plays an important role in the dynamics and evolution of nuclear "activity" in luminous infrared galaxies.

This work was supported by NSF grant AST 90-22996. I would like to thank Debbie Levine and Robert Hurt for helpful discussions.

REFERENCES

Athanassoula, E. 1992, *Mon. Not. R. Astr. Soc.*, **259**, 345.

Barnbaum, C., and Turner, J. L. 1993, in preparation.

Canzian, B. G., Mundy, L. G., and Scoville, N. Z. 1988, *Ap. J.*, **333**, 157.

Gusten, R. 1987, in IAU Symp. 136 on *The Center of the Galaxy*, ed. M. Morris (Dordrecht: Kluwer), p. 89.

Handa, T., Nakai, N., Sofue, Y., Hayashi, M., and Fujimoto, M. 1990, *P. A. S. J.*, **42**, 1.

Hurt, R. L., and Turner, J. L., *Ap. J.*, **377**, 763.

Ishizuki, S., Kawabe, R., Ishiguro, M., Okumura, S. K., Morita, K. -I., Chikada, Y., and Kasuga, T. 1990, *Nature*, **344**, 224.

Kenney, J. D. P., Wilson, C. D., Scoville, N. Z., Devereux, N. A., and Young, J. S. 1992, *Ap. J. Lett.*, **395**, L79.

Levine, D., Turner, J. L., and Hurt, R. L. 1993, in *Astronomy with Millimeter and Submillimeter Wave Interferometry*, ed. M. Ishiguro (San Francisco: ASP Conference Series), in press.

Langer, W. D., and Penzias, A. 1990, *Ap. J.*, **357**, 477.

Lo, K. Y., *et al.* 1984, *Ap. J. Lett.*, **282**, L59.

Meixner, M., Puchalsky, R., Blitz, L., Wright, M., and Heckman, T. 1990, *Ap. J.*, **354**, 158.

Sage, L. J., and Solomon, P. M. 1991, *Ap. J.*, **380**, 392.

Sanders, D. B., Soifer, B. T., Elias, J. H., Madore, B. F., Matthews, K., Neugebauer, G., and Scoville, N. Z. 1988, *Ap. J.*, **325**, 74.

Sanders, D. B., Solomon, P. M., and Scoville, N. Z. 1984, *Ap. J.*, **276**, 182.

Scoville, N. Z., and Sanders, D. B. 1987, in *Interstellar Processes*, eds. D. J. Hollenbach, and H. A. Thronson, Jr. (Dordrecht: Kluwer), p. 21.

Scoville, N. Z., Sargent, A. I., Sanders, D. B., and Soifer, B. T. 1991, *Ap. J. Lett.*, **366**, L5.

Turner, J. L., and Hurt, R. L. 1993, *Ap. J.*, **384**, 72.

Turner, J. L., Kenney, J., and Hurt, R. L. 1992, in preparation.

Young, J. S., and Scoville, N. Z. 1991, *Ann. Rev. Astr. Ap.*, **29**, 581.

Velocity Field Peculiarities in the Circumnuclear Regions of Seyfert Galaxies from Observations with the 6m Telescope

V. L. Afanasiev and A. I. Shapovalova

Special Astrophysical Observatory, Russian Academy of Sciences, Russia

ABSTRACT

Observational evidence for the existence of bars in the central parts of Seyfert galaxies Mkn 744, Mkn 573, and NGC 4151 is presented.

1 INTRODUCTION

Afanasiev (1981) investigated rotation curves of 28 Seyfert galaxies and found local maxima at distances 0.5–2 kpc from the centers in 20 objects. The rotation velocity declined with the increasing distance from the local maximum more rapidly than for the Keplerian law. This suggests a possibility of development of Kelvin-Helmholtz instability leading to essential turbulization of the circumnuclear disk (global gas instability in the disk) (Morozov 1979). Observed solid body rotation, from the center to the local maximum, indicates a possible presence of bars in the centers of Sy s.

2 GAS MOTION FEATURES IN THE BAR'S REGION

Gaseous disk in the region of the bar is thinner than the latter, has lower chaotic velocities and rotates faster than the bar $\Omega_d > \Omega_b$ (Fridman 1987). If the radius R_1, where the relative linear velocity ΔV_R reaches the sound speed in the gaseous disk $\Delta V_R = (\Omega_d - \Omega_b)R > V_s$, is within the bar region (R_b), then a shock may arise for $R_1 < R_b$. The shock front then will be observed on the rear side of the bar with respect to the direction of rotation, if the disk and the bar rotate in the same direction. Depending on the ΔV_R value, either strong or weak shocks will be observed in the optical band (mainly in the forbidden lines) as linear structures. Because of the finite thickness of the bar such linear structure will turn in the position angle (P.A.) with respect to the bar in the direction opposite to the rotation of the bar.

3 RESULTS OF OBSERVATIONS

3.1 The Velocity Field of Mkn 744

The smoothed isophotes obtained from the electron camera plates on the 60cm reflector of the *SAO* (scale 28" mm^{-1}) are displayed in Figure 1. The elongation of the outer isophotes coincides with the line of nodes, and the major axes of the inner isophotes ($R < 1.5$ kpc) are turned by 20° to the North with respect to the outer isophotes. Apparently it is a bar.

Spectra of Mkn 744 were obtained with a long slit (100" × 2") in the prime focus of the 6m telescope with an image tube in the wavelength range $\lambda\lambda6400$–6900ÅÅ with a dispersion 46Å mm^{-1} in position angles 0° − 180° every 10° − 15°. Figure 1*b* gives the radial velocities observed in different position angles found from Hα (filled circles) and from [N II] and [S II] (open circles) at a distance 3" from the center of the galaxy. The deviation of the observed radial velocities from purely circular motions (solid line) caused by the gas emitting in Hα line are shown at the bottom of Figure 1*b*. Large deviations from the circular velocity with a maximum amplitude of 100 km s^{-1} are seen in the P.A.~20° − 60° for the velocities found from the forbidden lines [N II] and [S II]. The amplitude of the noncircular motions is as large as 150 − 180 km s^{-1} at a distance of 5" from the center. The angular velocity measured in the forbidden lines exhibits a jump in the same direction as the P.A. of the bar.

Thus, the presence of the bar is suggested by:
1) elongation of the central isophotes coincident with the bar elongation;
2) coincidence of the boundaries of the central knot (~2 kpc) with the straight part on the rotation curve (solid body rotation);
3) deviations from circular motions in P.A.~20° − 60° that were found from the forbidden lines and caused by the shock at the edges of the bar;
4) increase in the amplitude of the velocity jump, measured in the forbidden lines, with the distance from the center. (V~100 km s^{-1} at R~3", and V~150 − 180 km s^{-1} at R~5"). The latter excludes the presence of a jet in this region.

3.2 Mkn 573

The spectral observations of the central part of Mkn 573 were carried out with a Multi-Pupil Fiber Spectrograph installed in the prime focus of the 6m telescope (Afanasiev and Sil'chenko 1991). About 80 spectra are registered simultaneously. The dispersion was 1.3Å px^{-1}, the seeing < 1.5", the spectral resolution 2–3Å, the spectral range $\lambda\lambda4800$–5200ÅÅ. The maps of the continuum brightness distribution in the central region of Mrk 573 (central wavelength — 5110Å, spectral window — 40Å) (Fig. 2*a*), and in the [O III]$\lambda5007$Å emission line peak (Fig. 2*b*) are presented. The continuum image of the central region has an oval shape elongated in the P.A.~62°. It is a mini-bar ~ 1 kpc. The isoptotes in the [O III] emission line peak are elongated in the P.A.~106° and coincide with the kinematic axes. Gaussian analysis of the Hβ and [O III]$\lambda5007$ contours obtained from the long-slit observations was performed for distances < 4". At each point of the slit, 2–3 components which differed in radial velocities were selected. Next, the brightness distributions in the wings of the line [O III] (high-velocity component) and near the center of [O III] (low-velocity

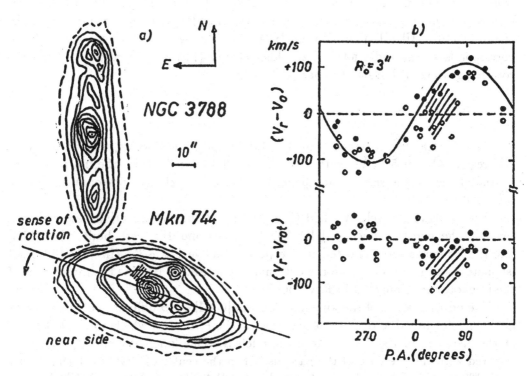

Fig. 1—(*a*) Isophotes of Mkn 744 and its companion NGC 3788; the solid line shows the line of nodes, the dashed line indicates the direction of the bar elongation in the central part; (*b*) Deviation of Hα (dark circles) and [N II] and [S II] (open circles) velocity field from the circular motion. The shock front region is hatched (Afanasiev and Shapovalova 1981).

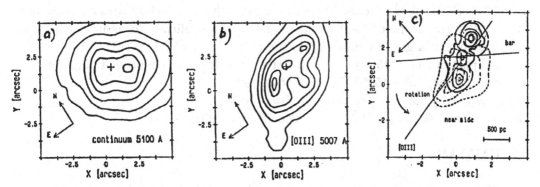

Fig. 2—The maps of the central region of Mkn 573 ($10'' \times 10''$): (*a*) in the continuum band λ5110Å; and (*b*) in the peak emission line [O III]λ5007Å; (*c*) The brightness distribution in [O III] line wings (the high-velocity component) is shown with the solid lines, and the low-velocity component — with the dashed lines.

component) were obtained (Fig. 2c).

The images in the wings display two condensations located at the same distances from the center as the condensations visible on the radio-maps . The P.A.~125° of the line running across the condensations coincides with the orientation of the radio structure (Ulvestad and Wilson 1984). So, the high-velocity component [O III] corresponds to the radio jet and the low-velocity (P.A.~105°) corresponds to the rotating gaseous disk inside the bar (P.A.~125°).

3.3 NGC 4151

The spectra were obtained with the Multi-Pupil Fiber Spectrograph in the spectral region $\lambda\lambda 4600-5300$ÅÅ with a dispersion of 1.8Å px^{-1} from the matrix 8×12 spatial elements, the size of each being $0.8'' \times 0.8''$. A 2D photon-counting system was used as a detector.

The maps of the velocity field (Fig. 3a, b) and the surface brightness distributions in the emission lines [O III]$\lambda 5007$ (Fig. 3c) and Hβ were built. In [O III] line the isophotes are elongated in the position angle P.A.~35° (from the internal isophotes) and P.A.~43° (from the external isophotes). In Hβ line the major axes of the isophotes are elongated in P.A.~50°.

The position angle of the kinematic axis in [O III] – Hβ difference of the velocity field (Fig. 3c) gives the velocity peculiarities in the P.A.~15°, which are ± 100 km s^{-1} at the distances of 100–150 pc from the nucleus. This kinematic effect may be associated with the presence of the gaseous bar in the center of NGC 4151 (Shlosman *et al.* 1990). This gaseous bar is observed in the P.A.~50° in Hβ. In [O III], the axes are turned with respect to the corresponding axes in Hβ by 10°. This may be due to the fact that as the disk rotates its gas impacts the inner gaseous bar, and shocks are formed at its edges. Thus the velocity [O III] – Hβ peculiarity in the P.A.~15°, which we observe, corresponds to the component [O III] associated with the shock excitation at the edges of the inner gaseous bar (Fig. 3d).

The velocity field (Fig. 3a, b) measured in the [O III] line indicates a slower rotation of the gas compared to Hβ velocity field. This confirms our assumption about the presence of a gaseous mini-bar in the central part of NGC 4151. Thus, in the central part of NGC 4151 the following structural features are present (Fig. 3d):
1) torus: 50 pc, P.A.~150° (Terlevich *et al.* 1991);
2) gaseous bar: 250–300 pc, P.A.~50° (this work);
3) radio jet: 500–600 pc, P.A.~ 77° (Pedlar *et al.* 1991).

4 SUMMARY

Hence, in the central parts of Seyfert galaxies (R ~1–2 kpc) a stellar bar may be formed as result of dynamical instabilities (*e.g.* Mkh 744, Mkn 573). This bar captures the interstellar medium forming a gaseous disk on the scale of hundreds of parsec, approximately along the minor axis of the bar. According to Shlosman *et al.* 1990), the cold gaseous disk may become unstable and form a gaseous bar (≥ 100 pc, *e.g.* NGC 4151) which would ensure the fueling of the nucleus. Subsequently, a gaseous torus with a rotation axis along the minor axes of the bar may form in the vicinity of the active nucleus (*e.g.* NGC 4151, Fig. 3*d*). Then the ionizing UV radiation cone's axis will point perpendicularly to the plane of the torus.

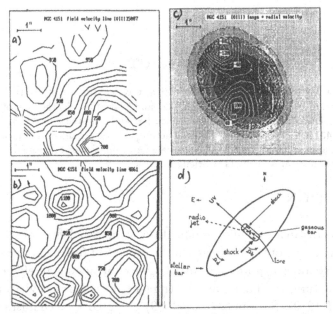

Fig. 3—The velocity field in (*a*) [O III]; and (*b*) Hβ lines; (*c*) The field velocity differences for [O III]–Hβ (white isophotes) superimposed on the image in [O III]λ5007Å; (*d*) The sketch of structural features in the main body of NGC 4151.

REFERENCES

Afanasiev, V. L. 1981, *Pis'ma v Astron. Zh.*, **7**, 390 (in Russian).

Afanasiev, V. L., and Shapovalova, A. I. 1981, *Astrofizika*, **17**, 403 (in Russian).

Afanasiev, V. L., and Sil'chenko, O. K. 1991, *Astrofiz. Issled.* (Izv. SAO), **33**, 88 (in Russian).

Fridman, A. M. 1987, *Vestnik AN SSSR*, **6**, 18 (in Russian).

Morosov, A. G. 1979, *Astron. Zh.* **56**, 498 (in Russian).

Pedlar, A., Kukula, M., Longley, D., Axon, D., Baum, S., and O'Dea, C. 1991, in Proc. Heidelberg Conference.

Shlosman, I., Begelman, M. C., and Frank, J. 1990, *Nature*, **345**, 679.

Simkin, S. M. 1975, *Ap. J.*, **200**, 567.

Terlevich, R., Portal, M. S., Diaz, A. J., Terlevich, E. 1991, *Mon. Not. R. Astr. Soc.*, **249**, 36.

Ulvestad, J. S., Wilson, A. S. 1984, *Ap. J.*, **278**, 544.

Bars within Bars, Spirals within Spirals

Johan H. Knapen and John E. Beckman

Instituto de Astrofísica de Canarias, E-38200 La Laguna, Tenerife, Spain

ABSTRACT

Using new high-resolution imaging at optical wavelengths, we study the morphology of the inner kpc of two barred spirals, NGC 4321 and NGC 4314. In the case of NGC 4321, we present evidence for the existence of a nested bar and a nested spiral structure. In the case of NGC 4314, the inner star formation activity is not organized in spiral arms, but has a ring-like appearance. The role of large scale bars and of possible nested bar structure in the fueling of nuclear starbursts is briefly discussed.

1 INTRODUCTION

The need to dissipate angular momentum of material flowing into the central region of a galaxy is one of the problems in the process of fueling active galactic nuclei (AGN). Non-axisymmetric potentials are capable of achieving an efficient dissipation of angular momentum, and bars may well trigger the nuclear activity of galaxies (see *e.g.* Simkin *et al.* 1980). Shlosman *et al.* (1989) proposed a mechanism where a set of nested bars can play an important role in this process. "Bars within bars", possibly of a similar kind to those proposed by Shlosman *et al.*, have been observed in a number of galaxies (see *e.g.* de Vaucouleurs 1974; Friedli and Pfenniger 1993 and references therein).

We show here preliminary results from an imaging study of two barred spiral galaxies which show a mild form of nuclear activity: a (circum-)nuclear starburst. These galaxies, NGC 4314 and NGC 4321, are well-known cases where strong formation of massive stars takes place in a region within about 1 kpc in radius from the center (see *e.g.* Pogge 1989). Since neither galaxy has a strong bulge, and NGC 4321 does not have a strong bar, effects such as those discussed above may be easier to observe, and these cases may be illustrative for a wider range of galaxies, including AGN hosts.

2 NGC 4314

NGC 4314 is a strongly barred spiral galaxy, which hardly shows any spiral arm structure outside the bar. The central region is characterized by a ring-like region of intense star formation. Pogge (1989) estimates from his Hα image of this galaxy that some 94% of the total Hα luminosity is emitted by the circumnuclear starburst. Benedict *et al.* (1992), in an extensive optical and NIR study of this galaxy, identify a nuclear bar, aligned in the same general direction of the main bar.

We have obtained new optical imaging of the central region of NGC 4314 at higher resolution then published before (subarcsec seeing conditions). In Figure 1 we show our $U - I$ color map, indicating in lighter shades areas of bluer color.

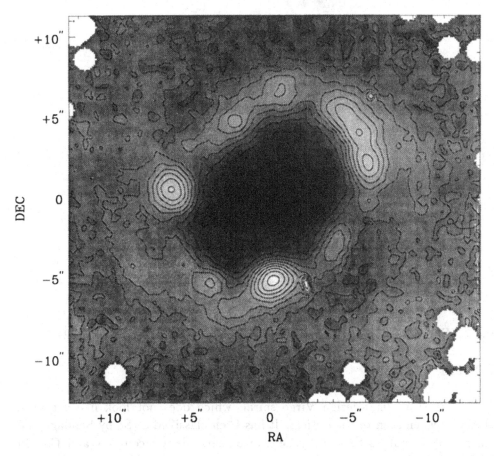

Fig. 1—$U-I$ color index map of the central region of the barred galaxy NGC 4314. White round areas near the edges of the image are instrumental effects.

It is easily seen that a number of blue regions, indicating zones of recent star formation, are lined up in a ring around the nucleus, without much spiral arm signature. Two spiral arms split off the ring, in the NW and SE parts of the ring. The relatively red part inside the star forming ring shows a SE–NW extension, which can be identified with the nuclear bar at PA= 136° as described by Benedict *et al.* (1992). The nature of this inner 'bar' is not clear, and projection effects may be important.

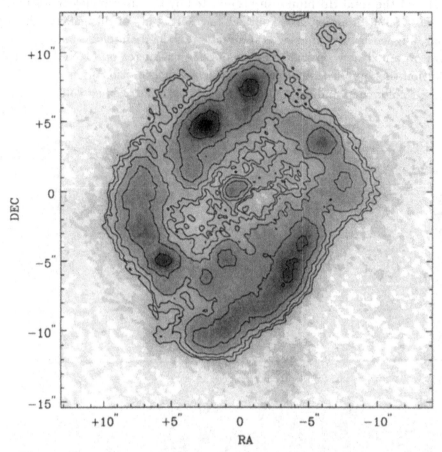

Fig. 2—Hα continuum-subtracted image of the central part of NGC 4321.

3 NGC 4321

NGC 4321 is a grand-design Virgo spiral, which does not look like a typical barred galaxy when seen in the optical. It has been classified as Sc by Sandage and Tammann (1987), but as SABbc by de Vaucouleurs, de Vaucouleurs and Corwin (1976). Pierce (1986) presents an *I*-band image where elongated contours are seen around the nucleus, indicating the existence of a bar. In a recent H I study (Knapen *et al.* 1993) we have presented evidence from the kinematics of the neutral hydrogen gas that a bar is indeed present on the scale of some 6 kpc. The contours of equal velocity show a change in the position angle of the kinematical major axis with radius, an effect that can also be seen in velocity-position diagrams along the minor axis. Comparison of the H I results with Hα and near-infrared images (see Knapen *et al.* 1993) shows that the effects of the non-axisymmetric potential can be seen in both the distribution of the old stellar population and of regions of recent massive star formation.

The central region of NGC 4321, of some 30″ diameter, shows an interesting morphology, especially in Hα. The strongly enhanced star formation around the center was first called a double-lobed ring (Arsenault *et al.* 1988), but was later classified as a four-armed spiral (Pogge 1989; Cepa *et al.* 1990) when Hα data of higher quality and improved resolution became available. We have obtained new Hα imaging using the 4.2m William Herschel Telescope at La Palma, at sub-arcsecond resolution. The final, continuum-subtracted image is shown here in Figure 2. It shows clearly the four spiral arm segments reported before in the literature, implying that the massive star formation does not occur in a 'ring' of star formation as in the case of NGC 4314.

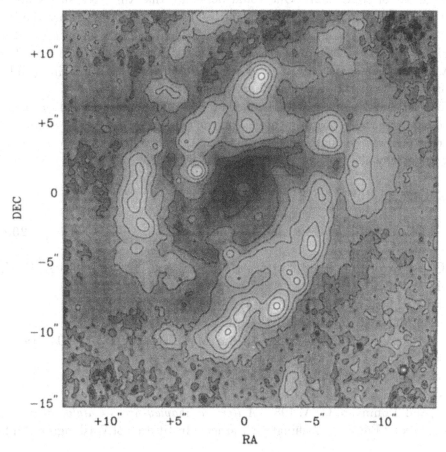

Fig. 3—$U - I$ color index map of the inner region of NGC 4321.

From our new optical broad-band observations of NGC 4321, obtained with the Auxiliary Port camera on the WHT, we have produced a $U - I$ color index map, shown here in Figure 3.

This image shows, in lighter shades, regions of relatively blue emission which coincide quite well with the regions of massive star formation seen in the Hα image.

In some cases, such as the region some 5″ to the NE of the nucleus, an H II region as apparent in the Hα map cannot be readily identified with a similarly strong blue region in the color map. This is probably due to the influence of dust extinction, which is much stronger in the U-band than in the spectral region of the Hα line (around 6,700Å). Careful comparison of the Hα and $U - I$ images thus gives a rough indication of the extent of dust extinction. Dust does not play an excessive role in the central zone, strong effects are noted only in a few regions such as the one mentioned before.

A striking feature in the color image is the red area inside the regions defined by the armlets. This area shows a bar-like feature, which is almost perfectly aligned with the large bar of NGC 4321. One could postulate that this area shows up as red only because of the presence of large quantities of dust, but in that case the Hα map would not give such good agreement with the color map in the inner area. It is more plausible that we are indeed seeing a nested bar structure, where the inner bar might be a continuation of the outer bar, given their alignment. NIR imaging will be needed to confirm the existence of this "bars within bars" structure.

From optical and Hα images it seems that two of the inner armlets connect through the bar to the outer arms in the disk of the galaxy. If that can be confirmed we have a case here of two continuing spiral arms, and possibly also one continuing bar, giving the galaxy the "self-mirrored" impression we observe.

REFERENCES

Arsenault, R., Boulesteix, J., Georgelin, Y., and Roy, J. -R. 1988, *Astr. Ap.*, **200**, 29.

Benedict, G. F., Higdon, J. L., Tollestrup, E. V., Hahn, J. M., and Harvey, P. M. 1992, *A. J.*, **103**, 757.

Cepa, J., and Beckman, J. E. 1990, *Astr. Ap. Suppl.*, **83**, 211.

Friedli, D., and Martinet, L. 1993, *Astr. Ap.*, in press.

Knapen, J. H., Cepa, J., Beckman, J. E., del Rio, M. S., and Pedlar, A. 1993, *Ap. J.*, in press.

Pierce, M. J. 1986, *A. J.*, **92**, 285.

Pogge, R. W. 1989, *Ap. J. Suppl.*, **71**, 433.

Sandage, A., and Tammann, G. A. 1987, *A Revised Shapley-Ames Catalog of Bright Galaxies*, Second Edition (Washington: Carnegie Institution of Washington Publ. 635).

Shlosman, I., Frank, J., and Begelman, M. C. 1989, *Nature*, **338**, 45.

Simkin, S., Su, H., and Schwarz, M. P. 1980, *Ap. J.*, **237**, 404.

Vaucouleurs, G. de 1974, in *Formation of Galaxies*, IAU Symp. 58, ed. J. R. Shakeshaft (Reidel: Dordrecht), p. 335.

Vaucouleurs, G. de, Vaucouleurs, A. de, and Corwin, H. G. 1976, *Second Reference Catalogue of Bright Galaxies* (RC2), University of Texas Press.

Accretion Disks in the Galactic Center

Wolfgang J. Duschl[1,2,3], Susanne von Linden[3], and Peter L. Biermann[3]

[1]Institut für Theoretische Astrophysik, Heidelberg, Germany
[2]Interdisziplinäres Zentrum für Wissenschaftliches Rechnen, Heidelberg, Germany
[3]Max-Planck-Institut für Radioastronomie, Bonn, Germany

ABSTRACT

We present a new model for the dynamics of molecular clouds in the innermost ≈ 200 pc from the Galactic Center. Our analysis allows us to determine the characteristic parameters of the accretion disk as well as the positions of individual clouds with respect to the Galactic Center. Finally, we show that the results are in good agreement with independent determinations of the same parameters and discuss a physical mechanism that allows for the theoretical understanding of these parameters.

1 OVERVIEW

We assume that the dynamics of accretion disks in the innermost ≈ 200 pc from the Galactic Center (GC) can be modelled in the framework of an accretion disk description. The two parameters that mainly determine the structure of the disk are the radial mass flow rate through the disk ($\dot{M}$) and the dynamical viscosity (ν). For a pair of these two parameters, one can calculate the radial velocities (v_s) in the disk with respect to the GC and, consequently, the observable quantity v_x, the radial velocity with respect to the observer (of course, after applying all the relevant corrections). In an iterative process, with this method, we determine the pair $\{\dot{M}, \nu\}$ that gives the best agreement between the observed and modelled values of v_x. Having found such a solution, we also determined the location of the individual clouds with respect to the GC. Thus, this technique allows us also to construct a map of the distribution of molecular clouds close to the GC. In the next section, we summarize our *ansatz* for describing the accretion disk. In section 3, we demonstrate our technique for one example, and in section 4, we summarize the results and compare them to parameters that have been determined independently. Finally, the last section is devoted to an overview of the process that allows for the physical understanding of the determined quantities.

2 THE ACCRETION DISK MODEL

We describe the dynamics of the molecular clouds in the vicinity of the GC in the framework of an accretion disk model. For this, we assume that the molecular clouds are the *only* prominent features in an otherwise smooth disk flow. This allows

105

us to use the observed radial velocities of the cloud as characteristic for all the disk at the galactocentric radius of the cloud, and thus to use the clouds as tracers of the gas flow in the disk.

In the following, we discuss only the case of stationary accretion disks with no other mass input than the radial one at the disk's outer edge. We model the disk as being geometrically thin and azimuthally symmetric. The disk is assumed to reside in a spherically-symmetric potential that is due to a prescribed mass distribution, $M_b(r)$. In the case of the GC, we take $M_b(r) = 1.14 \times 10^9 \; M_\odot (r/100 \; pc)^{5/4}$, where r is the radius in a spherical coordinate system (see, *e.g.*, Genzel and Townes 1987). This defines the azimuthal velocity, *i.e.*, the Keplerian velocity, v_ϕ.

The dynamical viscosity is assumed to follow a power law in radial direction: $\nu = \nu_0 (s/s_0)^\beta$, where s is the radius in a cylindrical coordinate system which is the appropriate choice for a geometrically-thin accretion system. Only in the plane of symmetry is $s = r$. We take s_0 as an arbitrarily chosen scaling radius.

Details of the model description are discussed in Linden *et al.* (1993). There, we also investigate further processes, namely the time dependent evolution of accretion disks, and the sinks and sources of the mass flow through the disk due to star formation and vertical infall of mass into the disk.

3 AN EXAMPLE: M-0.13-0.08

As an example, we have chosen one of the best observed molecular clouds in the GC area, namely M-0.13-0.08 and its vicinity. For validation of the method, we have taken a comparatively small set of seven observations. The observations are taken from the large sample of Zylka, Mezger, and Wink (1990). By matching observed and theoretically modelled radial velocities v_x, we get as the best fit pair of parameters $\dot{M} = 10^{-1.8} \; M_\odot$ yr, and $\nu(s = 115 \; pc) = 6 \cdot 10^{26} \; cm^2 sec^{-1}$. The radius of 115 pc is the average radius from the GC of the seven observations. This is thus the determined radial distance of M-0.13-0.08 from the GC. Details of the matching procedure and its accuracy may again be found in Linden *et al.* (1993).

4 THE PARAMETERS OF THE DISK

Obviously, the determination of one cloud position in the GC accretion disk does not allow us to determine radial variations of the viscosity. On the other hand, the determined radius of ≈ 115 pc from the GC is well within the limits of $30 \dots 200$ pc inferred from spectroscopy ($s > 30$ pc) and dynamics of the expanding molecular ring ($s < 200$ pc).

To get additional information about the radial variations of ν, we choose to include observations of Pauls *et al.* (1993) of material presumably much closer to the GC. We find that, indeed, this material is located at a GC distance of only ≈ 15 pc. Keeping the assumption of a stationary disk (*i.e.*, $\dot{M} = $ const.), this allows us to determine the radial derivative of ν. We find $\nu(s) = 5.7 \times 10^{26}$ cm^{2}sec$^{-1}(s/100$ pc$)^{0.4}$. (Biermann *et al.* 1993).

For the above mentioned radial mass flow rate, together with this viscosity, we have to choose the disk's inner radius at $\ll 15$ pc to ensure a) that numerical effects of the boundary condition do not influence the solution, and b) that the basic assumptions of the disk models (e.g., $v_s \ll v_\phi$) are still valid at the location of the clouds. To fulfill both conditions, in our final models, we put the inner disk boundary to 1 pc. It is important to note that this choice does not represent a physically mandated radius but one that guarantees numerical consistencies at the locations where the analysed structures are located.

It is also important to note that the viscosity that we have determined in this way is in good agreement with what one can independently estimate to be a reasonable value. Interpreting it as some kind of turbulent viscosity, we may estimate its value by taking the product of the observed velocity dispersion in the gas and the scale height of the gas distribution. In the case of the GC this, interestingly enough, leads to a number that agrees to within an order of magnitude with the one that we have determined through matching the radial velocities.

5 A MECHANISM GENERATING HIGH VISCOSITIES

Despite the fact that we find agreement between the characteristic velocities and lengths scales as observed in the GC, and the viscosity that our model calculations indicate, this kind of analysis does not provide a physical explanation of the viscosity.

Assuming a standard α accretion disk (Shakura and Sunyaev 1973) would immediately lead to a viscosity parameter α that is by several orders of magnitude larger than the usually assumed maximum value of 1. This maximum applies only to the case of a *closed system, i.e.,* to an accretion disk that has its only connection to the outside through its radiation.

If we allow for other processes, like interactions with bars or a stellar component, the classical α prescription breaks down. On the other hand, one can think of several other processes in the GC environment that would be capable of putting energy into the disk and thus would make (much) higher viscosities possible.

Linden *et al.* (1994) have proposed such a process and have shown that supernovae could very well put enough energy into the accretion disk to allow for such a high viscosity. Linden et al. have even succeeded in showing that a that there is a cyclic evolution possible that can maintain itself. Supernovae gives rise to high enough an energy input to explain the observed viscosity. The high viscosity, in turn, causes such large a mass flow rate that there is always sufficient mass replenished to ensure ongoing star formation and thus to ensure high enough a supernovae rate ...

Acknowledgements: WJD acknowledges travel support from the Deutsche Forschungsgemeinschaft DFG. WJD and SvL are greatful to the conference organizers for financial support.

REFERENCES

Biermann, P. L., Duschl, W. J., and Linden, S. v. 1993, *Astr. Ap.*, in press.
Genzel, R., and Townes, C. H. 1987, *Ann. Rev. Astr. Ap.*, **25**, 377.
Linden, S. v., Duschl, W. J., and Biermann, P. L. 1993, *Astr. Ap.*, **269**, 169.
Linden, S. v., Biermann, P. L., Schmutzler, T., Lesch, H., and Duschl, W. J. 1994,
    Astr. Ap., in press.
Pauls, T., *et al.* 1993, *Ap. J. Lett.*, **403**, L13
Shakura, N. I., and Sunyaev, R. A. 1973, *Astr. Ap.*, **24**, 337
Zylka, R., Mezger, P. G., and Wink J. E. 1990, *Astr. Ap.*, **234**, 133

HST Imagery of the Starburst Nucleus of M83

Jarita Holbrook[1], Sara Heap[1], Eliot Malumuth[2], Steven Shore[2], and Bill Waller[1]

[1]Laboratory for Astronomy and Solar Physics, Goddard Space Flight Center
[2]Astronomy Programs, Computer Sciences Corporation

ABSTRACT

We used *HST* Planetary Camera images of M83 in the Hα, U, V, and I filters to study the ionizing clusters in the nuclear starburst region. Our high resolution images revealed detailed structure, previously not visible in ground-based observations.

1 INTRODUCTION

M83 is a well-studied barred spiral galaxy due to its proximity and its near face-on orientation (i=24°). The estimated distance to M83 ranges from 3.75 Mpc (de Vaucouleurs 1976) to 7.5 Mpc (Lord 1991). The nuclear region is well resolved in ground-based observations (*e.g.* Gallais *et al.* 1991), but with the *HST* Planetary Camera it is possible to probe the region in much greater detail. Our *HST* images actually allowed us to resolve individual clusters within the starburst regions.

2 OBSERVATIONS

We obtained high-resolution images of the nuclear regions with the Planetary Camera (pixel size= 0."0436) on the Hubble Space Telescope *(HST)* on 14 December 1992. We chose U, V, and I filters which would reveal the ionizing star clusters in the bands. We were also able to obtain Hα images of M83 because its readshift ($v_{hel} = 504$ km s^{-1} RC3) places Hα in the [N II] filter, F658N, near its peak sensitivity. Since the images were taken 126 days after the most recent WFPC decontamination, it was necessary to correct for contamination especially in the U-band image.

3 REDUCTION

The primary processing of the data was done at the Space Telescope Institute, which inclused flat-fielding, bias removal, and "dark" image subtraction. The nuclear region of M83 was on PC6. Processing at Goddard Space Flight Center included absolute flux calibration and the removal of cosmic rays. Radiometric calibration involved correcting for WFPC decontamination, deriving WFPC standard magnitudes, and transforming the results over to the Johnson UBVRI magnitude system.

4 SUMMARY OF RESULTS

We compared our images to the near IR images of Gallais *et al.* (1991). Each knot presented in Gallais' paper is resolved into several clusters in our images, each having a FWHM size of 0."2 (4 pc) or larger. The absolute magnitudes of the observed clusters range from –9.46 to –14.63. Our data indicate that the nucleus is not a star-forming region as proposed by Gallais *et al.* instead the nucleus shows the characteristics of an old stellar population. Our Hα images show that the Hα flux is brightest in a region we labeled as region 8, but which corresponds to region A in Gallais *et al.* Region 6 has a companion knot in the Hα image that is not visible in the other bands not in the IR images, but the resolution of Gallais' images is such that it is difficult to be certain that flux from the second knot is not contributing to the total flux in region 6. An overlay of the Hα flux on the U and V band images shows the Hα flux to border the edges of the brightest starburst knots in regions 4, 5, 7, and, as previously mentioned, most intense around region 8. The Hα structure appears to be hollow with the star clusters prominent around the edges.

Fig. 1—Contour plot of an overlay of the V-band (thick) and the Hα (thin) images. The regions are numbered corresponding to Gallais *et al.* (1991). Region A is labeled as region 8 in our plot. The companion to source 6 is to the left of region 6. While the nucleus (region 1) appears to show Hα flux, that emission is due solely to continuum flux.

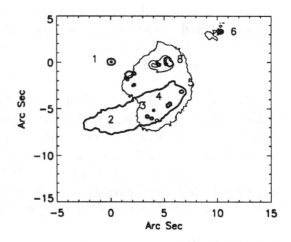

REFERENCES

de Vaucouleurs, G. 1979, *A. J.*, **84**, 1270.
Gallais, P., *et al.* 1991, *Astr. Ap.*, **243**, 309.
Lord, S. 1991, *Ap. J.*, **381**, 130.

Kinematics of the Nuclear Environment of NGC 5728

Santiago Arribas, and Evencio Mediavilla

Instituto de Astrofísica de Canarias. 38200-La Laguna, Tenerife, Spain

ABSTRACT

Here we summarize the main results derived from optical fiber observations (bidimensional spectroscopy) related to the kinematics of the circumnuclear region of NGC 5728. We present additional arguments supporting that the 'true' nucleus of this galaxy is displaced in velocity and space from the optical nucleus. In view of this result, the region of double-peaked emission line profiles is connected to the nucleus, and the red component probably represents outflowing gas.

1 INTRODUCTION

NGC 5728 is a spiral barred galaxy classified as Seyfert 2 (Véron-Cetty et al. 1982; Phillips et al. 1983), which shows some peculiar features in its circumnuclear environment. Here we can point out the presence of a region showing double-peaked emission line profiles in the vicinity of its nucleus, the asymmetric position (with respect to the optical nucleus) of a ring of blue stars and ionized gas (Rubin 1981; Schommer et al. 1988; Wagner and Appenzseller 1988), and the possible presence of a weak BLR misaligned with respect to the NEL maximum (Pecontal et al. 1990).

We have observed the circumnuclear region of NGC 5728 with a new observational arrangement based on the use of optical fibers (Arribas et al. 1991), which is similar to the one presented by Shapovalova (see Afanasiev and Shapovalova, these proceedings). The observations were performed on May 8, 1989, using the 4.2m WHT sited at the ORM. The detailed analysis and results on NGC 5728 can be found in Arribas and Mediavilla (1993).

2 RESULTS ON NGC 5728

The region showing double-peaked line profiles is separated from the optical nucleus. It was concluded that the blue component fits better the general kinematic behavior in the circumnuclear region (see Fig. 1), as well as the local kinematics.

It was found that the optical nucleus (maxima of the NEL) is not only displaced in velocity from the kinematic center, but also in space (by about 1 arcsec $\sim$ 260 pc). The good coincidence of the kinematic center with the radio peak suggests that the 'true nucleus' in NGC 5728 is obscured by dust. This explains the redshift of the

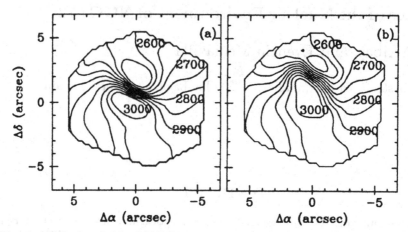

Fig. 1—Velocity field of NGC 5728 generated with the blue (*a*) and red (*b*) component of Hα in the double-peaked line region.

optical nucleus and the high ratio of the circumnuclear-to-nuclear flux, as well as the color effect when measuring the nucleus position, reported by other authors.

There are two additional results which support this interpretation. On the one hand, the kinematic center sits symmetrically within the ring of blue stars and ionized gas, according to the astrometry reported by Wagner and Appentzeller (1988). On the other hand, the kinematic center agrees better with the position of the weak BLR proposed by Pecontal *et al.* (1990).

With our spatial resolution, one edge of the region showing double-peaked line profiles is centered with the hidden nucleus (kinematic center) suggesting that the red component (with larger line fluxes and widths) is directly related to it. From geometrical considerations it is not possible to decide whether it represents inflow or outflow. The large widths of this component seem to support the outflow hypothesis.

REFERENCES

Arribas, S., and Mediavilla, E. 1993, *Ap. J.*, in press.

Arribas, S., Mediavilla, E., and Rasilla, J. L. 1991, *Ap. J.*, **369**, 260.

Pecontal, E., Adam, G., Bacon, R., Courtes, G., Georgelin, Y., and Monnet, G. 1990, *Astr. Ap.*, **232**, 331.

Phillips, M. M., Charles, P. A., and Baldwin, J. A. 1983, *Ap. J.*, **266**, 485.

Rubin, V. C. 1980, *Ap. J.*, **238**, 817.

Schommer, S. A., Nelson, C., Wilson, S. A., Baldwin, J. A., Phillips, M. M., Williams, T. B., and Turtle, A. J. 1988, *Ap. J.*, **324**, 154.

Afanasiev, V. L., and Shapovalova, A. I. 1993, these proceedings.

Véron-Cetty, M. P., Véron, P., Tarenghi, M., and Grosbol, P. 1982, *The Messenger*, **28**, 13.

Wagner, S. J. and Appenzeller, I. 1988, *Astr. Ap.*, **197**, 75.

Neutral Hydrogen Absorption in NGC 1068 and NGC 3079

Jack F. Gallimore[1,2], Stefi Baum[1], Chris O'Dea[1], Elias Brinks[3], and Alan Pedlar[4]

[1]Space Telescope Science Institute, 3700 San Martin Dr., Balto., MD 21218
[2]Astronomy Dept., University of Maryland, College Park, MD 20742
[3]NRAO, P.O. Box 0, Socorro, NM 87801
[4]NRAL, Jodrell Bank, Macclesfield, Cheshire, SK11 9DL, United Kingdom

ABSTRACT

Using the *VLA* in A-configuration we have obtained λ21cm absorption spectra of 12 Seyfert galaxies with bright, extended continuum radio emission. Currently, we have completed the data reduction for three galaxies: NGC 1068, Markarian 3, and NGC 3079. No absorption is detected in Mrk 3, but multiple and broad absorption lines were detected in NGC 1068 and NGC 3079. Here we present the preliminary analysis for these galaxies.

1 NGC 1068

The radio continuum emission from NGC 1068 is dominated by a 13″ radio triple extending SW–NE (*e.g.*, Wilson and Ulvestad 1987). We have detected H I absorption over the entire SW radio lobe and the southern half of the central source. Three kinematically distinct regions are apparent: the radio nucleus, the linear radio structure 1–2″ SW of the nucleus (the SW "jet"), and the SW radio lobe. Using the technique of Dickey, Brinks, and Puche (1992), we extracted optical depth spectra for each of these regions (Fig. 1). Broad (FWHM = 130 ± 25 km s^{-1}), double absorption lines are present in the nuclear region. Adopting v_{sys}(HI)= 1137 km s^{-1} (de Vaucouleurs *et al.* 1991, RC3), the lines are offset by $+55 \pm 11$ and -283 ± 11 km s^{-1}. We suspect that these lines may arise in a region of rapid rotation and streaming near the active nucleus. In the SW "jet" multiple absorption lines centered at v_{sys} are apparent. Since this region lies along the 2.2 μm stellar bar (Thronson *et al.* 1989), these multiple absorption lines may be due to gas streaming in the bar potential. The absorption profiles over the SW radio lobe are narrow (FWHM $\sim$ 20—40 km s^{-1}) and centered near v_{sys}. Absorption in this region likely arises in a nearly face-on, rotating disk. We also note that the distribution of the absorption line gas confirms that the NE (SW) radio lobe is situated on the near (far) side of the galaxy.

2 NGC 3079

We have detected broad absorption line gas which is slightly spatially resolved against the extended radio nucleus of NGC 3079. The $\tau(v)$ spectrum is provided in Figure 2. The neutral hydrogen column density exceeds $10^{22}(T_{spin}/100$ K) cm^{-2}. The primary absorption feature is broad, with FWHM = 310 ± 15 km s^{-1}, and has a deep

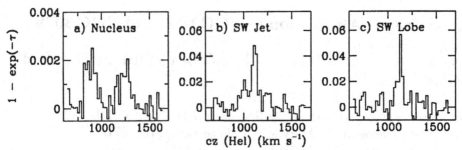

Fig. 1—Optical depth spectra for NGC 1068. These spectra were extracted from *a)* 0.5–1″ SE of the radio nucleus; *b)* 1–2″ SW of the nucleus (the SW "jet"); and *c)* ~ 6″ SW of the nucleus (the SW lobe).

core near $v_{sys} = 1,125$ km s^{-1} (RC3). This broad line component also has a slight blue asymmetry; the velocity difference between the narrow and broad line centers of a two component Gaussian decomposition is -42 ± 5 km s^{-1}. This asymmetry is probably due to gas outflow superimposed on an unresolved region of rapid rotation. In addition, there are red- and blueshifted absorption lines at $+107$ and -120 km s^{-1} relative to the centroid of the primary feature. The relative strengths of the shifted lines vary smoothly over the beam width. We suspect that these shifted features arise in a slightly resolved, rotating ring of neutral hydrogen. Assuming circular orbits and that the ring is just resolved with the *VLA* beam, the mass interior to this ring is $\sim 5 \times 10^8 / \sin i$ M$_\odot$ (i = ring inclination).

Fig. 2—Optical depth spectrum for NGC 3079. This spectrum was extracted by averaging spectra within 1.5″ of the radio nucleus.

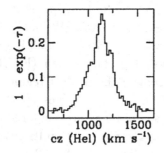

We wish to thank John Dickey for providing for us his τ–extraction code.

REFERENCES

de Vaucouleurs, G., de Vaucouleurs, A., Corwin, H. G., Buta, R. J., Paturel, G., and Fouque, P. 1991, *Third Reference Catalog of Bright Galaxies*, New York: Springer-Verlag (RC3).

Dickey, J., Brinks, E., and Puche, D. 1992, *Ap. J.*, **385**, 501.

Thronson, H. A. *et al.* 1989, *Ap. J.*, **343**, 158.

Wilson, A. S., and Ulvestad, J. S. 1987, *Ap. J.*, **319**, 105.

NGC 5506: Kinematics and Physics of the Extended NLR

Roberto Maiolino[1], Ruggero Stanga[1], Marco Salvati[2], and
Josè M. Rodriguez Espinosa[3]

[1]Dipartimento di Astronomia e Scienza dello Spazio, Università di Firenze, Italy
[2]Osservatorio Astrofisico di Arcetri, Italy
[3]Instituto de Astrofisica de Canarias, Spain

The galaxy NGC 5506 shows features intermediate between Sy 1 and Sy 2. We observed NGC 5506 with ISIS, at the 4.2m William Herschel Telescope. We obtained spectra at three slit inclinations (0° P.A., parallel to the disk minor axis, 68° and 93° P.A.), in the spectral ranges 4,675–5,475Å and 6,300–7,125Å. The relative proximity of this galaxy (z=0.006) allowed us to resolve the Narrow Line Region along the slit.

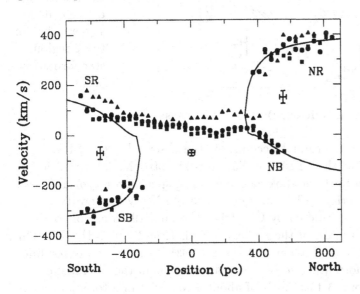

Fig. 1— Velocity field of the double peaks observed along 0° P.A.: circles for Hα, squares for [N II]6,584 and triangles for [O III]5,007. The curves are the results of the kinematical model.

The Hα, [N II]6,584 and [O III]5,007 lines show double peaks at distances larger than 300 pc from the center along 0° P.A. The velocity shift *vs.* position of the two peaks is plotted in Figure 1. We fitted the velocity curve of these peaks with a model similar to that by Wilson, Baldwin, and Ulvestad (1985) where the double peaked lines are emitted by gas flowing through two aligned and opposite cones; we obtain a flow velocity of 400 km s^{-1}, a cone aperture angle of 80° and an inclination angle of the cone axis with respect to the perpendicular to the line of sight of 13°, very close to galaxy inclination.

The relative intensity of the Hα double peaks and their [N II]/Hα ratio (Fig. 2) can be well explained assuming that the emitting clouds within the cones are *outflowing* and *optically thick* to Hα, so that the redshifted clouds (NR and SR), presenting us the illuminated side, show line ratios typical of the high ionization region, while the blueshifted clouds (NB and SB), presenting us the dark side, show line ratios typical of the transition (partially ionized) zone.

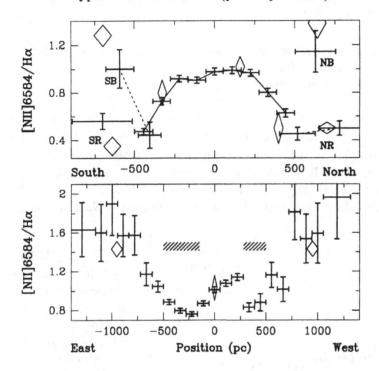

Fig. 2— [N II]/Hα ratio along 0° P.A. (top) and 68° (bottom). Diamonds are the ratios predicted by our model. Shaded marks indicate the regions where the data and the model are discrepant; we suggest that in these regions star formation occurs.

The observed UV emission cannot account for the observed Hβ and Hα luminosities, indicating that the ionizing radiation is anisotropic. Interpreting NGC 5506 as an obscured Sy 1, we took as the ionizing spectrum a typical Sy 1 spectrum (Mathews and Ferland 1987) normalized to the X-ray luminosity of NGC 5506. The line ratios predicted by CLOUDY (Ferland 1990) are in good agreement with the data in most of the nuclear regions (Fig. 2 for [N II]/Hα). Only in the shaded regions in Figure 2, along P.A. 68° and 93° P.A., the computed line ratios are too high. Star formation activity could well explain the discrepancy, diluting the AGN radiation. A total SFR of about 1 $M_\odot$ yr^{-1} in a toroidal region with 250 pc $\leq r \leq$ 500 pc is required to provide the necessary dilution, while the OB stars associated to the star forming regions are sufficient to produce the low UV emission observed by Bergeron *et al.* (1981).

REFERENCES

Bergeron, J., Maccacaro, T., and Perola, C. 1981, *Astr. Ap.*, **97**, 94.
Ferland, G. J. 1990, *OSU Astronomy Department Internal Report*, 90-02.
Mathews, W. G., and Ferland, G. J. 1987, *Ap. J.*, **323**, 456.
Wilson, A. S., Baldwin, J. A., and Ulvestad, J. S. 1985, *Ap. J.*, **291**, 627.

Two-Component Spectral Emission Lines in NGC 3227

Evencio Mediavilla and Santiago Arribas

Instituto de Astrofísica de Canarias. 38200 La Laguna, Tenerife, Spain

ABSTRACT

We have obtained bi-dimensional spectroscopy of intermediate spectral resolution of the circumnuclear region of NGC 3227. We found two-peaked emission line profiles which evidence the presence of different kinematical components. The classification of the line profiles after attempting their two-components Gaussian fitting is discussed.

1 INTRODUCTION

The presence of asymmetries in the profiles of the narrow emission lines (NEL) of Seyfert nuclei is a common result (see *e.g.* Veilleux 1991). In NGC 3227 the asymmetrical profiles of the NEL corresponding to the optical nucleus suggest the existence of a substructure. Observing with intermediate spectral resolution an extended region in the environment of the nucleus of this galaxy, we obtained profiles showing at least two components. We present here this result whose importance for kinematics is obvious.

Our instrumental set-up is based on an optical fiber bundle of 95 fibers disposed in an hexagonal lattice, covering a projected rectangle of 9 arcsec × 12 arcsec. The spatial sampling was ~ 1 arcsec, and the spectral resolution 2Å. The spectral range covered was (4600–5400Å) which includes the Hβ and [O III]$\lambda\lambda$4959, 5007 lines. We used this fiber bundle in combination with the ISIS spectrograph and the 4.2m WHT sited on the island of La Palma. The data were acquired on December 10, 1992. For details about bi-dimensional spectroscopy with optical fibers see Arribas, Mediavilla, and Rasilla (1991), Vanderriest (1993), or Shapovalova (these proceedings).

2 SPECTRAL COMPONENTS

Many of the spectra from the circumnuclear region of NGC 3227 show asymmetrical line profiles with traces of various components. One of them, located approximately 5 arcsec north and 1 arcsec west from the optical nucleus, presents two neatly resolved peaks in both the [O III]$\lambda\lambda$4959, 5007 lines (see Fig. 1a). This suggests a Gaussian decomposition of the line profiles in two components (blue and red components). After fitting the [O III] and Hβ lines we found three basic types of profiles:

i) asymmetrical [O III] lines with blue and red (narrower) components (Fig. 1*b*), ii) nearly symmetrical [O III] lines also with two components (Fig. 1*c*), and iii) possibly symmetrical [O III] lines found in spectra with Hβ/[O III]>1 (Fig. 1*d*). We do not found evidence of more than one component in the Hβ lines.

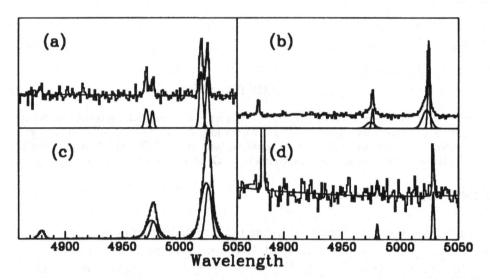

Fig. 1—Fits of the Hβ and [O III]$\lambda\lambda$4959, 5007 lines using two components. The four spectra are representative of (*a*) two-peaked lines; (*b*) asymmetrical [O III] lines; (*c*) nearly symmetrical [O III] lines (this is the spectrum of the optical nucleus after removing the Hβ broad component); and (*d*) symmetrical [O III] lines (Hβ/[O III]>1).

Using the profile decomposition of the [O III] lines we derive from the spectrum corresponding to the optical nucleus, heliocentric velocities of 1,142 km s^{-1} and 1,038 km s^{-1} for the red and blue components, respectively. It is worth noting the good agreement between the velocity corresponding to the red component and the systemic velocity obtained from 21 cm observations (Mirabel 1982) and, consequently, the strong blueshift of the blue component with respect to this velocity.

REFERENCES

Arribas, S., Mediavilla, E., and Rasilla J. L. 1991, *Ap. J.*, **369**, 260.
Mirabel, I. F. 1982, *Ap. J.*, **260**, 75.
Vanderriest, C. 1993, in *Fiber Optics in Astronomy II*, ed. P. M. Gray (San Francisco: ASP Conference Series), **37**, p. 338.
Veilleux, S. 1991, *Ap. J. Suppl.*, **75**, 383.

A Third Inner Ring in a Late-Phase Starburst Galaxy NGC 4736

Jie H. Huang[1], Qiu S. Gu[1], and Hong J. Su[2]

[1]Astronomy Department, Nanking University, China
[2]Purple Mountain Observatory, Nanking, China

A late-phase starburst galaxy NGC 4736 (Walker *et al.* 1988) has been noted to possess some peculiarities, including an expanding circumnuclear ring of H II regions at a radius of about 48″ and a faint outer ring of H I at ∼ 4′ radius. There is a controversy about the origin of this Hα ring (Buta 1988).

We have found a third nuclear ring at ∼ 5″ (∼ 150 pc) radius on deep-exposed images of this galaxy in V-band after reconstructing with Richard-Lucy method (Lucy 1974), shown in Figure 1. Same structures, shown in gradient of intensities in Figure 2, appeared after enhancing the V-band image with adaptive filter (Richter *et al.* 1991). The I-band image, however, does not show any kind of ring-like structure, no matter how you process the image, see Figure 3. Spectroscopic study of this region is under way.

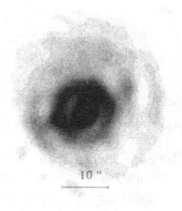

Fig. 1—This is a V-band image of NGC 4736, processed with the Richard-Lucy method. It shows a nuclear ring at ∼ 5″ (∼ 150 pc) radius. The central "bar" structure is also evident.

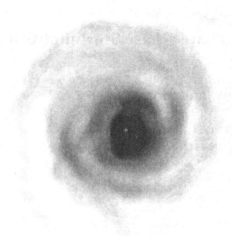

Fig. 2—This is a V-band image of NGC 4736, showing the structures in gradient intensities which look different from those shown on Figure 1. However, it does show the same structure.

Fig. 3— No ring structure appears on this I-band image processed with the Richard-Lucy method.

Acknowledgement: NOAO and STScI are thanked for providing IRAF and STS-DAS for our data reduction and analysis.

REFERENCES

Buta, R. 1988, *Ap. J. Suppl.*, **66**, 233.
Lucy, L. B. 1974, *A. J.*, **79**, 745.
Richter, G. M. *et al.* 1991, *Astron. Nachr.*, **312**, 6.
Walker, C. E., Lebofsky, M. J., and Rieke, G. H. 1988, *Ap. J.*, **325**, 687.

The FWHM-Critical Density Relationship in an AGN Sample

Brian R. Espey

Department of Physics and Astronomy, University of Pittsburgh, Pittsburgh, PA 15260, U. S. A.

ABSTRACT

We report on an analysis of a sample of objects observed to have a correlation between FWHM and critical density for de-excitation of forbidden emission lines. Included in the sample is a QSO of $M_V \approx -26$ which is the most luminous object to date known to show such a relationship. In addition, this QSO shows that the semi-forbidden lines which arise in the Broad Line Region (BLR) follow the same trend seen in the more distant Narrow Line Region (NLR), suggesting that a smooth velocity field joins both zones. A similar relationship between linewidth and critical density holds for all objects in the sample despite a range of $\approx 5000{:}1$ in luminosity.

1 INTRODUCTION

Previous studies of the velocity field in NLRs of low redshift AGNs suggested that the physical processes are at work in all objects (Fillipenko 1985). We have extended the early work by obtaining a sample of objects from the literature which show a correlation between linewidth and critical density and supplemented these with data for a QSO recently discovered to show the first direct evidence for such a relationship extending into the BLR itself.

2 RESULTS

From the available published data we have selected a sample of eight objects which show a correlation coefficient, r, for linewidths with $3.0 < \log(n_{\mathrm{crit}}) < 8.0$ of $r > 0.85$ (where n_{crit} is the line critical density in units of cm^{-3}). From a weighted mean of the slopes for each individual object we derive

$$\log(\mathrm{FWHM}) \propto (0.112 \pm 0.007)\log(n_{\mathrm{crit}}) \qquad (1)$$

There is a weak trend for slope to be correlated with $L(\mathrm{H}\beta)$ but this disappears if the most discrepant value (that for NGC 7213) is excluded. For the most luminous object in our sample the correlation holds into the BLR itself (as evidenced by the agreement of the FWHM of semi-forbidden lines with the value predicted from equation [1])) and there is tentative evidence that the FWHM of C IV 1549Å is consistent with its thermalization density (Rees, Netzer, and Ferland 1989).

The slope given in equation (1) is about half that expected in a simple model in which gas clouds of similar ionization level are in virial motion about a central photoionizing source (*e.g.* Fillipenko and Halpern 1984). It is possible that a more realistic distribution of virial cloud properties can explain the relationship given in equation (1), but the point we wish to stress here is the *similarity* of the slopes for objects which range from AGN of the Low Ionization Nuclear Emission-line Region (LINER) type through Seyfert 1s to QSOs.

There are no strong indications from the data of the direction of motion of the velocity field. While some observers have found evidence favouring infall models (Penston *et al.* 1990; Whittle 1992), theory suggests that infalling clouds will not be stable (Mathews and Veilleux 1989). An intriguing suggestion is that the observed correlation between linewidth and density is due to an outflowing wind shocking ISM gas in the AGN host galaxy (Contini 1988). If such is indeed the case then further work in which host galaxy type and inclination are included may prove revealing.

3 FUTURE WORK

We are planning to extend our data sample to search for possible differences between objects of different AGN types and also to determine the direction of motion of the emission line clouds through more detailed analysis of the emission line profiles. The motion of material in the BLR of AGNs has proven difficult to determine conclusively, although there is strong evidence for radial motion of gas with respect to the systemic velocity in both low and high luminosity AGN (*e.g.* Osterbrock and Cohen 1979; Carswell *et al.* 1991). The existence of a relationship between linewidth and critical density which extends from the NLR into the BLR has been suggested previously on the basis of statistical results (Heckman *et al.* 1984). If it can be proven to hold in more individual cases it will provide a powerful tool to decipher the velocity field in the nuclear region of AGNs through study of less complex NLR material.

REFERENCES

Carswell, R. F. 1991, *Ap. J. Lett.*, **381**, L5.
Contini, M. 1988, *Ap. J.*, **333**, 181.
Fillipenko, A. V. 1985, *Ap. J.*, **289**, 475.
Fillipenko, A. V., and Halpern, J. P. 1984, *Ap. J.*, **285**, 458.
Heckman, T. M., Miley, G. K., and Green, R. F. 1984, *Ap. J.*, **281**, 525.
Mathews, W. G., and Veilleux, S. 1989, *Ap. J.*, **336**, 93.
Osterborck, D. E., and Cohen, R. 1979, *Mon. Not. R. Astr. Soc.* **187**, 61p.
Penston, M. V., Croft, S., Basu, D., and Fuller, N. 1990, *Mon. Not. R. Astr. Soc.*, **244**, 357.
Rees, M. J., Netzer, H., and Ferland, G. J. 1989, *Ap. J.*, **347**, 640.
Whittle, M. 1992, *Ap. J*, **387**, 121.

Dust and Properties of BLR and NLR in AGNs

Ari Laor[1], and Hagai Netzer[2]

[1]Institute for Advanced Study, Princeton, U. S. A.
[2]School of Physics and Astronomy, Tel-Aviv University, Tel-Aviv, Israel

ABSTRACT

The apparent low covering factor of the narrow line region (NLR), the nearly "empty" intermediate region between the broad line region (BLR) and the NLR, and the size of the BLR, are all naturally explained if dust is embedded in the narrow line emitting gas. Such dust, together with the observed blue-excess asymmetry of the narrow line profiles, imply that the NLR gas has a net inflow motion.

1 THE BASIC QUESTIONS

In this paper we address the following three questions: 1) Why does $R_{\mathrm{BLR}} \sim 0.1 L_{46}^{1/2}$ pc? This scaling is indicated by the remarkably similar emission spectra of AGNs over a very large luminosity range. 2) What is the actual covering factor (C) of the cold gas in the NLR? Photon counting using the Hβ recombination line indicates $C \sim 2\%$, assuming optically thick clouds. The $3-30$ μm IR continuum, if due to dust reprocessing of the UV continuum, indicates $C \sim 20-40\%$. High resolution imagings of the NLR by the *HST* in nearby AGNs indicate, if the clouds are resolved, $C \geq 20\%$. 3). Why is there an apparent gap in the gas distribution between the BLR and the NLR? A gap is indicated most clearly in Seyfert 1.5 galaxies and some quasars where the permitted line profiles show distinct narrow and broad components.

2 THE EFFECTS OF DUST ON THE LINE EMISSION

Dust embedded with photoionized gas strongly suppresses the line emission when the ionization parameter $U \geq 10^{-2}$ (see Voit 1992; Laor and Draine 1993; Netzer and Laor 1993). Since U increases from $10^{-2} - 10^{-4}$ in the NLR to $\sim 10^{-1}$ in the BLR, we expect the following: at large distances dust has a small effect on the line emission. This is the standard Extended NLR. With decreasing distance, U increases and the line emission efficiency decreases. An increasing fraction of the incident ionizing flux is absorbed by dust, and an increasing fraction of the line emission is destroyed by the dust. The clouds gradually "disappear" into the intermediate region (see Fig. 1). This explains the presence of the apparent "gap", and the apparent low value of C deduced from Hβ photon counting. Below the critical radius, $R_{\mathrm{sub}} \simeq 0.2 L_{46}^{1/2}$ pc, all the dust sublimes. The line emission efficiency increases by $1-2$ orders of magnitude.

The clouds "reappear" in what we call the BLR. This explains both the absolute value of R_{BLR} and its scaling with luminosity. Further details and discussions are given in Netzer and Laor (1993).

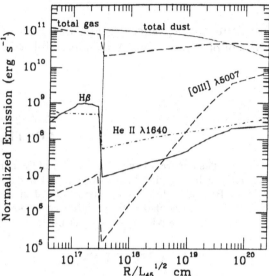

Fig. 1—Normalized line emission, per given solid angle, as a function of radius. A continuous cloud distribution with $U \propto R^{-1/2}$ is assumed. Internal dust is present at all distances outside $R = 0.1 L_{45}^{1/2}$ pc, and is absorbing the central source radiation and destroying line photons. This is the reason for the sharp drop in line emission at that radius. Dust absorption at larger distances is less pronounced because of the decrease in ionization parameter.

3 IMPLICATIONS AND PREDICTIONS

The presence of dust embedded in the narrow line clouds implies the following: 1) The narrow lines and the $3 - 30\ \mu$m IR emission originate from the same distribution of (molecular?) clouds. 2) The narrow line emission is not isotropic as generally assumed. 3) The blue asymmetry of the narrow line profiles implies that the NLR has a net *inflow* motion. 4) The $\sim 10^4$ K gas in the NLR does not originate from a cooling instability of gas in a hotter phase.

The following predictions will allow this idea to be tested: 1) The clouds in the "gap" should be observable in scattered optical-UV light. 2) The scattering is likely to result in a low level ($\sim 1 - 2\%$) optical-UV polarization. 3) Emission line surface brightness asymmetry is likely to be present for an axisymmetric cloud distribution (farther side will be brighter). 4) Significant molecular line emission is expected. 5) Depletion of heavy elements from the gas phase, in particular Fe, will affect the observed line ratios.

A. L. acknowledges support by the NSF grant PHY92-45317. H. N. acknowledges support by the US-Israel Binational Science Foundation grant 8900179.

REFERENCES

Laor, A. and Draine, B. T. 1993, *Ap. J.*, **402**, 441.
Netzer, H., and Laor, A. 1993, *Ap. J. Lett.*, **404**, L51.
Voit, G. M. 1992, *Ap. J.*, **399**, 495.

Compact Radio Cores in Seyfert Galaxies

R. P. Norris[1], and A. L. Roy[1,2]

[1]Australia Telescope National Facility, CSIRO
[2]School of Physics, University of Sydney

ABSTRACT

We find that Sy 2 galaxies have compact radio cores much more often than Sy 1 galaxies. This result is inconsistent with the popular unified models which explain the difference between Sy 1 and Sy 2 in terms of orientation. We propose a mechanism which is consistent with the observations while still supporting the unified model.

1 INTRODUCTION

The orientation unification scheme has enjoyed considerable success in accounting for the difference between Sy 1 and Sy 2 galaxies. This scheme suggests that the apparent differences between Sy 1 and Sy 2 galaxies are due simply to our viewing angle. It invokes a dense dusty torus which obscures our view of the broad-line region (BLR) when our line of sight lies close to the plane of the torus. In this case, only the lines from the more distant narrow-line region (NLR) are visible, and the galaxy then appears as a Sy 2. Light from the BLR is visible only when viewed from within a cone centred on the polar axis of the dust torus, and the galaxy then appears as a Sy 1.

This model has received support from observations such as those of Miller and Goodrich (1990), who found that the polarized emission from some Sy 2 galaxies has broad lines characteristic of a Sy 1 galaxy. They suggest that the light scattered from dust or electrons above and below the torus enable us to see the BLR in scattered light. This scheme predicts that, since the dust in the torus should be optically thin at radio wavelengths, Sy 1 and Sy 2 galaxies should appear identical at radio wavelengths.

It is already established that the total radio luminosity of the two types is similar (Ulvestad and Wilson 1989) but it is possible that the total flux density may be dominated by large-scale emission which is not directly related to the central core. However, high- resolution very long baseline interferometry ($VLBI$) observations are sensitive principally to non-thermal emission at 10^8 K from the Seyfert activity, and discriminate against the 10^4 K extended emission expected from H II regions and supernova remnants. Such observations therefore provide a sensitive test of the unified model.

125

2 OBSERVATIONS

We have selected two samples of Seyfert galaxies. One far-infrared (FIR) sample, containing 120 objects, was selected using *IRAS* flux densities, and the other optical sample, of 64 objects, was selected using optical and near-infrared properties. Each sample contains comparable numbers of Sy 1 and Sy 2 galaxies, and the sub-samples of Sy 1 and Sy 2 galaxies do not differ significantly in redshift or luminosity distribution.

All these galaxies were observed with the 275-km baseline Parkes-Tidbinbilla Interferometer (*PTI*) at 2.3 GHz, resulting in a survey with uniform sensitivity and high resolution. Preliminary results based on a subset of these samples have been published by Norris *et al.* (1992). Here we have increased the sample size, and refined the statistical treatment. A full account of this experiment is given by Roy *et al.* (in preparation).

3 RESULTS AND CONCLUSION

To simplify the statistics, we simply register a detection or non-detection for each source, without regard to the magnitude of the detected flux density. In our FIR sample, we obtained a detection rate of 47% on our sample of 78 Sy 2s, but only 24% on our sample of 42 Sy 1s. Similarly, in our optical sample, we obtained a detection rate of 50% on our sample of 26 Sy 2s, but only 29% on our sample of 38 Sy 1s. It is clear that Sy 1s have a much lower detection rate than Sy 2s in both samples. Roy *et al.* give a rigorous discussion of the statistics, and show that the level of significance of the effect in the combined samples is greater than 99%.

We suggest an explanation for this surprising result. Our model is based on that proposed by Norris *et al.* (1992). The dominant effect is that the NLR clouds are optically thick (from free-free absorption) at centimetre wavelengths, and so obscure the radio emission from the compact core when the galaxy is viewed down the cone of the NLR. Roy *et al.* discuss modifications to this model because of other effects, such as the location of the radio core relative to the BLR and NLR, the free-free self-absorption of NLR clouds, and the radio optical depth of the torus. A definitive test of the model will be to repeat the observations at a higher frequency.

REFERENCES

Miller, J. S., and Goodrich, R. W. 1990, *Ap. J.*, **355**, 456.
Norris, R. P., *et al.* 1992, in *Relationships between Active Galactic Nuclei and Starburst Galaxies*, ed. A. V. Filippenko (San Francisco: ASP-31), p. 71.
Ulvestad, J. S., and Wilson, A. S. 1989, *Ap. J.*, **343**, 659.

The Link Between Galaxy Disturbance and Emission Line Width in Seyfert Galaxies

Charles H. Nelson and Mark Whittle

. Department of Astronomy, University of Virginia

ABSTRACT

We have measured or compiled stellar velocity dispersions, σ_*, for a sample of ~ 80 Seyfert galaxies. The [O III] $\lambda 5007$Å emission line width correlates quite strongly with σ_*, suggesting ionized gas velocities result principally from motion in the host bulge potential. Here we concentrate on second order effects, looking for parameters which correlate with the <u>scatter</u> on the [O III] FWHM vs. σ_* relation. In decreasing order of clarity, we find that Seyferts with relatively broad emission lines (*e.g.* $V_{gas} > V_{stars}$) have strong linear radio sources, are disturbed, or have bars. Since these galaxies show no unusual scatter on the Faber-Jackson plot of σ_* vs M_{bul}, we conclude that radio luminous Seyferts and tidally disturbed Seyferts have unusual gas kinematics rather than unusual stellar kinematics.

The profiles of forbidden emission lines characterize the ionized gas kinematics in the Narrow Line Region (NLR) of Seyfert galaxies. What physical processes accelerate this gas, and are they related to the active nucleus or to the host galaxy? To explore the role of the host galaxy, we have measured stellar velocity dispersions, σ_*, for 78 objects using the cross-correlation method. We also include published measurements of σ_* (*e.g.* Terlevich, Dias, and Terlevich 1990; Whitmore *et al.* 1985) and complementary data on NLR and host properites, from Whittle (1992). In particular, we include the Perturbation Class, PC, which rates the degree of galaxy disturbance and/or interaction on a scale of 1 to 6 (based on the scheme of Dahari [1985]).

Figure 1 shows [O III] FWHM plotted against FWHM(stars) ($\sigma_* \times 2.355$). There is a moderately strong correlation (significance 99.992%) close to the dashed line $Y = X$, suggesting that the bulge gravitational potential is a primary factor in determining the gas velocity field in the NLR.

There is, however, considerable real scatter, suggesting that secondary factors influence the gas kinematics. We consider three possibilities for these other factors — radio source properties, galaxy disturbance, and bars. In Figure 1, Seyferts with luminous linear radio sources (LLRS), disturbed and/or interacting Seyferts (PC > 3), and barred Seyferts, are plotted with different symbols. Although not a dramatic separation, the flagged objects tend to fall to the right of the solid line, which is the best fit to the remaining objects (solid dots). We characterize these offsets in several

128   *Nelson and Whittle*

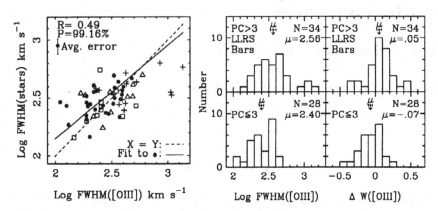

Fig. 1—FWHM([O III]) *vs.* FWHM(stars). + – Luminous linear radio sources (LLRS), △ – morphologically disturbed (PC > 3), □ – bars, • – normal Seyferts.

Fig. 2—Histograms of FWHM([O III]) and ΔW for the flagged and unflagged objects in Figure 1. The mean and its error for each distribution are indicated.

ways. First, counting objects left/right of the fit gives 3/11 (LLRS), 3/7 (disturbed), and 1/9 (barred), showing an unequal division with (binomial) significance levels 97%, 83%, and 99% respectively. Second, we compare the FWHM distributions of each flagged group with the unflagged group. Students t-test shows offsets to higher mean FWHM with (two tail) significance 99.96% (LLRS), 98.1% (disturbed), 85% (barred). The left hand panel of Figure 2 shows the combined subgroups (offset significance 99.8%). Third, to remove the influence of the primary correlation, we define ΔW, as the offset from $X = Y$ in Figure 1, *i.e.* $\Delta W = \log(\text{FWHM}_{[\text{O III}]}) - \log(\text{FWHM}_{\text{stars}})$. Comparing distributions of ΔW shows higher mean ΔW with significance 99.2% (LLRS), 94.6% (disturbed), and 95.8% (barred). The right hand panel of Figure 2 shows the combined subgroups (offset significance 99.5%).

While not statistically dramatic, we feel these results nevertheless indicate real effects. Excess linewidth in Seyferts with LLRS may result from interaction between the radio jet and the NLR gas. In disturbed, interacting, and possibly barred Seyferts, non-axisymmetric potentials seem to influence the kinematics of nuclear ionized gas.

Finally, we note that in a Faber-Jackson plot of σ_* against M_{bul}, the flagged objects do <u>not</u> stand apart from the others, suggesting that the offsets discussed above are due to differences in gas kinematics, rather than differences in stellar kinematics.

REFERENCES

Dahari, O. 1985, *Ap. J. Suppl.*, **57**, 643.
Terlevich, E., Diaz, A. I., and Terlevich, R. 1990, *Mon. Not. R. Astr. Soc.*, **242**, 271.
Whitmore, B. C., McElroy, D. B., and Tonry, J. L. 1985, *Ap. J. Suppl.*, **59**, 1.
Whittle M. 1992, *Ap. J. Suppl.*, **79**, 49.

Theoretical Line Profiles of Spatially Resolved Optically Thin Disks and Cones

H. Schulz[1,2] and A. Mücke[2]

[1]Radioastronomisches Institut der Universität Bonn, Auf dem Hügel 71,
 D-53121 Bonn, Germany
[2]Astronomisches Institut der Ruhr-Universität, D-44780 Bochum, Germany

ABSTRACT

A theoretical scheme is developed for computing line profiles from optically thin rotating and expanding disks and systems of emission line clouds moving radially inside cones. The models include turbulent motions, atmospheric seeing and effects induced by the size of the observing aperture. As an example, the asymmetric extranuclear Hα line profiles along the major axis of NGC 7469 are fitted by a rotating disk model.

1 INTRODUCTION

Disks and cones are basic geometries for systems of emission line clouds in active galaxies. Many lines, as *e.g.* forbidden emission lines, are formed in optically thin clouds so that the line profiles only depend on the geometrical and kinematical distribution of the clouds and radiative transfer effects can be neglected. Hence, infinitesimally spatially resolved profiles are calculated by integrating along that part of the line-of-sight that intersects the configuration. Those volume elements whose line-of-sight velocity component equals v contribute to the intensity $I(v)$.

For systems following radial power laws for both velocity and emissivity we derived analytical expressions for the profile functions. These results are used to check the pure numerical code whose working method is based on collecting and summing up all emissivities falling in a given velocity bin along the considered line-of-sight. Additional isotropic turbulent motions are taken into account by convolving the profiles with a normalized Gaussian of a given σ. The line-of-sight profiles are stored in a data cube (x, y, v). Spatial convolutions to mimic the effects of atmospheric seeing and finite observing apertur and pixel sizes are simply carried out with the *ESO* image processing system MIDAS.

2 DISKS

For an edge-on disk with emissivity law $j = r^\beta$ we obtain the following analytical solution assuming radial power laws for a circular velocity field $w_c = r^a$

$$I(v) = (\frac{v}{x})^{(\beta-a+3)/(a-1)} \left[x(a-1)\sqrt{(\frac{v}{x})^{2/(a-1)} - x^2} \right]^{-1}$$

and the implicit solution for a radial velocity $w_r = r^b$

$$v(y) = y(x^2 + y^2)^{(b-1)/2}$$

$$I[v(y)] = \frac{1}{(x^2 + y^2)^{(b-\beta-1)/2} + y^2(b-1)(x^2 + y^2)^{(b-\beta-3)/2}}$$

A possible inclination i of the geometry is taken into account by the transformation $v \to v/\sin i$.

Line-of-sight profiles from rotating disks turn out to be asymmetric.

3 CONES

For systems of clouds moving radially inside tilted cones (again obeying power laws for velocity and emissivity) we obtain the analytical solution

$$v(y) = [r(y)]^{(b-1)}[q(\Phi) + \zeta \tan \Phi]$$

$$I[v(y)] = \left\{ (b-1)r^{b-\beta-3}[y(1 + \tan^2 \Phi) + \frac{\zeta \tan \Phi}{\cos \Phi}][yq(\Phi) + \zeta \tan \Phi] + r^{b-\beta-1}q(\Phi) \right\}^{-1}$$

with $q(\Phi) = \cos \Phi + \sin \Phi \tan \Phi$, $\zeta = -y \sin \Phi + z \cos \Phi$ and $r(y) = [x^2+y^2+(\zeta/\cos \Phi+y \tan \Phi)^2]^{\frac{1}{2}}$, where Φ is the tilt angle of the cone measured from side-on and z the coordinate perpendicular to the line-of-sight coordinate y (for $\Phi = 0$ z-axis = cone-axis).

4 A NUMERICAL FIT

As a first result high resolution (45 km s^{-1}) Hα narrow-line profiles from the central 5 pixels (0.94″ px^{-1}) along the major axis of the Seyfert galaxy NGC 7469 are fitted by assuming a linearly rising rotation curve up to 1″ which becomes flat beyond. The line emissivity is constant in the innermost arcsec and then falls with the inverse square of the galactocentric distance. The disk inclination is 45°. The turbulence is 20% of the rotation velocity and the seeing 1″. With these parameters the observed asymmetries can be successfully fitted.

H. S. acknowledges support from DARA grant 50 OR 9102.

Star Formation in Barred Galaxies

Robert C. Kennicutt, Jr.

Steward Observatory, University of Arizona

ABSTRACT

Galaxy bars can be important triggering agents for star formation, radial gas flows, and nuclear activity. This paper reviews the observational evidence for bar-induced star formation and gas redistribution in spiral galaxies. Specific topics include the global star formation rates in barred vs normal galaxies, the spatial distribution and abundances of star forming regions in barred systems, and circumnuclear hotspots.

1 INTRODUCTION

Barred galaxies present one of the clearest cases of mass-transfer induced activity, and as such are valuable laboratories for understanding the triggering of starbursts and nuclear activity in a broader context. As reviewed by Athanassoula elsewhere in this volume, hydrodynamic simulations suggest that bars can trigger a wide range of phenomena, including large-scale gas compression, star formation, and radial transport of gas into the nuclear region (also see Sellwood and Wilkinson 1993).

This paper reviews the observational evidence for bar-induced star formation and circumnuclear activity. I begin by discussing the integrated properties of barred vs normal spirals, based on surveys in Hα, radio continuum, and the infrared (section 2). Section 3 summarizes the star formation properties of individual barred systems, with emphasis on the bars themselves and their surrounding disks. In section 4 I discuss the circumnuclear "hotspot" star formation regions, which are probably the most distinctive signatures of bar-induced activity. I conclude with a summary of outstanding questions and important areas for future work.

2 INTEGRATED PROPERTIES OF BARRED VS NORMAL SPIRALS

The first systematic comparisons of the global properties of barred and normal galaxies were based on radio continuum surveys (Cameron 1971; Dressel and Condon 1978; Dressel 1979; Heckman 1980). These data revealed no significant difference between the total radio fluxes of barred vs normal spiral, but showed indications of differences in their central radio sources. Hummel (1980, 1981) used high-resolution observations of a complete sample to show that the radio luminosity functions of spiral disks were indistinguishable for barred and nonbarred systems, but the central radio sources (interior to a few kpc) were roughly twice as strong in barred spirals. These results revealed that the main influence of the bars was in the central regions, though a physical connection to star formation was not yet firmly established.

The release of the *IRAS* survey in the mid-1980's stimulated a second wave of studies. Hawarden *et al.* (1986) compiled the *IRAS* colors of Shapley-Ames spirals, and found that roughly one third of the barred galaxies showed unusually strong emission at 25 μm. This mid-IR emission is usually associated with warm dust in the nuclear regions of galaxies, and Hawarden *et al.* hypothesized that the excess in barred systems was produced by luminous circumnuclear star forming regions. By contrast there was no significant difference between barred and normal spirals in the 60–100 μm range, where the bulk of the emission from disk star forming regions is found.

The case for circumnuclear star formation was indirect, because most galaxies were not spatially resolved in the *IRAS* survey. Subsequent high resolution observations have confirmed this interpretation, however. Puxley, Hawarden, and Mountain (1988) showed that the IR-bright barred systems often contain luminous central radio sources. Phillips (1993) has likewise shown that the mid-IR excess is often associated with visible circumnuclear hotspots, and there is a strong correlation between the excess IR and Hα emission. The most direct test comes from a 10 μm survey of the nuclei (roughly central kpc) of 133 nearby galaxies by Devereux (1987). Devereux found that 40% of *early-type* barred galaxies exhibit strong 10 μm excesses, which he attributed to nuclear activity (*e.g.,* Seyferts) or star formation. The distinction between early and late-type spirals is significant, and is probably tied to dynamical differences between the respective bar types (see section 3.1 and Devereux's paper, these proceedings). Although strong 10–25 μm emission is prevalent among early-type barred systems, a bar is not a necessary condition for such activity. For example H and K-band imaging for a large sample of IR-luminous galaxies shows that roughly half of the IR-bright spirals contain no discernable bar structure whatsoever (Pompea and Rieke 1990).

These radio and infrared surveys provide convincing evidence that the central regions of barred galaxies are unusually active, but they leave unanswered the original question of whether bars influence the *global* SFRs in disks. In order to address this question directly I compiled integrated Hα measurements of 217 spirals from the surveys of Kennicutt and Kent (1983) and Romanishin (1990), and subdivided the galaxies into barred (SB) and unbarred (SA, SAB) classes, according to published classifications. Figure 1 compares the distributions of Hα + [N II] equivalent width for the 161 normal and 56 barred galaxies, plotted as a function of Hubble type. The equivalent width is effectively the Hα luminosity of the galaxy normalized to its red continuum luminosity, and equivalently the star formation rate (SFR) per unit luminosity or mass. The line connects the median values for normal galaxies in each subtype, and shows that there is no significant difference between the integrated SFRs of the normal and barred systems. Ryder and Dopita (1993) reach the

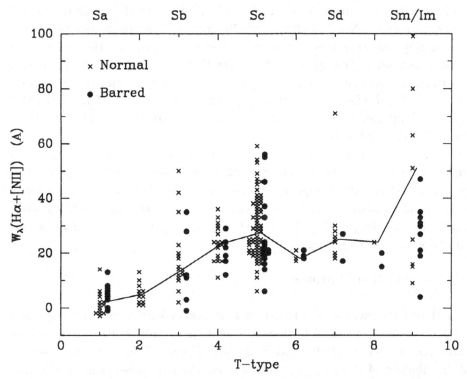

Fig. 1—Comparison of integrated Hα + [N II] equivalent widths for barred vs normal spirals, shown as a function of Hubble type. The line connnects the median values for normal spirals of each type.

same conclusion based on a sample of 24 southern spirals, though they do detect a difference in the central SFRs, as discussed in the next section. In summary, there is no detectable difference between the total SFRs of barred and normal spirals, but is a marked excess of star formation or other activities in the central regions of some (~30–50%) barred systems.

3 EXTRANUCLEAR STAR FORMATION

A casual inspection of a photographic atlas of barred galaxies will reveal a variety of unique star formation properties—star forming knots concentrated along the bar, in prominent spiral arms emanating from the bar, in a ring around the bar, in a pair of bright clumps at the ends of the bar, and/or in a bright circumnuclear ring within the bar. These striking patterns often lead to erroneous generalizations in the literature, for example that "star formation is triggered in bars" or just as often that "star formation is suppressed in bars"! The actual star formation properties of barred galaxies are much more complex, and before we discuss these properties in detail it is useful to examine actual images of a few examples. Figure 2 displays continuum-subtracted Hα images of 4 galaxies which illustrate the range in bar and star formation properties, NGC 1300, 1512, 7479, and 1073. As these images demonstrate, bars clearly influence the star formation in each case, but there is a large variety of disk responses. For instance it is clear that gas compression in the bars often does trigger strong star formation, as illustrated by the prominent bar H II regions in NGC 7479 and NGC 1073, but in other cases (*e.g.,* NGC 1300, NGC 1512) the the entire disk interior to the bar is devoid of star formation, with the prominent exceptions of the nuclear hotspots.

The diversity of structures so evident in Figure 2 must ultimately be tied to differences in the structure, dynamics, and gas content of the disks and bars themselves. Quantifying these relationships requires systematic observations of large samples of barred and normal galaxies. Thanks to several recent studies, most notably three Ph. D. studies by Feinstein (1992), Phillips (1993), and Ryder (1993), this observational foundation is now being laid.

3.1 Star Formation Distributions

Most of the current studies are based on Hα observations, which directly trace the SFR and have sufficient sensitivity and resolution to provide detailed spatially resolved information. Ryder (1993) and Ryder and Dopita (1993) have focussed on the radial distribution of the SFR. Outside the bar the Hα emission is usually well respresented by an exponential disk, as is the case in normal spirals (Kennicutt 1989). However in most cases, especially early-type systems, the emission within the bar radius drops well below the extrapolated exponential, suggesting that star formation has been suppressed (*e.g.,* NGC 1300 and NGC 1512 in Fig. 2). In many galaxies the Hα emission peaks again near the nucleus, often in a cluster of circumnuclear hotspots. Such circumnuclear emission is observed in approximately 25% of all barred galaxies, and 40% of early-type barred systems (Sérsic and Pastoriza 1967). The large increase in the circumnuclear SFR is often comparable to the net deficit elsewhere within the bar.

Phillips (1993) has undertaken the most comprehensive study of star formation in barred galaxies to date. He obtained deep Hα and continuum maps of 16 nearby barred systems, and used automated photometry programs to measure the fluxes of individual H II regions, and map the distribution of the SFR within the galaxies. This allowed him to test and quantify many of the generalizations that are often made about bar-triggered star formation. Data for three of his galaxies are shown in Figure 2. One of the most important results of his thesis is that most of the observed star formation properties of the barred galaxies can be subdivided into two distinct classes, which correlate with the photometric and kinematic properties of the underlying galaxies. Types SBb and earlier contain bars with flat stellar surface brightness profiles, flat rotation curves, strong rings, but no star formation in the bar region. Types SBc and later show exponential bars, rising rotation curves, relatively

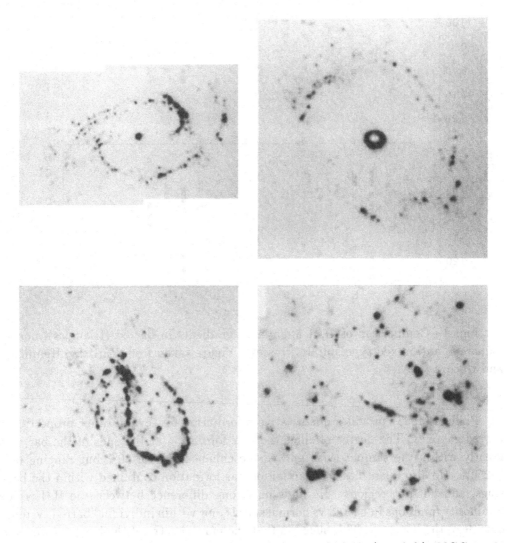

Fig. 2—Hα images of 4 barred spirals: NGC 1300, SBb (top left); NGC 1512, SBb (top right); NGC 7479, SBbc (lower left); NGC 1073, SBc (lower right). Note the bright circumnuclear emission regions in NGC 1300 and NGC 1512. NGC 7479 image provided by J. Kenney and J. Young, others by A. Phillips.

weak rings, and star formation enhancements in the bar. Bright circumnuclear regions may be found in both type ranges, though are especially frequent in the early-type systems (Sérsic and Pastoriza 1967; Devereux 1987; Phillips 1993). These results are in general accord with the dynamical models (Athanassoula, these proceedings), and hence provide a theoretical framework for understanding the diversity of star formation properties in barred galaxies.

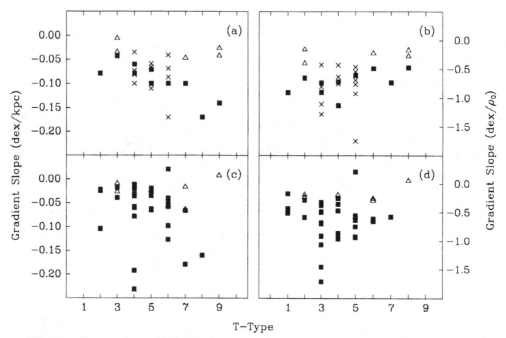

Fig. 3—Comparison of O/H abundance gradients in barred (triangles), normal (squares), and mixed type spirals (crosses). Figure taken from Zaritsky, Kennicutt, and Huchra (1993).

Phillips (1993) provides quantitative information on several other properties of barred galaxies. The degree of enhanced star formation at the ends of the bar vary greatly within the sample, with an average enhancement of 80% but ranging over 0–250%. He also quantifies the fraction of star formation contained within the bars, rings, and hotspot regions. He finds no strong difference between the H II region luminosity functions in barred vs normal spirals, nor within individual barred systems, outside of the hotspots. This suggests that the mass spectra of the bar-triggered star forming regions are not significantly different from those formed in normal disks.

3.2 Chemical Abundance Distributions

The abundance distributions in barred galaxies can impose valuable constraints on the gas flows and bar lifetimes, but until recently the available data were too sparse to be useful. The situation has changed markedly in the past year, with the publication of large H II region abundance surveys by Vila-Costas and Edmunds (1992), Zaritsky, Kennicutt, and Huchra (1993), and Martin and Roy (1993). All three studies show a striking difference between the abundance distributions in barred vs. normal disks.

Figure 3, taken from Zaritsky *et al.* (1993), summarizes the behavior of the O/H abundance gradients in barred and normal galaxies. The top two panels show data from Vila-Costas and Edmunds (1993), while the bottom panels show new data from the Zaritsky *et al.* study. Normal spirals are denoted by squares, barred galaxies by open triangles, and mixed types by crosses. The gradients, measured in units of both dex/kpc (left) and dex per isophotal radius ρ_0 (right), are plotted as a function of galaxy type. In all cases there is a marked separation between the barred and normal spirals, with the barred systems showing much weaker radial gradients or no gradient at all. The result is insensitive to the way in which the gradient is defined, or to the radial range over which it is measured. The dependence on bar type is clearly stronger than any underlying dependence on Hubble type or luminosity. Martin (these proceedings) and Martin and Roy (1993) find further evidence that the magnitude of the gradient among barred systems correlates with the relative strength of the bar.

The systematically weaker abundance gradients in barred disks is difficult to understand unless bars are very longlived structures, because the time scale for modifying the chemical abundance distributions should be comparable to the gas recycling time of disks, roughly 4–6 Gyr on average (Kennicutt 1983). If bar-induced gas flows are the explanation for the flat abundance distributions, as suggested by Martin and Roy (1993), then very efficient angular momentum transport and long bar lifetimes are required. Recent numerical simulations (*e.g.*, Pfenniger and Friedli 1991; Little and Carlberg 1991; Friedli and Benz 1993) suggest that such long lifetimes and high angular momentum transfer rates are not unreasonable, and it would be interesting to make a quantitative calculation of the expected abundance distributions in such models. Edmunds and Roy (1993), on the other hand, suggest that the barred systems may be evolving as normal closed-box systems, and that the anomalously steep abundance gradients in non-barred systems are produced by grand-design spiral patterns. It is even conceivable that external parameters, such as halo mass coupled to infall history, are responsible for the evolution of both the bar structure and the chemical enrichment histories of the disks. Whatever the explanation, understanding the chemical evolution of the barred galaxies may reveal fundamental insights in their dynamical evolution as well.

3.3 Gas Compression and Star Formation

Most of the work cited above has focussed on quantifying the observed star formation properties of barred galaxies. Understanding the physical mechanisms which are responsible for these differences requires coordinated studies of the gas distributions, kinematics, and star formation distributions in well resolved systems. To date only a handful of nearby galaxies have been studied in this manner, most notably M83 (Handa, Sofue, and Nakai 1991; Kenney and Lord 1991), NGC 3504 (Kenney, Carlstrom, and Young 1993), NGC 3992 (Cepa and Beckman 1990), and NGC 4321 (Cepa *et al.* 1992; Knapen *et al.* 1993). For illustrative purposes I will briefly summarize the work on M83, which is the best studied case to date.

M83 is one of the nearest barred spirals, and it possesses virtually all of the features attributed to barred galaxies, a two-armed grand design spiral pattern, pronounced clumps of star formation outside the ends of the bar, dust lanes and (faint) H II regions along the bar, and very bright circumnuclear regions (*cf.* Holbrook *et al.*, these proceedings). It has been mapped at high resolution in H I, CO, Hα, and in the stellar continuum at several visible and infrared wavelengths (see Kenney and Lord 1991 for a summary of the relevant literature). Handa *et al.* (1991) mapped the molecular gas distribution in the southern arm and bar of the galaxy, and combined this with a UV map of the galaxy to compare the SFR/density ratio in the disk, arm, bar, and circumnuclear regions. They found that this ratio was 5–10 times lower in the bar than in the surrounding disk and arm regions, suggesting a much less efficient conversion of gas to stars in the bar. Kenney and Lord (1991) obtained higher resolution CO maps, and used these to compare the distribution of gas and star formation with a simple dynamical model for the bar and disk. They showed that the most pronounced concentrations of gas and star formation coincide with the expected locations of orbit crowding between streamlines in the bar and the surrounding disk. It would be interesting to extend this comparison to a larger sample, to test whether the orbit-crowding model can reproduce the wide variety of star formation concentrations which are observed among barred systems (*e.g.,* Phillips 1993).

4 CIRCUMNUCLEAR STAR FORMATION, HOTSPOTS

The most spectacular star forming regions are the circumnuclear "hotspots" first identified by Morgan (1958) and cataloged in a series of papers by Sérsic and Pastoriza (1967). These bright knots generally lie from 100–500 pc of the nucleus, are are found almost exclusively in barred galaxies. A complete inspection of the Hubble plate collection by Sérsic and Pastoriza (1967) showed hotspots in 24% of barred galaxies (types SB and SAB) studied, but in none of the unbarred galaxies in the sample.

The photometric and spectroscopic properties of hotspots have been studied by numerous authors, including Osmer, Smith, and Weedman (1974), Alloin and Kunth (1979), Phillips *et al.* (1984), Wynn-Williams and Becklin (1985), Kennicutt, Keel, and Blaha (1989), Jackson *et al.* (1991), and Phillips (1993). Although the hotspots are often presumed to be H II regions, multiband imaging shows that they are a mixture of high surface brightness H II regions and very luminous star clusters (see Kennicutt *et al.* 1989 for examples). Infrared mapping (*e.g.,* Wynn-Williams and Becklin 1985) shows that extinction often obscures many of the brightest star forming regions in the visible, and further IR data will be needed to test the degree to which dust modulates the optical morphology of these regions. The integrated properties of the H II regions are typical of giant H II regions in the disk, with Hα luminosities of order 10^{39}–10^{41} ergs s^{-1}, and equivalent SFRs of order 0.01–0.1 M$_\odot$ yr^{-1} per hotspot (Kennicutt *et al.* 1989; Phillips 1993). The blue knots without emission lines are

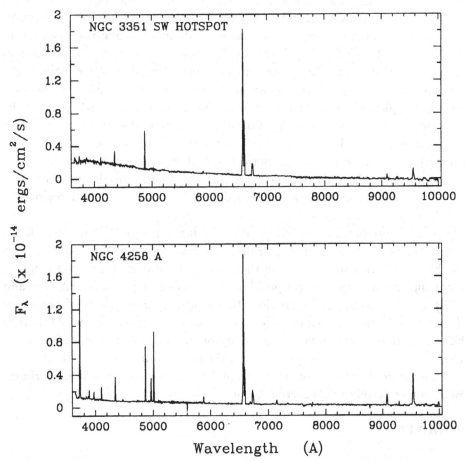

Fig. 4—Comparion of emission-line spectra of a typical circumnuclear hotspot (top) with a comparable luminosity disk H II region (bottom). Spectra were obtained at the MMT and the Steward Observatory 2.3m telescope.

presumably more evolved counterparts of the H II regions. The concentration of star formation in the circumnuclear disk is very high, typically 10–300 times the average SFR per unit area in the disk (Phillips 1993). In the hotspot galaxies studied by Phillips the circumnuclear regions contained 5–70% of all of the massive star formation in the respective galaxies.

Spectroscopic observations can provide more information about the physical conditions in the hotspots. Figure 4 (top) shows an optical spectrum (3,650–10,000Å) of the brightest hotspot in the SBb galaxy NGC 3351. For comparison the bottom panel shows the spectrum of a comparable luminosity H II region in the disk of the Sb galaxy NGC 4258. Several differences are immediately apparent. The forbidden lines in the blue are very weak in the hotspot, due to a combination of high reddening (typically 1–4 mag in optically identified hotspots) and unusually low nebular excitation. The latter implies a low electron temperature in the ionized gas ($T_e \leq$ 7,000 K) and a high heavy-element abundance ($Z \sim 1.5 - 3\ Z_\odot$). Other differences apparent in the spectrum are a very low ionization (note in particular the weakness of He I $\lambda 5876$, [O III]$\lambda 5007$, and [S III]$\lambda\lambda 9069, 9532$), and the strong stellar continuum relative to the emission lines. These properties are indicative of a soft ionizing radiation field, which could be produced by an evolved ionizing star cluster or one with an IMF which is depleted in very massive stars. The gas densities measured from the forbidden lines are also unusually high, of order a few hundred atoms per cm³ vs. a few tens in most disk giant H II regions. Some of these properties (*e.g.,* low excitation and ionization) are characteristic of all metal-rich H II regions, not just the hotspots, but other properties, such as the strong continua, high surface brightness, and high densities appear to be intermediate between disk H II regions and H II region-like galactic nuclei (Kennicutt *et al.* 1989).

The hotspots represent the consequence of the large gas inflow rates which are believed to occur in strongly barred galaxies. Several hotspot galaxies have been mapped at high resolution in CO, and in those cases the star forming rings coincide with massive molecular rings or 'twin peak' structures (Kenney *et al.* 1992, Roy and Belley 1993). The concentrations lie close to the expected locations of the inner Lindblad resonance in the bar, which may account for the ringlike structure (Kenney *et al.* 1992). It is interesting to note that many of the anomalous properties of the hotspots are also observed, on a larger scale, in the most luminous infrared starburst galaxies. Thus hotspots may offer an opportunity to study a bona-fide starburst environment in a relatively unobscurred region.

5 DESIDERATA

The problem of star formation in barred galaxies has only recently attracted the attention it deserves from observers, so our empirical understanding of the subject is still very immature. The main accomplishment of the past decade of observation has been to characterize the basic star formation properties of barred galaxies. We now have the basic numbers on star formation rates and distributions needed to fit theoretical models and simulations, and even some sense of the how these properties relate to the photometric and kinematic structures of the bars and their parent galaxies. However what is lacking is a deeper integration of the available data to probe the physical processes which account for the diversity of observed properties among barred galaxies. This hardly surprising— we do not understand large-scale star formation in *any* galaxy at this level— but the unique dynamical properties of barred galaxies offer unique opportunities to understand the underlying physics of the large scale star formation in general.

While preparing this review I was impressed by several important observations which have *not* been made, but are easily within reach of current instrumentation. We sorely need a large database of visible and near-IR surface photometry of nearby galaxies, in order to obtain accurate structural information and statistics on bar structure. In the age of CCDs and IR arrays our knowledge of barred galaxy frequency and structure still rests almost entirely on visual classifications of photographic plates by de Vaucouleurs and Sandage. This superb database is not likely to be replaced until the completion of the Sloan and 2MASS digital sky surveys, but in the meantime a modest imaging study of nearby galaxies could greatly improve our knowledge of the photometric parameters and frequency of bars. In a similar vein, further near-IR imaging of the dusty central regions of barred galaxies, especially those with intense star formation, would provide an unbiased database on the circumnuclear properties of these objects. Ideally this would include a combination of broadband (JHK) imaging and narrowband observations of key emission features such as Brγ.

Understanding the mechanisms that drive the star formation in barred galaxies requires a much broader attack, including high resolution observations in H I, CO, Hα, and the broadband visible–IR continuum at the very least. Relating these observations to physical models will almost certainly require spatially resolved observations of the velocity fields of both the gas and stars. The work on M83 described earlier provides a glimpse into how productive such an approach can be, but it needs to be extended to a much wider range of bar types and physical environments. Parallel studies of normal galaxies may be needed to separate bar-triggered phenomena from those which are simply manifestations of the normal star formation process.

The recent results on abundance gradients and hotspot H II regions illustrate the diagnostic power of optical and IR spectroscopy for this problem. Such spectra not only provide information on the chemical evolution and hence the gas flows in the barred galaxies, but also provide diagnostics of the densities, masses, energetics, and radiation fields in the star forming regions.

I am grateful to Jeff Kenney, Pierre Martin, Drew Phillips, and Stuart Ryder for discussing various aspects of barred galaxies, and for providing materials in advance of publication. I also thank Kenney and Phillips for permission to reproduce the images shown in Figure 2. This work was supported in part by the National Science Foundation through Grant AST90-19150.

REFERENCES

Alloin, D, and Kunth, D. 1979, *Astr. Ap.*, **71**, 335.

Cameron, M. J. 1971, *Mon. Not. R. Astr. Soc.*, **152**, 403.

Cepa, J., and Beckman, J. E. 1990, *Ap. J.*, **349**, 497.

Cepa, J., Beckman, J. E., Knapen, J. H., Nakai, N., and Kuno, N. 1992, *A. J.*, **103**, 429.

Devereux, N. 1987, *Ap. J.*, **323**, 91.

Dressel, L. L. 1979, *Ph. D.* thesis, University of Virginia.

Dressel, L. L., and Condon, J. J. 1978, *Ap. J. Suppl.*, **36**, 53.

Edmunds, M. G., and Roy, J. -R. 1993, *Mon. Not. R. Astr. Soc.*, **261**, L17.

Feinstein, C. 1992, *Ph. D.* thesis, University of La Plata.

Friedli, D., and Benz, W. 1993, *Astr. Ap.*, **268**, 65.

Handa, T., Sofue, Y., and Nakai, N. 1991, in *Dynamics of Molecular Cloud Distributions*, ed. F. Combes, and F. Casoli (Dordrecht: Kluwer), p. 156.

Hawarden, T. G., Mountain, C. M., Leggett, S., and Puxley, P. 1986, *Mon. Not. R. Astr. Soc.*, **221**, 41P.

Heckman, T. 1980, *Astr. Ap.*, **88**, 365.

Hummel, E. 1980, *Astr. Ap. Suppl.*, **41**, 151.

Hummel, E. 1981, *Astr. Ap.*, **93**, 93.

Jackson, J. M. *et al.* 1991, *Ap. J.*, **375**, 105.

Kenney, J. D. P., and Lord, S. D. 1991, *Ap. J.*, **381**, 118.

Kenney, J. D. P., Carlstrom, J. E., and Young, J. S. 1993, *Ap. J.*, **417**, in press.

Kenney, J. D. P., Wilson, C. D., Scoville, N. Z., Devereux, N. A., and Young, J. S. 1992, *Ap. J. Lett.*, **395**, L79.

Kennicutt, R. C. 1983, *Ap. J.*, **272**, 54.

Kennicutt, R. C. 1989, *Ap. J.*, **344**, 685.

Kennicutt, R. C., and Kent, S. M. 1983, *A. J.*, **88**, 1094.

Kennicutt, R. C., Keel, W. C., and Blaha, C. A. 1989, *A. J.*, **97**, 1022.

Knapen, J. H., Cepa, J., Beckman, J. E., Soledad del Rio, M., and Pedlar, A. 1993, *Ap. J.*, in press.

Little, B., and Carlberg, R. G. 1991, *Mon. Not. R. Astr. Soc.*, **250**, 161.

Martin, P., and Roy, J. -R. 1993, *Ap. J.*, submitted.

Morgan, W. W. 1958, *P. A. S. P.*, **70**, 364.

Osmer, P. S., Smith, M. G., and Weedman, D. W. 1974, *Ap. J.*, **192**, 279.

Pfenniger, D., and Friedli, D. 1991, *Astr. Ap.*, **252**, 75.

Phillips, A. C. 1993, *Ph. D.* thesis, University of Washington.

Phillips, M. M., Pagel, B. E. J., Edmunds, M. G., and Diaz, A. 1984, *Mon. Not. R. Astr. Soc.*, **210**, 701.

Pompea, S. M., and Rieke, G. H. 1990, *Ap. J.*, **356**, 416.

Puxley, P. J., Hawarden, T. G., and Mountain, C. M. 1988, *Mon. Not. R. Astr. Soc.*, **231**, 465.

Romanishin, W. 1990, *A. J.*, **100**, 373.

Roy, J. -R., and Belley, J. 1993, *Ap. J.*, **406**, 60.

Ryder, S. D. 1993, Ph.D. thesis, Australian National University.

Ryder, S. D., and Dopita, M. A. 1993, *Ap. J.*, submitted.

Sellwood, J. A., and Wilkinson, A. 1993, *Rep. Prog. Phys.*, **56**, 173.

Sérsic, J. L., and Pastoriza, M. 1967, *P. A. S. P.*, **79**, 152.

Vila-Costas, M. B., and Edmunds, M. G. 1992, *Mon. Not. R. Astr. Soc.*, **259**, 121.

Wynn-Williams, C. G., and Becklin, E. E. 1985, *Ap. J.*, **290**, 108.

Zaritsky, D., Kennicutt, R. C., and Huchra, J. P. 1993, *Ap. J.*, submitted.

Gas Dynamics and Star Formation in and around Bars

E. Athanassoula

Observatoire de Marseille, 2 Place Le Verrier, 13248 Marseille Cedex 04, France

ABSTRACT

In this paper I briefly review the flow of gas in and around the bars of early type, strongly barred galaxies. I discuss the formation and location of the shocks near the leading edges of bars and the parameters that influence them. Straight shock loci can also be loci of such high shear that no stars can form there, although they correspond to important density enhancements. The flows found in barred galaxies entail a considerable amount of inflow. If inner Lindblad resonances are absent, or if one or more secondary bars exist within the primary one, then this inflowing gas can come very near to the galactic center.

1 INTRODUCTION

Modelling the interstellar medium in order to follow the gas flow in and around bars is not a straightforward matter. Two families of approaches have been developed so far for that purpose:

1. Codes treating the cool dense clouds as ballistic particles with a finite cross section, often called "sticky particle" codes. Exactly when two such clouds are considered to collide and what happens in such a case varies from one code to the other (Taff and Savedoff 1972; Larson 1978; Schwarz 1979).

2. Codes considering a collection of these clouds as a fluid with a sound speed of the order of the velocity dispersion of the clouds, *i.e.* of the order of 5 to 10 km sec^{-1} (*cf.* Cowie 1980). This group of codes is composed of at least three subgroups:

i) Difference schemes, in which several different ways of solving the hydrodynamic equations are used (see *e.g.* Prendergast 1983 and references therein).

ii) the beam scheme (Sanders and Prendergast 1974).

iii) SPH (for Smooth Particle Hydrodynamics; Gingold and Monaghan 1977; Lucy 1977), where an ensemble of smoothed out particles is used to solve the hydrodynamic equations.

Given the intricacies and complexities of the real interstellar medium, these codes can be considered to give at best crude representations. More realistic, and therefore more elaborate, codes taking into account the multiphase nature of the gas and allowing for star formation, gas heating by young stars and cooling by radiation, would of course be a big improvement. Such codes have been developed (*e.g.* Chiang and Prendergast 1985) but have so far only been applied to the description of small patches of the interstellar medium. It would indeed necessitate large grids to resolve small features while describing objects the size of a galaxy.

Yet some of the simpler codes have proven to be surprisingly good in reproducing observed features of the gas flow. This argues that they are adequate for the task and that a more correct description of the local behaviour may not be necessary in all cases. In this paper I will review gas dynamics and star formation in and around bars, comparing different codes, and, in particular comparing their results with observations. For more general reviews on barred galaxies see Bosma (1992), and Sellwood and Wilkinson (1993).

2 GAS FLOWS WITHIN THE BAR REGION

Figure 1a shows the response of the gas to a barlike potential. The bar is at 45° to the x axis and rotates clockwise. The length of its semimajor axis, a, is 5 kpc, and the size of the box is 16 kpc. The greyscale coding is such that black denotes empty regions and white — the highest density ones. Note that there is little gas in the region of the bar and that most of it is concentrated near the center, at the ends of the bar and in two narrow lanes along the leading edges of the bar. Following the flow (Fig. 1b) or better still, considering the velocities along a line perpendicular to the density maxima, shows that these maxima are the loci of strong shocks as well as of considerable shear. The gas streamlines in a frame of reference corotating with the bar (Fig. 1b) follow the orientation of the x_1 orbits in the outer parts and that of the x_2 in the innermost ones. The gradual shift from one orientation to the other accounts for the form of the shock loci. The shape of these loci, which slit spectroscopy (Jörsäter 1984; Pence and Blackman 1984; Lindblad and Jörsäter 1985) has shown to be shocks, agrees very well with that of the observed dust lanes in real galaxies (Athanassoula 1984). In strong bars the dust lanes are known to be straight, parallel to each other, but at an angle to the bar major axis; for weak bars or ovals they are curved and concave towards the bar major axis. In the inner parts they curve around the nucleus in what is often termed a "nuclear spiral", or "nuclear ring" (*e.g.* NGC 1433, 4321, *etc.*). All these properties are well reproduced by the results of the hydrodynamical simulations. Plotting the density along the shocks one sees that it is not constant but shows a maximum not far from the innermost part, which could be associated with the "twin peaks" often found in the CO observations (Kenney, these proceedings and references therein).

In the model described above an x_2 family was present, while in the model shown in Figure 1c and 1d there is none. In other words the model in the upper panels has an inner Lindblad resonance (hereafter ILR), while the one in the lower panels has none. This explains the difference between the forms of the corresponding shock loci, which I will call "offset" in the former case and "centered" in the latter. Analysis of

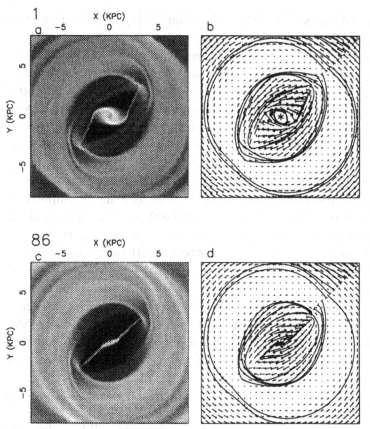

Fig. 1—Response of the gas to a bar potential. The left panels show the density and the right ones the velocity in a frame of reference corotating with the bar; the upper row corresponds to a case with an ILR, and the lower row to a case without.

over 200 runs shows that, in order for offset shocks to occur, the x_2 family must exist and must have a considerable extent.

Various model parameters influence the properties of the shock loci. In particular, for shocks to be offset the Lagrangian or corotation radius has to be at $r_L = (1.2 \pm 0.2)a$, which sets strict limits to the values the pattern speed of the bar may take. Both the axial ratio and the strength of the bar influence the form of the shock loci, in the sense that weak bars or ovals have curved shock fronts, with their concave part towards the bar major axis, while in strong bars they are straight, in good agreement with the observations. Finally, for models with low central concentration one sees no shocks and the gas flow can be modelled as a series of nested non-intersecting elliptical orbits. This is not true for the very centrally concentrated models which show offset shocks. The velocity jump observed at the shock can be sizeable, often exceeding 100 or 200 km sec^{-1}, and is more important for models with high mass and/or high axial ratio bars.

The results described so far have been obtained with a difference scheme hydro-

dynamical code (Athanassoula 1992*b*, hereafter A92*b*); it is of interest to compare
them with the results obtained with other codes. Several other difference scheme
codes, as well as the beam scheme, have given offset shock loci along the leading side
of the bar (see discussion and references in Prendergast 1983), although for some
of them the resolution was not good enough to allow detailed quantitative studies.
Sticky particle codes also show shock formation along the leading edges of the bar
(Schwarz 1984, 1985). During the initial phases of the simulation, when the bar is
still growing, the shock loci are very clear (*cf. e.g.* the left hand panel of Figure 7*d*
of Schwarz 1984). After that phase, however, as noted by Schwarz (1985), one needs
to introduce recycling of the gas in order to follow the gas dynamics, because of the
depopulation of the region within corotation. Even so, Schwarz's simulations, as well
as a couple of hundred of my own, show that the shock loci are most of the time not
very clearly delineated and certainly not as smooth as the observed ones. It is very
hard to reach some balance between the inflow and the recycling and the shocks do
not last long. Furthermore, if such a balance is achieved for a given run and a given
recycling constant, then this value will not work for another run. Thus the exercise
can become quite frustrating and the sticky particle method, which has proved so
useful for the study of outer rings, is not so well suited to the study of shocks. Nevertheless it is interesting to note that it is possible to recover, at least qualitatively,
a fair fraction of the results found with the hydrodynamic code, provided one knows
ahead of time what one is looking for.

All results discussed so far have been obtained from the response of the gas
to a given bar potential, so we should now ask ourselves what we have missed by
neglecting selfconsistency. The selfconsistency of the gas will be unimportant unless
the surface density of the gas is relatively high. For these cases one has to remember
that any ring (or part of a ring) of gas can be unstable and fragment into lumps. This
has been borne out by numerical simulations (*e.g.* Fukunaga and Tosa 1991; Heller
and Shlosman 1993, and these proceedings; Wada and Habe, these proceedings).
Unfortunately the time scales involved depend not only on dynamical quantities like
the halo to disc mass ratio, the dispersion of velocities *etc.*, but also on parameters like
the amount of heating provided by young stars, or on the various ways of modelling
star formation, or even on purely numerical parameters like the softening length or
the number of particles.

Stellar selfconsistency is also not included in the response calculations. The bar is
introduced gradually in a way loosely mimicking bar formation in N-body simulations,
i.e. by gradually taking mass out of an axisymmetric component and adding it to
the bar. After that, however, the bar, except for rotation, does not change, while the
bars in *N*-body simulations evolve in a slow secular way - their axial ratio decreases
(Sellwood 1981) while their pattern speed in purely stellar cases decreases (Sellwood
1981; Little and Carlberg 1991 *etc.*) and the central concentration of the galaxy
slightly increases. These changes, however, are slow enough so that the gas evolution
can be viewed as a sequence of responses to somewhat different potentials. Thus, as

the bar becomes thinner and/or more massive, the shock loci will gradually change shape, first from curved to straight offset and then to centered.

An important effect which is omitted in response calculations is the possibility of mode-mode interactions, which may well explain the asymmetries observed in the inner parts of late type barred galaxies and their offcentered bars, as well as the formation of secondary bars, as will be discussed in section 5.2.

3 STAR FORMATION

Observations show that in strong bars star formation occurs prolifically near the center and the bar's end (Kennicutt, these proceedings and references therein) but not at the locations of the dust lanes, which are very smooth and devoid of young stars and H II regions. This is not true for ovals, where star formation can occur at the dust lanes as well as the central parts and the ends of the bar. A good example is NGC 1566, where the H II region catalogue of Comte and Duquennoy (1982) shows that the northwestern dust lane is particularly well delineated by H II regions.

Simulations, on the other hand, show that the density of the gas is high in three regions: the ends of the bar, the center and along the narrow shock loci (see Fig. 1, or corresponding figures in Athanassoula 1991, hereafter A91, and A92b). The two first correspond, as expected, to regions of observed high star formation in real galaxies, but not the third one. The explanation lies in the very high shear present in the shocks. Indeed, a typical giant molecular cloud at the shock locus would not have time to collapse before it sheared out. On the other hand this is not true for the regions at the end of the bar, or for the shocks in ovals or weak bars, where star formation is observed (A92b). Thus the agreement between the models and the observations is again very satisfactory.

4 MASS INFLOW

Figures 1b and 1d, or similar figures from A92b show clearly that, if there are shocks, the gas flow is complicated, comprising both inflow and outflow, depending on the region considered. These figures show that before the shock the motion along a streamline is outwards and that after passing through the shock the gas moves inwards at a very high rate, often exceeding 100 km sec^{-1}. This, however, is the value at the position of maximum inflow, and it is much more representative to have a mean value. Such a density-averaged mean on a circle about the center is of the order of only a few km sec^{-1} (A92b).

An alternative way of measuring the mass inflow is to measure the mass within a given radius as a function of time. Figure 2, obtained for the same model as Figure 1a, shows that there is a time of very high inflow, linked to the growth of the bar. This phase is more obvious for stronger and/or more rapidly rotating bars. It is followed by a steady, and much less strong inflow, which I have followed for a few Gyrs. The

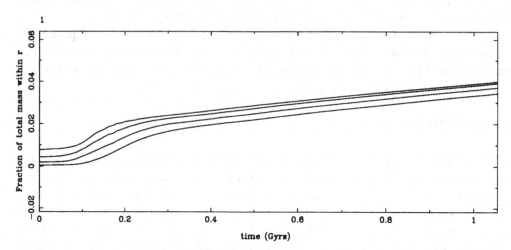

Fig. 2—Gas mass within a given radius as a function of time. The four curves correspond to radii 0.5, 1, 1.5 and 2 kpc respectively.

parameters of the model influence the rate of inflow during both stages in a similar way. Figure 3 gives the amount of inflow during the steady state as a function of the four basic parameters describing a bar potential, namely the axial ratio, quadrupole moment and pattern speed of the bar and the central concentration of the model (*cf.* Athanassoula 1992*a*, hereafter A92*a*). The rates were obtained by taking the slopes of $M(t)$ curves, similar to those of Figure 2, corresponding to the radius $r = 1$ kpc In each panel only one of these parameters is varied, while the other three are held fixed. Since what is interesting at this stage is the existence of trends, I normalised the inflow values so that the inflow is equal to 1 for a inhomogeneous bar model with $a/b = 2.5$, $r_L = 6$ kpc, $Q_m = 4.5 \times 10^{10}$ M$_\odot$kpc^2 and $\rho_c = 2.4 \times 10^4$ M$_\odot$/pc^2 (*cf.* A92*a* for definitions). We see that the inflow is highest for thin, slowly rotating bars in centrally concentrated potentials. If one plots the amplitude of the velocity jumps at the shocks as a function of the same four parameters one sees similar trends, so that one can conclude that highest inflow is found in models where the shocks are strongest, as expected. The non-monotonic behaviour of the mass inflow as a function of quadrupole moment can be understood as follows: since the total mass of all models is the same, models with higher bar mass have a lower "bulge" mass and are therefore less centrally concentrated. This, according to panel *c* (Fig. 3), entails a lower inflow rate, which counters the higher inflow rate expected from the higher bar mass cases.

The inflow rate increases also with the viscosity of the code (which in this case is set by the cell size). It also depends on the amount of recycling (set in these models by the recycling parameter α [A92*b*]), in the sense that models with more recycling show more inflow, as expected, since they have more gas available for it.

Similar results are found for the sticky particle code. In particular the trends

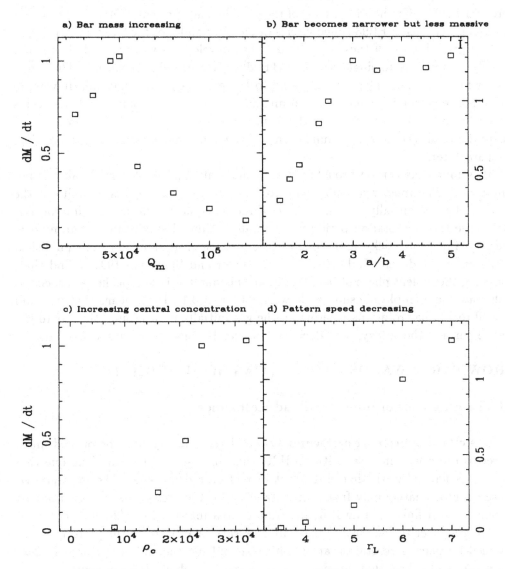

Fig. 3—Mass inflow as a function of *(a)* the bar quadrupole moment Q_m; *(b)* axial ratio a/b; *(c)* the model's central density ρ_c; and *(d)* Lagrangian radius r_L. For each panel the other three parameters are held fixed to the values $a/b = 2.5$, $r_L = 6$ kpc, $Q_m = 4.5 \times 10^{10}$ $M_\odot \mathrm{kpc}^2$ and $\rho_c = 2.4 \times 10^4$ $M_\odot \mathrm{pc}^{-2}$

shown in Figure 3 are also by and large found in the sticky particle results, except that there is more scatter, presumably due to the noise and to the fact that the flow is not really steady much of the time.

The distinction between the rapid inflow phase associated with bar formation and the lesser inflow during the subsequent quiet evolution can be seen in SPH simulations as well (*cf.* Figure 4 of Wada and Habe 1992 and Figure 13 of Friedli and Benz 1993, hereafter FB93). Furthermore, some of the global trends presented here are borne out by the simulations in FB93 (filled circles in panels *d* and *e* of their Figure 12). Their model E, which falls off these correlations, has developed two bars, and thus is not directly comparable to single bar runs (Friedli, private communication). The larger scatter found in these plots as compared to those of Figure 3 is natural. Indeed, in the response calculations only one parameter is varied in each panel, and the others are held fixed. On the other hand this is not true for the selfconsistent simulations, where the final results of two simulations differ in the values assigned to more than one parameter.

How much gas can be brought towards the central parts of barred galaxies per unit time? The most reasonable way to measure this quantity is in units of the fraction of mass initially within a circle of radius equal to the bar semimajor axis and to use the bar rotation period for the unit of time. Using results from my own hydrodynamic and sticky particle simulations, as well as numbers from published SPH results (Wada and Habe 1992; FB93; Heller and Shlosman 1993) I find that, excluding the violent phase of bar formation, this number is, for gas flows associated with reasonable amplitude shocks, between 0.01 and 0.1. That means that it would take 10 to 100 bar rotations to bring all the gas initially in or around the bar to the central parts of the galaxy, provided that the steady state regime lasts that long.

5 HOW FAR INWARDS CAN THE GAS BE BROUGHT?

5.1 The Existence of Inner Lindblad Resonances

It is clear from comparing Figures 1*a, b* and 1*c, d* that gas can be pushed much nearer to the center in cases with no ILRs than in cases with them. The question thus arises naturally whether ILRs exist in real barred galaxies. The existing evidence is unfortunately only fragmentary, and is, for the greatest part, based on the axisymmetric definition of an ILR, *i.e.* from comparing the $\Omega - \kappa/2$ curve, obtained by averaging over the azimuthal angle all necessary quantities, to the pattern speed, obtained by placing corotation at or somewhat outside the end of the bar. It does however seem to show that, at least in a fair fraction of bars, ILRs do exist.

5.1.1 Theoretical considerations

Linear theory shows clearly that, in order for a swing-amplified mode to exist, the galaxy must not have an ILR (Toomre 1981). This prompted Toomre (1981) to suggest that the existence of an ILR should be a way of stabilizing galactic discs. However nonlinear theory showed that this is not necessarily the case. Indeed, if a second mode (without an ILR) exists, then nonlinear mode coupling makes it possible for the cycle to close (Tagger *et al.* 1987; Sygnet *et al.* 1988). The same argument is valid in case of a WASER mode, although in this case a Q-barrier is also a possible way of closing the cycle (*e.g.* Mark 1977). Alternatively, the mode may be sufficiently unstable to saturate the resonances (Sellwood 1989).

5.1.2 Numerical simulations

The bars that form in numerical simulations sometimes have and sometimes do not have an ILR. The fact that the latter simulations are more numerous than the former is not an argument against the existence of ILRs. The reason is that centrally concentrated models require a shorter time step, and, unless one's code has an option for multiple time-steps (Sellwood 1985, 1987), they can become quite expensive to run. Yet several simulations producing bars with ILRs have been reported in the literature (Zang and Hohl 1978; Efstathiou *et al.* 1982; Sellwood 1985, 1989). Periodic orbits have so far been calculated in two potentials derived from numerical simulations. In one (Sparke and Sellwood 1987) the x_2 and x_3 families were found but their families had a small extent on the characteristic diagram, while in the other (Pfenniger and Friedli 1991) they were absent.

5.1.3 Observations

A simple way of estimating whether ILRs exist in certain types of barred galaxies is to look at the positions of their dust lanes. Most bars, and certainly all the prototypes, have offset dust lanes, as in Figure 1*a*. After looking through several atlases and plates, I could come up with two good candidates for centered dust lanes, NGC 4123 and 7479, both shown in the atlas by Sandage and Bedke (1988), where the dust lanes are very near the bar major axis.

After making assumptions about the mass to light ratio M/L of the galaxy and its geometry it is possible to obtain the potential of a barred galaxy, and from that, $\Omega(r)$ and $\kappa(r)$. Assuming further that the Lagrangian or corotation radius is equal to $1.2a$, where a is the bar semimajor axis (an assumption that seems to be borne out by several independent lines of evidence — see A91 and A92*b* for discussions), one can test whether an ILR exists or not. Assuming that M/L is constant and that the galaxy has an infinitesimal thickness, it was shown (A91; Athanassoula and Wozniak 1993) that all 11 galaxies with a thin bar from an SB0 sample have ILRs, while ovals are found to be more borderline (4 galaxies). Correcting for thickness (Athanassoula and Wozniak 1993) does not change this result for thin bars, but shows that ovals are probably devoid of ILRs.

The most convincing piece of evidence, however, comes from the work of Teuben *et al.* (1986), who showed, with the help of TAURUS observations of NGC 1365, that the x_2 family should be present in this galaxy in order to explain the inner parts of the observed velocity field.

5.2 Bars within Bars

The most promising scenario for bringing gas down to the center of the galaxy, to radii smaller that that of the ILR, is that of "bars within bars", proposed initially by Shlosman *et al.* (1989). According to this, if gas is pushed inwards by a bar it will accumulate at small radii and form a bar unstable disc, which will form a (secondary) bar within the primary one. This will also push gas inwards and the scenario may repeat itself, now at a smaller scale. A succession of such events, on ever-smaller scales, can be envisaged, bringing gas very near the galactic center. This was borne out beautifully, at least insofar as the first secondary bar is concerned, by the numerical simulations of Heller and Shlosman (1993, also these proceedings) and Friedli and Martinet (1993). The most promising way to look for such components in real galaxies is with the help of CO observations. Indeed, small secondary bars have been found in the molecular gas component of a number of galaxies. Good examples are NGC 3351 (Devereux *et al.* 1992) and M101 (Kenney *et al.* 1991), although the latter is not a barred galaxy. However these are not the only known examples of secondary bars, and several cases have been found in optical or infrared images. De Vaucouleurs (1975) discussed the prototype case NGC 1291, while for NGC 1326 he argued that there is yet a third bar within the secondary one (de Vaucouleurs 1974). These bars, however, are not made of gas, nor are they particularly young objects, so that one cannot argue that they have evolved out of a gaseous bar formed within the primary one. Nor can one rely on the formation of unstable disks from the accumulation of stars in the central parts. The most promising way to account for such bars is a nonlinear mode-mode interaction between the two bars (Tagger *et al.* 1987; Sygnet *et al.* 1988). This is most efficient if the resonant regions of the two bars overlap, and the two aforementioned publications give explicitly the case where the ILR of the outer bar coincides, or is very near to, the corotation resonance of the inner one, which is the most likely pair of resonances to overlap. It is not an easy matter to verify this prediction from observations, but it is straightforward to do so with the help of numerical simulations. It is thus very reassuring to see that the simulations of Friedli and Martinet (1993) indeed find that the corotation of the inner bar roughly coincides with the ILR of the outer one.

6 SUMMARY AND CONCLUSIONS

1) The gas streamlines within the bar region are related to the properties of the x_1 and x_2 orbits.

2) Depending on the parameters of the bar model
- there may be shocks offset towards the leading side of the bar
- there may be "centered" shocks (near the major axis of the bar), or
- there may be no shocks.

3) The shocks are offset only if the x_2 family exists and has a sufficient extent.

4) For strong bars the shock loci are straight, while for weak ones or ovals they are curved with their concave part towards the bar major axis.

5) In the inner parts there are nuclear "spirals" or "rings". The density along them is not constant and there are two strong clumps at roughly 180° from each other.

6) Star formation occurs mainly at the center and at the two ends of the bar. In particular, shear prohibits star formation for straight shock loci but not necessarily for curved ones.

7) Strict limits can be set to the main parameters describing a bar if it has offset dust lanes. In particular the Lagrangian radius $r_{\rm L} = (1.2 \pm 0.2)a$.

8) Gas flows associated with shocks necessarily imply important inflow, which is largest in the region of the shock.

9) The inflow is high during or after violent events like the formation of the bar, and much less during quasi-steady slow evolution of the bar.

10) Thinner, slower rotating bars in more centrally concentrated galaxies imply higher rates of inflow. The same is true for more recycling, or higher viscosity.

11) Most of the main results have been found by more than one of the methods used for modelling the interstellar medium, although quantitative differences of course do exist.

12) There are indications that ILRs exist in at least a fair fraction of cases.

13) "Bars within bars" help bring the gas near the center. Theoretical work predicts that their pattern speeds are linked, and that the corotation of the inner bar coincides with the ILR of the outer one; this has been verified by numerical simulations.

REFERENCES

Athanassoula, E. 1984, *Phys. Reports*, **114**, 320.

Athanassoula, E. 1991, in *Dynamics of Disc Galaxies*, ed. B. Sundelius (Gothenburg, Gothenburg University), p. 149. (A91)

Athanassoula, E. 1992a, *Mon. Not R. Astr. Soc.*, **259**, 328. (A92a)

Athanassoula, E. 1992b, *Mon. Not R. Astr. Soc.*, **259**, 345. (A92b)

Athanassoula, E., and Wozniak H. 1993, in preparation.

Bosma, A. 1992, in *Morphology & Physical Classification of Galaxies*, eds. G. Busarello *et al.* (Dordrecht: Reidel), p. 207.

Chiang, W. H., and Prendergast, K. H. 1985 *Ap. J.*, **297**, 507.

Comte, G., and Duquennoy, A. 1982, *Astr. Ap.*, **114**, 7.

Cowie, L. L. 1980, *Ap. J.*, **236**, 868.

De Vaucouleurs, G. 1974, *The Formation and Dynamics of Galaxies*, ed. J. R. Shakeshaft (Dordrecht: Reidel), p. 1.

De Vaucouleurs, G. 1975, *Ap. J. Suppl.*, **29**, 193.

Devereux, N. A., Kenney, J. D. P., and Young, J. S. 1992, *A. J.*, **103**, 784.

Efstathiou, G., Lake, G., and Negroponte, J. 1982, *Mon. Not. R. Astr. Soc.*, **199**, 1069.

Friedli, D., and Martinet, L. 1993, *Astr. Ap.*, submitted.

Friedli, D., and Benz, W. 1993, *Astr. Ap.*, **268**, 65. (FB93)

Fukunaga, M., and Tosa, M. 1991, *Publ. Astr. Soc. Jap.*, **43**, 469.

Gingold, R. A., Monaghan, J. J. 1977, *Mon. Not. R. Astr. Soc.*, **181**, 375.

Heller, C. H., and Shlosman, I. 1993, *Ap. J.*, in press; also these proceedings.

Jörsäter, S. 1984, *Ph. D. thesis*, Stockholm University.

Kenney, J. D., Scoville, N. Z., and Wilson, C. D. 1991 *Ap. J.*, **366**, 432.

Larson, R. B. 1978, *J. Comp. Phys.*, **27**, 397.

Lindblad, P. O. and Jörsäter, S. 1988, in *Evolution of Galaxies*, Proceedings of the X Regional Astronomy Meeting, ed. J. Palous, *Publ. Astr. Inst. Czech. Acad. Sci.*, **69**, 289.

Little, B., and Carlberg, R. G. 1991, *Mon. Not. R. Astr. Soc.*, **250**, 161.

Lucy, L. B. 1977, *A. J.*, **82**, 1013.

Mark, J. W.-K. 1977, *Ap. J.*, **212**, 645.

Pence, W. D., and Blackman, C. P. 1984, *Mon. Not. R. Astr. Soc.*, **207**, 9.

Pfenniger, D., and Friedli, D. 1991, *Astr. Ap.*, **252**, 75.

Prendergast, K. H. 1983, in *Internal Kinematics and Dynamics of Galaxies*, ed. E. Athanassoula (Dordrecht: Reidel), p. 215.

Sandage, A., and Bedke, J. 1988, *Atlas of Galaxies*, (Washington: NASA).

Sanders, R. H., and Prendergast, K. H. 1974, *Ap. J.*, **188**, 498.

Schwarz, M. P. 1979, *Ph. D. thesis*, Australian National University.

Schwarz, M. P. 1984, *Mon. Not. R. Astr. Soc.*, **209**, 93.

Schwarz, M. P. 1985, *Mon. Not. R. Astr. Soc.*, **212**, 677.

Sellwood, J. A. 1981, *Astr. Ap.*, **99**, 362.

Sellwood, J. A. 1985, *Mon. Not. R. Astr. Soc.*, **217**, 127.

Sellwood, J. A. 1987, *Ann. Rev. Astr. Astrophys.*, **25**, 151.

Sellwood, J. A. 1989, *Mon. Not. R. Astr. Soc.*, **238**, 115.

Sellwood, J. A., and Wilkinson, A. 1993, *Rep. Progr. Phys.*, **56**, 173.

Shlosman, I., Frank, J., and Begelman, M. C. 1989, *Nature*, **338**, 45.

Sparke, L. S., and Sellwood, J. A. 1987, *Mon. Not. R. Astr. Soc.*, **225**, 653.

Sygnet, J. F., Tagger, M., Athanassoula, E., and Pellat, R. 1988, *Mon. Not. R. Astr. Soc.*, **232**, 733.

Taff, L. G., and Savedoff, M. P. 1972, *Mon. Not. R. Astr. Soc.*, **160**, 89.

Tagger, M., Sygnet J. F., Athanassoula, E., and Pellat, R. 1987, *Ap. J.*, **318**, L43.

Teuben, P. J., Sanders, R. H., Atherton, P. D., and van Albada, G. D. 1986, *Mon. Not. R. Astr. Soc.*, **221**, 1.

Toomre, A. 1981, in *Structure and Evolution of Normal Galaxies*, ed. S. M. Fall and D. Lynden-Bell (Cambridge: Cambridge University Press), p. 111.

Wada, K., and Habe, A. 1992, *Mon. Not. R. Astr. Soc.*, **258**, 82.

Zang, T. A., and Hohl, F. 1978, *Ap. J.*, **226**, 521.

Nuclear Starbursts in Barred Spiral Galaxies

Nick Devereux

New Mexico State University, Department of Astronomy, Box 30001/Dept 4500. Las Cruces, NM 88003

ABSTRACT

The nuclear star formation rates have been measured for a complete sample of 156 spiral galaxies. The nuclear star formation rates of barred and unbarred galaxies are compared separately for early and late Hubble types. Significant differences are observed and critically evaluated within the context of a possible causal relationship between stellar bars and nuclear starbursts.

1 INTRODUCTION

Considerable interest surrounds the possible association between nuclear activity and galaxy morphology (Sersic and Pastoriza 1967; Heckman 1978; Simkin, Su, and Schwarz 1980). Indeed, the unambiguous association with a morphological feature, such as a stellar bar for example, may provide an important clue as to the origin of the activity. A number of studies have demonstrated the *frequency* of nuclear star formation to be enhanced in the nuclei of barred spiral galaxies. Heckman (1980) demonstrated a higher incidence of H II regions in the nuclei of barred spirals and Balzano (1983) noted an excess of barred spirals in an optically selected sample of starburst galaxies. The results discussed in this contribution, however, I believe are the first to demonstrate that the high mass star formation *rate* is enhanced in the nuclei of barred spirals when compared to unbarred spirals.

2 THE SAMPLE

A complete sample of 227 nearby ($15 \leq D[\text{Mpc}] \leq 40$), infrared luminous, $L(40 - 120 \ \mu\text{m}) \geq 3 \times 10^9 \ \text{L}_\odot$ spiral galaxies were selected upon comparison of the Infrared Astronomical Satellite (*IRAS*) Point Source Catalog with the Nearby Galaxies Catalog (Tully 1988). The sample is complete in the sense that it includes *all* spiral galaxies with far infrared luminosities, $L(40 - 120 \ \mu\text{m}) \geq 3 \times 10^9 \ \text{L}_\odot$, in the distance range 15—40 Mpc. The nuclear star formation rate was quantified for 156 galaxies in the complete sample by using the NASA Infrared Telescope Facility, located on Mauna Kea in Hawaii, to measure the 10 μm luminosity in a 5″ beam centered on the nucleus (Devereux 1987). The 5″ beam size corresponds to a linear diameter of 500—700 pc for most of the galaxies that were observed.

156   *Devereux*

3 RESULTS

3.1 The Central 10 μm Luminosity

The morphological assignments of de Vaucouleurs, de Vaucouleurs and Corwin (1976) were used to segregate the galaxies by Hubble type, in addition to bar morphology, in order to isolate potential differences between spirals of early and late-type that are not necessarily attributable to a stellar bar. A histogram showing the distribution of nuclear 10 μm luminosities that were measured for 57 early-type (Sb and earlier) spiral galaxies is presented in Figure 1. The barred galaxies are significantly different from the unbarred galaxies at the 97.5% confidence level based on a two tailed Kolmogorov-Smirnoff test. The difference is in the sense that 40% of the barred galaxies are associated with nuclear 10 μm luminosities that exceed the maximum of 6×10^8 L$_\odot$ measured for the unbarred galaxies.

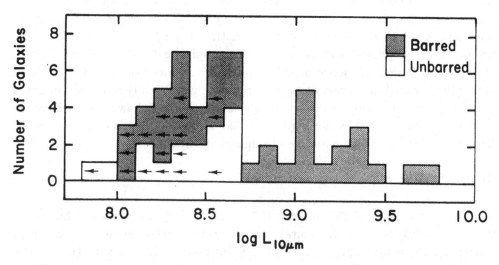

Fig. 1—Histogram showing the distribution of nuclear 10 μm luminosities that were measured for 57 early-type (Sb and earlier) barred (shaded histogram) and unbarred (unshaded histogram) spiral galaxies. Arrows indicate 2σ upper limits.

A histogram showing the distribution of nuclear 10 μm luminosities that were measured for 99 late-type (Sbc and later) spiral galaxies is presented in Figure 2. The distribution of nuclear 10 μm luminosities are similar for the barred and unbarred late-type spirals.

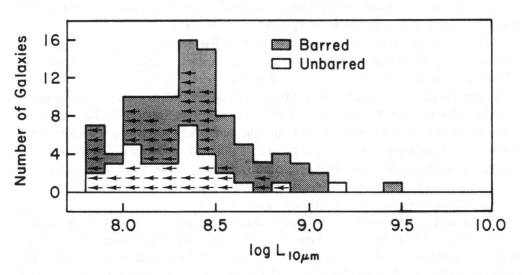

Fig. 2—Histogram showing the distribution of nuclear 10 μm luminosities that were measured for 99 late-type (Sbc and later) barred (shaded histogram) and unbarred (unshaded histogram) spiral galaxies. Arrows indicate 2σ upper limits.

4 DISCUSSION

The mean nuclear 10 μm luminosity of the early-type barred spirals is elevated by a factor of 3 compared to the early-type unbarred spirals. The association of the high central 10 μm luminosity with a nuclear starburst is supported by the detection of optical emission lines indicative of H II regions (Devereux 1989) and Hα emission line images showing a resolved nuclear starburst (Young and Devereux 1991). In the interests of brevity, the early-type barred spirals that are associated with nuclear 10 μm luminosities that exceed the maximum observed for the early-type unbarred spirals will hereafter be called starburst galaxies. The association is justified by the fact that the mean nuclear 10 μm luminosity of the starburst galaxies is comparable to that measured for a similar sized region in the nuclei of the archetype starburst galaxies M82 and NGC 253 (Rieke *et al.* 1980). Collectively, the results indicate that the nuclear star formation rate is enhanced in early-type barred spirals when compared to early-type unbarred spirals. Only 40% of the early-type barred spirals host a nuclear starburst however. Thus stellar bars may be *necessary*, but they are certainly *not sufficient* for a nuclear starburst. Clearly other parameters are important for a nuclear starburst to occur and one obvious requirement is an adequate supply of molecular gas, although confirmation of this hypothesis is beyond the scope of the present work.

No significant difference is observed between the nuclear 10 μm luminosities of barred and unbarred late-type spirals. Evidently, the relationship between bar morphology and nuclear 10 μm luminosity is different for early and late-type spirals which raises an interesting question. *Is there a mechanism which can explain why nuclear starbursts occur preferentially in early-type barred spirals?*

The association of nuclear starbursts with stellar bars strongly suggests that it arises as a consequence of a bar related phenomenon, such as a resonance for example. The inner Lindblad resonance, in particular, is expected to be the site of enhanced molecular cloud density and star formation in the central regions of barred spiral galaxies (Combes and Gerin 1985). The location of the inner Lindblad resonance depends on the form of the mass distribution and is expected to be closest to the nucleus in the galaxies with the highest central mass concentration. Thus, the stellar bulge, which distinguishes early-type spirals from late-type spirals, may be the reason why nuclear starbursts occur preferentially in early-type barred spiral galaxies. There are some problems with this hypothesis, however.

Numerical models indicate that molecular clouds and the associated star formation congregates in a ring at the location of the inner Lindblad resonance, thus one may expect to observe rings of molecular gas and star formation in the nuclei of the starburst galaxies. To date, only one of the early-type barred starburst galaxies, NGC 3504, has been studied in detail (Kenney *et al.* 1992). The observations show the molecular and ionized gas to be concentrated in a single peak centered on the nucleus interior to the inner Lindblad resonance. Thus there is no direct observational evidence, based on the morphology of the molecular and ionized gas, that the stellar bar had any role in initiating the nuclear starburst within NGC 3504.

Another problem is that although the starburst galaxies are classified as barred, upon closer inspection many show evidence for some sort of interaction, highlighting one of the disadvantages of statistical studies that are based on limited and perhaps misleading morphological information. Three of the starburst galaxies are so disturbed that they are included in the Arp Catalog of Peculiar Galaxies (Arp 1967). If the starburst galaxies showing evidence for interaction are excluded, the statistical significance of the association between nuclear starbursts and stellar bars, alluded to in section 3.1, would be weakened considerably. On the other hand, we know that interactions may lead to the formation of stellar bars (Noguchi 1987), so perhaps we should include them. How then are we to interpret the result concerning the association between nuclear starbursts and stellar bars? Are we to conclude that the starburst is causally related to the presence of a stellar bar or the result of an interaction? Perhaps the only way to resolve the dilemma is to conduct high angular resolution kinematic studies of the molecular and ionized gas within the starburst galaxies in order to elucidate the impact, if any, of the stellar bar on the interstellar medium and the origin of the nuclear starburst.

REFERENCES

Arp, H., 1966 *Atlas of Peculiar Galaxies*, California Institute of Technology.

Balzano, V. A., 1983, *Ap. J.*, **268**, 602.

Combes, F., and Gerin, M., 1985, *Astr. Ap.*, **150**, 327.

de Vaucouleurs, G., de Vaucouleurs, A., and Corwin, H. G. Jr. 1976, *Second Reference Catalog of Bright Galaxies* (Austin: University of Texas Press).

Devereux, N. A. 1987, *Ap. J.*, **323**, 91.

Devereux, N. A. 1989, *Ap. J.*, **346**, 126.

Heckman, T. M. 1978, *P. A. S. P.*, **90**, 241.

Heckman, T. M. 1980, *Astr. Ap.*, **88**, 365.

Kenney, J. D. P., Wilson, C. D., Scoville, N. Z., Devereux, N. A., and Young, J. S. 1992, *Ap. J.*, **395**, L79.

Noguchi, M. 1987, *Mon. Not. R. Astr. Soc.*, **228**, 635.

Rieke, G. H., Lebofsky, M. J., Thompson, R. I., Low, F. J., and Tokunaga, A. T. 1980, *Ap. J.*, **238**, 24.

Sersic, J. L., and Pastoriza, M. 1967, *P. A. S. P.*, **79**, 152.

Simkin, S. M., Su, H. J., and Schwarz, M. P. 1980, *Ap. J.*, **237**, 404.

Tully, R. B. 1988, *The Nearby Galaxies Catalog*, (Cambridge: Cambridge University Press).

Young, J. S., and Devereux, N. A. 1991, *Ap. J.*, **373**, 414.

Molecular Inflow towards Galactic Nuclei

John E. Beckman

Instituto de Astrofísica de Canarias, E-38200 La Laguna, Tenerife, Spain

ABSTRACT

The fact of starburst activity in a non-negligible fraction of galaxies implies a refuelling mechanism, since the star formation rates observed are sufficient to exhaust the gas in situ over relatively short timescales. A number of "gas bars" – elongated structures in which molecular gas is observed to flow in markedly non-circular orbits – has been observed in galaxies with circumnuclear star-forming activity. In this paper I discuss whether the observed properties of these flows fit the dynamical models which have been proposed for refuelling starbursts, concluding that they may well do so, but not according to some rather over-simplified previously postulated scenarios.

1 INTRODUCTION

Observations during the past five years of molecular gas in external galaxies on scales of order 100 pc, made possible by the commissioning of millimetre wave interferometers, have virtually put within our research the solution to the problem of how starbursts are refuelled. The need for refuelling is clear since in the majority of cases the observed star-formation rates, of order 0.1 to 1 $M_\odot$ per year, would exhaust the molecular gas observed within 100 pc of the nucleus in times between 10^8 and 10^7 years. On the other hand the observed association of starbursts and also Seyferts with the presence of bars in spirals, implies a causal link which would be satisfied naturally by the property of a bar to brake the rotational motion of interstellar gas, and thus allow it to flow down the gravitational potential gradient towards the nucleus. These considerations have led the molecular observers to search for evidence of the phenomenology of molecular inflow, with some success as I describe below.

However, it is important to point out the difficulty of inferring the true three-dimensional kinematic structure of a gas flow when struggling against the intrinsic problems of barely adequate signal to noise ratio, linear resolution just sufficient to resolve the structural elements of the flow, and the inevitable two-dimensionality of the observations. For a gaseous bar lying close to the major axis of a galaxy under study the projected component of any flow along the bar is vanishingly small, and essentially undetectable. For a bar aligned close to the minor axis, on the other hand, inflow causes compression, along the bar, of the isovelocity contours, which could also be caused by rapid rotation due to a condensed central mass. In addition we always face the classical ambiguity of deciding whether any non-circular flow observed is towards or away from the nucleus. Thus we must be cautious about claims to have measured inflow, particularly at velocities of tens of km s^{-1} which are at least an order of magnitude higher than the rates required to refuel the majority of observed starbursts.

My interest in this problem arose because during the course of a comprehensive analysis of the interstellar hydrogen: molecular, atomic, and ionized, in the Sc spiral NGC 4321 we discovered (Knapen *et al.* 1993) evidence for markedly non-circular motions in H I on scales of a couple of kpc around the nucleus, which led us to predict, and later find in the infrared, a bar on this scale. We subsequently analyzed the circumnuclear zone optically on a scale of < 1 kpc, finding a small circumnuclear bar aligned with the principal bar, as well as a spiral starburst structure, with dust-lanes oriented generally along the bar axis but swinging inwards across the bar towards the nucleus. Details are presented elsewhere (see Knapen *et al.*, these proceedings). This set of observations offers an apparently classical example of a "bars within bars" system of the type postulated by Shlosman *et al.* (1989). Here again notes of caution are required. Firstly the starburst in NGC 4321 is "no big deal" — an order of magnitude less vigorous than classical starbursts in its SFR. (However it has a very small bulge, and a weak bar, so we should not expect more; in fact these circumstances render the phenomena much easier to observe). Secondly the flow pattern originally detected in H I is most simply interpreted not as a direct inflow, but as counterflow: gas in highly elongated orbits with major axes along the bar, but flowing around it. Theorists had predicted that bars should give rise to counterflow, as a normal response of the gas to a bar potential (Roberts *et al.* 1979; Schwarz 1985), and it was indeed this pattern which induced us to look for and find a bar in the old stellar population of NGC 4321 using the near infrared, when no clear evidence was present in the visible. In the light of this experience I felt it to be instructive to review the observations allegedly supporting the presence of molecular inflows within 1 kpc of other galactic nuclei.

2 OBSERVATIONS OF GAS BARS

Here I will briefly summarize the observational results to date in those galaxies where elongated, bar-like structures have been observed, to either side of the galactic nucleus, in molecular gas, with resolution sufficient to detect features on scales of order 100 pc, or finer, at the distances of the galaxies. A more detailed description, with figures, can be found in Beckman (1993).

In the Sc starburst galaxy NGC 6946, following early two-beam interferometric work by Ball *et al.* (1985), Ishizuki *et al.* (1990) used the Nobeyama millimetric array to map the inner 2 kpc zone in the CO J= 2 $\rightarrow$ 1 line. Their map includes an elongated feature, symmetrical with respect to the nucleus, some 1500 pc long by 300 pc wide, in a slightly S-sharped curve. They also find a circumnuclear CO disc

300 pc in diameter, and dense, characteristic of a star formation zone. The radial velocity contours were interpreted by Ishizuki *et al.* as indicating rotation within the ring, and radial inflow along the bar, with an estimated amplitude of 60–80 km s^{-1}. The mass flow, from the measured column densities, corresponding to a vector of this magnitude, would be some 20 $M_\odot$ yr^{-1}, which is at least an order of magnitude in excess of the star formation rate inferred by DeGoia-Eastwood *et al.* (1984) from the observations. However the gas bar lies close to the minor axis so, as explained in section 1, estimates of the true non-circular velocity along the bar are subject to major systematic uncertainties. The mass of gas, on the other hand, is reliably estimable, and Ishizuki *et al.* find 3×10^8 $M_\odot$ of H_2 in the rotating disc, and claim that 25% of the total mass within 150 pc of the nucleus is in gaseous form. Such estimates are vital when modelling the kinematic response of the gas to the total gravitational potential.

The same observers, also using the Nobeyama array, mapped the central kpc of IC 342 in CO J= 1 $\rightarrow$ 0, finding an internal circumnuclear ring, of diameter 100 pc and a bar, in the form of two ridges, each of size 500 pc $\times$ 80 pc, leading outwards from opposite sides of the ring, though slightly offset. The total mass of H_2 within this zone had previously been estimated at 2×10^8 $M_\odot$ by Lo *et al.* (1984), who claimed the detection of gaseous counterflow along a bar-like structure. From radio emission, the presence of 3×10^4 OB stars can be inferred, leading to an estimated star formation rate averaged over some 10^4 years of 0.3 $M_\odot$ yr^{-1}. To resupply this at constant rates would imply inflow velocities of order 1 km s^{-1} at the observed H_2 column densities. More recently Turner and Hurt (1992) used the ^{13}CO J= 2 $\rightarrow$ 1 transition to obtain a more accurate mapping of the H_2 column density. They found an offset between the molecular gas bar and observed Hα ridges, which they cite as evidence for a spiral density wave as the agent in the formation of the (slightly curved) elongated gas structure. These authors claim that the absence of an observable stellar bar, even an oval distortion, even in the near infrared implies that in IC 342 the self-gravity of the gas gives rise to its own resonant structure. The "gas bar" in this case, would be really a pair of shallowly curved spiral arms, whose mass in the inner few hundred pc dominates that of the stellar component. This variant of the bar scenario does not contradict the more general idea that departures from axisymmetry stimulate the non-conservative forces required to cause inflow refuelling.

In the barred S(B)bc galaxy Maffei 2, a gas bar has been detected using both of the currently available interferometers: Nobeyama (Ishiguro *et al.* 1989), and OVRO (Hurt and Turner 1991). The Nobeyama group, using ^{12}CO, found a bar of length 1500 pc, and reported evidence for non-circular motions in the gas, whereas the OVRO observers did not find such clear evidence, although their ^{13}CO data do confirm an elongated structure. Combining the CO-based estimates of the mass of molecular gas in the bar, with radio (6 cm) determinations of the number of OB stars, a duration for the starburst of $\sim 5 \times 10^8$ years can be inferred. The net inflow velocities required to deliver the H_2 to the inner couple of hundred parsec are of order 1 or 2 km s^{-1}, assuming zero effective return of gas to the ISM during the wind and supernova phases of the massive stars. Hurt and Turner suggest that the gas bar in Maffei 2 is a self-stimulated resonant structure, and that the gas mass within 300 pc of the nucleus, some 40% of the total mass, is sufficient to maintain the surface density contrast observed.

In the SAB(c) galaxy M83, a map in CO J= 1 → 0 by Handa *et al.* (1990) with non-interferometric resolution of 15 arcsec revealed a gas bar of length 2 kpc, aligned with the optical bar, with ridges of high temperature along the leading edges of the latter, and with a velocity jump, across the bar, of 20–40 km s^{-1}. This is consistent with a large-scale counterflow pattern. Here the estimated gas mass constitutes only some 10% of the total dynamical mass in the central beam, corresponding to 300 pc radius. The optical kinematic data shows a high nuclear mass concentration, and all the shifts and line widths in CO can be explained in terms of this, without the need for rapid inflow. Higher angular resolution is required to obtain more detailed dynamical conclusions from the kinematics.

In the well known nearby starburst galaxy NGC 253 Canzian *et al.* (1988) made a CO map at OVRO, with a resolution of 90 pc × 50 pc finding an elongated structure centred on the nucleus, of dimensions 400 pc × 100 pc. Estimates of the H_2 mass give a gas: total mass ratio of order 10%, and the CO velocity field could be fitted by a central rigid rotator, together with an inflow component inferred by Canzian *et al.* as 8 (±2) km s^{-1}. Bearing in mind the possible sources of systematic error mentioned above, the limits quoted appear optimistic, but this result can be treated as *prima facie* evidence for inflow.

3 CONCLUSIONS

In all disc galaxies the total neutral gas distribution shows an exponential mean radial distribution of column density, and due to the well-known photochemical effects of dust, within the innermost scale-length (typically of order 3 kpc) the molecular component dominates. In starburst galaxies, however, there are additional concentrations of molecular gas towards the nucleus, and as we have seen in this brief survey, these are frequently not axisymmetric. Among the structures observed are bar-like forms, normally in a highly elongated S-shape with the nucleus at the point of inflexion, and often with a disc of gas around the nucleus itself. The details of the structure and dynamics of these bars and discs vary, above all according to the ratio of gaseous to stellar mass within the zones described.

Although the CO observations do have the effective velocity precision required ($\sim$ 1 km s^{-1}) to make valid judgments about inflow, although the linear resolution is typically only just enough to resolve these structures, even in nearby galaxies. While there is nearly always evidence for non-circular motions, the most characteristic signature is of a jump in velocity perpendicular to the bar, of order a few tens of km s^{-1}, and not radial inflow in this velocity range. Gaseous counterflow has

been predicted in various dynamical models of barred galaxies, notably by Roberts
et al. (1979), Schwarz (1989) and by Combes and Gerin (1985) who incorporated
molecular clouds and star formation in their scenarios. These models predict non-
circular motions of the gas in elongated orbits around the bar, *i.e.* counterflow,
with radial components of order tens of km s^{-1}. It is the dynamical friction in this
counterflow which may yield a net inflow towards the nucleus, with a corresponding
transfer of angular momentum along the bar. The rate of this inflow, equivalent to
mean velocities of up to 1 km s^{-1}, would be sufficient to maintain the star formation
rates observed in typical circumnuclear starbursts. Obviously episodes of more rapid
inflow, yielding more spectacular starbursts, are not excluded in this scenario, but
these would be difficult to observe, in general, because short-lived, while the steady
inflow component is extremely hard to measure directly since it represents a small
difference, of order 1 km s^{-1} at most, between counterflow velocities of tens of km
s^{-1}. However the counterflow signatures themselves represent important evidence for
the presence and the dynamical importance of non-axisymmetric, bar structures in
causing circumnuclear star-forming activity, and further molecular observations at
the best resolution and sensitivity possible, are clearly called for.

REFERENCES

Ball, R., Sargent, A. I., Scoville, N. Z., Lo, K. Y., and Scott, L. S. 1985, *Ap. J.*, **298**, L21.

Beckman, J. E. 1993, in *The Nearest Active Galaxies*, eds. J. E. Beckman, L. Colina and H. Netzer (C. S. I. C. Madrid), in press.

Canzian, B., Mundy, L. C., and Scoville, N. Z. 1988, *Ap. J.*, **333**, 157.

Combes, F., and Gerin, M. 1985, *Astr. Ap.*, **150**, 327.

DeGioia-Eastwood, K., Grasdalen, G. L., Strom, S. E., and Strom, K. 1984 *Ap. J.*, **278**, 564.

Handa, T., Nakai, N., Sofue, Y., Hayashi, M., and Fujimoto, M. 1990, *P. A. S. J.*, **42**, 1.

Hurt, R. L., and Turner, J. L. 1991, *Ap. J.*, **377**, 434.

Ishiguro, M. *et al.* 1989, *Ap. J.*, **344**, 763.

Ishizuki, S. *et al.* 1990, *Ap. J.*, **355**, 436.

Knapen, J. H., Cepa, J., Beckman, J. E., del Rio, M. S., and Pedlar, A. 1993, *Ap. J.*, in press.

Lo, K. Y. *et al.* 1984, *Ap. J. Lett.*, **282**, L59.

Roberts, W. W., Huntley, J. M., and van Albada, G. D. 1979, *Ap. J.*, **333**, 67.

Schwarz, M. P. 1985, *Mon. Not. R. Astr. Soc.*, **212**, 677.

Shlosman, I., Frank, J., and Begelman, M. C. 1989, *Nature*, **338**, 45.

Turner, J. L., and Hurt, R. L. 1992, *Ap. J.*, **384**, 72.

Bar-Driven and Interaction-Driven Starbursts in S0/Sa Galaxies

L. L. Dressel[1], and J. S. Gallagher III[2]

[1]Applied Research Corporation
[2]University of Wisconsin

ABSTRACT

Strong central starbursts occur most frequently in early-type disk galaxies, especially barred galaxies. We have taken U, B, R, I, and Hα images and long slit nuclear spectra of eight galaxies of type S0/SB0 to Sa/SBa with bright central starbursts, selected on the basis of *IRAS* colors and fluxes. Four of the galaxies have circumnuclear ring-like distributions of hot stars and gas. Some or all of these four galaxies may be cases of barred galaxies with inner Lindblad resonances (ILRs), in which gas has been channelled into a ring at an ILR and compressed into star formation. Another galaxy is clearly interacting with a companion Seyfert galaxy. The strong interaction has apparently lead to the producion of the bright nuclear and near-nuclear emitting regions and the complex of fainter star-forming clumps and filaments surrounding them. Two of the remaining galaxies have compact nuclear starbursts. We conjecture that these galaxies may not have ILRs, so that gas perturbed by a bar potential or by a companion has accumulated at the nucleus instead of in a ring. The final galaxy appears to have a very bright off-center clump of star formation, conspicuous at all wavelengths. Although the possibility of a centered but dust-obscured burst is discussed, the apparent asymmetry may be real and in need of explanation.

1 INTRODUCTION

Strong central starbursts are occurring in a significant minority of S0 and early-type spiral galaxies. The *IRAS* survey data can be used reliably to find complete samples of these galaxies, since on-going bursts produce far infrared radiation that is warmer than that produced by disks. Bright sources warmer than $F_{100\mu}/F_{60\mu} \approx 2.0$ are found in S0, Sa, and Sb galaxies; bright sources cooler than this are found in Sb and Sc galaxies. (See Figure 2 in Dressel 1988.) Supplemental optical, radio, and infrared data indicate that $F_{100\mu}/F_{60\mu} = 2.0$ is the approximate dividing line between central-starburst-dominated emission and disk-dominated emission. (Seyfert nuclei comprise a small minority of the warm population.) The warm central sources occur at a much higher rate in barred galaxies than in galaxies that do not have a bar. (See Table 1 in Dressel 1988.) The host galaxies of starbursts also tend to have peculiar optical morphologies, sometimes clearly caused by gravitational interaction with a companion galaxy. To study the evolution and determine the causes of central starbursts in galaxies with and without bars and companions, we are making a detailed optical study of several nearby bright S0 and Sa galaxies with bright warm infrared sources.

2 SAMPLE AND DATA

The observed galaxies were chosen from a list of candidates with $m_{pg} \leq 14.5$ mag, with Hubble type S0 to Sa (or SB0 to SBa) in Nilson (1973), and with $0° \leq \delta \leq 37°$. Consistent with observing constraints, the non-Seyfert galaxies with the brightest, warmest far-infrared sources were selected from this list. Some of the properties of the eight observed galaxies are listed in Table 1. The UGC, NGC, and Markarian numbers are given in the first three columns. *IRAS* fluxes and flux ratios are given in columns 4 and 5. The far infrared luminosity in column 6 has been calculated from the "FIR" flux from 43 μ to 123 μ as prescribed by Lonsdale *et al.* (1985). Column 7 gives the molecular hydrogen mass deduced from CO observations, corrected to a conversion factor of $N_{H_2}(\text{cm}^{-2}) = 3.5 \cdot 10^{20} I_{CO}$ (K km s^{-1}). The last column gives the Hubble type and information about companions. H_o is assumed to be 50 km s^{-1} Mpc^{-1} throughout.

TABLE 1
PROPERTIES OF THE OBSERVED GALAXIES

UGC #	NGC #	Mrk #	$F_{60\mu}$ (Jy)	$F_{100\mu}$ /$F_{60\mu}$	log L_{FIR} (L$_\odot$)	log M_{H_2} (M$_\odot$)	UGC Classification[1]
00089	23	545	8.77[2]	1.71[2]	11.1	10.4[4]	SBa pw 00094
00861	471		2.86[3]	1.29[3]	10.5		S0 brightest in mult sys
01157	632	1002	5.03[3]	1.29[3]	10.5	<9.9[5]	S0 pw 01153
01385		2	6.13[3]	1.30[3]	11.1		SBa brightest of 3
03201	1691	1088	7.23[3]	1.45[3]	10.9	9.7[6]	SB0/SBa
03265	1819	1194	7.36[3]	1.66[3]	11.0	10.4[7]	SB0
12575	7648	531	4.92[3]	1.50[3]	10.6		S0
12618	7679	534	7.79[3]	1.35[3]	11.1	10.0[8]	S0 pw 12622, disturbed

References: [1]Nilson 1973; [2]Soifer *et al.* 1989; [3]Knapp *et al.* 1989; [4]Young *et al.* 1989; [5]Jackson *et al.* 1989; [6]Krugel, Steppe, and Chini 1990; [7]Thronson *et al.* 1989; [8]Wikland and Henkel 1989.

Eight galaxies in Table 1 were observed with the 2.1m telescope at Kitt Peak in October 1991. U, B, R, I, and Hα images were taken on 3 nights with the CCD direct-imaging camera. The field of view was 2.5′ on a side, the pixel size was 0.19″, and seeing was typically 0.7 to 1.0″. Long slit spectra through the nucleus were taken with the Gold spectrograph on the following 3 nights. The 2.5′ × 2.5′ slit was oriented east-west. The spectra span 3,600Å to 4,500Å, with 3.7Å spectral resolution and 0.78″ spatial pixel size. All of the spectra have strong Balmer absorption lines and Balmer and [O II] emission lines, with rotation in the same direction for gas and stars.

3 RESULTS

3.1 Circumnuclear Rings

Four of the galaxies, UGC 00089, 00861, 03201, and 03265, have Hα emission in a bright clumpy circumnuclear ring-like structure. Two of these, UGC 03201 and 03265, have a very prominent optical bar. UGC 00089 may also have a bar, but conflicting classifications have been published. Several theoretical studies have convincingly shown that gas can be channelled into a ring at an inner Lindblad resonance (ILR) in a barred galaxy, where it can then be triggered into star formation by compression (*e.g.,* Casoli and Combes 1982; Combes and Gerin 1985). We do not have data adequate to find whether or where these galaxies have ILRs, but the long slit data are at least consistent with the gas having purely rotational motion in corotation with the stars in all four galaxies.

The continuum-subtracted Hα image of UGC 03265 is shown in Figure 1. The Hα ring is elongated perpendicular to the bar and has its brightest clumps at the farthest ends, where the ring is met by dust lanes in the continuum images. It is thus reminiscent of twin-peaked CO distributions, which have been explained in terms of streamline crowding near ILRs (Kenney *et al.*, 1992).

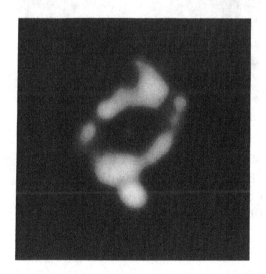

Fig. 1—Linear brightness display of the continuum-subtracted Hα emission of UGC 03265. North is up, east is to the right. Photo diameter = 18 arcsec = 8 kpc.

UGC 00861 is a smooth featureless galaxy in the continuum, except for one dust lane 4 arcsec from the nucleus which bears no obvious relation to the structure of the galaxy. The galaxy is not clearly barred, but its central isophotes are elongated and twisted relative to its outer isophotes. It is unique among the galaxies in having three small nearby companions (assuming that they have similar redshifts) within or near the Nilson (1973) radius. This may therefore be a case of the nearby passage of a companion triggering the formation of a temporary stellar bar, which then induces the formation of a star-producing ring of gas (Noguchi 1988).

3.2 Tidal Features

UGC 12618 = Arp 216 and its companion Seyfert galaxy three diameters east, UGC 12622, are both tidally disturbed (Arp 1966; Nilson 1973). UGC 12618 has been given a variety of classifications, including SB0 pec: (de Vaucouleurs, de Vaucouleurs, and Corwin 1976) and Sc/Sa (tides?) (Sandage and Tammann 1981). Figure 2 shows the continuum-subtracted Hα emission, with two bright clumps at and near the nucleus surrounded by a complex of fainter clumps and filaments. The east-west rotation curve is asymmetric, rising and falling more steeply in the west. Since the time scales of tidal features and starbursts are both cosmically short, interaction is implicated as the cause of both the morphological disturbances and the star formation.

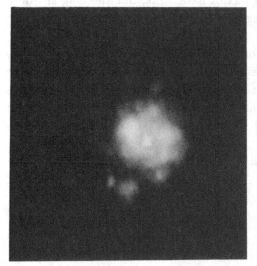

Fig. 2—Logarithmic brightness display of the continuum-subtracted Hα emission of UGC 12618. North is up, east is to the right. Photo diameter = 40 arcsec = 21 kpc.

Fig. 3—Logarithmic brightness display of the sum of the U, B, R, and I images of UGC 12575, with Hα removed from R. North is up, east is to the right. Photo diameter = 80 arcsec = 29 kpc.

3.3 Compact Central Bursts

UGC 01157 has a slightly resolved central Hα source with three faint extensions. One of the extensions lies along a bar-like structure in the I image; the other two are perpendicular to it. UGC 01385 has a slightly resolved elongated central Hα source, with the elongation similar in angle to the bar in the continuum images. Both galaxies have nearby bright companions. We suggest that these galaxies do not have ILRs, so that inflowing gas has accumulated at the nucleus rather than in a ring (Telesco, Dressel, and Wolstencroft 1993), possibly in response to the bar potential or to perturbations by the companions.

3.4 Off-Center Burst

UGC 12575 is unique among the galaxies in having a single bright clump distinctly off-center in all the images (Fig. 3). The centroid of the clump is the same in the U and Hα images: 1.8″, or 660 pc, from the center of the outer I isophotes. It moves slightly across the continuum images from U to I (from 1.8 to 1.2″ out). Complex structure in the continuum images indicates that dust is present, but the coincidence of the U and Hα centroids suggests that dust is not entirely responsible for the off-center appearance of the starburst. Shift in centroid from U to I could be due to increasing prominence of a red nucleus that is blended with the clump. The long slit spectrum of the clump is consistent with pure rotation; only the rising part of the rotation curve is seen, so the dynamical center cannot be determined. Perhaps significantly, this is the only galaxy observed that does not have either a bar, a comparably sized companion within 5 diameters, or a dwarf companion within its radius.

This work was supported in part by NASA grant NAG5-1630.

REFERENCES

Arp, H. 1966, *Atlas of Peculiar Galaxies* (Pasadena: California Inst. of Technology).

Casoli, F., and Combes, F. 1982, *Astr. Ap.*, **110**, 287.

Combes, F., and Gerin, M. 1985, *Astr. Ap.*, **150**, 327.

de Vaucouleurs, G., de Vaucouleurs, A., and Corwin, H. 1976, *Second Rererence Catalog of Bright Galaxies* (Austin: University of Texas).

Dressel, L. L. 1988, *Ap. J. Lett.*, **329**, L69.

Jackson, J. M., Snell, R. L., Ho, P. T. P., and Barrett, A. H. 1989, *Ap. J.*, **337**, 680.

Kenney, J. D., Wilson, C., Scoville, N. Z., Devereux, N. A., and Young, J. S. 1992, *Ap. J. Lett.*, **395**, L79.

Knapp, G. R., Guhathakurta, P., Kim, D.-W., and Jura, M. 1989, *Ap. J. Suppl.*, **70**, 329.

Krugel, E., Steppe, H., and Chini, R. 1990, *Astr. Ap.*, **229**, 17.

Lonsdale, C. J., Helou, G., Good, J. C., and Rice, W. 1985, *Catalogued Galaxies and Quasars Observed in the IRAS Survey* (Pasadena: Jet Propulsion Laboratory).

Sandage, A. and Tammann, G. A. 1981, *A Revised Shapley-Ames Catalog of Bright Galaxies* (Carnegie Institution of Washington Publication 635).

Nilson, P. 1973, *Uppsala General Catalogue of Galaxies, Nova Acta*, **1**, Sec. V: A.

Noguchi, M. 1988, *Astr. Ap.*, **203**, 259.

Soifer, B. T., Boehmer, L., Neugebauer, G., and Sanders, D. B. 1989, *A. J.*, **98**, 766.

Telesco, C. M., Dressel, L. L., and Wolstencroft, R. D. 1993, *Ap. J.*, **414**, in press.

Thronson, H. A., *et al.* 1989, *Ap. J.*, **344**, 747.

Wikland, T., and Henkel, C. 1989, *Astr. Ap.*, **225**, 1.

Young, J. S., Shuding, X., Kenney, J. D., and Rice, W. L. 1989, *Ap. J. Suppl.*, **70**, 699.

Gas Inflow due to Perpendicular Orbits in Barred Potentials

Françoise Combes

Radioastronomie, DEMIRM, Observatoire de Meudon
F-92 190 Meudon, France

ABSTRACT

Radial gas flows can be induced in a galactic disk by a bar potential, and its implied gravitational torques. All gas inside corotation is driven towards the center, and forms a nuclear ring at the inner Lindblad resonance (ILR). When the mass concentration is high enough, there exists two ILRs, and the existence of periodic orbits perpendicular to the bar makes the gas response to shift in phase with respect to the stellar main bar. This produces a strong torque on the gas, and drives a rapid nuclear gas flow inside ILR. With a more viscous gas, however, a second bar of stars and gas can decouple from the primary one, with a higher pattern speed. In this "bar within bar" configuration, the gas is in phase with the stellar component, and the gravity torques are minimised. The gas inside the second corotation flows slowly inwards. The nuclear bar is relatively long- lived, which explains its frequent occurence in observed barred spirals.

1 NUCLEAR BARS AND THEIR POSSIBLE INTERPRETATIONS

Bars are the way to redistribute angular momentum in a galaxy, and to reshape the mass distribution. The induced gas flow towards the center is the cause of starbursts, hot spots in nuclear rings, and may be of nuclear activity. A clue to the detailed mechanisms of the central gas flow is the observation of nuclear bars and central isophote twists. This phenomenon in barred spiral galaxies has been observed for a long time. Already de Vaucouleurs (1974) had noticed bars within bars, and Sandage and Brucato (1979) high surface brightness nuclear bars, as independent entities. Kormendy (1982) finds that isophote twists are preferentially found in SB galaxies, with strong central concentration, that the second small bar is aligned nearly perpendicular to the main bar, and proposes that the triaxiality of bulges is the cause of the projected twists. More recently, Buta and Crocker (1991) have shown that these twists are associated with nuclear rings. Sometimes the nuclear ring itself is almost perpendicular to the main bar. Shaw *et al.* (1993) have studied via near-IR photometry (1.2–2.2μ) the three nuclear bars galaxies NGC 1097, 4736 and 5728, and shown that the isophote twists cannot be interpreted in terms of a triaxial bulge; in these 3 galaxies, the nuclear bar is located just inside the nuclear ring (of about 1 kpc radius). They proposed two main interpretations for these isophote twists: first, in the presence of two ILRs, the family x_2 of perpendicular periodic orbits is the cause of phase shifts in the gas component; once enough gas is gathered in the center, a fraction of the stellar component follows its potential. Second, a nuclear bar can decouple from the main bar, with a larger pattern speed, as suggested by Shlosman *et al.* (1989), and demonstrated by Friedli and Martinet (1993).

2 NUMERICAL TECHNIQUES

To test the operational conditions of these two mechanisms, we have run self-consistent simulations containing gas and stars. The code used was the Particle-Mesh method with FFT (Combes *et al.* 1990), using a 2D exponential polar grid with 64 × 96 cells. The cell size is of the order of 100 pc in the central region, and the number of particles is 8×10^4. The gas was treated as an ensemble of 2×10^4 clouds suffering collisions, but with two different codes. In the first one, clouds are distributed according to a mass spectrum, between 10^3 and 10^6 $M_\odot$, they can exchange mass or coalesce in collisions; at the end of the mass spectrum, massive clouds have a finite life-time and disperse into small clouds, with re-injected energy to simulate star-formation (Combes and Gerin 1985). In the second one, clouds have all the same mass, and exchange only energy and momentum in inelastic collisions (absolute values of relative velocities are reduced by 25% in a collision). There is no energy re-injected by star-formation. We think that this second code is less viscous, in the sense that the equilibrium velocity dispersion of the gas particles is much lower.

The model galaxy is built with an analytic bulge and dark halo, while the disk is represented by particles. The bulge to disk mass ratio was varied to simulate early and late-types. The runs presented here have mass ratios of bulge:disk:halo of 1:6:10, and size ratios 1:12:54. The gas mass fraction was varied between 3 and 18%.

3 RESULTS: PERPENDICULAR ORBITS AND BARS WITHIN BARS

We find that the physical details of the gas code are essential to determine the dynamical behaviour of the barred galaxies. With the less viscous code, the gas settles in a phase-shifted bar, influenced by the existence of perpendicular periodic orbits, while in the more viscous code, a rapidly rotating nuclear bar of stars and gas decouples from the main bar.

3.1 Phase-Shifted Bar

This first model (Fig. 1) is run with the single-mass less viscous gas code, with a gas mass fraction of 6%. First the star-gas disk is unstable, and forms a barred spiral structure in a few 10^8 yrs; the bar ends at its corotation, with a pattern speed equal to 40 km s^{-1} kpc^{-1}. Two inner Lindblad resonances are present in the center, and consequently there exists the x_2 family of periodic orbits perpendicular to the main bar (Contopoulos and Papayannopoulos 1980). Gas clouds tend to follow these orbits, but collisions make them turn gradually from parallel to perpendicular to the bar. The gas then settles in a phase-shifted bar, leading by 30° the main stellar bar. The pattern speed of this secondary feature is the same as that of the main bar, as determined by Fourier transforms of the potential in function of radius and time.

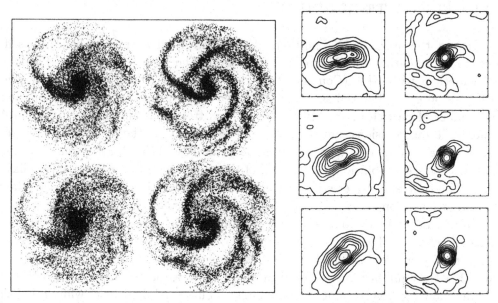

Fig. 1—First run (less viscous gas): particle plots of stars (left) and gas (right), and contours of the central parts. Note the phase-shift in the gas.

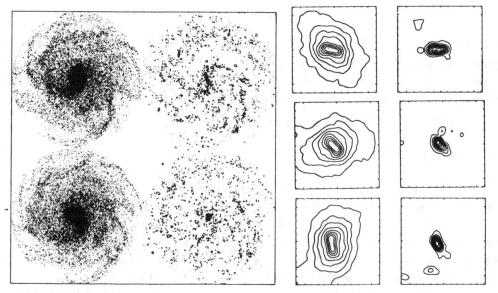

Fig. 2—Same for the second run (more viscous gas): 'bar within bar' configuration.

In this configuration, where the gas wave is in advance of the stellar bar, the gravitational forces of the stellar potential tends to slow down the gas such that the latter experiences a negative torque between the two ILRs. This produces strong gas flows towards the center.

3.2 Bars within Bars

The second model (Fig. 2) is run with the more viscous mass-spectrum gas code, with the same gas mass fraction of 6%. The primary bar forms as in the first run, but very quickly a secondary bar of stars and gas decouples from the first, with a pattern speed of 100 km s^{-1} kpc^{-1}, *i.e.* rotating about twice more rapidly than the primary bar. The corotation of this nuclear bar corresponds roughly to the inner Lindblad resonance of the primary one, suggesting that non-linear interactions between the two bars are at play (Tagger *et al.* 1987). The gas is entirely participating to the nuclear bar, and there is no phase shift between gas and stars. Gravity torques are therefore considerably reduced in this run with respect to the first one. This explains why the nuclear bar feature is relatively long-lived. It remains for about 40 rotations of the small bar, before dissolution occurs, by concentration of gas towards the center.

To better understand the relative role of gas and stars in the second bar decoupling, we have run a simulation without gas, in exactly the same conditions. The first bar appears as in the two first runs, but no secondary bar ever decouples. This can be interpreted by the higher central velocity dispersion in the pure stellar run, as is shown in Figure 3. The presence of only 6% gas mass prevents such a large heating of the stellar component.

We have also repeated the second run with higher gas mass fractions. The nuclear bars are even more clearer, and gas flows somewhat more efficiently, with relatively shorter-lived bars. Figure 4 summarises the efficiencies of gas flows.

4 CONCLUSION

Gas flows in barred galaxies occur in several steps. First, gravity torques drive the gas within corotation of the primary bar towards the inner Lindblad resonance to form a nuclear ring (*e.g.* Schwarz 1984; Combes and Gerin 1985). If the mass concentration is not enough to allow two ILRs, the gas can remain steady there for a few 10^8 yrs. When the mass concentration has built up, and two ILRs exist, there can be two possibilities, according to the viscosity of the gas. First, the presence of periodic orbits perpendicular to the main stellar bar makes the response of the gas to shift in phase with respect to the stellar potential. The strong gravitational torques then felt by the gas which leads the potential, drive a nuclear gas flow inside ILR. Second, a secondary nuclear bar of stars and gas decouples from the primary one, gravity torques inside corotation can take over the gas flows, but now more slowly, since the gas is in phase with the potential. Third, dynamical friction could play a large role in the final gas flow, since the characteristic decaying time-scale for a GMC of 10^7 $M_\odot$ is 10^9 yrs at 1 kpc, but varies with the square of distance from the center; it is 3×10^7 yrs at r = 200 pc.

In the nuclear bar formation, the gas-star coupling is essential. In the pure stellar run, but also in a pure gas run, no such bar develops. The "bar within bar" phenomenon is relatively long-lived, at least 20 rotations of the second bar. This might explain the high frequency of its observation.

REFERENCES

Buta, R., and Crocker, D. A. 1991, *A. J.*, **102**, 1715.

Combes, F., and Gerin, M. 1985, *Astr. Ap.*, **150**, 327.

Combes, F., Debbasch, F., Friedli, D., and Pfenniger, D. 1990, *Astr. Ap.*, **233**, 82.

Contopoulos, G., and Papayannopoulos, T. 1980, *Astr. Ap.*, **92**, 33.

de Vaucouleurs, G. 1974, in *IAU Symp.* **58**, *The Formation and Dynamics of Galaxies*, ed. J. R. Shakeshaft (Dordrecht: Reidel), p. 335.

Friedli, D., and Martinet, L. 1993, *Astr. Ap.*, in press.

Kormendy, J. 1982, *Ap. J.*, **257**, 75.

Sandage, A., and Brucato, R. 1979, *A. J.*, **84**, 472.

Shaw, M. A., Combes, F., Axon D. J., and Wright, G. S. 1993, *Astr. Ap.*, **273**, 31.

Schwarz, M. P. 1984, *Mon. Not. R. Astr. Soc.*, **209**, 93.

Shlosman, I., Frank, J., and Begelman, M. C. 1989, *Nature*, **338**, 45.

Tagger, M., Sygnet, J. F., Athanassoula, E., and Pellat, R. 1987, *Ap. J. Lett.*, **318**, L43.

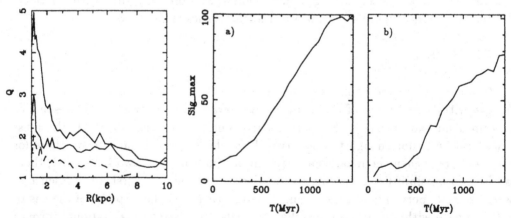

Fig. 3—Distribution of the stellar Toomre Q parameter, *vs* radius. Dash (initial conditions); full line, bottom (6% gas); top (pure stellar run).

Fig. 4—Gas surface density inside 1 kpc as function of time: *a*) with phase-shifted gas response, *b*) with two bars.

Neutral Gas in the Barred Starburst Galaxy Maffei 2

Robert L. Hurt

Department of Astronomy, UCLA

1 RESULTS

Maffei 2, a nearby SBb *pec* galaxy behind the Galactic plane, is the site of a nuclear burst of star formation as evidenced by near infrared (Rickard and Harvey 1984, *A. J.*, **89**, 1520; Ho *et al.* 1989, *Ap. J.*, **344**, 135), Brackett line (Ho *et al.* 1990, *A. J.*, **349**, 57), and radio continuum emission (Seaquist, *et al.* 1976, *Astr. Ap.*, **48**, 413; Turner and Ho 1993, preprint). Its proximity (∼5 Mpc) and large angular extent (∼10′) make it an ideal candidate for high resolution neutral gas studies which are necessary for understanding the processes driving the nuclear starburst.

The integrated H I emission $\int I(v)dv$ is shown in greyscale in Figure 1. These *VLA* observations combine C and D configurations and, with tapered uniform weighting, achieve 20″ resolution. The peak column densities of 3.7×10^{21} cm^{-2} can be seen to lie in two arms around the nucleus, coincident with enhanced 20 cm continuum emission (Fig. 1, left). This continuum emission traces the spiral arms of Maffei 2 as seen in the near infrared (Hurt, *et al.* 1993, *A. J.*, **105**, 121). Very little H I is present in the nucleus, although the region between the dashed lines is strongly affected by extinction/emission from foreground Galactic H I clouds. There is a strong N/S asymmetry in the H I distribution that reflects the observed peculiar morphology in the radio continuum near infrared.

The H I velocity field (Fig. 1, right) exhibits deviations from symmetric circular rotation that are characteristically seen in barred galaxies. The large velocity gradients perpendicular to the bar are what would be expected from gas moving in elliptical streamlines with cusps along the bar. Notable is the *lack* of any other large scale perturbations of the velocity field which might be expected in the case of a dynamically recent tidal encounter.

There is a striking discrepancy in the rotation curves of Maffei 2 derived from ^{13}CO (Hurt and Turner 1991, *Ap. J.*, **377**, 434) and H I observations (Fig. 2). The ^{13}CO rotation curve derived from major axis slices rises steeply within the inner 2″, turns over at 5″ and stays flat out to a radius of 20″. While consistent with the H I, there must be a second steep rise in rotational velocity between 20″ and 50″ to connect the two dynamical regions. However, the systemic velocity determined from the two maps yield notably different values: -21 ± 1 km s^{-1} for the H I and -30 ± 1 km s^{-1}.

This result must be evaluated carefully since the ^{13}CO emission samples only an azimuthally limited region of the galaxy's velocity field near its major axis.

The neutral gas emission in Maffei 2 hints at the possibility of a dynamically recent tidal encounter which could have stimulated the nuclear star formation and produced the peculiar morphology of the galaxy. While there are no strong perturbations seen in the velocity field like those of some other interacting galaxies, the distorted morphology and variations in the dynamics of the nuclear and disk gas may well be the products of such an encounter.

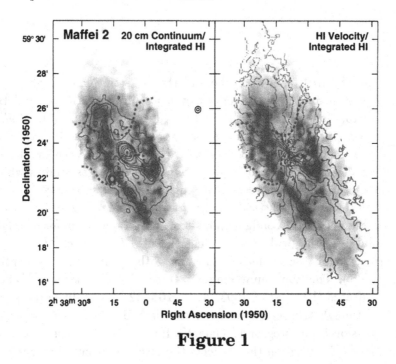

Figure 1

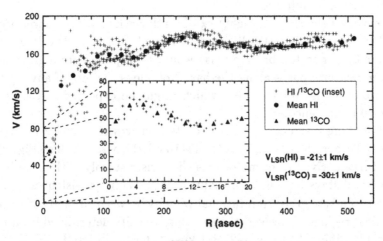

Figure 2

The Abundance Gradients in Barred Galaxies: The Role of Radial Flows

Pierre Martin

Steward Observatory, University of Arizona, Tucson, AZ 85721, U. S. A.

ABSTRACT

We present a study of the O/H gradients derived from large numbers of H II regions in a sample of barred galaxies. It is shown that the slopes of the O/H gradients in galaxies with a barred structure are related to the bar strength: the stronger is the bar, the flatter the abundance gradient becomes. This result is in agreement with a scenario of mixing of the interstellar gas by radial flows across the disks of barred galaxies. Thus we concluded that a bar acts as a homogenizer on the chemical composition in a spiral galaxy.

1 INTRODUCTION: RADIAL FLOWS

Observations and numerical simulations have shown that strong radial flows are present in barred spiral galaxies (*e.g.* Sellwood and Wilkinson 1993). Although the exact origin of such flows is unclear, their amplitudes seem to be related to the bar properties and to the presence of radiative shocks induced when the gas passes through the bar. A possible important consequence of radial flows in a spiral galaxy is a radial redistribution of the chemical composition across the disk. Such mechanism was suggested to explain why the O/H gradients in few barred galaxies (*e.g.* NGC 1365) are shallower than gradients observed in normal spirals.

Review works by Vila-Costas and Edmunds (1992, VE) and Zaritsky *et al.* (1993, ZKH) have confirmed this difference. However, their large samples of galaxies included very few barred spirals where the O/H gradient is derived from a large number of H II regions. Moreover, they used the morphological type alone to compare the gradient properties in SA and SB galaxies; a more quantitative parameter is needed to describe the amplitude of a bar in a galaxy. Thus to have a more reliable picture of the influence of radial flows on the chemical composition of spiral galaxies, a larger sample of barred galaxies with well-established O/H gradients and a better parametrisation of the strength of bars are needed.

2 DATA AND PARAMETRISATION OF THE BAR STRENGTH

Using monochromatic imagery with interference filters, we obtained new O/H data in 6 galaxies showing a barred structure (*e.g.* Martin and Roy 1993). This technique allowed us to derive the O/H gradients from at least 50 H II regions per galaxy. Other data were taken from papers by VE and ZKH. Bar strength is defined by the axis ratio of the bar as $E_B = 10(1 - b/a)$ where a and b are the major and the minor axis respectively. Thus E_B defines a continuously changing morphology going from a normal galaxy ($E_B = 0$) to a very strong barred galaxy ($E_B = 8$).

3 RESULTS

Figure 1 presents the relationship between the slopes of the O/H gradients in spiral galaxies (SAB and SB) showing a barred morphology against the bar strength E_B. A very strong correlation is observed ($R = 0.82$): *the stronger is the bar, the flatter the abundance gradient becomes!*

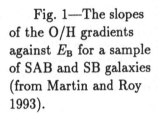

Fig. 1—The slopes of the O/H gradients against E_B for a sample of SAB and SB galaxies (from Martin and Roy 1993).

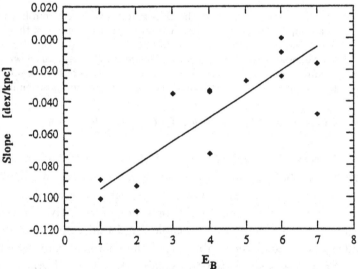

This trend suggests that a more efficient mixing occurs across the disks of barred spirals when the axis ratio b/a decreases. This is in agreement with recent models of large-scale radial flows by Friedli and Benz (1993) showing that the amount of gas accreted in the central region of a barred galaxy increases when b/a decreases. Thus a bar acts as a homogenizer on the chemical composition in a spiral galaxy and, the importance of this effect is strongly related to the bar strength.

Acknowlegements: this work was funded by the NSERC of Canada, the Québec FCAR-centre program and in part by the NSF through Grant AST-9019150.

REFERENCES

Friedli, D., and Benz, W. 1993, *Astr. Ap.*, **268**, 65.
Martin, P., and Roy, J.-R. 1993, *Ap. J.*, submitted.
Sellwood, J. A., and Wilkinson, A. 1993, *Rep. Prog. Phys.*, **56**, 173.
Vila-Costas, M. B., Edmunds, M. G. 1992, *Mon. Not. R. Astr. Soc.*, **259**, 121. (VE)
Zaritsky, D., Kennicutt, R. C., and Huchra, J. P. 1993, in preparation. (ZKH)

Bar-Induced Non-Circular Molecular Gas Motions in M82

Alice C. Quillen[1], Connie Walker[2], and T. G. Phillips[3]

[1]Astronomy Department, Ohio State University
[2]Astronomy Department, University of Texas
[3]Downs Laboratory, Caltech

ABSTRACT

We model the kinematics of the molecular gas in the nearly edge-on disk in M82, by considering velocity and surface density perturbations caused by a possible rotating kpc long bar consistent with the angle of the bar observed from K (2.2 μm) isophotes. A model with a bar that has an Inner Linblad Resonance (ILR) at $r \sim 10'' \sim 150$ pc fits the molecular observations of the inner torus. The clouds have a cloud-cloud velocity dispersion of less than ~ 30 km s^{-1}.

The nearby "starburst" galaxy M82 is one of the most powerful infrared sources which is the result of a high star formation (SF) rate. There is a high concentration of molecular gas ($\sim 5.5 \times 10^7$ M$_\odot$) in the vicinity of the central starbursting region. The double lobed molecular structure observed is thought to be in the form of a rotating ring with a radius of approximately 250 pc (*e.g.* Weliachew *et al.* 1984). In the near infrared K or 2.2 μm band, there is a plateau of emission which is interpreted as evidence for a kpc long bar (Telesco *et al.* 1991) which may have caused the molecular gas to sink into the nuclear region. This mechanism for fueling a starburst has been predicted by numerical simulations of galaxy-galaxy collisions which include gas dynamics (*e.g.* Barnes and Hernquist 1991). The wide profile shapes observed at high spatial resolution (Brouillet and Schilke 1993), the double peaks in the channel maps of ^{12}CO(1-0) (Carlstrom 1988), and a twist observed in the contours of the projected velocity field (*e.g.* Sofue *et al.* 1992) suggest that there are non-circular motions in the molecular gas, possibly induced by the bar.

By modeling the kinematics of the molecular material we attempt to see if peculiarities of the velocities observed in the molecular material of the torus can be caused by bar induced non-circular motions. Because the disk of M82 is edge on and direct observation of the bar is not possible, we use the simplest of assumptions for the rotating bar. We assume that the bar is a rotating quadrupolar perturbation to the potential. Gas clouds are expected to settle onto closed orbits so first order velocity and surface density perturbations were computed. We compare our models to ^{12}CO(3-2) maps of M82 observed at the Caltech Submillimeter Observatory.

We do a simulated observation in which a model velocity field is convolved with the beam of the telescope. Two rotation curves were explored, with and without an

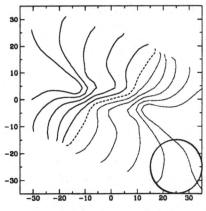

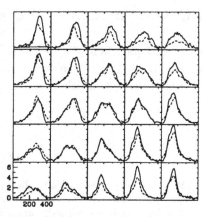

1*a*. Model velocity map. 1*b*. Model and Data.

Fig 1—(*a*) Line-of-sight velocities on the sky. The contour interval is 20 km s⁻¹. Axes are in arcsecs. Solid lines are at velocities greater than the systemic velocity and dotted lines are at velocities lower than the systemic velocity; (*b*) ^{12}CO(3-2) observations are the solid lines; the model is dotted. Each plot is a spectrum. Spectra are 12″ apart on the sky. The beam is 20″ FWHM. Velocities are in km s⁻¹ VLSR.

ILR. The rotation curve without an ILR did not fit the data which suggests that M82 has a stellar bulge. Since the disk in M82 is close to edge-on we had to model the vertical structure of the disk. We find that the wide profile shapes and the twist in the velocity field can be accounted for by bar induced non-circular molecular gas motions. The gas has a low velocity dispersion of $\sigma_g < 30$ km s⁻¹ implying that the high SF rate is due to the short time scales of gravitational instability (Larson 1987).

REFERENCES

Barnes, J. E., and Herquist, L. E. 1991, *Ap. J. Lett.*, **370**, L65.

Brouillet, N., Combes, F., and Baudry, A. 1991, IAU Symp. 146 on *Dynamics of Galaxies and their Molecular Cloud Distributions*, eds. F. Combes, and F. Casoli (Dordrecht: Kluwer), p. 347.

Carlstrom, J. E. 1988, *Ph. D.* thesis, University of California, Berkeley.

Larson, R. 1987, in *Starbursts and Galaxy Evolution*, eds. T. Thuan, T. Montmerle, and T. Tran Thahn Van (Paris: Editions Frontieres), p. 467.

Sofue, Y., Reuter, H. P., Krause, M., Wielebinski, R., and Nakai, N. 1992, *Ap. J.*, **395**, 126.

Telesco, C. M., Campins, H., Joy, M., Dietz, K., and Decher, R. 1991, *Ap. J.*, 369, 135.

Weliachew, L., Fomalont, E. B., and Grieson, E. W. 1984, *Astr. Ap.*, 137, 335.

Star Complexes and Evolution of Disk Galaxies

Arthur D. Chernin and Yury N. Efremov

Sternberg Astronomical Institute, Moscow University, 119899 Moscow, Russia

ABSTRACT

Major features of star complexes as basic "building blocks" of disk galaxies are presented and their role in the galactic formation and evolution is discussed.

Star complexes are the largest aggregates of individual stars, stellar associations and clusters, together with interstellar gas clouds. They have been recently recognized as universal and ubiquitous "building blocks" of disk galaxies and the contemporary star formation is concentrated within the complexes (Efremov 1979, 1988, 1989).

A typical complex is a kpc-size system containing about 10^7 $M_\odot$, mostly in H I and H_2 clouds. Young star clusters and OB-associations are strongly confined within the star complex and usually fill in only a small part of its volume. Star formation lasts 50–70 Myr in a common complex, and it is probable that after 100 Myr the complexes are disrupted. Super-associations may be considered as a rare kind of complexes with violent star formation over the whole complex: there are more than 200 star complexes in M31 and only one super-association.

The well known Gould's Belt, the Local system, is the only star complex whose structure and dynamics we are able to study. It contains 8 small clusters, 3 OB associations, at least 3 Cepheides, a dozen other late supergiants, and surely a lot of main sequence stars. The oldest stars and clusters in the Local system complex are about 60 Myr old. Tayler *et al.* (1987) estimated that the total stellar, H I, and molecular masses of the complex are 0.5, 1.0, and 0.4 $\times$ 10^6 $M_\odot$, respectively. Other complexes in the Galaxy are probably more massive, especially those concentrated within the Car – Sgr arm (Alfaro *et al.* 1992).

The great majority of the high luminosity stars and about 90% of the associations and young clusters in our Galaxy, M31, M33 and most probably elsewhere are formed inside complexes; they evidently are the general features of all disk galaxies though not always easily detected (Efremov 1989).

The first detailed discussion of the connection between the star complexes and the giant gas clouds was made by Elmegreen and Elmegreen (1983). They suggest that the galactic disks fragment spontaneously into the superclouds of 10^6 $M_\odot$, which subsequently fragment into several molecular clouds forming OB-associations and clusters. Observations indicate that complexes are generally confined to the main spiral arms where the formation of superclouds is much easier (Elmegreen 1987).

Taking into account these data and the above considerations, we propose a scenario for the formation and evolution of disk galaxies based on the concepts of star complexes as fundamental building blocks of galactic structure. We assume that some 15–18 Gyr ago all the barionic matter of a Milky Way-type galaxy was dispersed over a space volume of about 100 kpc across in the form of a rarified gas cloud. The dark non-barionic matter coexisted within the same volume. Both matter components formed a proto-galaxy which separated out from the general cosmological background at $z = 3 - 10$. The first stage of the evolution of the protogalaxy was the free-fall contraction of the central denser region of the gaseous cloud which could involve 10 – 30% of its mass. The collapse and fragmentation of this region would lead to the formation of the spherical (bulge/halo) subsystem of the galaxy.

The rest (and most) of the cloud contracted as well, but on a longer time scale, about 4–6 Gyr, which was determined by the cooling processes. The cooling time was longer than the free-fall time for the less dense outer part of the protogalaxy. Initial rotation of the gas cloud led to the formation of a flat rotating disk around the bulge. Subsequently, the disk fragmented into the huge proto-complexes: because the time scale of formation of superclouds — proto-complexes is much shorter than for protogalaxy disk formation, the complexes are expected to be the first structures appearing in the disk. Because their masses have been at least tenfold larger than those of giant molecular clouds, the complexes have to determine the collective processes in the galactic disk, such as the ones leading to the increase of stellar velocity dispersion and development of the major elements of large-scale galactic structure, and first of all the spiral arms (Chernin and Efremov 1993).

Some features of this scenario seem to be in agrement with recent observational studies of the youngest objects in the universe (Isotov *et al.* 1990; Dressler *et al.* 1993).

REFERENCES

Alfaro, E. J., Cabrera-Cano, J., and Delgado, A. J. 1992, it Ap. J., **399**, 576.

Chernin, A. D., Efremov, Yu. N. 1993, *Astron. Nachr.*, **314**, 7.

Dressler, A., *et al.* 1993, *Ap. J. Lett.*, **404**, L45.

Efremov, Yu. N. 1988, *Stellar complexes*, (London: Harwood).

Efremov, Yu. N. 1989, *Centers of Star Formation in Galaxies*, (Moscow: Nauka), in Russian.

Elmegreen, B. G., and Elmegreen D. M. 1983, *Mon. Not. R. Astr. Soc.*, **203**, 31.

Elmegreen, B. G. 1987, *Ap. J.*, **312**, 626.

Isotov, V. A., *et al.* 1990, *Sov. Astron.*, **67**, 682.

Taylor, D. K., *et al.* 1987, *Ap. J.*, **315**, 104.

Interstellar Gas Flows in the Gravitational Potential Well of Density Waves

Leonid Marochnik, Vladimir Berman, and Gwyn Fireman

Computer Sciences Corporation, Science Programs

1 OVERVIEW

One–dimensional numerical simulations of a large–scale flow of interstellar gas in galactic spiral density waves were produced using the approach described by Roberts (1969), Shu *et al.* (1972, 1973), Woodward (1975), Marochnik *et al.* (1983), Lubow *et al.* (1986), and Berman *et al.* (1990).

We have employed a two–phase model for the initial state of the interstellar medium (ISM). Thermal processes and self–gravitation of the gas were taken into account. Magnetic fields were not included in this version of the mathematical model. A discussion on the role of the magnetic field and status of the two–phase model of ISM versus Supernova–dominated model can be found in Marochnik and Suchkov (1991).

We have used the Schmidt model of the galaxy and the following parameters of its spiral structure: $\Omega_p = 23$, 24, and 13.5 km s^{-1} kpc^{-1} (the angular velocity of the spiral pattern), $F = 10\%$ (the gravitational field of a spiral arm).

The right-hand side of the energy equation is $L = n(n\Lambda - \Gamma)$, where $n\Lambda(T)$ and Γ are the standard cooling and heating functions, respectively (Penston 1970).

Numerical simulations of the interstellar gas flow have been produced for a number of galactocentric distances, $R = 4, 5, 6, 8, 10$ kpc and two different values of the initial density of the interstellar gas, $n_0 = 0.05$ and 0.5 cm^{-3}. Figure 1 demonstrates one of the examples simulated.

2 CONCLUSION

Numerical simulations show that a large–scale flow of the interstellar gas in spiral density waves can produce either a small–scale (on the order of a few pc) cold dense phase, formation of multiphase flows, galactic shocks, or shockless flows, depending on the characteristics of the galactic model, the galactic spiral structure's model, and the parameters of the interstellar medium.

This study demonstrates the necessity of performing fully three-dimensional numerical simulations of a large–scale interstellar gas flow in the gravitational well of galactic spiral arms, with magnetic fields, thermal processes, and self–gravitation

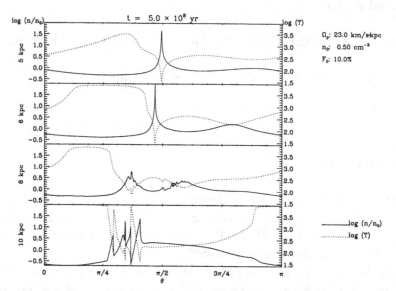

Fig. 1—Gas density and temperature behavior under the flow in the gravitational potential well of a spiral arm. n_0 = initial density; n = modelled density; T = temperature (K); R = galactocentric distances at 5, 6, 8, and 10 kpc.

taken into account.

This work has been supported by the High Performance Computing Laboratory of Computer Sciences Corporation. The authors would like to express their gratitude to H. Rudd for assistance in preparing this paper.

REFERENCES

Berman, V. G., Marochnik, L. S., Mishuzov, Y. N., and Suchkov, A. A. 1990, in *Galactic and Intergalactic Magnetic Fields*, ed. R. Beck *et al.*, p. 159.

Lubow, S. H., Balbus, S. A., and Cowie, L. L. 1986, *Ap. J.*, **309**, 496.

Marochnik, L. S. and Suchkov, A. A. 1991, in *Evolutions of Interstellar Matter and Dynamics of Galaxies*, eds. J. Palous *et al.* (New York: Cambridge University Press), p. 234.

Marochnik, L. S., Berman, V. G., Mishuzov, Y. N., and Suchkov, A. A. 1983, *Ap. Space Sci.*, **89**, 177.

Penston, M. V. 1970, *Ap. J.*, **162**, 771.

Roberts, W. W. 1969, *Ap. J.*, **158**, 123.

Shu, F. H., Milione, V., Gebel, W., Yuan, C., and Cuzzi, J. N. 1972, *Ap. J.*, **173**, 557.

Shu, F. H., Milione, V., and Roberts, W. W. 1973, *Ap. J.*, **183**, 819.

Woodward, P. R. 1975, *Ap. J.*, **195**, 61.

Massive Gas Rings in the Nuclei of Barred Spiral Galaxies

M. Shaw[1], F. Combes[2], D. Axon[3], and G. Wright[4]

[1]Department of Physics, University of Sheffield, United Kingdom
[2]D. E. M. I. R. M., Observatoire de Meudon, France
[3]N. R. A. L., Jodrell Bank, United Kingdom
[4]Joint Astronomy Center, Hawaii, U. S. A.

ABSTRACT

Near–IR photometry within the nuclei of 3 barred spiral galaxies reveals 63–100° isophote twists between the innermost regions and the PA's of the major bars. New N-body simulations indicate that such twists can be best described as the response of the stars to the influence on the potential of the gas within circumnuclear rings.

1 INTRODUCTION AND OBSERVATIONS

The clearest consequence of barred potentials is manifest in those galaxies undergoing enhanced nuclear star formation, particularly those possessing circumnuclear rings (CNRs). These rings probably form in the region of the inner Lindblad resonances (ILRs), the accumulation of molecular gas acting to enhance star formation. Isophote twists have been noted previously in the centres of barred potentials, both in the stars (Jarvis *et al.* 1988) and the gas (Devereux *et al.* 1992), and a link has been suggested between these twists and CNR's.

We have obtained near–IR (1.2–2.2 μm) photometry within the central regions of the barred spirals NGC 1097, NGC 4736 and NGC 5728. Twists of 63°–100° are observed between the innermost isophotes and the known position angles of the bar/lens components, corresponding to the most unambiguous and extreme such detections in any barred spiral later than S0/a. In NGC 1097 and NGC 5728 such twists are immediately interior to the radii of known CNRs. We have considered three possibilities by which such twists could arise.

In weak–intermediate strength bars, there exist two ILRs. The (stable) x_2 orbital family dominate between the ILRs and are orthogonal to the main bar. Unfortunately, the observed isophote twists do not merely reflect motions of stars along the x_2 orbits since projection effects fail to account for the twists seen in all cases. Furthermore, no purely stellar N-body simulations of bars conducted to date yield stable twists due to rapid heating of the stars within the barred potential.

In the presence of a central massive object, gas interior to the ILR can form a disc–like structure. This structure can decouple from the main bar if it has sufficient mass, and can develop independent bar instabilities (*e.g.* Shlosman *et al.* 1989) —

the angular speeds of the two bars differing by a factor 3–10. However, no kinematic evidence exists for any rapidly rotating component in NGC 5728 (Wagner and Appenzeller 1988). Indeed, in our N–body simulations, the pattern speed of the twisted isophotal structure is *identical* to that of the main bar.

As there exists no evidence for current or recent star formation interior to the radii of the CNRs in NGC 1097 and NGC 5728, or within the central $\sim 1'$ of NGC 4736, the presence of M supergiants formed from radial gas flows of material from the CNR similarly fails to account for the observed twists.

2 N–BODY SIMULATIONS, CONCLUSIONS AND IMPLICATIONS

We have modelled each of the galaxies concerned using a 2D FFT scheme with a polar grid providing an unprecedented nuclear resolution of 70 pc (see Combes — these proceedings — for further details). In our simulations, clear gaseous CNRs are formed, the orientations of which are misaligned with that of the main stellar bars. This almost certainly reflects gas motions along the x_2 orbits, the non–orthogonality resulting from cloud–cloud collisions. This twist in the gas alone is unlikely to account for the near–IR morphology, but the stars in such simulations also show an elongated morphology aligned with that of the gaseous CNR.

We therefore believe the observations identify the response of the stars to the potential of the gas in the CNR. The gas follows the x_2 orbits, and has sufficient mass to amplify the tendency of the stars to do the same. The gas suppresses any stellar heating which may take place: our purely stellar dynamical simulations show no evidence for twisted isophotes. The twists are essentially a gas dynamical phenomenon so we rarely expect to see exactly orthogonal isophotes. Since the presence of 2 ILRs is a direct result of the central mass concentration, we expect twisted isophotes to predominate in early–type barred spirals. The gas component loses angular momentum to the bar potential, suggesting a complete depopulation of gas between the two ILRs over $\sim$ few $\times 10^9$ yrs. Subsequently, the stars should dynamically "heat up" in the barred potential, slowly reducing any evidence for isophote twists.

As $\sim 80\%$ of SBab to SBc galaxies may possess CNRs (Kormendy 1979), more extensive near–IR imaging surveys should yield many other objects showing similar morphology.

REFERENCES

Devereux, N., Kenney, J., and Young, J. 1992, *A. J.*, **103**, 784.
Jarvis, B., Dubath, P. Martinet, L., and Bacon, R. 1988, *Astr. Ap. Suppl.*, **74**, 513.
Kormendy, J. 1979, *Ap. J.*, **227**, 714.
Shlosman, I., Frank, J., and Begelman, M. C. 1989, *Nature*, **338**, 45.
Wagner, S., and Appenzeller, I. 1988, *Astr. Ap.*, **197**, 75.

Pattern Speeds and Time Evolution in Ringed Galaxies from Observational and Simulational Databases

G. Byrd[1], P. Rautiainen[2], H. Salo[2], R. Buta[1], and D. Crocker[1]

[1]University of Alabama, Tuscaloosa, AL 35487, U. S. A.
[2]University of Oulu, Oulu, Finland

ABSTRACT

We have made new N-body simulations of ringed, barred spiral galaxies that extend and improve upon the previous work of M. P. Schwarz. Using a comprehensive database of multicolor images of a large sample of ringed galaxies, we are able to morphologically "match" galaxies to simulation frames that differ in bar pattern speed and time step. From these matches, we can reliably identify both low and high pattern speed galaxies.

1 INTRODUCTION

It is now well-established that the rings commonly observed in disk galaxies are probably caused by orbital resonances with a bar or bar-like potential in the disk (see Buta and Crocker 1991, *A. J.*, **102**, 1715, hereafter BC; 1993, *A. J.*, **105**, 1344, for recent discussions). In a pioneering N-body investigation, Schwarz (1981, *Ap. J.*, **247**, 77) simulated the behavior of gas clouds in two dimensional rotating bar potentials, and identified the outer Lindblad resonance (OLR), the inner 4:1 resonance (UHR), and the inner Lindblad resonance (ILR) as the resonances responsible for the features known as outer, inner, and nuclear rings, respectively. To better understand ring formation and evolution, we have carried out new simulations at many more bar pattern speeds (Ω_p) than Schwarz with more particles and an improved treatment of cloud collisions. We also take advantage of a recently acquired database of BVI CCD images of 150 ringed galaxies for comparison.

Our simulations can be divided into three different Ω_p "Domains". In Domain 1, Ω_p allows an OLR but is high enough to disallow a UHR and an ILR; in Domain 2, Ω_p allows an OLR and a UHR but disallows an ILR; and in Domain 3, Ω_p is low enough that OLR, UHR, and ILR all exist. Figure 1 illustrates three galaxies which fit into these domains and their matching simulation frames. ESO 509−98 (top, Domain 1) includes only a bar and an R_1R_2' outer pseudoring (see BC). ESO 575−47 (middle, Domain 2), has prominent inner and outer pseudorings but no nuclear ring. Finally, NGC 6782 (bottom, Domain 3) shows all three ring types. The close matches demonstrate that we can reliably identify low and high pattern speed galaxies.

Supported by the Finnish Academy of Sciences (GB) and NSF grants RII8996152 (EPSCoR) and AST 9014137 to the University of Alabama.

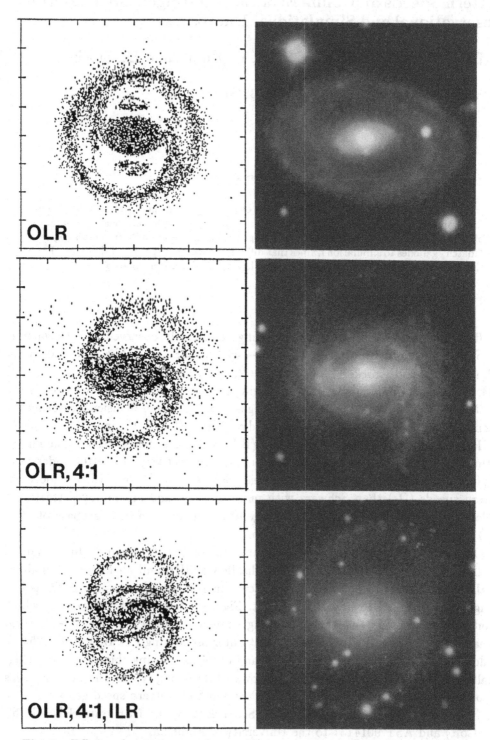

Fig. 1—Effects of pattern speed on galaxy morphology (simulations at left)

Large–Scale Gravitational Instability and Galactic Viscosity

Susanne v. Linden[1], Harald Lesch[1], and Francoise Combes[2]

[1]Max-Planck Institut für Radioastronomie, Bonn, Germany
[2]Observatoire de Meudon, Paris, France

ABSTRACT

The large-scale dynamics of the interstellar medium in spiral galaxies is significantly determined by molecular cloud physics. Inelastic cloud-cloud collisions, coupled with gravitational instabilities generated by the gas cooling, give rise to a net flux of angular momentum which is equivalent to a (gravitational) "viscosity". By using N–body model calculations of a self-gravitating disk we study the efficiency of collisions coupled with gravitational waves, and its importance for large-scale mass accretion in disk galaxies.

1 THE IDEA

The mass of the interstellar medium of galaxies is essentially included in an ensemble of clouds, whose physics on a small-scale dominates the dynamical behaviour. In a non-axisymmetric potential, the gas clouds are submitted to gravitational torques, which transfer angular momentum outwards, inducing an accumulation of gas in the central regions. Combes and Gèrin (1985) developed a collision model for the interstellar medium, including a mass spectrum of molecular clouds. If on the other hand, differential rotation is sufficient, the gas is submitted to viscous torques, which also accumulates gas in the central regions (Lesch *et al.* 1990). It is the aim of our investigation to connect these different approaches, by identifying the angular momentum transport in a non-axisymmetric potential with a viscosity. A study of the exchange of angular momentum and energy between clouds will provide an understanding of the nature of the gas "viscosity".

To understand this scenario we use self-consistent 2D simulations with gas (19,048 particles) and stars (38,000 particles) (Linden *et al.* 1993). With the FFT method on 128×128 grids self-gravity is computed. Clouds are simulated as spheres of constant density and they have masses between 10^4 and 10^7 $M_\odot$ ($M_{gas} = 6 \times 10^9$ $M_\odot$; 6% M_{tot}). Small clouds grow by collisions until they become giant molecular clouds, which have a life-time of 4×10^7 yrs (Casoli and Combes 1982).

Initially, we distribute the mass of the galaxy as 7×10^{10} $M_\odot$ for the halo and 3×10^{10} $M_\odot$ for the disk. The scale length for the disk is 3 kpc.

After a few time steps a spiral structure develops in the gas. Since the gaseous component exhibits considerably smaller velocity fluctuations than the stellar component, the self-regulation mechanism leads, via collisions, to a Toomre-Q-parameter of ~ 1. Thus, the gas disk is subject to a gravitational instability in the form of the $m = 2$ spiral wave. These waves generate a spiral instability in the stellar component and after a few more time steps the stellar disk also shows a spiral structure. We note that a ring develops at the inner Lindblad resonance ($R = 5$ kpc). This ring appears and disappears for $10 < t/1.2 \times 10^7$ yrs. < 70 on a time scale of $\sim 5 \times 10^7$ yr. Afterwards the ring is stable.

To understand the physical nature of the viscosity we concentrate on the "microscopic" physics of the cloud-cloud-interactions:

- mass transfer from one cloud to another depending on the relative velocity
- elasticity of collisions
- exchange of angular momentum
- influence of large scale perturbations like bars and oval distortions on the cloud dynamics

1.1 Results up to Now

An estimate for the viscosity ν produced by a non-axisymmetric perturbation with a pattern speed Ω_p and m arms is given by (Norman 1987): $\nu \simeq \epsilon R v_R$ with $v_R \simeq m^2(\Omega_p/\Omega)^2(R/\tau_{\text{diss}})(\Omega - \Omega_p)/\Omega$, where $\tau_{\text{diss}} \simeq 2\pi/\Omega\alpha$ presents the cloud-cloud collision time (v_R is the radial velocity). For a rough estimate we assume that a cloud collides during one rotation period α—times. Our simulation give $\alpha = 5$.

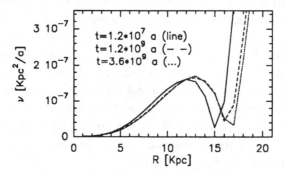

The Figure shows the calculated viscosity ν during 3 different time steps ($m = 2$, $\epsilon = 0.1$). The values of ν are just compatible with observed gas flows towards galactic center regions.

REFERENCES

Casoli, F., and Combes, F. 1982, *Astr. Ap.*, **110**, 287.

Combes, F., and Gèrin, M. 1985, *Astr. Ap.*, **150**, 327.

Lesch, H., *et al.* 1990, *Mon. Not. R. Astr. Soc.*, **225**, 607.

Linden von, S., Lesch, H., and Combes, F. 1993, in *Theory of Accretion Disks – 2*, eds. W. J. Duschl, *et al.* (Dordrecht: Kluwer), in press.

Norman, C. 1987, in *Galactic and Extragalactic Star Formation*, eds. R. . Pudritz and M. Fich (Dordrecht: Kluwer), p. 495.

H I and H₂ in Luminous Interacting Galaxies

N. Scoville[1], J. E. Hibbard[2], M. S. Yun[1], and J. H. van Gorkom[2]

[1]Astronomy Department, California Institute of Technology
[2]Astronomy Department, Columbia University

ABSTRACT

The interstellar gas plays a fundamental role in the interacting galaxy systems: it is dissipative and thus responds irreversibly to strong dynamical perturbations; it is also the active component which fuels starbursts and may fuel AGNs in the central regions of the galaxies. In this article we review new high resolution aperture synthesis observations of the atomic and molecular gas in luminous interacting galaxies.

Due to the extended nature of the atomic gas in disk galaxies, the H I observations bear most critically on the long range tidal perburbations and therefore serve to elucidate the interaction history. Here we summarize the H I observations of six systems: the M81/M82 group and five interacting pairs with long tidal tails. In all of these systems the impact of geometry is evident from the spatial and kinematic structure of the atomic gas within the extended tidal features. The six systems may be crudely placed in a temporal sequence for galactic merging — in the early stages large amounts of H I still exist within the galactic disks and the star formation is spread throughout the disks as evidenced by Hα emission while in the final stages the H I is almost entirely contained within the tidal features beyond the merger body and the molecular gas is mostly found in the central remnant.

Molecular line observations have largely concentrated on the intermediate evolutionary phase in which the galaxies become strongly emitting in the far infrared. Molecular line surveys of luminous infrared bright galaxies sampled by the *IRAS* survey have shown these galaxies to be extraordinarily rich in molecular gas as evidenced by the CO emission. High resolution aperture synthesis data are presented for two of these systems — also representing a temporal sequence in the intermediate phases of merger induced activity. In each case the molecular gas is found to be much more highly concentrated than in normal galaxies, probably a result of the viscous dissipation of energy and angular momentum in the ISM as a result of the strong dynamical perturbations. Within the central few hundred parsecs as much as 10^{10} $M_\odot$ of H₂ is found. Such concentrations are surely the source of nuclear starbursts, possibly also rejuvinating nuclear nonthermal activity. It is suggested that interaction induced activity in the interstellar gas of galaxies may play a fundamental role in galactic evolution – leading to star formation rates elevated by at least an order of magnitude for a period $\sim 2 \times 10^8$ years. Since such activity is presently occurring in approximately 2% of all spiral galaxies, it is clear that galactic interactions can be critical to both the long term stellar populations of galaxies and to the most luminous, energetic phases for galactic evolution.

1 INTRODUCTION

The important role of interstellar gas in galactic interactions has been recognized only in the last few years. It has long been suggested that interaction induced gas

191

inflow may play an important role in the enhanced central activity (see discussions by Kennicutt and Heckman, these proceedings), but the dynamical importance of interstellar gas in galactic interactions has been recognized only recently. The critical difference between the gas and the stars is that the gas clouds have a relatively short mean free path and are highly dissipative. The gas is also the medium from which luminous young stars form. The atomic hydrogen in the extended tidal features provides us with a unique record of the galactic encounter from its spatial distribution and kinematics.

The last few years have seen an explosive growth in high resolution observations of both the low density atomic and high density molecular (via CO) phases of the ISM in interacting galaxy systems. Much of the interest in these systems was spurred by follow-up studies from the *IRAS* survey revealing a strong correlation between infrared luminous galaxies ($L_{\rm IR} > 10^{10}$ $L_\odot$), presumably undergoing bursts of star formation, and optical morphologies suggesting a past or ongoing galactic interaction. Under the assumption that most of the luminosity seen in the far infrared originates as UV and optical radiation from young stars which has been absorbed and reradiated by dust, the far infrared luminosity provides an indication of the present day star formation rates, albeit with uncertainties related to the form of the stellar inital mass function. Recent high resolution near infrared imaging has pinpointed the centers of activity — most often in the nuclei of the merging galaxies. Relative to non-interacting galaxies of similar interstellar gas content, the inferred star formation rates are elevated by an order of magnitude and occasionally by two orders of magnitude.

The importance of understanding interacting galaxy systems arises from the dramatic effect these interactions may play in determining large scale galactic structure and in resuscitating a most luminous phase in galactic evolution. Mergers were probably more frequent in the past and they might be responsible for much of the present structure of galaxies. In this sense the study of galactic mergers can shed light on protogalactic evolution and hierarchical formation of galaxies (*e.g.* White and Rees 1978; Peebles 1980).

In this review we present HI and H_2 data for a sample of interacting systems which range in sequence of elevated activity and probably also temporal evolution of galactic merging. The HI data often reveal extended tidal features not at all apparent in optical images while the CO observations reveal massive concentrations of gas which are presumably the sites of interaction-induced starbursts. As the interaction progresses the atomic gas resides predominantly in the tidal features outside the remnant body, while the molecular gas appears concentrated in the nuclei of the merging systems. Using canonical conversion factors from CO to H_2, the molecular gas often accounts for more than 30% of the indicated dynamical mass in the central regions. Since this is much higher than the mass fractions found in the presumed progenitors ($< 10\%$), it is a strong indication that gaseous dissipation has played a role, and perhaps a dominant one, during the merger (Kormendy and Sanders 1992).

2 H I — THE INTERACTION RECORD

Neutral hydrogen studies are particularly useful for disentangling the geometry of interactions involving disk galaxies due to two facts: (1) the tidal features are drawn primarily from the outer regions of the disks (*cf.* Toomre and Toomre 1972) which are rich in atomic gas (Wevers *et al.* 1986); and (2) this outer gas has not undergone more than 1–2 rotations since the encounter started and therefore retains spatial and kinematic signatures of the encounter (Toomre and Toomre 1972).

In this section we describe the observational results for six systems: the M81/M82 group, Arp 295, NGC 4676 (the "Mice"), NGC 520/UGC 957, NGC 3921, and NGC 7252 (the "Atoms for Peace" galaxy). For the M81/M82 system, a numerical model has been constructed for the encounter. For the other systems, detailed models have not yet been generated, but the signature that the interaction has left on the outer gas is clearly seen.

2.1 The M81/M82/NGC 3077 Group

At a distance of only 3.3 Mpc the M81/M82/NGC 3077 group is the closest of those studied at high resolution. It differs from those discussed later in that it is dominated by a single massive galaxy, M81, with two smaller companions. The entire group has recently been mapped with the *VLA* D array at 60″ resolution by Yun, Ho, and Lo (1993a), and the mosaic H I image of twelve fields is shown in Figure 1. For comparison at the same scale an optical photographic image is reproduced in Figure 1b. The comparison most clearly demonstrates the unique ability of the H I emission to trace extended tidal features well beyond the optical disks. The total atomic hydrogen mass in the group is approximatley 6×10^9 M$_\odot$ with 60% contained in the disk of M81 and 20% concentrated near NGC 3077 to the southeast. The remainder of the H I, approximately 10^9 M$_\odot$, is contained in the extended spiral/tidal arm reaching from M81 through NGC 3077 and up towards M82 to the north.

Yun (1992) modeled these data by tracing the response of the disk particles under the influence of the potentials representing three orbiting galaxies. As shown in Figure 2, the hyperbolic passage of M82 from the south and the parabolic passage of NGC 3077 from the northeast result in tidally driven spiral arms in M81 and severe disruption of the disks of M82 and NGC 3077. Although most of the H I seen outside the disks of the galaxies probably originated from M81, one exception is the apparent tidal tail extending from NGC 3077 to the northwest, tracing the orbit of M82. This feature would be hard to understand unless the gas has originated from NGC 3077, and thus all three galaxies probably had substantial amounts of gas prior to the interaction.

The major effects of the interactions among the three galaxies in this group are: the stimulation of the strong spiral density wave in the disk of M81; a complete disruption of the disk of NGC 3077; and the central concentration of gas in the disk

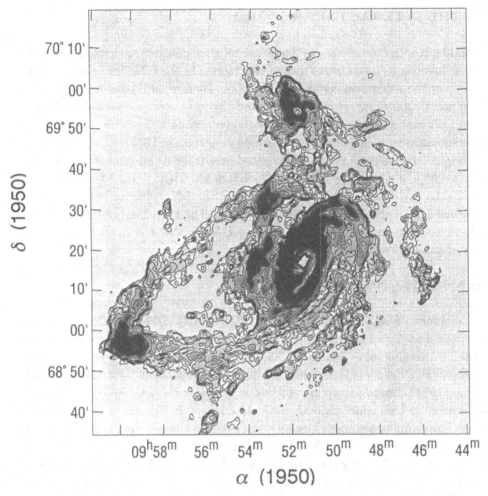

Fig. 1*a*—A *VLA* D array map of the integrated H I emission in the M81 group is shown in grey scale. In the region mapped, the total H I mass is 6×10^9 M$_\odot$, 60% of which is contained in the disk of M81.

of M82. With regard to the strong spiral pattern in M81, it is interesting to note that the presence of four spiral arms is probably consistent with there having been two interactions (*i.e.* from both NGC 3077 and M82). The numerical simulation required a nearly simultaneous interaction involving all three galaxies to reproduce the tidal features found around NGC 3077.

The higher resolution H I observations of M82 using the *VLA* C array (Yun, Ho, and Lo 1993*b*) found an amorphous disk and a pair of prominent tidal tails extending out of the stellar disk, indicating a severe tidal disruption of its own H I disk rather than accreting gas captured from M81. The H I kinematics suggests the presence of a bar potential in M82, which may be responsible for channeling the gas into the nuclear region. The extraordinary effects on M82 are best seen in molecular line

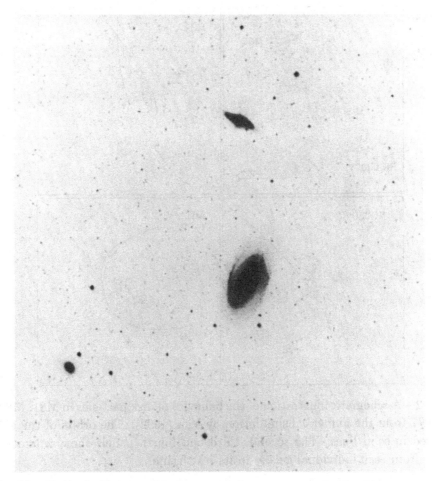

Fig. 1*b*—Optical photographic image of the same area in the M81 group shown in the Figure 1*a* H I image.

tracers and in the infrared. In this galaxy, the total molecular gas mass may be as much as 25% of the dynamical mass and the dense gas is highly concentrated in the center (*cf.* Young and Scoville 1984). The nucleus of M82 has been mapped at high resolution in CO, HCN and HCO⁺ by Lo *et al.* (1987), Nakai *et al.* (1987), and Carlstrom (1990). These studies reveal a torus of gas at radius $\sim$ 150 pc and mass $\sim 10^8$ M$_\odot$. The relatively strong emission from high dipole moment molecules such as HCO⁺ and HCN (Carlstrom 1990) suggests that the physical conditions in this gas are quite different from those in typical Galactic giant molecular clouds (GMCs). The derived kinetic temperatures are 50 K and the volume densities within the clouds are approximately 10^5 cm^{-3}. The fact that the bulk of the near infrared stellar light, presumably from relatively young giant stars, occurs *within* the torus has led to the suggestion that the starburst is progressing from the center outwards. Carlstrom

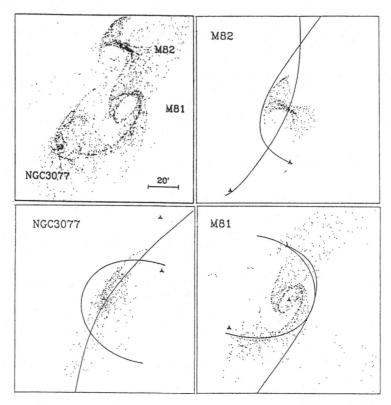

Fig. 2—A schematic illustration of the behavior of disk particles in M81, M82, and NGC 3077 from the numerical simulation by Yun (1992). The orbits of the galaxies are traced in solid lines. The second, third and fourth panels show separately the particles from each individual galaxy in its rest frame.

(1990) estimates from the 3 mm free-free emission that the UV production rate is 6×10^{53} sec^{-1}, corresponding to 3×10^5 O8 stars. This ionizing photon production rate deduced for the center of M82 (R < 500 pc) is in fact somewhat greater than that of the entire Milky Way. A recent analysis of the cosmic ray heating for the moleuclar gas in M82 by Suchkov, Allen, and Heckman (1993) suggests that elevated temperatures seen in the molecular gas can be understood if the cosmic ray density is approximately 300 times that of the Milky Way as would be expected in such a starburst.

2.2 H I in Strongly Interacting Systems

The systems studied at high resolution with the *VLA* by Hibbard *et al.* (1993a) and Hibbard and van Gorkom (in preparation) were chosen for the appearance of their extended tidal tails. Four of the five galaxies were taken from Toomre's (1977)

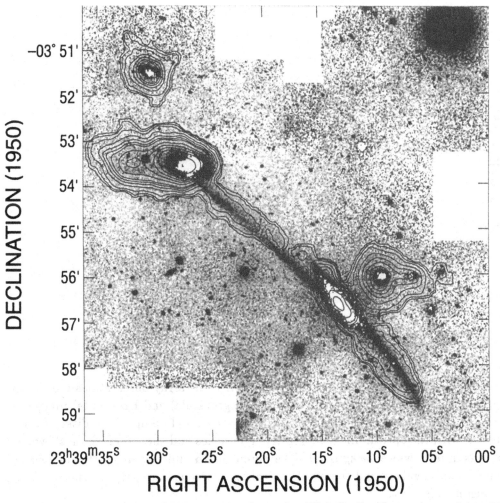

Fig. 3—The integrated H I intensity from the *VLA* C+D observations of Arp 295 is shown superposed as contours on a negative grayscale of the *KPNO* 2.1m *R*-band image with all pixels containing Hα emission set to white.

proposed sequence of approximately equal mass disk galaxies in progressive stages of merging. The goal of the study is to quantify the atomic gas distribution and star formation activity as one progresses in a sequence through more advanced interactions and merging.

Though they were selected purely on the basis of their optical appearance, all of these systems turned out to be IR luminous, with IR luminosities in the range of $10^{10} - 10^{11}$ L$_\odot$ ($H_0 = 75\,\mathrm{km\,s^{-1}\,Mpc^{-1}}$ is used throughout), and all show signs of central activity: all are radio continuum sources, and all but Arp 295B and NGC 7252 have optical spectra typical of LINERs (Keel *et al.* 1985; Dahari 1985).

Added to the beginning of the Toomre (1977) sequence is Arp 295, at a distance

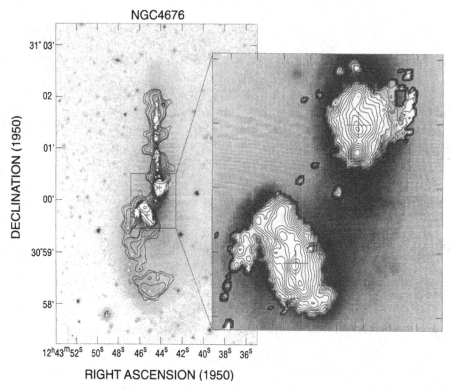

Fig. 4—H I intensity contours from the *VLA* C+D array observations are shown superposed on a *KPNO* 2.1m optical (negative greyscale) and Hα (white) images of the NGC 4676 system, "The Mice". The inset shows a close-up of the disks of both galaxies, with the optical image in negative greyscales and the Hα emission shown as contours upon a white background. The plume of Hα emission along the minor axis of the northern galaxy and the offset optical and gaseous bar of the southern galaxy are seen in this inset.

of 88 Mpc. In Figure 3 the integrated H I intensity is shown contoured on a negative greyscale of the *R*-band optical light with white pixels indicating Hα emission. The system has both a bridge and a tail indicating at least one pass of the two galaxies about each other, but both disks are still distinct indicating that the interaction is not too far advanced. The H I observations reveal that this system is part of a loose group, with nine smaller galaxies detected in H I, three of which are seen in Figure 3.

Already at this relatively early stage of interaction we find approximately 30% of the total H I ($M_{\rm tot,HI} = 3 \times 10^{10}$ M$_\odot$) outside of the optical bodies of the galaxies. The bridge extending approximately 120 kpc between the two galaxies is continuous in the optical continuum from stars but is broken in H I. The Hα emission is widely distributed through both disks and there is no evidence of extensive star formation occurring within the tidal features. Plumes of Hα emission are also seen along the minor axis of the northern disk galaxy.

The next system in the sequence, NGC 4676 (the "Mice"), is at approximately the same distance as Arp 295. The optical continuum, Hα, and neutral hydrogen emission are shown in Figure 4. In Figure 4*a*, extended tidal tails are seen on both sides of the two galaxies and a luminous bridge of material flows between them. Approximately 65% of the total HI mass (9×10^9 M$_\odot$) appears outside of the optical disks. The Hα is widely distributed over both galactic disks and tidal tails and along the minor axis of the northern companion. The details of the disks are seen in Figure 4*b*, with Hα shown contoured on a white background, and the optical light in negative greyscales. This Figure shows very clearly the Hα along the minor axis of northern galaxy, as well as a bar in the southern galaxy at approximately PA∼ 30°, both in the gas and the stellar light. The gaseous bar leads the stellar bar as seen in the numerical calculations of Barnes and Hernquist (1991) who point out that the offset stellar bar will torque the gas, leading it to lose its angular momentum.

The system consisting of NGC 520 and UGC 957 is at a distance of approximately 30 Mpc. This system is a more advanced merger than the previous systems, and a near infrared (2 μm) map clearly shows the presence of two nuclei (Stanford and Balcells 1991) — a primary one coinciding with the center of the dust lane, and a secondary one to the northwest, still within the main body of the NGC 520. At the location of the primary nucleus is a strong extended radio-continuum source (Condon *et al.* 1982) and a molecular gas disk (Sanders *et al.* 1988*a*). This gas concentration has a column density that is ten times that of typical Galactic GMCs, and comprises 24% of the indicated dynamical mass. The Hα emission is distributed throughout the main body of NGC 520: at both nuclei, along the minor axis of the primary nucleus, and within the inner tidal loops and tails.

UGC 957 is seen much further to the northwest (∼ 55 kpc) within the center of a long HI arm/tail, much like NGC 3077 in the M81 system (Fig. 1). But unlike in M81, the kinematics of the outermost HI in NGC 520 is dominated by rotational motion, rather than swinging tidal motions (*cf.* van Gorkom 1993). At present 60% of the HI (9×10^9 M$_\odot$) is outside the main optical disk, and much of this gas has no optical counterpart.

NGC 3921 (=Mkn 430) consists of a single nucleus with two tidal systems — one stellar and one gaseous — suggesting that this system may be the result of a merger of a spiral and a gas poor S0 galaxy. That the stellar light distribution is well described by an $r^{\frac{1}{4}}$ law (Stanford and Bushouse 1991) is an indication that the system is in a late stage of merging and has relaxed in the inner regions. This is supported by spectroscopy of the stellar light from the nucleus, which is consistent with a starburst approximately 10^9 years ago (Schweizer 1990). The presence of a large star-forming cluster in the southern tail suggests that self gravitating systems may well be formed in some of the tidal tails of interacting galaxies (*cf.* Hibbard and van Gorkom 1993). In NGC 3921, 75% of the HI (7×10^9 M$_\odot$) is located within the tidal features, and the kinematics suggests that perhaps all of it is outside of the main body of the remnant. The Hα emission in the body is more regular than in the earlier stages, and extensive ongoing star formation is seen along the southern tail even at this late stage.

NGC 7252 (the "Atoms-for-Peace" galaxy) is probably an advanced merger of two disk systems. The optical light distribution indicates an $r^{\frac{1}{4}}$ law, and the merged galaxy may be well along towards forming an eliptical galaxy core (Schweizer 1982). This conclusion is supported by the ageing burst population evident in the optical spectra (Schweizer 1982). NGC 7252 has also been mapped in CO by Wang, Schweizer, and Scoville 1992, who find that most of the molecular gas (80% of the total CO emission) is in the very central regions. This is in contrast to the 5×10^9 M$_\odot$ of atomic gas, all of which is contained in the tidal features (Hibbard *et al.* 1993*b*). Hα emission is seen not only in the nucleus but also filling in the gap between the CO and the innermost H I. Recent *ROSAT* observations reveal X-ray emission from NGC 7252, possibly indicating the presence of hot gas (Hibbard *et al.* 1993*b*).

For this system, detailed observations suggest that the atomic gas appears only in the tidal tails for two reasons: the outer H I is sent to these regions by the tidal forces experienced during the merger, and any inner H I is converted to other forms. The central molecular-gas coincides with the inner star-forming disk and accounts for 45% of the indicated dynamical mass, suggesting that dissipation has played a dominant role during the merger (Kormendy and Sanders 1992). The only ongoing star formation in the tails coincides with large concentrations of atomic gas and optical light, possibly being assembled out of the tidal material, and the blue colors seen throughout most of the tails indicate a young stellar population.

Briefly summarizing the qualitative changes as one goes from the less advanced to the more advanced interacting systems: 1) a larger fraction the H I appears outside the optical bodies in the extended tidal features and the most advanced mergers show virtually no H I left within the optical bodies; 2) the Hα is at first widely distributed through the galactic disks, and frequently along the minor axis of the galaxies. At intermediate stages there are signs of increased activity in the central regions, and star formation is seen more widely in the tidal tails. Finally, in the latest stages, the star formation is centrally concentrated towards the merger nucleus, and the tails have blue colors; 3) in the later stages, large concentrations of molecular gas are collected in the remnant body, which has relaxed into a $r^{\frac{1}{4}}$ law in its optical light and exhibits a post-starburst spectrum.

Apparently large quantities of both gas and stars are sent to large radii during such interactions, and large quantities of gas are brought into the central regions where it forms stars. The form that this material takes during its sojourn into intergalactic space and where it ends up after a few dynamical times will have important repercussions for the future evolution of the remnants. We find star formation taking place abundantly in the tails, and bound subsystems may be forming from the tidal debris as originally suggested by Zwicky (1956). The peak column densities of atomic hydrogen within the tidal tails averaged over the resolution are at the threshold column density at which molecular hydrogen usually dominates atomic hydrogen within the disks of normal spiral galaxies, and systematic searches for molecular emission within the tidal tails might prove fruitful.

3 MOLECULAR GAS — THE ACTIVE PHASE

In recent years a major direction of extragalactic molecular line research has been in studies of infrared bright galaxies highlighted as a result of the *IRAS* Survey. At the same time major ground based follow-up studies have been done in the optical, near infrared, and radio continuum (*e.g.* Lawrence *et al.* 1986). The *IRAS* bright galaxy survey (Soifer *et al.* 1987) included 238 objects in the northern hemisphere. Optical imaging of this sample reveals a strong correlation between the infrared luminosities of the galaxies and the existence of morphological features suggestive of a strong galactic interaction or merger (*e.g.* extended tidal tails or double nuclei; Sanders *et al.* 1988*b*). At the lowest luminosities, $10^{10} - 10^{11}$ L$_\odot$, the large majority of the *IRAS* galaxies appear isolated and undisturbed whereas at $10^{11} - 10^{12}$ L$_\odot$ approximately 30% show evidence of interactions and above 10^{12} L$_\odot$, $\geq 90\%$ appear to have undergone a recent strong interaction.

The ultimate source of the IR luminosity is likely trapped UV radiation from either a central starburst or AGN, which heats the dust that has been widely distributed during the encounter. Using *VLBI* radio observations, Norris *et al.* (1990) concluded that when an AGN was responsible for the central activity, it generally made itself known through other means, such as the optical emission lines (*cf.* Sanders *et al.* 1988*b*). Thus, some IR galaxies are probably powered by an AGN, but while others, especially the less luminous ones, by starbursts.

Molecular line observations of the most luminous *IRAS* galaxies are relevant for understanding of the effects of interactions on the molecular gas phase. In order to access the total molecular gas content of the most luminous *IRAS* galaxies, Sanders, Scoville, and Soifer (1991) obtained single dish measurements for a sample of 89 galaxies from the *IRAS* bright galaxy survey. The resulting H$_2$ masses, assuming the same CO emissivity per unit mass as for the Milky Way GMCs, are shown in Figure 5, plotted as a function of infrared luminosity. Also shown are the equivalent masses and luminosities of "normal" galaxies — the Milky Way and M51.

It is clear from Figure 5 that the most luminous *IRAS* galaxies are extremely rich in molecular gas with total H$_2$ masses in the range $2 - 50 \times 10^9$ M$_\odot$, that is 1–20 times the H$_2$ content of our Galaxy. It is also clear from Figure 5 that these galaxies are not simply scaled up versions of normal galaxies since the infrared luminosity to H$_2$ mass ratio is usually elevated considerably from the value of ~ 4 L$_\odot$/M$_\odot$ seen in normal galaxies.

On the basis of their optical morphologies the galaxies in the CO survey sample were placed in one of four possible categories: isolated spirals, spirals with a disturbed appearance, galactic pairs with non-overlapping disks, and advanced mergers. The latter category was distinguished from single isolated galaxies on the basis of extended

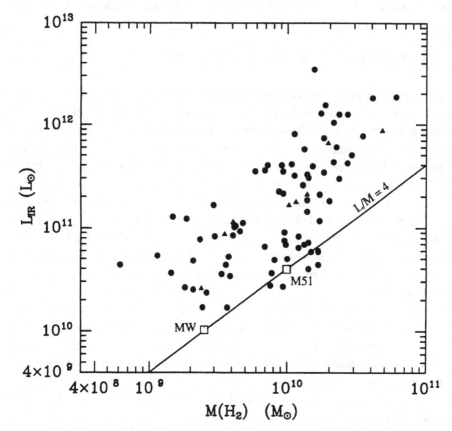

Fig. 5—The infrared luminosity ($\lambda = 8 - 1,000$ μm) is shown as a function of the global molecular gas mass for a sample of 89 bright infrared galaxies (Sanders, Scoville, and Soifer 1991; Tinney *et al.* 1990). Comparable data are also shown for the Milky Way and M51.

tidal tails in the optical or double nuclei seen in the near infrared (*e.g.* Carico *et al.* 1992). In Table 1 the mean infrared luminosity-to-molecular mass ratios are given as a function of infrared luminosity and interaction classification. The number of galaxies within each category is shown in parentheses. For the purpose of the present discussion the most important result is summarized in the last column. Averaging over IR luminosity it can be seen that both the galactic pairs and advanced merger categories have approximately an order of magnitude higher IR luminosity-to-gas mass ratio than the normal galaxies. It is not totally clear from this sample that the overall level of energetic activity is ten times higher in strongly interacting systems than in isolated galaxies since this sample is biased towards bright infrared galaxies. It will be important for a future study to investigate the IR luminosity-to-gas mass ratio in a sample which is not biased towards infrared bright galaxies.

A more surprising result of the molecular line observations in the interacting

TABLE 1. LUMINOSITY–TO–H$_2$ MASS RATIOS

L_{IR}/M_{H_2} (#Gal)	L_{IR} ($10^{11}L_\odot$)			
	$0.2-1$	$1-10$	$10-30$	all
Morphology				
Isolated	6 (15)	–	–	6 (15)
Disturbed	12 (13)	22 (11)	–	17 (24)
Non-overlapping Disks	20 (4)	30 (7)	75 (1)	30 (12)
Merger Pairs	32 (5)	34 (23)	74 (7)	42 (35)

$\}$ 39(47)

L_{IR} is the infrared luminosity at $\lambda = 8 - 1,000$ μm and M_{H_2} is the H$_2$ mass assuming a Galactic CO–to–H$_2$ conversion ratio of 3×10^{20} cm^{-2} (K km s^{-1})$^{-1}$. The units are $L_\odot M_\odot^{-1}$ and the number of objects in each group sampled by Sanders, Scoville, and Soifer (1990) and Tinney *et al.* (1990) is given in parenthesis.

galaxies highlighted as a result of the *IRAS* survey has been the finding that the molecular gas is not only more abundant, but also more concentrated than in normal galactic disks. To date, approximatley 30 of the high luminosity galaxies have been imaged with millimeter interferometers in the CO line. In Table 2, observational results are summarized for 19 systems observed with the Owens Valley array. The galaxies listed in Table 2 span the luminosity range $2 \times 10^{10} - 3 \times 10^{12}$ L$_\odot$($\lambda = 8 - 100$ μm). Ten galaxies, some of which are in the Owens Valley sample, have also been imaged using the Nobeyama array (Okumura *et al.* 1991). The significant results born out in both studies are: a significant fraction of the total molecular gas, usually greater than 50%, is typically concentrated in a nuclear source less than or equal to 1 kpc in radius; and secondly, the gas mass fractions in the center (M_{H_2}/M_{dyn}) often exceed 50% (as compared with typical values of 5% near the centers of normal galaxies). Below, we describe in detail two cases — NGC 4038/39 and Arp 220 — which serve to illustrate molecular gas morphology in intermediate and high luminosity interacting galaxies.

3.1 NGC 4038/39 — The Antennae

At a distance of 21 Mpc, the "antennae" is one of the best examples of a strong interaction in progress where excellent spatial resolution can be achieved. Stanford *et al.* (1990) imaged two 1′ fields in CO at 6″ resolution. The integrated intensity maps are shown in Figure 6, superposed on a 6500Å image of the inner disks. Three major CO concentrations are evident — two centered near the galaxy nuclei and a third northeast of NGC 4039, where the two galaxies overlap. To within the position uncertainties, the two nuclear CO peaks coincide with 2 μm peaks. The third CO complex, in the overlap region, coincides with a number of radio continuum, Hα, and 10μm emission knots (Hummel and van der Hulst 1986; Rubin, Ford, and D'Odorcio 1970; Wright *et al.* 1988). The CO concentration at the nucleus of the northern

TABLE 2. OWENS VALLEY OBSERVATIONS OF *IRAS* GALAXIES

Object	$\langle cz \rangle$ km s⁻¹	D Mpc	Radius " (kpc)	L_{IR} 10¹¹ $L_\odot$	Nuclear M_{H_2} 10⁹ $M_\odot$	M_{H_2}/M_{dyn}	CO morphology
Mrk 231	12660	174	~ 0.9 (0.7)	34.7	18	(5.6)	100% nulear source
IRAS 17208 − 0014	12850	171	< 1.5 (1.2)	27.0	56.0		100% nuclear source
Arp 220	5452	77	1 (0.3)	15.5	16.3	0.90	70% nuclear source
Mrk 273	11350	150	1.1 (0.8)	13.0	11.0	1.4	33% nulear source
VII Zw 31	16245	221	2.5 (2.7)	8.7	29.4		60% nuclear source
IRAS 10173 + 0828	14680	196	3.5 (3.3)	6.0	9.0		40% nuclear source
NGC 6240	7285	101	~ 1.5 (0.7)	6.6	20	0.9	60% nuclear source
IC 694 (Arp 299)	3030	42	1.3 (0.3)	4.1	3.6		triple source
VV114	6028	78	1.8 (0.7)	4.2	10.0		double source
NGC 1614	4847	62	2 (0.6)	4.0	6.0		30% nuclear source
IC 883 (Arp 193)	6941	98	1.7 (0.8)	4.0	18	0.7	70% nuclear source
Arp 55	11957	163	4 (3.2)	3.9	17.3	0.6	double source
NGC 1068	1137	18	1.5 (0.13)	1.5	4.5		ring+nuclear source
NGC 7469 (Arp 298)	4963	66	2.5 (0.8)	2.6	7.4	0.7	30% nuclear source
ZW 049.057	3900	52	1.5 (0.4)	1.7	4.0		40% nuclear source
NGC 828	5359	72	2.5 (0.9)	2.1	11.8	0.31	interacting pair
NGC 2146	838	21	4/13 (0.4)	1.2	4		
NGC 520 (Arp 157)	2261	29	2.5 (0.4)	0.6	3.2		interacting pair
NGC 3079	1137	24	3/7 (0.3)	0.7	5		
NGC 4038/39 (Arp 244)	1550	21	3.5 (0.4)	0.2	0.8	0.2	triple source

REFERENCES: (1) Scoville *et al.* (1989); (2) Planesas, Mirabel, and Sanders (1991); (3) Wang, Scoville, and Sanders (1991); (4) Sargent and Scoville (1991); (5) Sanders *et al.* (1988*a*); (6) Planesas, Scoville, and Myers (1991); (7) Young, Claussen, and Scoville (1988); (8) Young *et al.* (1988); (9) Stanford *et al.* (1990); (10) Scoville *et al.* (1991); (11) Brvant and Scoville (1993); (12) Yun and Scoville (1993)

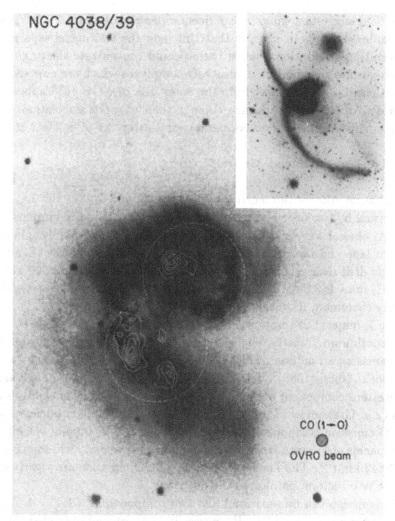

Fig. 6—CO emission contours are shown superposed on a 6500Å image of the NGC 4038/39 system. In the upper right a deep, larger scale image showing the tidal tails is also shown. Three massive concentrations of molecular gas are seen: at the centers of the two galaxies and at a possible interaction interface where the two galaxies overlap (Stanford *et al.* 1990).

galaxy, NGC 4038, is approximately 1 kpc in diameter and the integrated CO intensity suggests a molecular mass of approximately 8×10^8 M$_\odot$. At the NGC 4039 nucleus, an H$_2$ mass of 2×10^8 M$_\odot$ is found.

The massive gas concentration in the overlap region is probably the result of cloud-cloud collisions within the disks of the two colliding galaxies. Such collisions will be highly inelastic due to rapid cooling in the dense gas, and cancellation of angular momentum may result in the eventual accumulation of much of this gas near one of the galactic centers or in a merged nucleus. The N-body simulation of the interaction by Barnes (1988) suggests a time since pericenter passage of about 2×10^8 years, with the two disks brushing their edges at that time and the two nuclei separated by 20 kpc. The resulting gravitational perturbation could concentrate the originally more tenuous disk gas into the giant off-nuclear CO complexes which are now seen. Indeed the long northern tail pulled away from the outer disk of NGC 4039 appears to lead to the inner disk of NGC 4038, passing through the CO super association. Thus, gas in the disks might have been shocked, perhaps initiating the off-nuclear starburst.

3.2 Arp 220 – An Advanced Merger

Arp 220 has an infrared luminosity at $\lambda = 8 - 1,000$ μm of 1.5×10^{12} $L_\odot$, exceeding that in the visual by two orders of magnitude and placing it in the luminosity regime of quasars. At optical wavelengths, the galaxy appears approximately spherical with a central dust lane and faint tidal tails extending up to 70 kpc away (Sanders *et al.* 1988*b*). Single dish measurements show the CO emission spanning 900 km s^{-1} and the derived H_2 mass is 3.5×10^{10} $M_\odot$, approximately a factor of 15 greater than that of the Galaxy (Solomon, Radford, and Downes 1990).

In Figure 7, maps at $2''$ resolution (750 kpc) are shown from Scoville *et al.* (1991). The 3 mm continuum (mostly dust emission) and CO line emission are shown in the upper panels as a function of displacement coordinates from the cm-wave radio continuum peak (Norris 1988). Both the CO and 3 mm continuum peak are within $0.5''$ of the western component of the nucleus seen in high resolution near infrared and radio maps (*e.g.* Graham *et al.* 1990). In addition to the compact nuclear source, an extended CO emission component can be seen in maps with smaller velocity ranges. In the lower panels of Figure 7, the emission is shown for velocity windows centered at 5,330 and 5,642 km s^{-1}. The low level contours of CO emission are clearly extended along a NE–SW direction, parallel to the dust lane.

The CO emission can be separated into two components: a $1.4'' \times 1.9''$ core and $7'' \times 15''$ extended component. The H_2 masses in the core and extended components are 0.9 and 1.8×10^{10} $M_\odot$ respectively. The mean diameter of the core component ($\sim 1.7''$) corresponds to a radius of 315 pc and the mean H_2 density over the volume is 2,900 cm^{-3}. The dynamical mass for the core is similar to the H_2 gas mass there.

In Figure 7 (lower panels), a velocity gradient in the core and the extended components may be seen — lower velocity emission displaced to the SW, higher velocity emission to the NE. Direction of this velocity gradient is parallel to the major axis (PA$= 52 - 63°$) of the two CO components and the dust lane seen in the optical. Thus, the molecular gas has partially relaxed to a disk-like configuration with angular momentum playing a significant role in the gas distribution, probably inhibiting further collapse towards the center. The upper limit to the emission measure of ionized gas derived from the 3 mm continuum permits only 10^{11} $L_\odot$ for a population of only O5 stars. If a starburst is to account for the total luminosity, its duration must exceed 10^9 yr or the IMF must be deficient at ≥ 15 $M_\odot$ (Scoville and Soifer 1991).

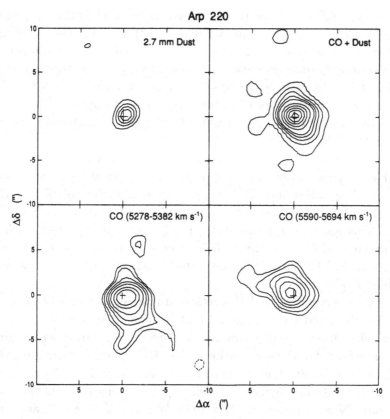

Fig. 7—The $\lambda = 2.7$ mm dust emission and integrated CO emission are shown in the upper panels as a function of offsets from the centimeter wave non-thermal radio peak in Arp 220 (Scoville *et al.* 1991). In the lower panels the CO emission is shown separately for a blueshifted and redshifted veloctiy ranges. The synthesized beam is $1.9'' \times 2.1''$.

3.3 Later Evolutionary Phases

There are a number of objects mapped at high resolution in molecular lines which have not recently undergone an interaction, but which may be reasonable analogs of the late evolutionary phases for interacting galaxy systems — that is an ageing nuclear starburst or a dust obscured AGN.

NGC 3079 is an edge-on spiral galaxy exhibiting radio lobes emerging from its nucleus and a high concentration of molecular gas within the central 1 kpc. The galaxy has recently been studied at high resolution in CO (Sofue and Irwin 1992). Within the central 200 pc a rapidly rotating core of gas can be seen with total mass approximately 1.5×10^9 M$_\odot$ and a gas mass fraction of approximately 50%. At slightly larger radii they see approximately 10^{10} M$_\odot$ of molecular gas situated in a ring feature. Sofue and Irwin (1992) also found molecular gas along the periphery of one of the minor axis radio lobes and they suggest that the nuclear radio jet is possibly focused by molecular gas. Tilanus and Veilleux (1993) compare the distribution of CO and HCN emission in the center of NGC 3079 — the latter having a higher dipole moment preferentially arises from molecular gas at density ~ 100 times greater than that sampled in CO. They find a strong gradient in the ratio of HCN to CO, with the latter more centrally concentrated in the core. Thus, it appears that the volume densities within molecular clouds are considerably higher at small radii, *i.e.* similar to the situation found in M82 (section 2.1).

In NGC 1068 (Seyfert 2), the molecular gas is mainly distributed in two massive spiral arms which originate from the ends of a stellar bar seen at 2 μm. Between these spiral arms at 1.4 kpc radius and the nucleus, there exists relatively little molecular gas, but on the nucleus there is possibly 6×10^7 M$_\odot$ of H$_2$ (Planesas, Scoville, and Meyers 1991). This gas (and associated dust) may hide the hypothesized Seyfert 1 nuclei (see Antonucci and Miller 1985). It is interesting to note that the molecular gas and young stars in NGC 3310 (Kikumoto *et al.* 1992) exhibit similar distributions to those in NGC 1068.

A number of observational and theoretical studies (*e.g.* Sanders *et al.* 1988*b*; Norman and Scoville 1988) have suggested that in some cases strong galactic interactions in gas-rich galaxies might ultimately lead to the formation and fueling of a central AGN or quasar. On the observational side, the optical emission lines seen in approximately 50% of the ultraluminous infrared galaxies have intensity ratios similar to those seen in LINER and Seyfert 1 nuclei. In fact, strong CO emission has now been seen in four UV-excess quasars: Mrk 1014, IZw 1, PG 0838+77, and Mrk 876 (Sanders, Scoville, and Soifer 1988; Barvainis, Alloin, and Antonnucci 1989; Alloin *et al.* 1992). The Owens Valley array has been used recently for a sensitive search for CO emission in the radio loud quasar 3C 48 (Scoville *et al.* 1993). A probable emission feature was detected at the redshift of the narrow optical emission lines ($z = 0.3695$) and the width of this feature is approximately 300 km s^{-1}. The integrated CO line flux implies a H$_2$ mass of 7×10^{10} M$_\odot$. 3C 48 is a particularly significant object inasmuch as it was the first optically identified quasar (Matthews and Sandage 1963) and yet it has many other characteristics similar to those of the ultraluminous infrared galaxies. It has a strong far infrared excess (Neugebauer, Soifer and Miley 1985), extended emission line gas and a possible double nucleus (*cf.* Stockton and Ridgway 1991). The detection of strong Balmer absorption lines in the extended nebulosity also indicates a luminous young stellar population (Boroson and Oke 1984), possibly an ageing starburst.

Probably the most extreme object with respect to its interstellar medium mass is the recently detected object *IRAS* 10214+4724 at $z = 2.3$ (*cf.* Brown and Vanden Bout 1991). Here the mass of molecular hydrogen exceeds 10^{11} M$_\odot$ and the far infrared luminosity is approximately 3×10^{14} L$_\odot$. Recent near infrared images obtained by Matthews *et al.* (1993) with the Keck telescope shows several near infrared peaks, possibly indicating a compact cluster of interacting galaxies.

4 GALACTIC EVOLUTION RESULTING FROM INTERACTIONS

In the previous sections we have attempted to show the important role played by the interstellar medium during galactic interactions — both in terms of the record of the encounter left in the extended tidal tails seen in H I and the dissipative/starburst role played by the molecular gas. We now address the issue of the long term effects which interactions of gas-rich galaxies may have on galactic evolution. We have seen from the molecular line data that an enormous mass of molecular gas may be concentrated at the center of the interacting system, while the diffuse atomic gas is relagated to the extended tidal tails. In the nucleus the star formation will be both more concentrated and much more rapid as judged from the infrared spatial distribution and the IR luminosity-to-gas mass ratios which are observed. These two facts make it possible to create a remnant with a higher central phase-space density than the precursors, removing one of the objections to creating ellipticals from mergers of spirals.

As discussed by Scoville and Soifer (1991), the ratio of space densities at the present epoch of infrared interacting galaxies and optical spiral galaxies is $\sim 2 \times 10^{-3}$. It is reasonable to believe that the duration of the infrared/starburst phase is approximately equal to the dynamical time ($\sim 2 \times 10^8$ years). Since two galaxies must take part in the interaction, the present day frequency of interaction-induced starbursts must be such that approximatley 2% of the present spiral galaxies would have passed through this phase in the last 10^9 years. And since the IR luminosity to gas mass ratio (see section 3) is enhanced by an order of magnitude during the starburst we conclude that a significant fraction (10-20%) of the stars now seen in spiral galaxies could have formed during interaction-induced starbursts. Including the effects of cosmological evolution on galaxy interaction rates makes this conclusion that much stronger.

A second major effect of galactic interactions is the redistribution and mixing of the ISM at different radii. Thus, ultimately the metalicity gradients in galaxies may be limited by galactic encounters such that more uniform metalicities will be observed in galaxies which have undergone relatively recent interactions. In addition, although the initial effect of the interaction on the low metalicity outer disk gas is to remove this gas into extended tidal features, much of this gas may eventually be absorbed back into the disk after several orbited periods. The result will be the gradual replenishment of the inner disk with low metalicity outer disk gas. This replenishment of the inner galactic disk both during and long after the galactic interaction may ultimately lead to a highly dissipative, "cold" disk out of which the next generation of post-interaction disk population stars will form. Clearly, the usual argument that interaction can not have played a significant role in the recent evolution of disk galaxies since the interaction would produce a vertically extended, hot disk population is no longer a strong argument if the majority of the thin disk stellar population was formed after the last interaction out of the resulting thin, gaseous disk.

Acknowledgement: This research was supported in part by NSF grants AST 90-16404 (NS and MSY) and AST 89-17744 and AST 90-23254 (JEH and JHvG). It is a pleasure to thank Alicia Rodriguez for the preparation of this manuscript.

REFERENCES

Alloin, D., Barvainis, R., Gordon, M. S., and Antonucci, R. R. J. 1992, *Astr. Ap.*, **265**, 429.

Antonucci, R. R. J., and Miller, J. S. 1985, *Ap. J.*, **297**, 621.

Barnes, J. E. 1988, *Ap. J.*, **331**, 699.

Barnes, J. E. and Hernquist, L. 1991, *Ap. J. Lett.*, **370**, L65.

Barvainis, R., Alloin, D., and Antonucci, R. 1989, *Ap. J. Lett.*, **337**, L69.

Boroson, T. A., and Oke, J. B. 1984, *Ap. J.*, **337**, 535.

Brown, R. L., and Vanden Bout, P. A. 1991, *A. J.*, **102**, 1956.

Bryant, P., and Scoville, N. Z. 1993, (in preparation).

Carico, D. P., Graham, J. R., Matthews, K., Wilson, T. D., Soifer, B. T., Neugebauer, G., and Sanders, D. B. 1990, *Ap. J. Lett.*, **349**, L39.

Carlstrom, J. 1990, unpublished *Ph. D.* thesis, Univesity of California (Berkeley).

Condon, J. J., Condon, M. A., Gisler, G., and Puschell, J. J. 1982, *Ap. J.*, **252**,102.

Dahari, O. 1985, *Ap. J. Suppl.*, **57**, 643.

Graham,J. R., Carico,D. P., Matthews,K., Neugebauer,G., Soifer,B. T., and Wilson,T. D. 1990, *Ap. J. Lett.*, **354**, L5.

Hibbard, J. E., and van Gorkom, J. H. 1993, in *The Globular Cluster — Galaxy Connection*, eds. G. H. Smith, and J. P. Brodie (San Francisco: ASP Conference Series), **48**, p. 619.

Hibbard, J. E., van Gorkom, J. H., Kasow, S., and Westpfahl, D. J. 1993*a*, in *Proceedings of the III Tetons Summer School* on *The Evolution of Galaxies and their Environment*, eds. J. M. Shull, and H. A. Thronson (Dordrecht: Kluwer), in press.

Hibbard, J. E., Guhathakurta, P., van Gorkom, J. H., Schweizer, F. 1993*b*, *A. J.*, submitted.

Hummel, E., and van der Hulst, J. M. 1986, *Astr. Ap.*, **155**, 151.

Keel W. C., *et al.* 1985, *A. J.*, **90**, 708.

Kikumoto, T., Tanigushi, Y., Suzuki, M., and Tomisada, K. 1992, in *Millimeter and Submillimeter Interferometry*, ed. M. Ishiguro (San Francisco: ASP Conference Series), in press.

Kormendy, J., and Sanders, D. B 1992, *Ap. J. Lett.*, **390**, L53.

Lawrence, A., Walker, D., Rowan-Robinson, M., Leech, K. J., and Penston, M. V. 1986, *Mon. Not. R. Astr. Soc.*, **219**, 687.

Lo, K. Y., Cheung, D. W., Masson, C. R., Phillips, T. G., Scott, S. L., and Woody, D. P. 1987, *Ap. J.*, **312**, 574.

Matthews, K., Soifer, B. T., Graham, J. R., and Neugebauer, G. 1993, in preparation.

Matthews, T. A., and Sandage, A. R. 1963, *Ap. J.*, **138**, 30.

Nakai, N. Hayashi, M., Handa, T. Sofue, Y., Hasegawa, T. 1986, *P. A. S. J.*, **38**, 603.

Neugebauer, G., Soifer, B. T., and Miley, G. R. 1985, *Ap. J. Lett.*, **295**, L27.

Norman, C. A., and Scoville, N. Z. 1988, *Ap. J.*, **332**, 124.

Norris, R. P. 1988, *Mon. Not. R. Astr. Soc.*, **230**, 345.

Norris, R. P., Allen, D. A., Sramek, R. A., Kesteven, M. J., and Troup, E. R. 1990, *Ap. J.*, **359**, 291.

Okumura, S. K., Kawabe, R., Ishiguro, M., Kasuga, T., Morita, K. I., and Ishizuki, S. 1991, in IAU Symposium No. 146, *Dynamics of Galaxies and their Molecular Distributions*, ed. F. Combes (Dordrecht: Kluwer), p. 425.

Peebles, P. J. E. 1980, *The Large-Scale Structure of the Universe*, (Princeton: Princeton University Press).

Planesas, P. Mirabel, I. F., and Sanders, D. B. 1991 *Ap. J.*, **370**, 172.

Planesas, P., Scoville, N. Z., and Myers, S. T. 1991, *Ap. J.*, **369**, 364.

Rubin, V. C., Ford, W. K., and D'Odorico, S. 1970, *Ap. J.*, **160**, 801.

Sanders, D. B., Scoville, N. Z., and Soifer, B. T. 1988, *Ap. J. Lett.*, **335**, L1.

Sanders, D. B., Scoville, N. Z., and Soifer, B. T. 1991, *Ap. J.*, **370**, 158.

Sanders, D. B., Scoville, N. Z., Sargent, A. I., and Soifer, B. T. 1988a, *Ap. J. Lett.*, **324**, L55.

Sanders, D. B., Soifer, B. T., Elias, J. H., Madore, B. F., Matthews, K., Neugebauer, G., and Scoville, N. Z. 1988b, *Ap. J.*, **325**, 74.

Sargent, A. I., and Scoville, N. Z. 1991, *Ap. J. Lett.*, **366**, L1.

Schweizer, F. 1982, *Ap. J.*, **252**, 455.

Schweizer, F. 1990, in *Dynamics and Interactions of Galaxies*, ed. R. Wielen (Heidelberg: Springer), p. 60.

Scoville, N. Z., Sanders, D. B., Sargent, A. I., Soifer, B. T., and Tinney, C. G. 1989, *Ap. J. Lett.*, **345**, L25.

Scoville, N. Z., Sargent, A. I., Sanders, D. B., and Soifer, B. T. 1991, *Ap. J. Lett.*, **366**, L5.

Scoville, N. Z., Padin, S., Sanders, D. B., Soifer, B. T., and Yun, M. Z. 1993, *Ap. J. Lett.*, **415**, L75.

Scoville, N. Z., and Soifer, B. T. 1991, in *Massive Stars and Starbursts*, Space Telescope Science Institute Symposium Series No. 5, eds. C. Leitherer, N. R. Wallborn, T. M. Heckman, and C. A. Norman (Cambridge: Cambridge University Press), p. 233.

Soifer, B. T., Sanders, D. B., Madore, B. F., Neugebauer, G., Danielson, G. E., Elias, J. H., Lonsdale, C. J., and Rice, W. L. 1987, *Ap. J.*, **320**, 238.

Sofue, Y., and Irwin, J. 1992, *P. A. S. J.*, **44**, 353.

Solomon, P. M., Radford, S. J. E., and Downes, D. 1990, *Ap. J. Lett.*, **348**, L53.

Stanford, S. A., and Balcells, M. 1991, *Ap. J.*, **370**, 118.

Stanford, S. A., Bushouse, H. A. 1991, *Ap. J.*, **371**, 92.

Stanford, S. A., Sargent, A. I., Sanders, D. B., and Scoville, N. Z. 1990, *Ap. J.*, **349**,

492.

Stockton, A., and Ridgway, S. E. 1991, *A. J.*, **102**, 488.

Suchkov, A., Allen, R. J., and Heckman, T. M. 1993, preprint.

Tilanus, R. P., and Veilleux, S. 1993, in preparation.

Tinney, C. G., Scoville, N. Z., Sanders, D. B., and Soifer, B. T. 1990, *Ap. J.*, **362**, 473.

Toomre, A. 1977, in *The Evolution of Galaxies and Stellar Populations*, eds. B. M. Tinsley, and R. B. Larson (Yale Univ. Observatory), p. 401.

Toomre, A., and Toomre, J. 1972, *Ap. J.*, **178**, 623.

van Gorkom, J. H. 1993, in *Proceedings of the III Tetons Summer School* on *The Evolution of Galaxies and their Environment*, eds. J. M. Shull, and H. A. Thronson (Dordrecht: Kluwer), in press.

Wang, Z., Scoville, N. Z., and Sanders, D. B. 1991, *Ap. J.*, **368**, 112.

Wang, Z., Schweizer, F. P., and Scoville, N. Z. 1992, *Ap. J.*, **396**, 510.

Wevers, B. H. M. R., van der Kruit, P. C., and Allen, R. J. 1986, *Astr. Ap. Suppl.*, **66**, 505.

White, S. D. M., and Rees, M. J. 1978, *Mon. Not. R. Astr. Soc.*, **183**, 341.

Wright, G. S., Joseph, R. D., Robertson, N. A., James, P. A., and Meikle, W. P. S. 1988, *Mon. Not. R. Astr. Soc.*, **233**, 1.

Young, J. S., Claussen, M. J., and Scoville, N. Z. 1988, *Ap. J.*, **324**, 115.

Young, J. S., Claussen, M. J., Kleinman, S. G., Rubin, V. C., and Scoville, N. Z. 1988, *Ap. J. Lett.*, **331**, L81.

Young, J. S., and Scoville, N. Z. 1984, *Ap. J.*, **287**, 153.

Yun, M. S., Ho, P. T. P., and Lo, K. Y. 1993*a*, *Nature*, submitted.

Yun, M. S., Ho, P. T. P., and Lo, K. Y. 1993*b*, *Ap. J. Lett.*, in press.

Yun, M. S. 1992, unpublished *Ph. D.* thesis, Harvard University.

Yun, M. S., and Scoville, N. Z. 1993, in preparation.

Zwicky, F. 1956, *Ergebnisse der Exakten Naturwissenschaften*, **29**, 344.

HIFI Results on the Superbubble of NGC 3079

S. Veilleux[1,2], G. Cecil[3], R. B. Tully[1], J. Bland-Hawthorn[4], and
A. V. Filippenko[5]

[1]Institute for Astronomy, University of Hawaii
[2]Kitt Peak National Observatory, NOAO
[3]Department of Physics and Astronomy, University of North Carolina
[4]Department of Space Physics and Astronomy, Rice University
[5]Department of Astronomy, University of California, Berkeley

ABSTRACT

The Hawaii Imaging Fabry-Perot Interferometer (*HIFI*) was used to produce a large data cube of the edge-on SBc galaxy NGC 3079 covering Hα + [N II] $\lambda\lambda$6548, 6583. The complete two-dimensional coverage of the Fabry-Perot data allowed us to derive the general flow pattern of the nuclear gas making up the superbubble in this object. Comparisons of our results with the well-known outflows in the Seyfert galaxy NGC 1068 and the starburst galaxy M82 indicate that the mass of entrained material is similar in these three galaxies, but that the kinetic energy involved in the outflow of NGC 3079 is at least an order of magnitude larger than in NGC 1068 and M82. The active nucleus in NGC 3079 is probably powering some of the outflow.

1 INTRODUCTION

Recent observations suggest that a violent outflow is taking place in the core of the edge-on SB(s)c galaxy NGC 3079. The optical line emission in the nucleus is LINER-like (Heckman 1980), and Hα presents faint, broad wings (Stauffer 1982; Keel 1983) reminiscent of low-luminosity Seyfert 1 galaxies. On closer inspection, however, the line emission responsible for the broad wings in the Hα profile is produced by a complex of extranuclear high-velocity clouds (Heckman, Armus, and Miley 1990 [HAM]; Filippenko and Sargent 1992 [FS]) which coincides in position with a loop-like structure first discovered in Hα + [N II] images (Ford *et al.* 1986).

There is very strong evidence for the presence of an active galactic nucleus (AGN) and intense star formation in the nuclear region of NGC 3079 (*e.g.,* Duric and Seaquist 1988 [DS]; Irwin and Seaquist 1988; Young, Claussen, and Scoville 1988). The relative importance of the AGN and the nuclear starburst in powering the outflow in NGC 3079 is still unclear. One reason for this uncertainty is the kinematic complexity of the outflowing gas and, therefore, the difficulty in determining accurately the kinetic energies involved in this process. The optical long-slit studies of HAM and FS have had only partial success in defining the geometry and kinematics of the outflowing gas due to a lack of spatial coverage. In this paper, we present the results of a study which used the Hawaii Imaging Fabry-Perot Interferometer (*HIFI:* Bland and Tully 1989) to produce complete line profile grids of Hα and [N II] $\lambda\lambda$6548, 6583 over the full extent of NGC 3079. A finesse 60 etalon with free spectral range of

92Å at Hα was used in its 71st and 72nd orders to yield a velocity resolution of 70 km s^{-1} (Nyquist sampling). The data were reduced using the method described in Bland and Tully (1989), Veilleux, Tully, and Bland-Hawthorn (1993), and Veilleux *et al.* (1993).

2 RESULTS

A nearly circular shell of line-emitting gas with an apparent diameter of 13″ or 1.1 kpc at the distance of 17.3 Mpc (Tully, Shaya, and Pierce 1992) is clearly seen on the east side of the nucleus. The presence of this bubble-like structure was first pointed out by Ford *et al.* (1986). The fainter detection limit of our data shows that the shell is contiguous at low intensity levels.

A survey of profiles across the bubble showed that most of them appear to be composed of 3 distinct velocity systems. The blue- and red-wing "bubble" components appear to have near-constant FWHM ≈ 350 km s^{-1}, while the "disk" component has FWHM ≈ 130 km s^{-1} (after accounting for our instrumental FWHM of 70 km s^{-1}). The different components were isolated using a multiple Gaussian decomposition technique. We fit 3 components to each line in the [N II] λλ6548,6583 + Hα complex, fixing the [N II] doublet ratio to its quantum value, and using the same line widths for the blue and red components. Stable fits are found across most of the bubble, and satisfactorily reproduce the complex variations. The distributions of the velocity centroids are shown in Figure 1.

Our velocity decomposition quantifies the asymmetric "bubble" pattern that was first described by FS, based on their long-slit spectra. Our datacube now shows that blueshifted velocities generally exceed those on the red wing, and that both coexist. Throughout most of the bubble, the velocity and surface brightness of the wing components are anti-correlated as in planetary nebulae. Velocity splitting is largest along the mid-axis of the bubble (panels *c, d, e,* and *f* in Fig. 1).

These patterns support FS's contention that we are observing an optically thin bubble, with blue and red components arising from the front and back volumes. However, our data show velocity patterns that are not observed in planetary nebulae, where gas expands radially from the central star. Instead, in NGC 3079 velocities increase almost linearly with nuclear radius until about 3/4 of the diameter of the bubble; the high-velocity gas disappears very rapidly beyond this point. This decrease is less abrupt in the SE quadrant of the bubble (left hand panels in Fig. 1). As FS have noted, this flow is more consistent with a radially increasing rate of expansion from the galaxy nucleus.

An ovoidal bubble with the more pointed extremity centered on the nucleus was found to best match the observed morphology of the emission-line structure. Good fits to the observed velocity distribution along the bubble mid-axis were found for $n = 1.5-2.5$, where $V_{outflow} \sim R^n$. With this flow field the abrupt decrease in velocities near the top of the bubble arises when the rapidly increasing velocity vector swings

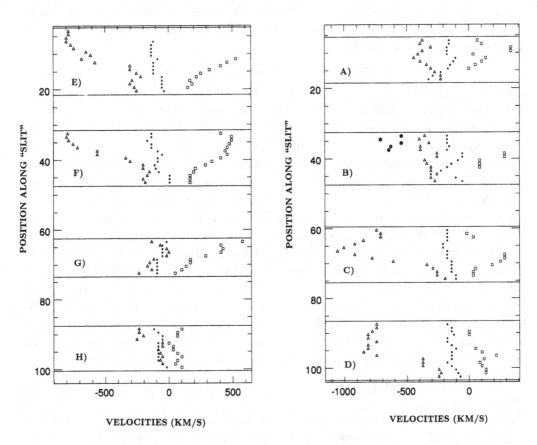

Fig. 1—Centroids of the emission-line components in the bubble. Each panel corresponds to a different "slit" position with respect to the bubble. The offset between the panels is 2″ in declination, with panel *a)* being at the extreme northern edge of the bubble. In each panel, the horizontal axis indicates the measured radial velocities of the components in km s⁻¹ and the vertical axis, the positions (right ascensions with arbitrary offsets, 1 unit = 0.″ 57) along each "slit". Three components are observed at most points in the bubble: two blueshifted (open triangles) and redshifted (open squares) "bubble" components, and a systemic disk components (small filled circles). In panel *b)*, a fourth component is also observed (large filled circles).

into the sky plane, rather than from a rapid deceleration. The bubble axis can only be tipped toward us by 5° − 9° to produce the observed asymmetry in maximum absolute velocities on the blue and red profile wings.

3 DISCUSSION

3.1 Time Scales, Masses, and Energies

The dynamical time scale of the superbubble can be estimated from the deprojected velocity of the entrained material derived from our kinematic model and the linear dimensions of the bubble:

$$t_{\text{dyn}} \approx 5 \times 10^5 \; R_{\text{bubble,kpc}} \; V_{\text{bubble,2000}}^{-1} \quad \text{yr}, \tag{1}$$

where $V_{\text{bubble,2000}}$ and $R_{\text{bubble,kpc}}$ are in units of 2,000 km s^{-1} and kpc, respectively. This time scale is comparable to the results found by DS from their wind model.

Assuming Case B recombination conditions at 10^4 K and gas density N_e, the ionized mass involved in the outflow of NGC 3079 is of order $10^7 \; N_e^{-1} \; M_\odot$. A conservatively small reddening correction was used to correct the observed fluxes for the presence of dust. An upper limit on the average gas density of 150 cm^{-3} is obtained from the [S II] $\lambda 6731/\lambda 6716$ ratio derived from complementary long-slit data. This mass is similar to those involved in the well-known outflows in the starburst galaxy M82 ($\sim 2 \times 10^5 \; M_\odot$; Bland and Tully 1988; HAM) and the Seyfert galaxy NGC 1068 ($\sim 2.1 \times 10^5 \; M_\odot$; Cecil, Bland, and Tully 1990). By deprojecting the observed velocities, we derive a bulk kinetic energy summed across the superbubble of a few $\times 10^{56} \; N_e^{-1}$ ergs. This energy is at least an order of magnitude larger than the kinetic energies involved in the outflows of M82 ($\lesssim 2 \times 10^{53}$ ergs; Bland and Tully 1988; HAM) and NGC 1068 ($\sim 4.4 \times 10^{53}$ ergs; Cecil, Bland, and Tully 1990).

The input rate of kinetic energy in the outflow of NGC 3079 can be estimated by assuming a constant rate over the dynamical time scale of the bubble:

$$\frac{dE_{\text{kin}}}{dt} \simeq \frac{E_{\text{kin}}}{t_{\text{dyn}}} > 6 \times 10^{42} \; E_{\text{kin,56}} V_{\text{bubble,2000}} R_{\text{bubble,kpc}}^{-1} \; N_e^{-1} \gtrsim 3 \times 10^{43} \; N_e^{-1} \quad \text{erg s}^{-1}. \tag{2}$$

The inequalities in equation (2) come from the facts that the rate does not include the kinetic energy of the wind material and a conservatively small reddening correction was applied to the observed fluxes. Neither does this rate include the contribution from the outflow on the west side of the nucleus (*cf.* Veilleux *et al.* 1993).

3.2 Energy of the Superbubble: Starburst or AGN?

The input rate of kinetic energy derived in the previous section can be compared with the injection rate *predicted* by the starburst scenario. In the starburst model, both stellar winds and supernovae are contributing to the pool of kinetic energy available to power the bubble. The rate at which this kinetic energy is injected in the surrounding medium is related to the star formation rate through the slope of the IMF and the upper mass limit. Using the calculations of Elson, Fall, and Freeman (1989; *cf.* also HAM), we obtain

$$\frac{dE_*}{dt} = 7 \times 10^{42} \; L_{\text{ir,11}} \quad \text{erg s}^{-1} \lesssim 3 \times 10^{42} \quad \text{erg s}^{-1}. \tag{3}$$

The infrared luminosity used in this calculation (Soifer *et al.* 1987) includes a contribution from the galactic disk of NGC 3079. This energy injection rate therefore is an *upper* limit to the injection rate of kinetic energy available to drive the nuclear superbubble.

These results suggest that the AGN present in the core of this galaxy may drive some of the outflow, but its exact importance depends critically on the density of the entrained material. In this black hole scenario, the starburst may still contribute to the outflow or was simply triggered by the expanding bubble-shock that propagates through the plane of the galactic disk (*e.g.*, Schiano 1985).

4 CONCLUSIONS

The violent outflow in NGC 3079 involves an ovoidal bubble of mass $\sim 10^7$ M$_\odot$ N_e^{-1} and kinetic energy $\sim$ a few $\times 10^{56} N_e^{-1}$ ergs with $N_e < 150$ cm^{-3}. This kinetic energy is at least an order of magnitude larger than in the outflows in NGC 1068 and M82. Energy arguments combined with the published radio data suggest the presence of a central AGN powering some of the outflow. A more detailed analysis of these results is presented in Veilleux *et al.* (1993).

This research is supported by NSF grants to the U. Hawaii and North Carolina. S. V. acknowledges the financial support of the Natural Sciences and ERC of Canada, and NASA (grant HF-1039.01-92A) awarded by the STScI.

REFERENCES

Bland, J., and Tully, R. B. 1988, *Nature*, **334**, 43.
_____. 1989, *A. J.*, **98**, 723.
Cecil, G. N., Bland, J., and Tully, R. B. 1990, *Ap. J.*, **329**, 38.
Duric, N., and Seaquist, E. R. 1988, *Ap. J.*, **326**, 574. (DS)
Elson, R. A. W., Fall, S. M., and Freeman, K. C. 1989, *Ap. J.*, **336**, 734.
Ford, H. C., Dahari, O., Jacoby, G. H., Crane, P. C., and Ciardullo, R. 1986, *Ap. J. Lett.*, **311**, L7.
Heckman, T. M. 1980, *Astr. Ap.*, **87**, 152.
Heckman, T. M., Armus, L., and Miley, G. K. 1990, *Ap. J. Suppl.*, **74**, 833. (HAM)
Irwin, J. A., and Seaquist, E. R. 1988, *Ap. J.*, **335**, 658.
Keel, W. 1983, *Ap. J.*, **269**, 466.
Schiano, A. V. R. 1985, *Ap. J.*, **299**, 24.
Stauffer, J. R. 1982, *Ap. J.*, **262**, 66.
Tully, R. B., Shaya, E. J., and Pierce, M. J. 1992, *Ap. J.Suppl.*, **80**, 479.
Veilleux, S., Cecil, G., Bland-Hawthorn, J., Tully, R. B., and Filippenko, A. V. 1993, *Ap. J.*, submitted.
Veilleux, S., Tully, R. B., and Bland-Hawthorn, J. 1993, *A. J.*, **105**, 1318.
Young, J. S., Claussen, M. J., and Scoville, N. Z. 1988, *Ap. J.*, **324**, 115.

Low Surface Brightness Galaxies: Evolution Without Mass Transfer

Gregory D. Bothun

Department of Physics, University of Oregon

ABSTRACT

On scales larger than 1 Mpc, Low Surface Brightness (LSB) galaxies are found in the same environment as the general population of disk galaxies. However, in a region of phase space defined by projected radius 0.5 Mpc and relative velocity = 500 km s^{-1}, LSB galaxies are extremely isolated. In addition, the average distance to a nearby galaxy of comparable mass is 1.7 times farther for LSB galaxies than for conventional disks. Since it is this small scale environment which determines the frequency of tidal interactions, the data argue that LSBs have not experienced a mass transfer event in the last Hubble time. The lack of such interactions clearly give these disks a different star formation history than their high surface brightness brethren and further implies that mean galactic surface brightness is a function of small-scale environment. To add further complexity, we have also identified a particular class of large-scale length LSB galaxy that, although isolated, invariably hosts a Seyfert 1 nucleus.

1 INTRODUCTION

Most conferences on topics in extragalactic astronomy are an entertaining mixture of apparent observational data which gives rise to theoretical conjecture followed by rampant folklore, wishful thinking and/or just plain rejection of the data as being relevant. This allows most theories to remain relatively unconstrained. For instance, the role that environment plays in the evolution of galaxies remains a contentious issue. To be sure, the present arrangement of galaxies into clusters, low density but large scale walls, or shells surrounding large scale voids means that a wide range of environments do exist. The morphology-density relation (Dressler 1980; Postman and Geller 1984) and the Butcher-Oemler effect (Butcher and Oemler 1978; see also Bothun and Dressler 1986; Dressler and Gunn 1990; Lavery *et al.* 1992) are the two most obvious examples of environmental influences on galaxy evolution. The physics of this influence, as well as its duration and at which redshift it is most effective, however, is not at all clear from extant data. Hence it is important to begin with a list of what has been established via observation. In short, this is what we know:

• Galaxy interactions are facilitated in a low velocity dispersion environment, that is, those encounters in which the encounter velocity is like that of the internal rotation velocity (150—300 km s^{-1}) will do the most damage.

• The interaction time is $\approx$ twice the dynamical time or 5% of a Hubble time.

• 3–5% of the General Disk Galaxy Population (GDGP) are currently interacting. This is based on the space density of objects in the Arp or Arp–Madore catalogs in comparison with galaxies in *ZCAT*.

• The GDGP is an exclusive club of High Surface Brightness (HSB)galaxies — just look at the pictures in the Arp catalog! While LSB tidal features are prevalent, the main portions of the interacting galaxies are quite HSB.

• All members of the GDGP are in some kind of structure (group, wall, sheet, cluster, void) where the velocity dispersion is usually low. Interacting galaxies are commonly found in these structures. For instance, even the Bootes void now contains several examples of HSB interacting galaxies (see Aldering *et al.* 1993).

• Most HSB disks have some low mass companion galaxies associated with them (see Zaritsky 1990).

On the basis of these established observations it seems reasonable to conclude that all HSB disk galaxies have had at least one tidal encounter (an probably more) over a Hubble Time. The impact of these encounters on the overall evolution is unknown hence leaving the door wide open for speculation.

2 FROM OBSERVATIONS TO FOLKLORE

Based on the observations cited above, here is a list of what we *pretend* to know:

• Interactions make starbursts. While this certainly explains why interacting galaxies are of HSB, the reality is that a truly wide range of present day star formation rates (SFRs) can be measured in almost any sample of interacting galaxies (see Kennicutt *et al.* 1987). Still the perception remains, enhanced primarily by detecting starburst galaxies in the far-infrared (*e.g.* Lonsdale *et al.* 1984), that interactions invariably cause a burst of star formation in one or both of the galaxies depending upon the amount of gas which is available. Detailed modeling (*e.g.* this conference or see Mihos *et al.* 1993) of the response of the gas to these interactions and the subsequent intensity and distribution of star formation also serve to validate this perception.

• Mergers make QSOs/Seyferts. Since the energy for these nuclear events most likely comes from mass infall onto massive black holes, then mergers are an appealing delivery mechanism. Indeed, the evidence presented by Heckman *et al.* (1986) that most powerful radio galaxies have merger morphology coupled with the widely publicized work of Sanders *et al.* (1988) involving ultra-luminous *IRAS* sources, seems to place this conjecture on firm observational footing (see also Bothun *et al.* 1982). If so, then galaxy interactions can clearly play a major role in determining galaxy evolution.

• Galaxies form via hierarchical clustering or merging of sub-units — well we all hope that gravity is involved!

• The morphology-density relation is a manifestation of formation time scales, which in turn is a measure of $\delta\rho/\rho$. That is, low $\delta\rho/\rho$ objects will have long formation timescales and will be disrupted if they try to form in relatively dense environments.

• If *isolated* 0.5–1 σ peaks really exist, then these would collapse (slowly) to form the least biased tracer of the mass distribution. The key here is isolation. In many N-body simulations, these common low σ peaks often sit on the shoulders of the rarer, denser peaks, and ultimately get incorporated into that larger structure.

From this list of things that we *pretend* to know we can conclude that interactions clearly influence galaxy evolution, perhaps in a dramatic manner. Moreover, the interacting strength and/or frequency may well be pre-determined by the initial conditions of galaxy formation. This then begs the question, are there examples of galaxies which did form in isolation? The answer is yes, but unfortunately, those galaxies are invisible, *e.g.* Malin 1 (Bothun *et al.* 1987).

3 WHAT IS A LSB GALAXY AND WHY ARE THEY RELEVANT?

A convenient definition of an LSB galaxy is one with an extrapolated central disk surface brightness B(0) fainter than 23.0 mag arcsec^{-2} . This is a full 4σ away from the constant B(0) disk of Freeman (1970) and is equal to the blue-band night sky brightness at a good site. According to Impey *et al.* (1988) the faintest galaxies that can be discovered from the ground have B(0) $\approx$ 27.0 mag arcsec^{-2} or 2% peak contrast with respect to the night sky background. In other words, these galaxies are essentially invisible.

There are three types of LSB galaxies that have been discovered by Impey *et al.* (1988, 1993) or Schombert *et al.* (1992). These types are distinguished only by disk scale length, α^{-1}, and circular velocity, not by B(0), and serve to show that LSB disk galaxies span the entire range of the Galactic Mass Spectrum. These LSB types are:

• Dwarfs: $\alpha^{-1} \leq 1$ kpc; $v_{\text{circ}} \leq 100$ km s^{-1}
• Disks: $1 \leq \alpha^{-1} \leq 5$ kpc; $100 \leq v_{\text{circ}} \leq 300$ km s^{-1}
• Crouching Giants: $\alpha^{-1} \geq 5$ kpc; $v_{\text{circ}} \geq 300$ km s^{-1}

The scanty evidence which is available (*e.g.* McGaugh 1992) suggests that M/L for LSB Disks (*not* dwarfs) is the same or even less than that for HSB disks. This certainly implies a lower volume mass density which suggests collapse from a rather low $\delta\rho/\rho$ perturbation. In addition, LSB disks have normal global H I masses but a) reduced peak H I column densities, b) very low molecular gas content (see Schombert *et al.* 1990) and c) very low H II region abundances for their mass. From this it is reasonable to conclude that LSB disk galaxies have had a relatively whimpy SFR averaged over a Hubble time. This is most likely the result of reduced surface H I density which is below the threshold for molecular cloud formation and the subsequent metal enrichment from massive stars born in such cloud complexes. Indeed, it may well be possible that LSB disks represent the evolutionary path taken by galaxies that form stars in atomic instead of molecular material. This now brings us to the main topic of this conference and the relevance of these newly discovered LSB galaxies.

Recent work by Bothun *et al.* (1993) and Mo *et al.* (1993) have shown that LSB disk galaxies are extremely isolated on small scales. In particular, LSB disk galaxies exhibit a correlation function with the same overall shape as that of HSB galaxies (*e.g.* it goes as $r^{-\gamma}$ where γ is 1.7) but has an amplitude which is 50% that of HSB galaxies. Moreover, the correlation function turns over on small scales. In fact, Bothun *et al.* (1993) show that the distribution of nearest neighbor distances for LSB and HSB galaxies are highly dissimilar and that, on average, the distance to the nearest galaxy of comparable mass is 1.7 times farther in the LSB than the HSB sample. Thus, LSB disk galaxies are not located in an environment which is conducive for tidal interactions. Furthermore, van der Hulst *et al.* (1993) show that the H I distribution in LSB galaxies is usually everywhere below the critical value for star formation. Thus, without an interaction to help in clumping the gas, the evolution of these galaxies will remain quiescent.

Hence, available redshift data on LSB galaxies indicates that, although their spatial distribution on scales ≥ 1 Mpc is indistinguishable from that of HSB galaxies, on smaller scales they are much less strongly clustered. The probability of tidal interactions to boost the SFR is dramatically lowered and one is left with a population of passively evolving galaxies, which are the least biased tracers of the mass but which are the most absent from any optically based redshift survey. In short, the discovery of LSB disk galaxies has given us a new window in to galaxy evolution, one which specifically shows the evolutionary history of disks which occurs when no mass transfer is involved. The primary implication of this is somewhat staggering, namely that the mean surface brightness of a galaxy, not its bulge-to-disk ratio, not its circular velocity, not its morphological type, is a function of local galaxy density.

We close by considering one final curiosity about the LSB galaxy population. In general, these galaxies are not active, indeed not a single one in a sample of ≈ 500 well studied disks shows evidence for an AGN. This in itself is a remarkable statistics. What other galaxy selection criteria could one define such that a sample of 500 objects would not contain one example of an AGN? However, of the three truly huge objects (i.e. $\alpha^{-1} \geq 12$ h^{-1} kpc) that we have discovered to date (see Bothun *et al.* 1987, 1990; Sprayberry *et al.* 1993), all of them have a Seyfert 1 nucleus as manifested by the FWZI of Hα of 5,000—10,000 km s^{-1}. The overall luminosity of the nucleus, however, is relatively low which probably indicates a low fueling rate. All three of these galaxies are quite isolated and have a total dynamical mass of $\approx 10^{12}$ solar. All three of them also have a normal bulge, which just happens to be surrounded by an immense, very diffuse, relatively gas-rich disk. We are completely unsure what this means but the existence of these three galaxies clearly shows that AGN activity in isolated galaxies is possible and is in marked contrast to the rest of the LSB galaxy population in which nuclear activity is non-existent.

Acknowledgments: I would like to thank C. Impey, S. McGaugh, H. Mo, J. Schombert, D. Sprayberry and T. van der Hulst, who actually have done all the real work that I have reported on.

REFERENCES

Aldering, G., Bothun, G., Marzke, R., and Kirshner, R. 1993, in preparation.

Bothun, G., Mould, G., Heckman, T., Schommer, R., Balick, B., and Kristian, J. 1982, *A. J.*, **87**, 1621.

Bothun, G., and Dressler, A. 1986, *Ap. J.*, **301**, 57.

Bothun, G., Impey, C., Malin, D., and Mould, J. 1987, *A. J.*, **94**, 23.

Bothun, G., Schombert, J., Impey, C., and Schneider, S. 1990, *Ap. J.*, **360**, 427.

Bothun, G., Schombert, J., Impey, C., Sprayberry, D., and McGaugh, S. 1993, in press.

Butcher, H., and Oemler, A. 1978, *Ap. J.*, **219**, 18.

Dressler, A. 1980, *Ap. J.*, **236**, 351.

Dressler, A., and Gunn, J. 1990, in *Evolution of the Universe of Galaxies: Edwin Hubble Centennial Symposium*, ed. R. Kron (San Francisco: Astronomical Society of the Pacific), p. 200.

Freeman, K. 1970, *Ap. J.*, **160**, 811.

Heckman, T. *et al.* 1986, *Ap. J.*, **311**, 526.

Impey, C., Bothun G., and Malin, D. 1988, *Ap. J.*, **330**, 634.

Impey, C., Irwin, M., Sprayberry, D., and Bothun, G. 1993, preprint.

Kennicutt, R. *et al.* 1987, *A. J.*, **93**, 1011.

Lavery, R., Pierce, M., and McClure, R. 1992, *A. J.*, **104**, 2067.

Lonsdale, C., Persson, E., and Matthews, K. 1984, *Ap. J.*, **287**, 95.

McGaugh, S. 1992, *Ph. D.* Thesis, University of Michigan.

Mihos, C., Bothun, G., and Richstone, D. 1993, in press.

Mo, H., McGaugh, S., and Bothun, G. 1993, preprint.

Postman, M., and Geller, M. 1984, *Ap. J.*, **281**, 95.

Sanders, D. B., Soifer, B. T., Elias, J. H., Madore, B. F., Matthews, K., Neugebauer, G., and Scoville, N. Z. 1988, *Ap. J.*, **325**, 74.

Schombert, J., Bothun, G., Impey, C., and Mundy, L. 1990, *A. J.*, **100**, 1523.

Schombert, J., Bothun, G., Schneider, S., and McGaugh, S. 1992, *A. J.*, **103**, 1107.

Sprayberry, D. *et al.* 1993, in press.

van der Hulst, T., Skillman, E., Smith, T., Bothun, G., and McGaugh, S. 1993, preprint.

Zaritsky, D. 1990, *Ph. D.* Thesis, University of Arizona.

Bulge Formation by Starbursts in Young Galaxies

A. Habe[1], K. Wada[1], and Y. Sofue[2]

[1]Department of Physics, Hokkaido University, Sapporo, Japan
[2]Institute of Astronomy, The University of Tokyo, Mitaka, Japan

ABSTRACT

We have studied the bulge formation process by starbursts in young disk galaxies whose disks and halos are gas-rich. If such galaxies tidally encounter another galaxies, large starbursts are easily induced and create galactic superwinds. We study the interaction between the superwind and the halo gas by using a similarity solution and show that a massive, radiativelly-cooled, gaseous shell is formed and becomes gravitationally unstable. In this way, we expect that shells of stars are formed. In order to study further evolution of these shells and their interaction with the disk, we model both the shell and the disk by using an N-body code. Our numerical results show that a large bulge with de Vaucouleurs' density profile is formed from the shell. We also show that the disk is thickened due to the interaction with the shell. The large bulges and thick disks are very similar to these found in S0 galaxies.

1 INTRODUCTION

Starburst galaxies release huge energy by frequent supernovae. In some starburst galaxies, hot gas and molecular outflows are observed. These outflows are called superwinds (Heckman *et al.* 1990; Tomisakak and Ikeuchi 1988; Mac Low and McCray 1989). Since some starburst galaxies are interacting galaxies, it was proposed that the starbursts occurs due to gravitational interaction between galaxies (*e.g.* Noguchi 1988). Such interactions induce bar formation in a galaxy, and the subsequent gas inflow towards the galactic center, as a result of gas-bar interaction and due to the self-gravity in the gas (*e.g.* Wada and Habe 1990).

In an early universe, since galaxies are young and have much more gas than now, galactic encounters induce even more violent star formation, causing large energy releases. We investigate the effects of these starbursts on the host galaxies. In section 2, we investigate the evolution of a superwind produced by violent starbursts, by using the similarity solution, and demonstrate that the superwind can produce a large-scale cold gaseous shell in the halo and can become gravitationally unstable. We can expect the star formation in this gaseous shell. In section 3, we show the numerical simulation of a shell of stars produced by the gravitational instability in the gaseous shell. We demonstrate that the shell evolves into a large bulge-like structure and gravitational interaction between the shell and the disk changes the structure of the disk which fattens. These numerical results suggest the explanation for the distribution of S0 galaxies in clusters, if starbursts favor denser regions and if the shells of stars are formed from large gaseous shells driven by superwinds.

2 SUPERWINDS PRODUCED BY STARBURSTS

We show the propagation of a shock induced by energy release from many supernovae in a starburst galaxy. The similarity solution is given by Umemura and Ikeuchi (1987). In this solution, there is a critical energy over which the halo gas is blown out,

$$L_{\rm w,cr} = 8.9 \times 10^{42} \left(\frac{n_c}{0.01~{\rm cm}^{-3}}\right) \left(\frac{r_c}{10~{\rm kpc}}\right)^{1/2} ~{\rm erg~s}^{-1},$$

where the halo gas is assumed to have $n_g(r) = n_c(r/r_c)^{-1/2}$. This energy corresponds to 1 supernova per year with $\epsilon = 0.1-$ the efficiency of supernovae-to-outflow energy conversion factor, and is larger than the energy released in nearby starburst galaxies. With this energy, the shock wave propagates the distance of $R_s = 7.6(t/10^8~{\rm yr})^{2/3}$ kpc, where t is age of the shock wave. We adopt this case as a typical one. Next, we consider the radiative cooling effects on the shock wave. The condition $t_{\rm cool} < t$ for the shocked gas is, $t > 0.7 \times 10^6 (n_c/0.01~{\rm cm}^{-3})^{-0.58}(r_c/10~{\rm kpc})^{-0.29}$ yr. where $t_{\rm cool}$ is the radiative cooling time of shocked gas at the shock front. If this condition is satisfied, a radiatively-cooled shell is formed and the halo gas is swept-up. When this shell becomes massive enough, it will be gravitationally unstable. The energy condition of self-gravitational instability is given by Ostriker and Cowie (1980), $E_{\rm tot} < 0$, where $E_{\rm tot}$ is the total energy of a part of the shell in the comoving frame, and is defined by $E_{\rm tot} = E_k + E_g + E_T$, where E_k, E_g and E_T are kinetic, self-gravitational, and thermal energy in this part of the gas shell. Hence,

$$t_{\rm cr} = 1.6. \times 10^8 \left(\frac{n_c}{0.01~{\rm cm}^{-3}}\right)^{-3/4} \left(\frac{r_c}{10~{\rm kpc}}\right)^{-3/8} ~{\rm yr},$$

$$M_s(t_{\rm cr}) = 2.6 \times 10^{10} \left(\frac{n_c}{0.01~{\rm cm}^{-3}}\right)^{-1/4} \left(\frac{r_c}{10~{\rm kpc}}\right)^{-1/8} ~{\rm M}_\odot,$$

and

$$R_s(t_{\rm cr}) = 10.7 \left(\frac{n_c}{0.01~{\rm cm}^{-3}}\right)^{-1/2} \left(\frac{r_c}{10~{\rm kpc}}\right)^{-1/4} ~{\rm kpc}.$$

Under these conditions the large shell becomes gravitationally unstable. Although the star formation efficiency is uncertain, the massive shell of stars is expected to be formed. Note, that a high efficiency of star formation has been suggested in the central star cluster (Lada *et al.* 1985). After stellar shell formation, it falls onto the disk, changing its shape and going through relaxation process. If the shell is massive enough, its gravity affects the galactic structure. Below, we investigate these dynamical effects.

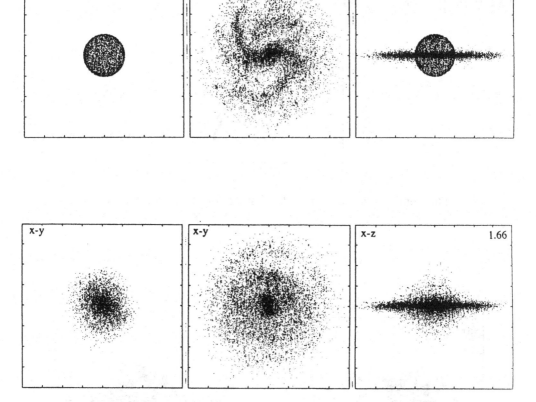

Fig. 1—Time evolution of a shell of stars and a disk. Left, center and right panels show the shell projected onto the disk, the shell component and both the shell and disk components, respectively.

3 *N*-BODY SIMULATION OF A SHELL OF STARS AND A LIVE DISK

In order to investigate the evolution of stellar shell formed by superwinds in the starburst galaxies as well as the effect of shell's gravity on the disk, we simulate both the shell of stars and the stellar disk. Our main assumptions are as follows:
Initially, the disk is thin and the central bulge is absent. We set a stellar disk to be stable according to Ostriker and Peebles criterion (Ostriker and Peebles 1973), by assuming a massive dark halo as a static potential and subsequently confirm that it is stable by using *N*-body simulations. Initial shell of stars is spherical and its mass

is $0.2 - 0.5 M_d$. We use $N_s = 2,000 - 5,000$ for the shell component and $N_d = 10,000$ for the disk component.

We use a workstation with the GRAPE-3 board, which is a special purpose hardware for calculating gravitational interactions between particles by the direct method. Its peak speed is 15 Gflops. We use the Runge-Kutta method for time integration. The softening length is 100 pc.

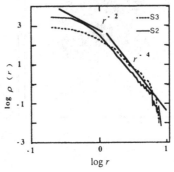

Fig. 2—The density distribution of final state for the fat stellar component.

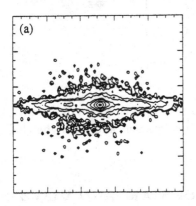 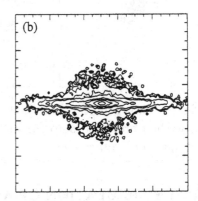

Fig. 3—The final isodensity contours for our models. Left panel shows only disk component; right panel shows disk and shell components.

We show one of our results in Figure 1. Initially, there is a weak spiral wave in the disk. Figure 1 shows the density wave in disk component induced by the gravity of the shell after it passes through the disk. Finally, a large fat stellar system and a thick stellar disk are formed. Figure 2 shows the density distribution of the fat stellar system. The fat stellar system rotates, as a result of gravitational interaction with the disk component. As in numerical simulation of a violent relaxation of a stellar system, core and halo structure are found. The density distribution of fat stellar system is similar to the 1/4 law of de Vaucouleurs (1948). We show the isodensity contours of our numerical results in Figure 3.

4 SUMMARY AND DISCUSSION

We study the effect of the starburst on the galactic structure in young galaxies. Since the galaxy encounters are frequent during cluster of galaxies formation, the starbursts are expected to be frequent as well. These starbursts are very energetic, since the young galaxies retain much gas in the disk and the halo.

We show that the energy released by many supernovae can create a superwind, and a large gaseous shell is formed due to the interaction between the superwind and the gas in the halo. We examine gravitational stability of the gaseous shell in a radiative cooling stage. We show that the gaseous shell can be gravitationally unstable and a shell of stars is formed. We simulate the evolution of both the large shell of stars and the stellar disk. We show that the evolution leads to a large bulge and a thick disk. Final structures of model galaxies resemble the S0 galaxies. If starbursts are induced by galaxy encounters, the process must be more efficient in a galaxy-crowded region. This process can explain the density-dependent morphology (Dressler 1980).

Acknowledgement: This work was supported in part by the Grant-in-Aid for Scientific Research (C) (04640261) of the Japanese Ministry of Education, Science and Culture.

REFERENCES

Dressler, A. 1980, *Ap. J.*, **236**, 351.

Heckman, T. M., Armus, L., and Miley, G. K. 1990, *Ap. J. Suppl.*, **74**, 833.

Lada, C. J., Margulis, M., and Dearborm, D. 1985, *Ap. J.*, **285**, 141.

Mac Low, M. M., and McCray, R. 1989, *Ap. J.*, **337**, 141.

Noguchi, M. 1988, *Astr. Ap.*, **203**, 259.

Ostriker, J. P., and Peebles, P. J. E. 1973, *Ap. J.*, **186**, 467.

Ostriker, J. P., and Cowie, L. L. 1981, *Ap. J. Lett.*, **243**, L127.

Sofue, Y., and Habe, A. 1992, *P. A. S. J.*, **44**, 325.

Sugimoto, D., Chikada, Y., Makino, J., Ito, T., Ebisuzaki, T., and Umemura, M. 1990, *Nature*, **345**, 33.

Tomisaka, K., and Ikeuchi, S. 1988, *Ap. J.*, **330**, 695.

Umemura, M., and Ikeuchi, S. 1987, *Ap. J.*, **319**, 601.

de Vaucouleurs, G. 1948, *Ann. Rev. Astr. Ap.*, **11**, 247.

Wada, K., and Habe, A. 1992, *Mon. Not. R. Astr. Soc.*, **258**, 82.

CO Observations of Nearby Active Galaxies

Nora Loiseau[1] and Françoise Combes[2]

[1]Instituto de Astrofísica de Canarias, Spain
[2]Observatoire de Meudon, France

ABSTRACT

Southern nearby active and starburst galaxies are being mapped in the $^{12}CO(1\text{-}0)$ and $^{12}CO(2\text{-}1)$ lines with the *SEST* telescope in order to investigate possible peculiarities in the dynamics, content and distribution of the gas which could lead to one or another type of activity, as was proposed by some recent models. Large-scale stellar bars, rings or closely interacting companions seem to be present in most active galaxies, being obviously related with the mechanism of gas transport from the disc into the nuclear or circumnuclear regions. But given that many of the barred or interacting galaxies do not show enhanced nuclear activity, other relevant parameters must exist, like the total gas content, the strength of the bar, or the impact parameters of the interactions. A signature of these processes would also be nuclear elongated features or "mini-bars" that are being observed in many galaxies, mainly in the infrared. We present CO data obtained for five active interacting and barred galaxies: NGC 134, NGC 986, NGC 4027, IC 1623 and IC 2554.

We mapped the nearby starburst and active galaxies described in Table 1 in the $^{12}CO(1\text{-}0)$ and $^{12}CO(2\text{-}1)$ lines with the 15m *SEST* telescope, with resolutions of 45″ and 22″ respectively. The galaxies in our sample were selected for their angular size, morphology and significant FIR luminosity, which is an indication of their activity. All the galaxies studied happen to have companions, sometimes clearly interacting.

TABLE 1. PARAMETERS OF THE OBSERVED GALAXIES

Galaxy	Morph. Type	Type of activ. (region)	S_{FIR} (100 μm/60 μm)	Size (arcmin)	T_{MB}[a] (mK)	Δv[a] (km s^{-1})
NGC 134	SAB(s)bc	Sy-like nucleus	54/19	8.5 x 2	60	150
IC 1623	S pec	STB (spread)	30/22	1.1 x 0.8	55	380
NGC 986	SB(rs)ab	STB (nucl.)	49/23	3.9 x 3	160	100
IC 2554	SB(r)b pec	Liner + STB (sp)	31/16	5 x 3	40	200
NGC 4027	SB(s)dm	weak STB (nucl.)	28/10	3.2 x 2.4	25	150

[a]Parameters of the $^{12}CO(1\text{-}0)$ central profiles.

NGC 134 is characterized by its bright nucleus, some intense blue starforming regions, and a plume in the direction opposite to the companion NGC 131. Both CO transitions show intense narrow lines ($\sim$50 km s^{-1}) associated with the north-eastern starforming region. The emission is weaker in the center and shows broader and more intense lines around it.

IC 1623 A/B (VV114) is a very disturbed merging pair of galaxies and an ultraluminous FIR source. The ^{12}CO(1-0) and ^{12}CO(2-1) spectra are very broad and show large variations from point to point. Our total intensity distribution is elongated in the North-South direction, with the maximum away from the nuclei, in the north side of the dusty region between IC 1623A and B. It does not coincide either with two compact southern concentrations recently detected interferometrically by Scoville *et al.* (1989). The flux of these compact components is less than 30% of the total CO flux, so they can not be clearly distinguished in our total intensity map.

NGC 986: This asymmetric barred spiral has a bright bulge and marked dust lanes in its bar, which end in a spiral or isophotal twist in the nuclear region (see Shaw *et al.* 1993). The ^{12}CO(1-0) spectra show a very strong, but rather narrow ($\sim$100 km s^{-1}) central profile. This suggests high starburst activity in the circumnuclear region, without large perturbations or peculiar motions. The spectra are broader in the bar region, where the gas appears to be streaming along the bar towards the nucleus at a deprojected velocity of 20 km s^{-1}).

IC 2554 is a little-studied nearby southern merging galaxy with remarkable dust lanes and numerous emission nebulae. Our ^{12}CO(1-0) and ^{12}CO(2-1) spectra show strong point to point variations, even at the highest resolution, indicating the clumpiness of the molecular gas. It is also noticeable that the ratio between the two transitions is different for the different components, probably associated with different members of the pair seen along the line-of-sight. A more uniform component is associated with a long north-eastern dust lane ending near the nucleus.

NGC 4027 is a prototypical late-type asymmetric barred galaxy (*i.e.* de Vaucouleurs and Freeman, 1973), well studied at many frequencies except in CO! Our observations show very extended CO emission. There are intense and narrow CO lines ($\sim$50 km s^{-1}) associated with the starforming regions of the arms, and broader (two component) profiles along the bar. The total intensity map is very asymmetric with the highest intensities associated with dusty regions 20″ north of the optical center (this galaxy has no strong dust lanes associated with the bar itself). The constant velocity contours show perturbations associated with the bar.

This research has made use of the NASA/IPAC Extragalactic Database (NED). NL acknowledges financial support from the FAPESP foundation (Brazil) for an observational trip to Chile.

References

de Vaucouleurs, G., and Freeman, K. C. 1973, *Vistas Astronomy*, **14**, 163.

Scoville, N. Z. *et al.* 1989, *Ap. J. Lett.*, **345**, L25.

Shaw, M. A., Combes, F., Axon, D. J., and Wright, G. S. 1993, *Astr. Ap.*, (in press).

Starburst Evolution on the IRAS–Color Diagram

Yoshiaki Taniguchi

Astronomical Institute, Tohoku University, Japan

ABSTRACT

In order to study effects of the starburst activity on far-infrared (FIR) colors, we have constructed starburst models which are able to trace FIR color evolution of starburst activity.

1 INTRODUCTION

The *IRAS* mission discovered that FIR emission properties of galaxies are significantly altered by starburst activity. Therefore, the FIR properties of galaxies are used to study the nature of starburst activity, in principle. Since, however, it is difficult to know properties of dust grains unambiguously, little effort has been made to construct starburst models which are able to reproduce FIR properties of galaxies (*cf.* Rowan-Robinson 1992). We here present a simple starburst model capable for calculating FIR emission of starburst galaxies.

2 MODEL AND RESULTS

The infrared continuum emission comes from dust grains embedded in ionized gas and from those in molecular clouds. The large grains are in thermal equilibrium with the radiation field and radiate as a blackbody with an emissivity $Q_\lambda \propto \lambda^{-1}$. On the other hand, small grains are heated transiently by single UV photons. In our model, emission at 12 and 25 μm mainly comes from small grains in the molecular clouds and large grains in the ionized gas, while emission at 60 and 100 μm comes from large grains in the molecular clouds. The basic concept of our model was given in Mouri and Taniguchi (1992). Our new model is now able to calculate FIR emission from 12 to 100 μm directly and to trace the full phase of starburst evolution. Detailed description will be given elsewhere.

In Figure 1, we show the model results together with *IRAS* color plots for normal spiral, starburst, Wolf-Rayet, and luminous infrared galaxies. Note that the sample galaxies are selected to cover the full phase of starburst evolution from pre- through on-going to post-starburst.

3 DISCUSSION

1) The on-going starburst galaxies are located along the model starburst line with a considerable scatter. Since the life time of the starburst activity is usually considered to be short (*i.e.*, $\sim 10^7$ years), this scatter is attributed to the scatter in star formation rate in the sample starburst galaxies.

2) It is remarkable that the majority of FIRGs have colder color in $\alpha(25,60)$, in spite of their blue colors of α (60, 100). These FIRGs are considered to be mainly powered by early B or later-type stars rather than O-type ones. The steeper decrease in α (25, 60) with respect to $\alpha(60,100)$ is due to the following evolutionary effect: since O stars first would die because of their shorter lifetime, emission at 25 μm becomes to fall off at first. In this phase, on the other hand, B stars still heat up dust grains which re-emit their energy mainly around 60 - 100 μm, resulting smaller changes in α (60, 100).

3) The above argument suggests that some FIRGs are in a post-starburst phase. Alternatively, it is possible to consider that B-star enhanced luminous starburst phenomena might occur in these galaxies. Since, however, their FIR properties are well described with blackbody radiation, we cannot rule out a possibility that heavy extinction may cause the observed steeper color index of α (25, 60).

REFERENCES

Mouri, H., and Taniguchi, Y. 1992, *Ap. J.*, **386**, 68.
Rowan-Robinson, M. 1992, *Mon. Not. R. Astr. Soc.*, **258**, 787.

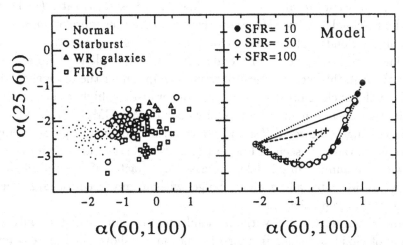

Fig. 1— Comparisons of the *IRAS* colors between the observations and the models. The model results are plotted with a time step of 5×10^6 years. Star formation rates are normalized to that in normal phase. The upper and lower masses are 40 and 5 $M_{\odot}$, respectively. The IMF slope is fixed at -2.5 during all the three stages.

Cloud Collisions and Bulge Formation in Disk Galaxies

Mario Klarić

Department of Physics and Astronomy, University of Alabama

ABSTRACT

I present results of several simulations of a flat self-gravitating disk composed of gas clouds and stars, to show that cloud collisions cause bulge formation.

1 THE CODE

I have performed a dozen runs using N-body code with a self-gravitating "zero-thickness" disk containing stars and/or gas clouds, and imbedded inside a spherical halo. Clouds could either collide or not. I have developed a new elaborate cloud collision routine, allowing clouds to either coalesce, fragment, or undergo star formation. The type of cloud collisions depends on their relative distance, while the relative speed and total mass of a colliding pair determine the outcome of a collision.

2 RESULTS

Bulge formation occurs in all runs with a low halo/disk mass ratio (H/D < 3) containing colliding clouds. A low H/D allows clouds to coalesce more often via cloud collisions into bigger clouds. These bigger clouds then act as very effective scattering agents, throwing stars and some other clouds out of the disk plane. The scattered particles make the disk thicker, and most of them end up, under the influence of halo potential, around the nucleus creating a bulge. Longer lasting, higher cloud collision rates, especially those between giant molecular clouds (GMC–GMC collisions), cause creation of bigger bulges. Lower H/Ds and larger initial cloud radii produce higher GMC–GMC collision rates, as seen in Figure 1 (clouds and stars, and bigger clouds).

In a run with the number of particles increased by a factor of 4, I obtain similar effect of higher collision rates and a bigger bulge, without any need for larger initial clouds (see Fig. 1 – more particles).

Bulges form in neither the run without clouds nor the run with stars only, confirming the role of cloud collisions in bulge formation, as shown in Figure 1 (stars only and no collisions).

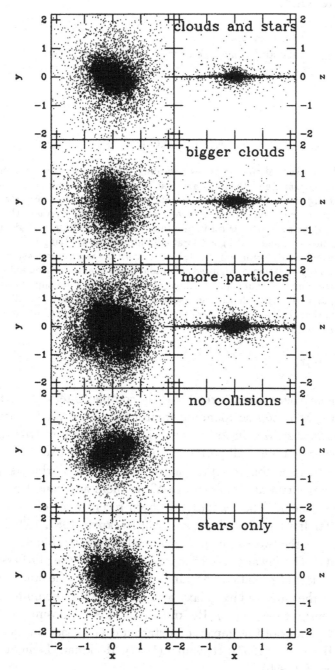

Fig. 1—Stellar components of five runs, each after 6 disk-edge rotations. Bulges form in runs with colliding gas clouds (top three panels), while runs without clouds or with non-colliding clouds end up without bulges.

Starbursts, Quasars, and their Environments

Timothy M. Heckman

Department of Physics and Astronomy,
Johns Hopkins University and Space Telescope Science Institute

ABSTRACT

I briefly discuss several specific issues regarding the possible inter-relationships between starbursts, quasars, and their extranuclear environments. First, I will argue that the case for fueling starbursts from the extranuclear environment is very strong. The luminosities of extreme starbursts are so large that they essentially require the complete conversion of a galaxy's interstellar medium into massive stars within a single dynamical time. Such starbursts should make good local laboratories for studying the processes involved in galaxy (spheroid) formation. Next, I will discuss the recent proposal by Terlevich and Boyle that the population of high-redshift radio-quiet quasars can be understood as the post-starburst-cores of young/proto elliptical galaxies (with no supermassive black holes required), and will argue that it has a serious energetics problem. Finally, I will describe the effects that the mechanical energy released by starbursts (and possibly quasars) has on their gaseous environments.

1 INTRODUCTION

The subject I have been asked to review — the complex inter-connections between starbursts, quasars, and their surrounding environment — is far too broad in scope to adequately summarize here in its entirety. I will therefore restrict my review to discussing some specific topics, all pertaining in some way to two general issues.

The first general issue concerns the connection between starbursts and quasars. How can we tell starbursts and quasars apart? Is there any causal or evolutionary connection between them? These issues have been recently reviewed by several different authors (Blandford 1992; Filippenko 1993; Heckman 1987, 1991). The second general issue concerns the two-way communication between starbursts or quasars and their environments. The inward flow of material can 'fuel' the starburst or quasar, while the outward flow of radiation and kinetic energy from the starburst or quasar will strongly effect the surrounding galaxy and inter-galactic medium. The 'fueling' issue has been recently reviewed (e.g. Heckman 1990, 1992; Shlosman 1990; Stockton 1990), and considerations of the impact of a starburst or quasar on its surroundings can be found in Begelman (1993), Heckman (1993), Heckman, Lehnert, and Armus (1993a, b), and Krolik (1990).

In the first part of my talk in section 2, I will sketch out a very general argument that luminous starbursts are events requiring the global rearrangement of a galaxy's interstellar medium, and the subsequent conversion of a significant fraction of this gas into stars on a single dynamical time for the system. In section 3, I will discuss and critique the provocative recent proposal by Terlevich and Boyle (1993) that radio-quiet quasars are the young cores of massive elliptical galaxies forming at high redshifts. Finally, in section 4 I will summarize the impact that the mechanical energy produced by starbursts (and possibly also by quasars) can have on their surrounding inter-stellar and inter-galactic media.

2 FUELING RATES AND DYNAMICAL TIMESCALES

2.1 The Method and the Data

On simple grounds of causality, it seems physically reasonable to posit that the maximum possible star–formation rate in a self-gravitating system is that correspond-ing to consuming all the gas in the system within one dynamical time (*e.g.* a crossing time or freefall time) for that system. An upper bound on the gas mass will be the total baryonic mass, and the gas fraction of the total mass M_{tot} can be parameterized as f_{gas}. This yields:

$$dM/dt_{max} = M_{tot}f_{gas}t_{dyn}^{-1} \approx M_{tot}f_{gas}(G\rho)^{1/2} \approx 115 f_{gas}\Delta v_{100}^3 \ M_\odot \ yr^{-1}, \qquad (1)$$

where Δv_{100} is the internal velocity dispersion of the self-gravitating system in units of 100 km s^{-1}. If mass is processed into stars at a rate dM/dt_{max} for a time $\approx 10^7$ to 10^9 years with a Salpeter Initial Mass Function (IMF) extending from 0.1 to 120 M$_\odot$, the resulting bolometric luminosity will be:

$$L_{max,Salp} \approx 7 \times 10^{11} f_{gas}\Delta v_{100}^3 \ L_\odot \qquad (2)$$

The corresponding expression for an IMF that contains only massive stars (defined here to be stars whose lifetimes are less than the duration of the starburst, so that the full amount of nucleosynthetic energy available is extracted during the burst) is:

$$L_{max,high-mass} \approx 6 \times 10^{12} f_{gas}\Delta v_{100}^3 \ L_\odot \qquad (3)$$

Thus, there is a simple predicted relationship between the maximum luminosity pos-sible for a starburst, and the velocity dispersion within the self-gravitating region that contains the gas to fuel the starburst. Note here that in extreme starbursts, the dynamical time and the gas fraction in equation (1) will not be the corresponding quantities appropriate for an entire galaxy. Rather, they are the dynamical time and gas fraction characterizing the region in which the velocity dispersion is measured by Δv. For example, in a starburst like M82, the CO $\lambda 2.6$ mm emission-line profile measures Δv in the region of the starburst (where the bulk of the molecular gas is located). The gas fraction within this region is much larger than gas fractions in the disks of normal galaxies, and the dynamical time is more than an order-of-magnitude shorter than in the disks of normal spiral galaxies (owing to the small size of the starburst).

Let us now consider the relevant data (bolometric luminosities and velocity dispersions) for samples of normal and infrared-luminous galaxies. The sample of normal galaxies is taken from Bothun, Lonsdale, and Rice (1989) who studied the mid and far-infrared properties of the galaxies in the UGC. The sample of infrared-bright galaxies is taken from the surveys of the molecular (CO $\lambda 2.6$ mm) and atomic (HI $\lambda 21$ cm) gas in *IRAS*-selected galaxies by Mirabel *et al.* (1990), Mirabel and Sanders (1988), and Sanders, Scoville, and Soifer (1991).

I have taken Δv to be half of the HI or CO emission-line profile width as measured at a level of 20% of the line profile's peak intensity (since this quantity is most often tabulated). I have applied no inclination correction to the line widths, even though in most cases the gas is probably largely rotationally supported. This is because the most extreme galaxies (and hence the objects of most interest in the present context) are usually highly disturbed systems for which the inclination of any presumed starburst disk can not be guessed without using (generally unavailable) high resolution maps of the gas in the starburst. Projection effects will therefore produce a sparsely populated tail in the distribution of $\log[\Delta v]$ toward low values at fixed $\log L$ (caused by relatively rare face-on systems). More quantitatively, $\approx 13\%$ ($\approx 4.5\%$) of the galaxies will have an observed value for Δv that is less than than one-half (one-third) the true deprojected value, assuming randomly oriented thin disks supported fully by rotation. This caveat should be kept in mind in the discussion to follow.

Luminosities have been computed assuming $H_0 = 75$ km s^{-1} Mpc^{-1}, and the bolometric luminosity has been approximated by the luminosity in the 8 to 1,000 μ band. This latter has been computed from the *IRAS* data using the corrections adopted in the original references given above. This quantity should be a reasonable approximation of the bolometric luminosity for the set of extreme objects that are most germane to the present discussion. It will also be a rough measure of the average star-formation rate for the most actively-star-forming subset of the normal galaxies (*e.g.* excluding those galaxies whose *IRAS* emission is dominated by a cool IR-Cirrus component — *cf.* Bothun, Lonsdale, and Rice 1989).

2.2 Results and Implications

The result — in the form of a plot of $\log[\Delta v]$ versus $\log[L_{IR}]$ — is shown in Figure 1. The principal conclusions to be drawn from this plot are as follows:

1). Normal (optically-selected) galaxies fall far below the maximum allowed luminosities. Adopting $f_{gas} = 0.1$ and a Salpeter IMF, a typical optically-selected galaxy has an IR luminosity that is only about 1% of the maximum allowed. Since the dynamical time for an ordinary disk galaxy is of-order 10^8 years, the galaxy can produce stars at its current rate for about a Hubble time (in agreement with direct estimates of the gas-depletion times in typical late-type galaxies — *cf.* Kennicutt 1990).

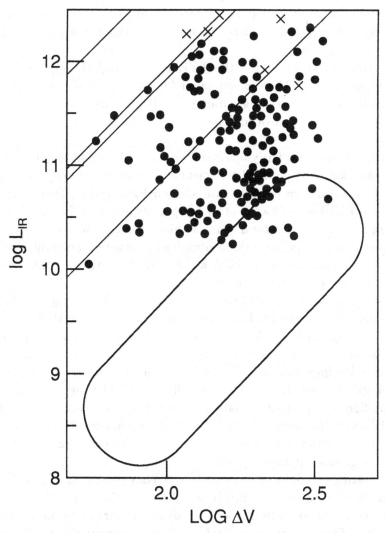

Fig. 1—A plot of the log of the internal velocity dispersion (taken to be half of the H I or CO profile's width at 20% of the profile's peak intensity) versus the log of the infrared (8–1,000 μ) luminosity for the sample of *IRAS*-selected galaxies described in the text. Those galaxies with quasar nuclei are indicated by crosses. The location of optically-selected galaxies (taken from Bothun, Lonsdale, and Rice 1989) is indicated schematically by the diagonal region near the bottom of the plot. The four diagonal lines are (in order of luminosity at a given value of Δv the upper bounds to luminosity predicted by eqs. (2) or (3) for: a Salpeter IMF with $f_{gas} = 0.1$ (appropriate for typical conditions of star-formation in ordinary galaxies); a 'massive-stars-only' IMF for $f_{gas} = 0.1$; a Salpeter IMF for $f_{gas} = 1.0$; and a 'massive-stars-only' IMF for $f_{gas} = 1.0$ (an absolute upper-bound to L).

2). The *IRAS*-selected galaxies range upward from the locus of normal galaxies and spill well over the relationship predicted by equation (2) for $f_{gas} = 0.1$.

3). The upper bound to the locus of the *IRAS*-selected galaxies is in approximate agreement with the upper bound predicted by either equation (2) with $f_{gas} = 1.0$ or by equation (3) with $f_{gas} = 0.1$.

4). For galaxies with the $\log[L] > 11.7$, the distribution of $\log[\Delta v]$ implies that few of the galaxies lying above the line for equation (2) and $f_{gas} = 0.1$ do so because of projection effects (see the preceding paragraph). Below this luminosity, there is a low-end tail in the distribution of $\log[\Delta v]$ at fixed $\log L$ that may be caused by projection effects.

The implications of this are fairly obvious: the most IR–luminous galaxies for a given velocity dispersion — if powered by starbursts — must not only be forming stars at close to the maximum possible rate allowed by causality, they must also preferentially produce high-mass stars and/or be systems in which the gas comprises an unusually large fraction of the total mass within the region where it is located. There have been arguments that both these situations ('massive-stars-only' IMF and nearly self-gravitating molecular gas disks) are generic to extreme starbursts (*cf.* Bernlohr 1993; Rieke *et al.* 1993; Scoville and Soifer 1991).

Perhaps of most importance to this meeting, Figure 1 strongly suggests that extreme starbursts must involve the wholesale rearrangement of the interstellar medium of a galaxy on a truly global scale. At this point, the only model that can plausibly accomplish this is one invoking a major galaxy merger (*cf.* Hernquist 1993).

Another interesting implication of Figure 1 concerns the relevance of starbursts to understanding galaxy formation at high redshifts. The old Eggen, Lynden-Bell, and Sandage paradigm for spheroid formation involves the efficient conversion of gas into stars on a galaxy freefall timescale. Insofar as the extreme local starbursts seem to correspond to this situation, they are — at least in this regard — reasonable local anologs to primeval galaxies at high redshift.

It is also interesting to note that none of the galaxies plotted in Figure 1 come close to violating the absolute upper limit imposed by equation (3) with $f_{gas} = 1.0$. There is therefore no strong and simple energetic/dynamical argument for favoring a model in which most of the infrared emission from the extreme galaxies in Figure 1 is powered by a 'shrouded quasar' (*e.g.* Sanders *et al.* 1989). Of course, this is not the same as saying that all the objects in Figure 1 must be starbursts. In fact six of the objects plotted in Figure 1 are classified as quasars based on the optical properties of their nuclei (and are so denoted in the Figure).

Finally, it is interesting to ask where *IRAS* FSC10214+4724 (also known as 'the most luminous object in the universe') would fall in Figure 1. For $H_0 = 75$ km s^{-1} Mpc^{-1} and $q_0 = 0$, it has a bolometric luminosity of 5×10^{14} L$_\odot$ (Rowan-Robinson *et al.* 1993). Its CO (J=6–5 and J=3–2) profiles yield $\Delta v = 170$ km s^{-1} (Solomon, Downes, and Radford 1992). Inspection of Figure 1 then shows that in order for this object to be powered by a starburst it must be turning all its gas into stars in a dynamical time, the gas must comprise 100% of the gravitating mass in the region probed by the CO emission, only massive stars must be forming, and the CO emission must be produced in a disk-like region viewed nearly face-on (we require an inclination correction corresponding to $\sin i = 0.40$ to place FSC10214+4724 on the uppermost of the four limits shown in Figure 1). There is an a priori probability of about 8% that a randomly oriented disk will be viewed this face-on.

3 QUASARS AND THE FORMATION OF ELLIPTICAL GALAXIES

3.1 Summary of the Terlevich and Boyle Model

In a bold and provocative recent paper, Terlevich and Boyle (1993, hereafter TB) have proposed that the population of radio-quiet quasars (which reaches its peak co-moving space density at $z \gtrsim 2$) corresponds to a population of post-starburst systems associated with the formation of the cores of elliptical galaxies. Their hypothesis is a further development of the 'quasars without black holes' model — Terlevich and Melnick (1985, 1987, 1988), Terlevich, Melnick, and Moles (1987), Terlevich (1989, 1990a, 1990b, 1992), Terlevich, Diaz, and Terlevich (1990), Terlevich *et al.* (1992) — in which quasars and other radio-quiet AGNs are powered by a coeval starburst (massive stars and supernovae) occurring in a nuclear region of high metallicity and high gas density.

In this model, a nearly instantaneous (delta-function in time) burst of star-formation occurs involving the central-most region in a forming elliptical galaxy. During the first several million years, the 'proto-quasar' is powered directly by nucleosynthesis in massive stars and would resemble an ultraluminous H II region. This early stage is actually the most luminous one, and is assumed to be shrouded in dust (since young starbursts with luminosities exceeding those of quasars have not been found in optical surveys). As the system evolves, the bolometric and especially the ionizing luminosity of the stellar population drops as the most massive stars evolve to their endpoints. Meanwhile, the mechanical energy supplied by type II supernovae becomes increasingly important, so that by a time of 8 Myr, it is argued that the core enters a 'QSO phase' dominated by type II supernovae and their remnants. By this time it is also supposed that much of the dusty shroud has been dispersed so that the core can be clearly seen in the optical and ultraviolet. This stage ends at 60 Myr when the least massive type II supernova progenitors explode.

By hypothesis, the kinetic energy associated with the type II supernovae is converted (via a blast wave interaction with a dense circumstellar shell around each supernova) into X-ray and EUV radiation. Some of the EUV radiation is further reprocessed via its interaction with dust and dense gas to produce the thermal IR continuum and the Broad Line Region respectively. In earlier versions of this model, it was argued that the optical and near-UV (non-ionizing) continuum was also produced by supernovae, but TB attribute this continuum to stellar photospheres.

Having given a brief overview of the model, let me now describe the specific assumptions and calculations made by TB. First, they assume that the event giving rise to a high-redshift quasar involves the innermost 5% of the mass of the young elliptical galaxy. This value may be regarded as an adustable parameter in their model, subject to the constraint that the angular size of the quasar can not exceed the observed upper limits. The size is also constrained by a more indirect argument. If the duration of the star-forming event is too long, a 'QSO phase' dominated by type II supernovae will not occur (since new very massive stars will be continuously formed). Quantitatively, the existence of their 'QSO phase' requires that the duration of star-formation must be short compared to the lifetime of type II supernova progenitors (*e.g.* $<<$ few $\times 10^7$ years). This is why they refer to the core as a 'coeval' stellar population. By the causality arguments given in section2 above, such a short burst duration requires that the starburst be physically small.

Taking the local ($z = 0$) optical luminosity function for elliptical galaxies and adopting a value for the mass/light ratio for the cores of ellipticals then allows them to calculate the co-moving space density of the elliptical cores as a function of mass at any redshift. They then use stellar evolution models to predict the mass/light ratio for a coeval stellar population with an age of 8 to 60 Myr (appropriate to the QSO phase). Combining the co-moving space density vs. mass relation with the light/mass ratio predicted by the stellar evolution models then yields the optical luminosity function for the cores. This luminosity function must be normalized in co-moving density by multiplying it by the QSO duty cycle at high redshifts. This is just the ratio of the lifetime of the QSO phase in any individual core (52 Myr by hypothesis) to the duration of the 'QSO epoch' which they take to be the time elapsed between $z = 10$ and $z = 2$. Thus, the duty cycle is a few percent.

The outcome of this is a predicted optical luminosity function for quasars (actually a set of functions for different choices of H_0, q_0, and the different shapes adopted for the local elliptical galaxy luminosity function). The goal is to reproduce the shape of the quasar luminosity function at $z \approx 2.0$ to 2.9 and its normalization in both co-moving density (the y–coordinate) and in luminosity (the x–coordinate). The model has some notable successes in this regard. First, the overall shapes of the predicted and observed luminosity functions are similar. This reflects the fact that the shapes of the adopted luminosity functions for elliptical galaxies at $z = 0$ (modulated by the slow dependence of M_{core}/L_{gal} on L_{gal}) are similar to the shape of the quasar luminosity function. Second, the normalization in co-moving density is good. Note that this is not really an adjustable parameter in the model since it is constrained to within a factor of about two by the 'duty cycle' argument given above. Finally, the normalization in luminosity is also good. This is by design, since it is set by the adjustable parameter that specifies the fraction of the mass of the elliptical that participates in the event (hence the 5% value adopted by TB).

3.2 Problems with the Model

While the model has an appealing elegance and simplicity, and the agreement between the model and data is surprisingly good, I believe that there are very serious problems with this model, which I will now explain.

To begin, the angular sizes predicted by the TB model already exceed the upper limits to the angular sizes of high-redshift quasars observed with the *HST*. The 'snapshot survey' of high-z quasars (Maoz *et al.* 1993 and references therein) — which was designed to look for gravitational lenses — observed 354 quasars, about half of which were in the specific redshift range modeled by TB ($z = 2.0$ to 2.9). These are all more luminous than $M = -25.5, -27.0, -26.4, -27.9$ for $(H_0, q_0) = (100, 0.5), (50, 0.5), (100, 0), $ and $(50, 0)$ respectively. In the TB model, quasars this luminous would correspond to the cores of massive elliptical galaxies (*e.g.* galaxies that are more than 1, 2, 2, and 3 magnitudes brighter today than a present-day L_* elliptical, for the four respective cosmological models given above). Using the scaling adopted by TB, the implied core diameters for the four cosmological models are greater than 1, 2, 2, and 4 kpc respectively, corresponding to angular diameters of greater than 0.22, 0.22, 0.29, and 0.29 arcsec. In contrast, a conservative upper limit to the angular sizes of the quasars imaged in the snapshot survey is < 0.1 arcsec (D. Maoz, private communication).

A much more serious problem (*e.g.* few orders-of-magnitude rather than a factor of several) is one of energetics. TB assume that the optical and near-UV continuum in high-z quasars is produced predominantly by starlight (this is implicit in their calculation of the light/mass ratios of the cores during the QSO phase). This can not be correct for at least two reasons. First, the optical-near-UV continuum of quasars is known to be time-variable: typical radio-quiet quasars at high redshift vary by about 0.5 magnitudes on timescales of 10 years or so (Pica and Smith 1983). This is clearly incompatible with the continuum being produced by an ensemble of millions of stars distributed over a kpc-scale core. Indeed, Terlevich himself has argued in detail elsewhere that the optical-UV continuum is produced by supernova remnants, and has shown how his model can reproduce both the correlated variability in the optical-UV continuum and the BLR (Terlevich *et al.* 1992) and the inverse relationship between variability amplitude and optical luminosity (Terlevich 1990*b*).

An equally serious problem with ascribing the UV continuum to starlight is the lack of any intrinsic Lyman edge in the spectra of quasars. Antonucci, Kinney, and Ford (1989) and Koratkar, Kinney, and Bohlin (1992) showed that any intrinsic Lyman edge must be less than 15% in amplitude in typical quasars, and more recent work by Tytler and Davis (1993) has lowered this limit to $< 4\%$. In contrast, the type of 8–60 Myr-old stellar population invoked for the QSO phase by TB will have an extremely strong Lyman edge (greater than a factor of 20 in amplitude — *cf.* Leitherer and Heckman 1993).

The firm conclusion then is that the optical-UV continuum in quasars can not be starlight, as TB implicitly assumed. Rather (and as Terlevich has argued elsewhere), the model requires that this continuum be powered by the thermalization of supernova kinetic energy rather than by nucleosynthesis. Since the X-rays are also produced in this way in the TB model, and since the IR is simply reprocessed optical and UV light, this is tantamount to saying that most of the bolometric luminosity of quasars must be powered by supernova remnants. This then leads to several severe energetics problems.

The first is one of self-consistency. Simple arguments show that nucleosynthesis will dominate supernova explosions in terms of the overall energetics of a coeval population of massive stars with an age of 8–60 Myr. In fact, detailed models (Leitherer and Heckman 1993) show that for an instantaneous burst of star-formation, the ratio of kinetic energy input from supernovae and stellar winds to the bolometric radiant luminosity due to nucleosynthesis ranges from about 1% at 8 Myr to 11% at 50 Myr. Thus, if the optical-UV continuum is produced by radiation from supernova remnants (even with 100% efficiency of conversion of kinetic energy into radiation), self-consistency requires that 10 to 100 times as much radiation is being produced in the form of starlight. Where is all this missing light? It can not be hidden in the IR, since the IR and optical-UV continua carry rough equal amounts of the overall energy budget in quasars (*cf.* Sanders *et al.* 1989).

Even more troubling, the relatively low 'yield' available in the form of supernova kinetic energy (10^{51} ergs per supernova implies $E \approx 10^{-4} \, Mc^2$ for a 'massive stars only' IMF and $\approx 10^{-5} \, Mc^2$ for a normal Salpeter IMF) means that explaining the tremendous amount of radiation that has been produced by quasars entails the processing of staggeringly large amounts of mass through supernovae. Let me be quantitative here, by computing the average amount of quasar radiant energy and the implied number of supernovae per elliptical galaxy in the TB model.

There have been several recent estimates of the total amount of radiant energy per unit present-day volume element produced by quasars over the history of the universe (Padovani, Burg, and Edelson 1990; Chokshi and Turner 1992). These estimates are especially useful because they are independent of H_0 and q_0. The former authors include only those quasars brighter than $B = 22.5$ and at redshifts < 2.2. I estimate (using the results in Boyle 1993) that their value should be increased by about a factor of two to extend down to $B = 25$ and out to $z = 5$. The results for the total amount of quasar radiant energy per unit present-day volume element are 8.3×10^{58} erg Mpc^{-3} for the revised Padovani, Burg, and Edelson calculation and about half this value for Chokshi and Turner. Recent results on the cosmic X-ray background can also be used to estimate this quantity. Taking the results from Zamorani (1993) that the monochromatic surface brightness of the X-ray background at 1 keV is 2.2×10^{-8} erg cm^{-2} s^{-1} ster^{-1}, that $80 \pm 20\%$ of this is due to quasars, and that the mean redshift of these quasars is ≈ 1.5, and then assuming a Bolometric correction of a factor of 190 at 1 keV for X-ray selected quasars (Giacconi and Burg 1992; Sanders *et al.* 1989), yields a quasar radiant energy production per unit present-day volume element of 1.1×10^{59} erg Mpc^{-3}. I will therefore adopt the mean of the X-ray, Chokshi and Turner, and Padovani, Burg, and Edelson values (7×10^{58} erg Mpc^{-3}).

The present-day luminosity density in optical light due to elliptical galaxies is about 1.4×10^7 L$_\odot$ h_{75} Mpc^{-3} (where h_{75} is H_0 in units of 75 km s^{-1} Mpc^{-1}, and where I have used the total galaxy optical luminosity density given by Loveday *et al.* [1992] and then assumed that 15% of this light is due to ellipticals). Dividing the quasar energy density by the elliptical galaxy luminosity density then implies that the amount of quasar radiant energy per elliptical galaxy with luminosity L_E is 5.0×10^{61} h_{75}^{-1} [$L_E/10^{10}$ L$_\odot$] erg. To get this much energy from thermalizing the kinetic energy of supernovae requires the formation of a total stellar mass of 6.2×10^{12} h_{75}^{-1} [$L_E/10^{10}$] M$_\odot$ for a normal Salpeter IMF extending from 0.1 to 120 M$_\odot$ and 'only' 8×10^{11} h_{75}^{-1} [$L_E/10^{10}$] M$_\odot$ for an IMF with massive stars only (*e.g.* no stars below 8 M$_\odot$).

These huge masses can be compared to the present-day masses of the cores of ellipticals. Adopting the parameters in TB, the core mass is given by $M_{core} = 7 \times 10^9$ $h_{75}^{1.69}$ [$L_E/10^{10}$]$^{1.35}$ M$_\odot$. The ratio of the mass required by quasar energetics to the present-day mass of a core is then 900 $h_{75}^{-2.69}$ [$L_E/10^{10}$ L$_\odot$]$^{-0.35}$ for a normal Salpeter IMF and 110 $h_{75}^{-2.69}$ [$L_E/10^{10}$ L$_\odot$]$^{-0.35}$ for a 'massive stars only' IMF. I would conclude that there is a severe energetics problem with the TB model, especially since a 'massive stars only' IMF is logically inconsistent with the model (since it is supposed to explain the formation of the cores of ellipticals, explicitly including the low mass stars that dominate these cores today).

4 THE EFFECTS OF STARBURSTS AND QUASARS ON THEIR ENVIRONMENTS

4.1 General Considerations and Global Estimates

In order to assess whether the energy produced by starbursts and quasars could have played a significant role in the formation and evolution of galaxies, it is useful to compile an inventory of their contributions. Following Songaila, Cowie, and Lilly (1990), the amount of radiant energy per Schecter L_* galaxy associated with the production of the metals currently present in the universe is:

- $E_{metals} \approx 2 \times 10^{61}$ h_{75}^{-1} ergs per L_* galaxy for the metals locked $-$ up inside galaxies
- $E_{metals} \approx 4 \times 10^{61}$ $h_{75}^{-2.5}$ ergs per L_* galaxy for the metals expelled from galaxies (based on the ratio of the metals in the Intra-Cluster Medium to the integrated optical light in great clusters, assuming this ratio is universal).

I have adopted here (and in the similar estimates to follow) an average galaxy density in the present universe of $\Phi_* = 0.0059$ h_{75}^3 Mpc^{-3} from Loveday *et al.* (1992). Note that if I include nucleosynthesis with an end-product of Helium, I need to multiply the above by a factor of $[1 + dY/dZ]$, where dY/dZ is at least ≈ 3 inside galaxies (Pagel 1993).

Following the discussion in section 3 above, the corresponding amount of radiant energy per L_* galaxy produced by quasars over the history of the universe is:

• $E_{\text{quasar}} \approx 9 \times 10^{60} \ h_{75}^{-3}$ ergs per L_* galaxy

These figures ignore type 2 Seyfert galaxies or the dust-shrouded 'ultraluminous' galaxies. Locally at least, these AGN classes produce at least as much radiant energy per unit co-moving volume as do the quasars. Thus, I conclude that $E_{\text{AGN}} > 10\% E_{\text{metals}}$.

The amount of kinetic energy per L_* galaxy produced by the stellar winds and supernovae associated with the metal-producing stars is:

• $KE_{\text{metals}} \approx 3 \times 10^{59} \ h_{75}^{-1}$ ergs per L_* galaxy for metals locked − up inside galaxies

• $KE_{\text{metals}} \approx 6 \times 10^{59} \ h_{75}^{-2.5}$ ergsper L_* galaxy for the metals expelled from galaxies.

In the above, I have used the calculations presented in Leitherer, Robert, and Drissen (1992) and Leitherer and Heckman (1993) to relate the kinetic energy to the radiant energy for a typical population of massive stars. While some of the kinetic energy supplied by massive stars may be dissipated and then radiated away, observations of local starburst galaxies imply that a significant fraction of the kinetic energy in these systems remains available to drive large-scale galactic 'superwinds', as I will briefly describe below (see also Heckman, Armus, and Miley 1990; Heckman, Lehnert, and Armus 1993a, b).

The kinetic energy produced by AGNs is more difficult to estimate. The BAL QSO phenomenon (*cf.* Weymann *et al.* 1991), is most naturally explained if essentially all radio-quiet quasars drive powerful sub-relativistic winds. Recent work suggests that the kinetic (wind) energy output of a typical quasar is not much lower than the total radiant energy output (Voit, Weymann, and Korista 1993). Thus, it may be that $KE_{\text{quasar}} \approx E_{\text{quasar}}$ (see above).

The jets associated with radio-loud AGNs may also carry out substantial amounts of kinetic energy. Integrating the radio source counts over flux density and frequency yields the total energy density in the extragalactic radio sky. The data given in Peacock (1993) imply that the surface brightness between 10 MHz and 100 GHz is 6.3×10^{-10} erg cm^{-2} s^{-1} ster^{-1}. To convert this into an estimate of jet kinetic energy requires knowing the conversion efficiency of jet kinetic energy into radio synchrotron emission (*e.g.* Begelman, Blandford, and Rees 1984). Parameterizing this ignorance as the quantity epsilon(rad) (which is probably in the range 10^{-1} to 10^{-2}), and taking the mean redshift at which an observed AGN radio synchrotron photon originates to be $< z > \approx 1$ to 2, I get:

• $KE_{\text{jet}} \approx 3 \times 10^{57} \ \epsilon_{\text{rad}}^{-1} \ h_{75}^{-3}$ ergs per L_* galaxy

For reasonable values of ϵ_{rad}, it appears that jets are less important global sources of kinetic energy than either massive stars or (possibly) BAL QSOs. On the other hand, if I assume that only elliptical galaxies produce strong radio emission, then the corresponding energy is:

- $KE_{\text{jet}} \approx 2 \times 10^{58}\ \epsilon_{\text{rad}}^{-1}\ h_{75}^{-3}$ ergs per L_* elliptical galaxy.

All these figures may be compared to the gravitational binding energy of an L_* galaxy:

- $E_{\text{bind}} \approx 4 \times 10^{59}\ h_{75}^{-1}$ ergs.

I therefore conclude that (integrated over the Hubble time), the kinetic energy supplied by massive stars and (possibly) AGNs per L_* galaxy is at least comparable to a typical galactic binding energy.

The question is then whether the similarity of the kinetic energy supplied by massive stars and AGNs to the gravitational binding energy is mere numerology, or whether it instead represents an important clue to the process of galaxy (spheroid?) formation. Others have directly explored related issues elsewhere under the rubrics of 'self-regulated galaxy formation', 'feedback', or 'explosive formation of galaxies' (*cf.* Ostriker and Cowie 1981; Ikeuchi 1981; Silk 1987; White and Frenk 1991).

4.2 Galactic Superwinds

A specific and well-established process by which starburst galaxies can effect their environment is via the large-scale outflow of kinetic and thermal energy in the form of a galactic superwind. I have recently reviewed the observational evidence in considerable detail elsewhere (Heckman, Armus, and Miley 1990; Heckman, Lehnert, and Armus 1993*a, b*), and will confine my remarks here to a general discussion of the effects such outflows can have on their environments.

We expect a superwind to occur when the kinetic energy in the ejecta supplied by supernovae and winds from massive stars in a starburst is efficiently thermalized. This means that the collisions between stellar ejecta convert the kinetic energy of the ejecta into thermal energy via shocks, with little energy subsequently lost to radiation. This situation is expected to arise when the (suitably normalized) kinetic energy input rate is so high that most of the volume of the starburst's ISM is filled by hot, tenuous supernova-heated gas (*e.g.* McKee and Ostriker 1977; MacLow and McCray 1988). The collective action of the supernovae and stellar winds then creates a 'bubble' of very hot (T up to 10^8 K) gas with a pressure that is much larger than that of its surroundings.

As the overpressured bubble expands inside a disk-like ISM (as expected in a typical starburst), it will expand most rapidly along the direction of the steepest pressure gradient (the disk's minor axis). Once the bubble has a diameter that is a few times the scale-height of the disk, it will evolve into the 'blow-out' phase as the swept-up shell of gas accelerates outward and fragments via Rayleigh-Taylor instabilities. The wind can then propagate out into the galactic halo as a moderately collimated bipolar flow where it will inflate a much larger and more tenuous bubble (*cf.* Tomisaka and Bregman 1993; Suchkov *et al.* 1993). Depending upon the extent of the halo and its density and upon the wind's mechanical luminosity and duration, the wind may ultimately blow out through the halo and into the inter-galactic medium.

A starburst characterized by star-formation that occurs for a timescale of 10^7 to 10^8 years (*cf.* Berhlohr 1993; Rieke *et al.* 1993), will produce kinetic energy via supernovae and stellar winds at a rate that is about 1 to 2% of the bolometric (radiant) luminosity of the starburst (*cf.* Leitherer and Heckman 1993). Thus, for a typical starburst like M82, the implied kinetic energy injection rate is $dE/dt \approx 2 \times 10^{42}$ erg s^{-1}. If the wind driven by this energy injection expands into a uniform galactic halo with density n (cm^{-3}) for a time t_{15} (10^{15} sec ≈ 32 Myr) the radius of the wall of the bubble and its expansion speed are given by:

$$r = 10 \ [dE/dt/2 \times 10^{42}]^{1/5} \ [n/10^{-2}]^{-1/5} \ t_{15}^{3/5} \ \text{kpc} \tag{4}$$

$$v = 190 \ [dE/dt/2 \times 10^{42}]^{1/5} \ [n/10^{-2}]^{-1/5} \ t_{15}^{-2/5} \ \text{km s}^{-1}, \tag{5}$$

where I have assumed the bubble is adiabatic. Thus, for typical starburst parameters, the size of the region influenced by the wind is galactic-scale and the characteristic expansion velocity of the accelerated ambient gas is similar to galaxy orbital velocities. The characteristic pressure associated with the wind-blown bubble (roughly the total kinetic energy injected divided by the bubble volume) is just:

$$P/k = 1.2 \times 10^5 \ [dE/dt/2 \times 10^{42}]^{2/5} \ [n/10^{-2}]^{3/5} \ t_{15}^{-4/5} \ \text{K cm}^{-3} \tag{6}$$

and this may be compared with typical interstellar pressures in normal galaxies of $P/k \approx 10^4$ K cm^{-3}. Thus, a modest starburst can significantly over-pressure its environment on a truly galactic scale. If the wind breaks out of the galactic halo and expands freely thereafter, it's ram pressure will be given by:

$$P(r)/k = 8 \times 10^5 \ [dE/dt/2 \times 10^{42}] \ [r/\text{kpc}]^{-2} \ \text{K cm}^{-3} \tag{7}$$

If the wind then impinges on another galaxy, it will be able to strip gas from this galaxy provided that its ram pressure exceeds the gravitational force per unit area exerted by the galaxy on its ISM:

$$P(r)/k > 2\pi G \sigma_* \sigma_{\text{gas}}/k = 1.2 \times 10^5 \ [\sigma_*]/100] \ [\sigma_{\text{gas}}/10] \ \text{K cm}^{-3} \tag{8}$$

where the stellar and gas surface mass densities in the galactic disk (σ_* and σ_{gas}) are given in units of M$_\odot$ pc^{-2}. Irwin *et al.* (1987) show that the wind from NGC 3079 (which I estimate has $dE/dt \approx 2 \times 10^{42}$ erg s^{-1}) can strip gas from the outer disk of its companion galaxy NGC 3073 (where $\sigma_* \sigma_{\text{gas}} \approx 10$ M$_\odot^2$ pc^{-4}).

Starburst-driven superwinds will have especially devastating consequences when they occur in dwarf galaxies (see Marlowe *et al.* 1993 for observations of this phenomenon). Suppose there was a (proto)galaxy in which all the baryons were initially in the gas phase. Turning only 1% of this gas into stars with normal IMF would return 5×10^{13} ergs in kinetic energy (from stellar winds and supernovae) per gram of remaining gas. This specific kinetic energy corresponds to a velocity of ≈ 100 km s^{-1}, which is at least comparable to the escape velocity for a dwarf galaxy. Thus, in principle, a burst involving only 1% of baryons could blow all the other baryons out of the galaxy (*cf.* Silk, Wyse, and Shields 1987; Babul and Rees 1992). Forming stars with a top-heavy IMF is even more efficient in turning mass into kinetic energy. If this mass were ejected impulsively and more than half the total mass were ejected, the galaxy would become unbound and 'dissolve'. This is rather unlikely (*cf.* DeYoung and Heckman 1993), but even if the galaxy were not destroyed by the mass loss, the mass ejection of much of the ISM could effectively shut-off all further star-formation, thereby allowing the galaxy to rapidly fade. This may offer a possible way out of the dilemma posed by the existence of a plethora of faint, blue galaxies at relatively moderate redshift (Broadhurst, Ellis, and Glazebrook 1992; Cowie *et al.* 1993).

The ejection of highly metal-enriched material by superwinds also offers a very natural explanation for the long-standing puzzle that the intra-cluster medium in clusters of galaxies contains at least as much mass in the form of metals as do the galaxies in the cluster (*e.g.* Sarazin 1986). This idea has been explored by many authors, including Larson and Dinerstein (1975), DeYoung (1978), and (more recently) White (1991) and David, Forman, and Jones (1991).

On long timescales ($>>$ the duration of the starburst), the 'fossil' of the wind-blown bubble can be treated as an explosion (with a total energy of 2×10^{57} ergs for a mild starburst like M82 and perhaps 10^{59} to 10^{60} ergs for an ultraluminous starburst or a forming galaxy). Neglecting the expansion of the universe — for illustrative purposes only — the adiabatic Sedov solution yields the maximum radius achieved by the superwind-inflated cavity in a Hubble time:

$$r_{\text{Hubble}} \approx 3.3 \ [E/10^{59}]^{1/5} \ [\Omega_{\text{IGM}}/0.01]^{-1/5} \ h^{-4/5} \ \text{Mpc}, \qquad (9)$$

where Ω_{IGM} is the density of the intergalactic medium into which the bubble expands in units of the closure density of the universe. Since r_{Hubble} is comparable to typical distances between galaxies, and superwinds may have carried out an average of $E \approx 10^{59}$ to 10^{60} ergs per galaxy (see section 4.1 above), this suggests that the structure of the IGM may be regulated by superwinds and the cavities they create (*cf.* Ostriker and Cowie 1981).

Miralda-Escude and Ostriker (1990) and Shapiro (1989) have recently evaluated the possible sources for heating and ionizing the IGM at high-redshifts. Both conclude that photoionization by quasars or hot stars is probably inadequate. On the other hand, the kinetic energy provided by superwinds should be able to shock-heat and ionize the IGM. Using the results from section 4.1 on the kinetic energy associated with the metals ejected from galaxies I find:

$$T_{\text{IGM}} \approx 6 \times 10^6 \ [\Omega_{\text{IGM}}/0.01]^{-1} \ h^{-3/2} \ \text{K}, \qquad (10)$$

assuming that all the ejected kinetic energy is used to heat the IGM, and ignoring subsequent radiative or abiabatic (cosmological-expansion) cooling of the IGM.

I conclude that galactic superwinds (particularly the extremely energetic events that were likely associated with galaxy formation) may have played key roles in the evolution of galaxies and the inter-galactic medium.

REFERENCES

Antonucci, R., Kinney, A., and Ford, H. 1989, *Ap. J.*, **342**, 64.

Babul, A., and Rees, M. 1992, *Mon. Not. R. Astr. Soc.*, 255, 346.

Begelman, M. 1993, in III Tetons Meeting on *The Evolution of Galaxies and their Environments*, ed. S. M. Shull and H. Thronson (Dordrecht: Kluwer), in press.

Begelman, M., Blandford, R. D., and Rees, M. J. 1984, *Rev. Mod. Phys.*, **56**, 255.

Bernlohr, K. 1993, *Astr. Ap.*, **268**, 25.

Blandford, R. D. 1992 in *Relationships Between Active Galactic Nuclei and Starburst Galaxies*, ed. A. V. Filippenko (San Francisco: ASP-31), p. 455.

Bothun, G. D., Lonsdale, C. J., and Rice, W. A. 1989, *Ap. J.*, **341**, 129.

Boyle, B. 1993, in III Tetons Meeting on *The Evolution of Galaxies and their Environments*, eds. S. M. Shull and H. Thronson (Dordrecht: Kluwer), in press.

Broadhurst, T., Ellis, R. and Glazebrook, K. 1992, *Nature*, **355**, 55.

Chokshi, A., and Turner, E. 1992, *Mon. Not. R. Astr. Soc.*, **259**, 421.

Cowie, L., Gardner, J., Hu, E., Wainscoat, R., and Hodapp, K. 1993, *Ap. J.*, in press.

David, L., Forman, W., and Jones, C. 1991, *Ap. J.*, **369**, 121.

DeYoung, D. S. 1978, *Ap. J.*, **223**, 47.

DeYoung, D., and Heckman, T. 1993, in preparation.

Filippenko, A. 1993, in *Physics of Active Galactic Nuclei*, eds. S. Wagner and W. Duschl (Berlin: Springer), in press.

Giacconi, R., and Burg, R. 1992, in *The X-Ray Background*, eds. X. Barcons and A. Fabian (Cambridge: Cambridge University Press), p. 3.

Heckman, T. 1987, in *Starbursts and Galaxy Evolution*, ed. T. Thuan, T. Montmerle, and J. Tran Thahn Van (Paris: Editions Frontieres), p. 467.

Heckman, T. M. 1990, in IAU Coll. 124 on *Paired and Interacting Galaxies*, eds. J. Sulentic, W. Keel, and C. Telesco (NASA CP-3098), p. 359.

Heckman, T. M. 1991, in *Massive Stars in Starburst Galaxies*, ed. C. Leitherer, T. Heckman, C. Norman, and N. Walborn (Cambridge: Cambridge University Press), p. 289.

Heckman, T. 1992, in *Testing the AGN Paradigm*, eds. S. Holt, S. Neff, and C. M. Urry (New York: AIP), p. 595.

Heckman, T. 1993, in *The Epoch of Galaxy Formation*, ed. M. J. Rees, (Pontifical Acad. of Sciences), in press.

Heckman, T., Armus, L., and Miley, G. 1990, *Ap. J. Suppl.*, **74**, 833.

Heckman, T., Lehnert, M., and Armus, L. 1993a, in III Tetons Meeting on *The Evolution of Galaxies and their Environments*, eds. S. M. Shull and H. Thronson (Dordrecht: Kluwer), in press.

Heckman, T., Lehnert, M., and Armus, L. 1993b, in *The Nearest Active Galaxies*, ed. J. Beckman, in press.

Hernquist, L. 1993, these proceedings.

Ikeuchi, S. 1981, *P. A. S. J.*, **33**, 211.

Irwin, J., Seaquist, E., Taylor, A., and Duric, N. 1987, *Ap. J. Lett.*, **313**, L91.

Kennicutt, R. 1990, in *The Evolution of the Universe of Galaxies*, ed. R. Kron, (San Francisco: ASP), p. 141.

Koratkar, A., Kinney, A., and Bohlin, R. 1992, *Ap. J.*, **400**, 435.

Krolik, J. 1990, in II Tetons Meeting on *The Interstellar Medium in Galaxies*, eds. H. Thronson and S. M. Shull (Dordrecht: Kluwer), p. 239.

Larson, R. B., and Dinerstein, H. L. 1975, *P. A. S. P.*, **87**, 911.

Leitherer, C., and Heckman, T. 1993, in preparation.

Leitherer, C., Robert, C., and Drissen, L. 1992, *Ap. J.*, **401**, 596.

Loveday, J., Peterson, B. A., Efstathiou, G., and Maddox, S. 1992, *Ap. J.*, **390**, 338.

MacLow, M., and McCray, R. 1988, *Ap. J.*, **324**, 776.

Maoz, D., Bahcall, J., Schneider, D., Bahcall, N., Djorgovski, S., Doxsey, R., Gould, A., Kirhakos, S., Meylan, G., and Yanny, B. 1993, *Ap. J.*, **409**, 28.

Marlowe, A., Heckman, T., Wyse, R., and Schommer, R. 1993, in preparation.

McKee, C., and Ostriker, J. 1977, *Ap. J.*, **218**, 148.

Mirabel, I. F., and Sanders, D. B. 1988, *Ap. J.*, **335**, 104.

Mirabel, I., Booth, R., Garay, G., Johansson, L., and Sanders, D. 1990, *Astr.Ap.*, **236**, 327.

Miralda-Escude, J., and Ostriker, J. 1990, *Ap. J.*, **350**, 1.

Ostriker, J., and Cowie, L. 1981, *Ap. J. Lett.*, **243**, L127.

Padovani, P., Burg, R., and Edelson, R. 1990, *Ap. J.*, **353**, 438.

Pagel, B. 1993, preprint.

Peacock, J. 1993, in *Extragalactic Background Radiation*, ed. M. Livio (Cambridge: Cambridge University Press), in press.

Pica, A., and Smith, A. 1983, *Ap. J.*, **272**, 11.

Rieke, G., Loken, K., Rieke, M., and Tamblyn, P. 1993, *Ap. J.*, **412**, 99.

Rowan-Robinson, M., *et al.* 1993, *Mon. Not. R. Astr. Soc.*, **261**, 513.

Sanders, D., Phinney, E. S., Neugebauer, G., Soifer, B. T., and Matthews, K. 1989, *Ap. J.*, **347**, 29.

Sanders, D., Scoville, N., and Soifer, B. 1991, *Ap. J.*, **370**, 158.

Sarazin, C. 1986, *Rev. Mod. Phys.*, **58**, 1.

Scoville, N. Z., and Soifer, B. T. 1991, in *Massive Stars in Starburst Galaxies*, eds. C. Leitherer, N. Walborn, C. Norman, and T. Heckman (Cambridge: Cambridge University Press), p. 233.

Shapiro, P. R. 1989, in Fourteenth Texas Symposium on Relativistic Astrophysics, *Ann. N. Y. Acad.*, ed. E. J. Fenyves.

Shlosman, I. 1990, in IAU Coll. 124 on *Paired and Interacting Galaxies*, eds. J. Sulentic, W. Keel, and C. Telesco (NASA CP-3098), p. 689.

Silk, J. 1987, in *Dark Matter in the Universe*, eds. J. Kormendy and G. Knapp, (Dordrecht: Reidel), p. 335.

Silk, J., Wyse, R., and Shields, G. 1987, *Ap. J. Lett.*, **322**, L59.

Solomon, P., Downes, D. and Radford, S. 1992, *Ap. J. Lett.*, **398**, L29.

Songaila, A., Cowie, L., and Lilly, S. 1990, *Ap. J.*, **348**, 371.

Stockton, A. 1990, in *Dynamics and Interactions of Galaxies*, ed. R. Weilen (Dordrecht: Kluwer), p. 440.

Suchkov, A. , Balsara, D., Heckman, T., and Leitherer, C. 1993, *Ap. J.*, submitted.

Terlevich, E., Diaz, A., and Terlevich, R. 1990, *Mon. Not. R. Astr. Soc.*, **242**, 271.

Terlevich, R. 1989, in *Evolutionary Phenomena in Galaxies*, ed. J. Beckman, and B. Pagel (Cambridge: Cambridge Univ. Press), p. 149.

Terlevich, R. 1990*a*, in *Windows on Galaxies*, eds. G. Fabbiano, J. Gallagher, and A. Renzini (Kluwer: Dordrecht), p. 87.

Terlevich, R. 1990*b*, in *Structure and Dynamics of the Interstellar Medium*, eds. G. Tenorio-Tagle, M. Moles, and J. Melnick (Berlin: Springer), p. 343.

Terlevich, R. 1992, in *Relationships Between Active Galactic Nuclei and Starburst Galaxies*, ed. A. Filippenko (San Francisco: ASP), p. 133.

Terlevich, R., and Boyle, B. 1993, *Mon. Not. R. Astr. Soc.*, **262**, 491.

Terlevich, R., and Melnick, J. 1985, *Mon. Not. R. Astr. Soc.*, **213**, 841.

Terlevich, R., and Melnick, J. 1987, in *Starbursts and Galaxy Evolution*, eds. T. Thuan, T. Montmerle, and J. Tran Thahn Van (Paris: Editions Frontieres), p. 393.

Terlevich, R., and Melnick, J. 1988, *Nature*, **333**, 239.

Terlevich, R., Melnick, J., and Moles, M. 1987, in *Observational Evidence for Activity in Galaxies*, eds. E. Khachikian, J. Melnick, and K. Fricke (Reidel: Dordrecht), p. 499.

Terlevich, R., Tenario-Tagle, G., Franco, J., and Melnick, J. 1992, *Mon. Not. R. Astr. Soc.*, **255**, 713.

Tomisaka, K., and Bregman, J. 1993, *P. A. S. J.*, in press.

Tytler, D., and Davis, C. 1993, private communication.

Voit, G. M., Weymann, R., and Korista, K. 1993, *Ap. J.*, **413**, 110.

Weymann, R., Morris, S., Foltz, C., and Hewett, P. 1991, *Ap. J.*, **373**, 23.

White, R.E. III 1991, *Ap. J.*, **367**, 69.

White, S., and Frenk, C. 1991, *Ap. J.*, **379**, 52.

Zamorani, G. 1993, in *Extragalactic Background Radiation*, ed. M. Livio (Cambridge: Cambridge University Press), in press.

Nuclear Fueling in Two-Component Star-Gas Disks

Masafumi Noguchi

Astronomical Institute, Tohoku University, Japan

ABSTRACT

Theoretical studies of gas dynamics in disk galaxies are reviewed in relation to the fueling of nuclear activity. Importance of self-gravitational effects in the interstellar gas component is emphasized.

1 FUELING PROBLEM

Recent observations have revealed that some galaxies show an unusual level of activity near the nuclear regions. One type of activity is the enhanced star formation in the central few kpc regions around galaxy nuclei. Another is the non-thermal activity which originates from sub-parsec region, which is called active galactic nuclei (AGN). This type of activity is believed to be powered by mass accretion onto supermassive black holes located in the galactic nuclei (*e.g.*, Begelman *et al.* 1984). Both types of activity require adequate source of fuel. In typical AGNs, a gas supply rate of ~ 1 M$_\odot$ yr^{-1} is needed to generate the radiation of the observed amount. CO line observations (*e.g.*, Kenney 1990) have detected large accumulation of a potential fuel in the form of molecular gas in the central kpc in many starburst galaxies.

Possible candidate sources of fuel are divided into local and global ones according to their spatial distribution. The difficulties for local sources such as compact star clusters have been pointed out by several works (*e.g.*, Shlosman *et al.* 1990). In this article we concentrate on global sources, especially the interstellar gas in the galactic disks. Because interstellar gas is distributed on the 10 kpc scale, one problem is how the material can be channelled into the nuclear region with a sufficiently high efficiency to maintain the observed level of activity. This fueling problem is stringent especially for AGN, which require gas supply to a region several orders of magnitude smaller than the size of a galactic disk.

From the dynamical point of view, disks of spiral galaxies can be regarded as two-component thin disks comprised of stars and gas. Stellar component is considered to be collisionless and dissipationless whereas energy dissipation (cooling) and collisions (in the case of a cloudy interstellar medium) characterize the gas component. These two components are coupled to each other gravitationally while obeying their own physical laws respectively. The problem of fueling is closely connected to the dynamical instability of galactic disks, because unstable disks develop a wide variety of non-axisymmetric structures which gravitationally drive the redistribution of the interstellar gas. We shall see later that the gas component does not only serve as a fuel for activity but also can affect the disk dynamics and its stability.

2 THREE STAGES OF MODELLING

The interstellar gas component, whose main constituents are hydrogen molecular clouds and H I clouds, usually occupies only a minor fraction of the total mass of a galaxy (*e.g.*, Knapp 1990). However this does not necessarily means that the gas component is dynamically unimportant. Theoretically, the importance of galactic gas dynamics can be understood within the following classification which is based on the dynamical contribution of the gas.

In the first category (1), the gravity of the gas is completely neglected. In the next stage of approximation, the category (2), we consider the self-gravitational effect, that is, the effect of the gas gravity on the gas itself. It is, however, assumed that the stellar component is not affected by the gas. In the third category (3), we take into account the gravity of the gas *and* the stars and treat the dynamical coupling of the two components in a self-consistent way. In the following, we briefly describe the theoretical studies related to the fueling problem in each category.

2.1 Non-Selfgravitating Gas Dynamics

In this category, the kinematics of the gas component is governed by the gravitational field made by the stellar component. It is well known that non-axisymmetric structures in the stellar component, such as spiral arms and bars, play an important role in redistributing the interstellar gas. Among these structures, the importance of bars is widely acknowledged. Extensive theoretical studies (mainly numerical) have been carried out so far to investigate the response of the gas to an imposed bar potential. Qualitative behavior of the gas changes depending on the bar parameters (*e.g.*, Sanders and Tubbs 1980).

Existence of inner Lindblad resonances (ILRs) seems to be important in relation to the fueling problem. Theoretical studies have indicated that the gas initially located between ILR and co-rotation points is deprived of its angular momentum by the bar torque and accumulates in a ring near the ILR (*e.g.*, Schwarz 1984). This gas transfer takes place on the dynamical timescale, giving rise to a rapid gas inflow with a rate of up to ~ 10 $M_\odot$ yr^{-1}. Because the ILRs are located very close to the nucleus (< 1 kpc) for any realistic disk models, bars with ILRs provide one promising tool to move the interstellar gas to the direct vicinity of galactic nuclei. Simkin *et al.* (1980) have pointed out that their gas cloud simulation in a barred potential successfully provides the observed link between the active nuclei of Seyfert galaxies and the wide range of spiral, ring, and amorphous features in their disks.

Non-self-gravitating models are considered to be inapplicable if the gas component becomes dense. For example, the ultimate fate of the gas gathered at the ILR ring is not clear within the framework of non-self-gravitating gas dynamics.

2.2 Self-Gravitating Gas Dynamics in Given Gravitational Potentials

The gravitational field due to the stellar component is still given as an external field in this category. However the gas component is not only influenced by the stellar gravitational field but also by the gravitational field arising from the gas itself.

Effect of gaseous gravity on the global gas response to the bar-like potentials has been investigated by Huntley (1980). The pitch angle of the spiral arms outside the bar increases as the gas mass fraction increases, which is consistent with the observed trend that late-type (therefore gas-rich on the average) galaxies have arms which are more open.

In relation to the fueling problem, one interesting possibility is that the circumnuclear gas ring formed by bar forcing becomes gradually gravitationally unstable as more and more gas accumulates, and breaks up into several clumps (fragmentation). Numerical studies by Fukunaga and Tosa (1991) and Wada and Habe (1992) show that the gas clumps thus formed interact and collide between themselves and induce further inflow from the original ring to the nucleus. The ILRs are not always required. The rings formed secularly by the viscous effect show this behavior also (Fukunaga and Tosa 1991, see Fig. 1).

The minimum mass fraction of the gas for this mechanism to operate is ~ 10 percent and well within the observed range for late type spiral galaxies. Some barred galaxies actually display clumping in their nuclear molecular rings (Kenney *et al.* 1992), reinforcing the similarity to the numerical simulations.

Another interesting possibility is the so-called bars-in-bars mechanism. Shlosman *et al.* (1989) have analytically shown that under suitable conditions the gas inflow induced by a bar builds up enough gas concentration on a smaller scale so that a new small bar develops. Though not a numerical simulation, this work suggests a possibility that bar instability propagates towards the nucleus as a runaway process over several orders of magnitude in radius. Observational evidence for bars-in-bars is not certain at present. Gaseous bars inclined with respect to the larger stellar bars (Kenney *et al.* 1992) observed in some barred galaxies may be ascribed to this mechanism. It should however be noted that such a misaligned configuration is relatively easily realized in the gas response to a single bar *without* invoking self-gravity if there is ILR (Athanassoula 1992*b*). We do not know how bars-in-bars actually look like.

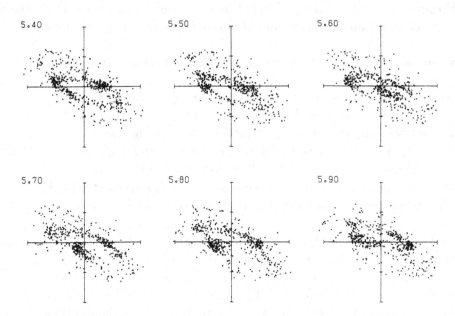

Fig. 1—Fragmentation and collapse of a gas cloud ring shown by Fukunaga and Tosa (1991).

2.3 Self-Gravitating Gas Dynamics in Live Stellar Potentials

In this category the stellar component is also allowed to respond freely to the change of the total gravitational field. In other words, the stellar and gaseous components are treated without discrimination as far as the gravity is concerned.

Pfenniger and Norman (1990) have shown that the growth of a significant central mass concentration in a barred galaxy induces broad radial resonance regions which can act to enhance the fueling. Thus bar-induced gas inflow can be a runaway process. A central mass concentration, in cooperation with bar perturbation, will also puff up the disk through the vertical resonances, possibly leading to the bulge formation. However, further growth of the central component will be limited because it will move the ILR towards the bar's end and destroy the main family of orbits which support the bar (Hasan and Norman 1990). Though these treatments are not fully self-consistent, the results mentioned above suggest that the growth of central gas concentration by bar-induced inflow may be important in global shaping of disk galaxies.

2.3.1 Dynamics of Two-Component Disks

Shlosman and Noguchi (1993) have investigated dynamical evolution of two-component galactic disks by means of 3D self-consistent simulations. Their disk models consist of collisionless stellar particles and collisional gas cloud particles. Disks are embedded in spherical halo components made of collisionless particles. Gravitational interactions between all the particles have been calculated by the tree code (Barnes and Hut 1986). Collisions between gas cloud particles are assumed to be inelastic with some portion of kinetic energy dissipated away. Fundamental parameters which specify the model are the total mass of the disk, f_d, the mass of gas component, f_g, and the restitution coefficient, f_{coll}. The radial component of relative velocity of two colliding particles is multiplied by $f_{coll}(< 1)$ and reversed, while the tangential component is kept unchanged. Total mass of the galaxy is set to be unity so that the halo occupies the mass $1 - f_d$ and the mass of the stellar disk is $f_d - f_g$. Initial Q–values (Toomre 1964) for the stellar and gaseous disks are set 1.5 initially for all the models.

Three families of models have been run with different values of f_d. The disk mass f_d is 0.2, 0.5 and 1.0 for model families A, B, and C, respectively. The purely stellar model (*i.e.* $f_g=0$) in A-family is stable against bar formation, whereas its counterparts in B and C are violently bar unstable.

Models from B-family provide a good example of how the gaseous gravity affects disk evolution. When f_g is small (< 0.05), the gas has negligible effect on the disk evolution. The stellar disk develops a strong bar within a few rotation periods (Fig. 2), just as in the case of the purely stellar model of the same family. Kinematics of the gas clouds is governed by the stellar gravitational field. Gas inflow to the nuclear region is induced by the bar potential.

When f_g exceeds ~ 0.05, the disk evolution is drastically changed. The stellar disk does not form any appreciable bar structure but remains nearly axisymmetric except transient spiral structures (Fig. 3). Many massive clumps are observed in the gas disk. The clump mass is typically few times 10^7 $M_\odot$ which could be compared to the giant molecular associations such as observed in M51 (Rand and Kulkarni 1990). Formation of massive gas clumps is closely related to the suppression of bar instability as shown in Figure 4, where the clumping in the gas disks is indicated as a function of f_g and f_{coll}. When f_g is small and/or f_{coll} is large, no clumps are formed (open circles) or only transient clumps appear (triangles). A stellar bar develops in this domain. When f_g is large and f_{coll} is small, gas clumps are stable and maintain their identity for a long time (filled squares). In this case, growth of gravitationally unstable perturbations is aided by efficient energy dissipation. The domain of stable clumps agrees with that of no bar instability.

The mechanism of the bar formation is not fully understood yet, despite much dispute (*e.g.*, Lynden-Bell 1979). Analysis of dominant orbits in strong bars reveals their alignment with the bar major axis (*e.g.*, Athanassoula 1992a). It is inferred that the growing correlation between increasingly elongated orbits is a prerequisite for bar formation. Shlosman and Noguchi (1993) have argued that massive clumps, as observed in Figure 3, will scatter stellar particles in a random manner and thus destroy coherent behavior required for bar growth. In other words, the clumps heat

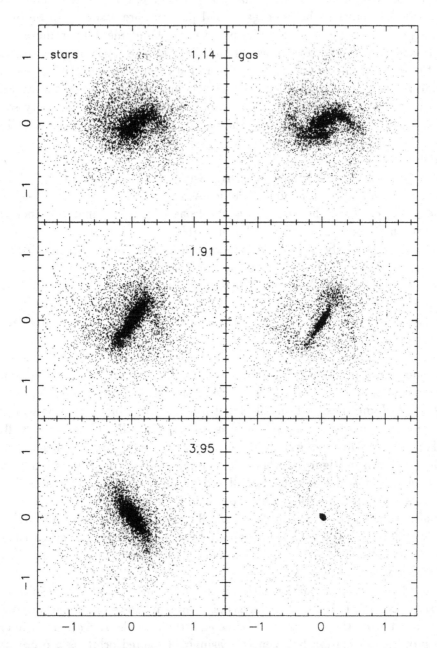

Fig. 2—Morphological evolution of a gas-poor disk model in which the stellar and the gas disks occupy 49% and 1% of the total galaxy mass, respectively (Shlosman and Noguchi 1993). The disk is rotating counter-clockwise. The time in units of the disk rotation period is given in each panel.

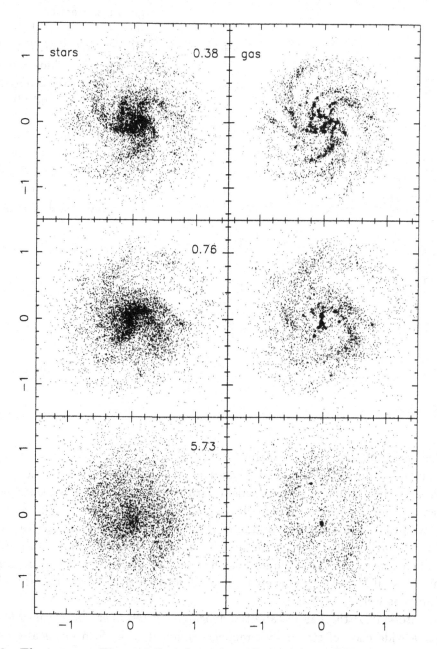

Fig. 3—The same as Figure 2, but for a gas-rich model in which the stellar and the gas disks occupy 40% and 10% of the total mass, respectively (Shlosman and Noguchi 1993).

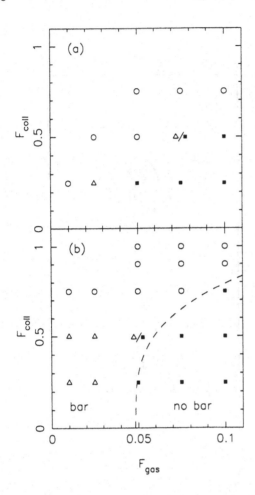

Fig. 4—Clumping behavior as a function of the gas mass fraction, F_{gas} (f_g in the text), and the restitution coefficient, F_{coll} (f_{coll} in the text) (Shlosman and Noguchi 1993). Open circles, triangles, and filled squares indicate the formation of no clumps, transient clumps, and stable clumps, respectively. Dashed line delineates the boundary between bar instability and stability.

the stellar disk component dynamically.

Gas inflow induced by bar forcing does not take place in the gas-rich cases. However, a significant inflow of gas clouds towards the nuclear region is seen (Fig. 3). Shlosman and Noguchi ascribe this infall to dynamical friction. The gas clumps, being much heavier than stellar particles and moving a little faster than stellar particles due to asymmetric drift, suffer from dynamical friction and gradually spiral to the disk center. The dynamical-friction-induced gas flow is very efficient, with the maximum infall rate of ~ 10 $M_\odot$ yr^{-1}. This new type of inflow occurs in the gas-rich models of the A-family as well.

The criterion for the bar instability advocated by Ostriker and Peebles (1973) works for a wide class of stellar disk models though it lacks firm physical ground. Using this criterion, Shlosman and Noguchi have derived a simple condition for bar instability in the two-component disks as follows. The Ostriker-Peebles' criterion for bar instability is given by

$$t = T_{rot}/(-W) > 0.14, \tag{1}$$

where T_{rot} is the total energy of rotational motion of the disk and W is the total potential energy of the system. Another form convenient here is

$$t \sim \frac{f_d}{2(2\alpha^2 + 1)} > 0.14, \tag{2}$$

where $\alpha (= \sigma / <v_\phi>)$ is the velocity dispersion in the disk, σ, normalized by the typical rotational velocity, $<v_\phi>$ (Frank and Shlosman 1989).

Effect of self-gravitating gas added to the system is two-fold. First, active disk mass is reduced by f_g because the gas, due to its collisional nature, cannot participate in bar formation process which requires coherent alignment of elongated orbits. Second, the gas dynamically heats the stellar disk by forming massive clumps which randomly scatter the stellar particles.

The heating process by massive clumps was calculated by Lacey (1984):

$$\frac{d\sigma^2}{d\tau} \sim \gamma \sigma^{-2}, \tag{3}$$

where τ is the time, σ is the *stellar* velocity dispersion, and the coefficient γ is defined as

$$\gamma = 2N_{cl} M_{cl}^2 \omega F(\beta) \ln\Lambda. \tag{4}$$

Here, N_{cl} and M_{cl} are the surface number density and individual mass of clumps, ω is the vertical epicyclic frequency, F takes the value between 0 and 3.1 depending on the rotation law $\beta = 2\Omega/\kappa$, where Ω and κ are orbital and radial epicyclic frequencies, respectively. Λ is the Coulomb logarithm.

Assuming that the gas clumps in the present galaxy models are due to the local gravitational instability,

$$M_{cl} = \frac{\pi^5 \Sigma_g^3}{\kappa^4}, \tag{5}$$

where Σ_g is the gas surface density. Noting that $N_{cl} M_{cl} = \eta \Sigma_g$, where η is of order unity, and integrating equation (3) and using equation (2),

$$t \sim \frac{f_d - f_g}{2[2(\alpha_o^4 + K f_g^4 \tau_{bar})^{1/2} + 1]}, \tag{6}$$

where $K \sim 2,000$ is a quantity not sensitive to the model parameters. α_o is the initial *stellar* velocity dispersion normalized by rotational velocity, τ_{bar} is the time scale of bar growth. In other words, $\tau_{bar} \sim$ orbital time, is the time available for gas clumps to heat the stellar disk before the bar will fully develop. By taking $t \sim 0.14$, equation (6) is solved to give the critical gas fraction, $f_{g,crit}$, required to suppress the stellar bar instability:

$$f_{g,crit} = [(\frac{f_d - f_{g,crit}}{0.56} - \frac{1}{2})^2 - \alpha_o^4]^{1/4} / (K\tau_{bar})^{1/4}. \tag{7}$$

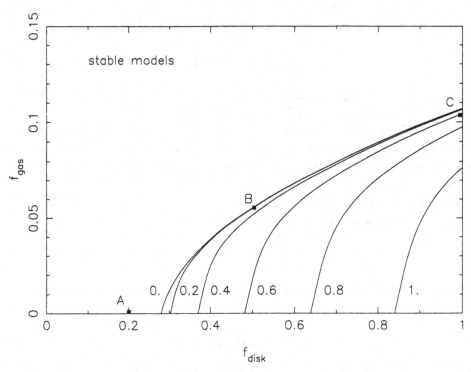

Fig. 5—The gas mass fraction required to suppress bar instability, f_{gas} ($f_{g,crit}$ in the text), plotted as a function of the disk mass fraction, f_{disk} (f_d in the text) (Shlosman and Noguchi 1993). Each line corresponds to the given value of α_o, *i.e.*, the intitial stellar velocity dispersion normalized by the rotational velocity. Critical gas mass fractions numerically determined for model families A, B, and C are indicated by filled squares.

Figure 5 plots $f_{g,crit}$ against f_d for various values of α_o. Although an increasingly large f_g is required to suppress the bar instability as f_d increases, the gas of ~ 10 percent of the total mass is sufficient even when $f_d = 1$, *i.e.*, when there is no halo at all. The numerical results for model families A, B, and C agree with this criterion.

The main results by Shlosman and Noguchi, namely the suppression of stellar bar instability and the emergence of new type gas infall driven by dynamical friction in gas-rich galactic disks, indicate that the gas component must have played a dynamically important role in the evolution of disk galaxies in the early cosmological epoch. The heating of stars by massive gas clumps is not a new idea. Forty years ago, Spitzer and Schwarzschild (1953) have proposed that the gravitational scattering by the interstellar clouds has caused a gradual increase of the stellar velocity dispersion for increasingly older populations observed in the solar neighborhood. The main point in Shlosman and Noguchi is that the effect of this mechanism is not necessarily of the first order, but can be large enough to affect the global structure of the galactic disks. Theoretically predicted value of $f_{g,crit}$ is well in the range of gas mass fraction observed in the late-type spiral galaxies at the present epoch. Rich interstellar gas may be working as a stabilizing agent in addition to massive haloes in these galaxies.

3. STEPS TOWARD MORE REALISTIC TREATMENTS

We have seen that the self-gravitating interstellar gas can cause many interesting dynamical phenomena even if its mass fraction relative to the total galaxy mass is small. The behavior of actual interstellar gas will be more complicated due to star formation and related processes. Giant molecular clouds and giant molecular associations are considered to be sites of active star formation. Energy input from stellar winds and supernova explosions may eventually destroy these gas clumps. Depletion due to star formation will make the gas component less important dynamically. To avoid overestimating the dynamical effect of gas clumps, the star formation process must be correctly incorporated in the models.

Numerical treatment of star formation process in galactic dynamics is still in a preliminary stage. Nevertheless the recent study by Heller and Shlosman (1993, and these proceedings) suggests that inclusion of star formation process changes the dynamics of two-component disks and brings about a variety of behavior in the central region of disk galaxies, sometimes in the presence of supermassive black holes there. These results suggest the importance of further investigation into the role of self-gravitating and star forming interstellar gas in galaxy evolution on both the global and nuclear scales.

Acknowledgement: The author thanks the SOC of this conference and the Japanese Ministry of Education, Science and Culture for financial support.

REFERENCES

Athanassoula, E. 1992a, *Mon. Not. R. Astr. Soc.*, **259**, 328.

Athanassoula, E. 1992b, *Mon. Not. R. Astr. Soc.*, **259**, 345.

Barnes, J. E., and Hut P. 1986, *Nature*, **324**, 446.

Begelman, M. C., Brandford, R. D., and Rees, M. J. 1984, *Rev. Mod. Phys.*, **56**, 255.

Frank, J., and Shlosman, I. 1989, *Ap. J.*, **346**, 118.

Fukunaga, M., and Tosa, M. 1991, *P. A. S. J.*, **43**, 469.

Hasan, H., and Norman, C. 1990, *Ap. J.*, **361**, 69.

Heller, C., and Shlosman, I. 1993, *Ap. J.*, in press, also these proceedings.

Huntley, J. M. 1980, *Ap. J.*, **238**, 524.

Kenney, J. D. P. 1990, in *The Interstellar Medium in Galaxies*, eds. H. A. Thronson, Jr., and J. M. Shull (Dordrecht: Kluwer), p. 151.

Kenney, J. D. P., Wilson, C. D., Scoville, N. Z., Devereux, N. A., and Young, J. S. 1992, *Ap. J. Lett.*, **395**, L79.

Knapp, G. R. 1990, in *The Interstellar Medium in Galaxies*, eds. H. A. Thronson, Jr., and J. M. Shull (Dordrecht: Kluwer), p. 3.

Lacey, C. G. 1984, *Mon. Not. R. Astr. Soc.*, **208**, 687.

Lynden-Bell, D. 1979, *Mon. Not. R. Astr. Soc.*, **187**, 101.

Ostriker, J. P., and Peebles, P. J. E. 1973, *Ap. J.*, **186**, 467.

Pfenniger, D., and Norman, C. 1990, *Ap. J.*, **363**, 391.

Rand, R. J., and Kulkarni, S. R. 1990, *Ap. J. Lett.*, **349**, L43.

Sanders, R. H., and Tubbs, A. D. 1980, *Ap. J.*, **235**, 803.

Schwarz, M. P. 1984, *Mon. Not. R. Astr. Soc.*, **209**, 193.

Shlosman, I., Begelman, M. C., and Frank, J. 1990, *Nature*, **345**, 679.

Shlosman, I., Frank, J., and Begelman, M. C. 1989, *Nature*, **338**, 45.

Shlosman, I., and Noguchi, M. 1993, *Ap. J.*, **414**, 474.

Simkin, S. M., Su, H.-J., and Schwarz, M. P. 1980, *Ap. J.*, **237**, 404.

Spitzer, L., and Schwarzschild, M. 1953, *Ap. J.*, **118**, 106.

Toomre, A. 1964, *Ap. J.*, **139**, 1217.

Wada, K., and Habe, A. 1992, *Mon. Not. R. Astr. Soc.*, **258**, 82.

Radio Loud Far-Infrared Galaxies

Arjun Dey[1], and Wil van Breugel[2]

[1]University of California at Berkeley
[2]Institute of Geophysics and Planetary Physics,
  Lawrence Livermore National Laboratory

ABSTRACT

We discuss the results from a study in progress of radio–loud *IRAS* galaxies. We have discovered a class of gas-rich AGN which are characterized by large IR luminosities, and which are intermediate in radio luminosity, IR colour, and optical spectral class between the *IRAS* ultraluminous galaxies and powerful quasars. These objects form an important sample in studies of the AGN–starburst connection and may be the evolutionary link between the ultraluminous galaxies and quasars.

1 INTRODUCTION

The relationship between nuclear activity and galaxy evolution is poorly understood. Speculation on the subject ranges between ideas that Seyfert activity is due to nuclear starbursts, to the theory that the ultraluminous ($L_{\mathrm{FIR}} \gtrsim 10^{12}$ L$_\odot$) far-infrared galaxies like Arp 220 are dust enshrouded, young quasars (Sanders *et al.* 1988). Unfortunately there exists no sample of active galaxies that can be used to test these various hypotheses since most samples used in these studies are selected either on the basis of optical emission-line equivalent width, or *IRAS* colour selection criteria, *IRAS* luminosity, *etc.* There is no ideal sample, but a good way of studying possible nuclear activity/evolution relationships is to select gas–rich galaxies with nuclear activity. We have selected sources from the *IRAS* survey (which is biased toward dusty, gas–rich objects) by using a completely independent indicator of galaxy activity: strong non-thermal radio emission. This sample of gas–rich radio galaxies also allows us to investigate the complex relationships between the ambient interstellar medium, the radio source, and the active nucleus.

2 SAMPLE DEFINITION AND OBSERVATIONS

The sample of galaxies was selected by cross-correlating the entire *IRAS* Faint Source Data Base (Moshir *et al.* 1992) and the Texas 365 MHz radio catalogue (Douglas *et al.* 1980). A search for objects with positive detections in the *IRAS* 60 μm band and with *IRAS*–radio positional coincidence less than 40″ resulted in a total list of 255 sources with |b| > 15°. The optical astrometry for this sample was done using the Guide Star astrometric facility at STScI. We searched published lists of active

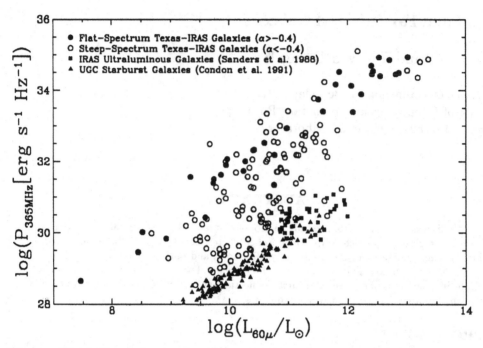

Fig. 1—*IRAS* 60 μm luminosity *vs.* 365 MHz radio power for the sample objects. The solid and open circles are the objects with flat and steep radio spectral indices resectively. Also shown for comparison are the ultraluminous galaxies of Sanders *et al.* and the UGC starburst galaxies from Condon *et al.*

objects and the recent literature for identifications with our sources and found that most of the sources with 60 μm flux densities greater than ∼ 3 Jy were known active objects (*e.g.* Arp 220, NGC 6240) or nearby bright galaxies listed in the NGC. In addition, about 25% of the fainter (*i.e.* $f_{60\mu} \lesssim$ 3 Jy) sources were also known active galaxies (*e.g.* BL Lac, OJ287, 3C48, 3C273, 3C120, 3C459, Mkn 348). We found no published information on the remaining 75% of the sources.

We mapped these remaining sources at the *VLA* to obtain morphological and radio spectral index information. Thus far we have snapshot observations at at least one frequency for the entire sample, and spectral index information for more than 50% of the sample. We also obtained optical images of all the fields using the 1m Nickel reflector at Lick Observatory and the 0.9m at CTIO. Optical spectroscopy of the candidates were then obtained using telescopes at Lick Observatory, ESO and CTIO.

3 RESULTS

There are a total of 123 matches (*i.e.* where the *IRAS* source and the radio source appear to be associated with the same object), of which we have good spectroscopic information on 102 sources. The median redshift of the sample is 0.1 for the objects with $f_{60\mu} < 3.16$ Jy, of which roughly 18% are BL Lac objects, 19% are quasars or Seyfert 1s, 16% are Seyfert 2s and 36% are LINERs. In addition, 11% of the sample are Post–Starburst AGN objects which exhibit both a Seyfert 2 emission line spectrum and also show evidence for a recent burst of star formation. A large fraction of the sample ($\sim 80\%$) have galaxy–scale radio sources (<8 kpc), and about 60% of these have steep ($\alpha < -0.4$) radio spectra. Nearly all of these sources belong to the class of Compact Steep Spectrum sources which generally have radio sources embedded in dense environments.

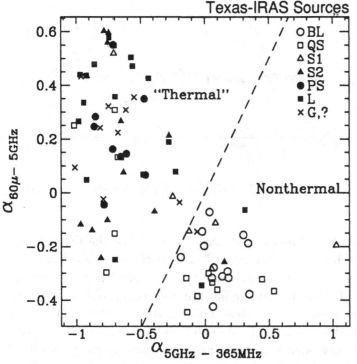

Fig. 2—Radio spectral index (4.85 GHz–365 MHz) vs. the IR–radio spectral index (60 μm – 365 MHz). The dashed line represents a pure power-law spectrum. The symbols represent the optical spectral classifications. BL: BL Lac, Q: Quasar, L: LINER, PS: Post-Starburst AGN, S1: Seyfert 1, S2: Seyfert 2, and ×: galaxies with only weak emission and as yet unidentified sources.

Although the sample appears very diverse, the objects in the sample appear to straddle an intermediate range in their infrared and radio properties between powerful AGN and the ultraluminous galaxies of Sanders *et al.* Figure 1 shows the *IRAS* 60 μm luminosity plotted versus the 365 MHz radio power for the sample objects. Also shown are the *IRAS* ultraluminous galaxies (from Sanders *et al.* 1988) and normal UGC starburst galaxies (Condon *et al.* 1991). The ultraluminous galaxies appear to be just the high luminosity counterparts of the normal starburst galaxies. The tight relationship seen in the starburst galaxies is just the low radio frequency counterpart of the well-known 6 cm—60 μm correlation (de Jong *et al.* 1985; Helou *et al.* 1985). Although there is no apparent correlation between the 365 MHz power and the 60 μm luminosity for our sample galaxies (the diagonal spread being largely due to Malmquist bias), it is of note that most of the objects in the sample appear to scatter inbetween the envelopes defined by the *IRAS* ultraluminous galaxies and the flat radio spectrum, radio powerful BL Lacs and quasars.

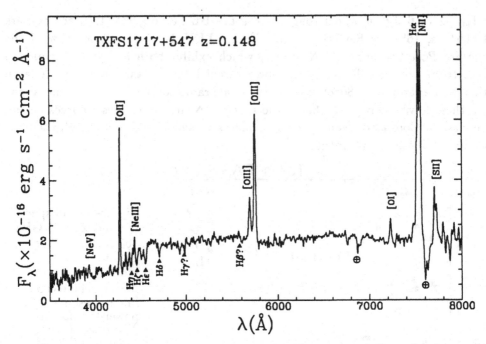

Fig. 3—Optical spectrum of TXFS1717+547, a $z = 0.148$ Post–Starburst AGN.

The objects in the sample can be divided into two groups on the basis of their spectral indices. Figure 2 shows a "colour-colour" plot of the radio spectral index plotted versus the far-infrared–radio spectral index. The dashed line represents a pure power-law spectrum. It is seen that the flat radio spectrum objects lie below this line (*i.e.* their infrared fluxes are consistent with a non-thermal origin) whereas the steeper radio spectrum objects ($\alpha_{\mathrm{radio}} < -0.4$) all have infrared excesses. Optical observations of the sample have shown that the flat spectrum objects are QSRs or objects that show continuous, featureless spectra and are probably BL Lac objects. On the other hand the steep spectrum sources are LINERs, Seyfert 2s and Post–Starburst AGNs.

The Post–Starburst AGNs are characterized in general by small radio sources with steep radio spectra, and compact optical morphologies. The spectra of these objects show both strong Balmer absorption lines as well as Seyfert 2 or LINER-like emission lines (*e.g.* Fig. 3). Some of these objects show no emission or absorption at the wavelength of Hβ although they show strong Hα emission. In this respect these objects show similar rest frame optical spectra to F10214+4724 (Elston *et al.* 1993). In addition, the radio powers of these objects are comparable to that of F10214+4724.

We have also discovered five new steep-spectrum *IRAS* quasars with redshifts greater than 1.5. All of these have asymmetric radio structures, exhibit infrared excesses and appear to have associated absorption in Lyα and C IV which indicate the presence of gas and dust in the vicinity of these objects. The broad-band continuum spectra of these quasars are similar to F10214+4724 in the optical and infrared and they have comparable bolometric luminosities. However, the radio power of these quasars is about two orders of magnitude greater than that of F10214+4724. These quasars may be objects similar to F10214+4724 in which the AGN is in the process of being unveiled.

4 CONCLUSIONS

We have found a class of gas–rich AGN with large far-infrared luminosities which are intermediate in radio luminosities, far-infrared colour, and optical spectral class between the *IRAS* ultraluminous starbursts and powerful quasars. These objects have very compact optical and radio morphologies and appear to have undergone a recent burst of star formation. In addition, we have discovered 5 new high redshift ($z > 1.5$) steep spectrum *IRAS* quasars which appear to have far-infrared thermal excesses, associated absorption, and asymmetric radio morphologies, all of which point to the presence of large quantities of gas around these objects.

We are now beginning a comprehensive observational program of spectropolarimetry, mm-line and VLBI observations of our steep spectrum sample to determine the physical characteristics of the AGNs, starbursts and their local environments in these intermediate type objects. This will allow us to conduct a systematic study of possible evolutionary connections between starburst and galaxy activity, and of AGN unification models.

We thank Jim Condon for allowing us to use his UGC galaxy data and Daniel Golombek for assistance with the astrometry. This project was performed at IGPP/ LLNL under the auspices of the U. S. Department of Energy under contract W-7405-ENG-48.

REFERENCES

Condon, J. J., Frayer, D. T., and Broderick, J. J. 1991, *A. J.*, **101**, 362.
de Jong, T., Klein, U., Wielebinski, R., and Wunderlich, E. 1985, *Astr. Ap. Lett.*, **147**, L6.
Douglas, J. N., Bash, F. N., Torrence, G. W., and Wolfe, C. 1980, *Univ. Texas Publ. Astron.*, **17**.
Elston, R., McCarthy, P. J., Eisenhardt, P., Dickinson, M., Spinrad, H., Jannuzi, B. T., and Maloney, P. 1993, *A. J.*, submitted.
Helou, G., Soifer, B. T., and Rowan-Robinson, M. 1985, *Ap. J. Lett.*, **298**, L7.
Moshir, M. *et al.* 1992, *Explanatory Supplement to the IRAS Faint Source Survey*, **2**, JPL D-10015 8/92 (Pasadena: JPL).
Sanders, D. B., Soifer, B. T., Elias, J. H., Madore, B. F., Matthews, K., Neugebauer, G., and Scoville, N. Z. 1988, *Ap. J.*, **325**, 74.

Massive Central Black Holes as Generators of Chaos and Drivers of Large-Scale Dynamics

Daniel Friedli

Geneva Observatory, CH-1290 Sauverny, Switzerland

ABSTRACT

Very dense clusters or massive black holes (MBH) located in the nuclei of disc galaxies generate inner Lindblad resonances (ILR) and chaos which then strongly modify the global dynamics. MBH with mass $M_{bh} \lesssim M_{bh}^{lim} \approx 0.02 M_d$, where M_d is the total stellar disc mass, round the central isodensity contours of barred galaxies, and the bar strength decreases with increasing M_{bh}. Near the horizontal ILR, gas can form a nuclear ring where star formation takes place and reduces the bar-driven gas inflow. Initial MBH with $M_{bh} \gtrsim M_{bh}^{lim}$ are sufficient to prevent an otherwise forming bar. If they slowly increase or are added in barred potentials, they dissolve the bar as soon as $M_{bh} \gtrsim M_{bh}^{lim}$, stopping the large scale gas fueling. MBH candidates for the upper end of the black hole mass function must preferably be searched in S0, Sa galaxies, some of them being turned-off quasars.

1 INTRODUCTION

Observations of nearby galaxies indicate non negligible (dark) mass in their nuclei (see *e.g.* Kormendy 1992) interpreted either as very dense clusters or MBH. The latter hypothesis is sustained by the widespread idea that MBH can be the engine powering AGN. However, there are less AGN at present time than at high redshifts z and it remains to be understood why some AGN are now turned-off. Whereas the influence of MBH on the local dynamics is indisputable, their effects on the global dynamics of multi-component systems remain to be explored in a self-consistent way. Indeed, although numerous previous studies have been devoted to this subject, they mainly concerned orbital studies in rigid triaxial potentials either non-rotating (Norman *et al.* 1985; Gerhard and Binney 1985; Gerhard 1986; Pfenniger and de Zeeuw 1989) or rotating (Martinet and Pfenniger 1987; Hasan and Norman 1990; Udry 1991; Hasan *et al.* 1993), collisionless N-body simulations (Norman *et al.* 1985), or the fate of dissipative particles in barred potentials (Pfenniger and Norman 1990). These studies indicate significant changes in the orbital structure around the nucleus and some of them predict possible modifications at larger scale such as bulge growth or bar dissolution. This assertion is corroborated by the fact that bar-driven gas fueling can finally lead to the destruction of the bar (Friedli and Benz 1993). Below, by means of self-consistent 3D numerical simulations with stars and gas, we explore in more detail the dynamical effects at different scales of such mass concentrations in nuclei of isolated disc galaxies. The MBH – bar – fueling connections are also underlined.

2 ACCRETION OF MASS INTO GALACTIC CENTERS

2.1 Basic Ingredients

As a result of dissipative and angular momentum transport processes, mass inflow has proved to be a quite common phenomenon in the universe. Gas represents the basic ingredient since it can both suffer from strong shocks when it is on self-intersecting orbits ($E_{\text{kinetic}} \rightarrow E_{\text{thermal}}$) and it can radiate ($E_{\text{thermal}} \rightarrow E_{\text{radiant}}$). If the gas has some angular momentum, it has to be efficiently removed and transported outwards by a torque, which corresponds to another essential ingredient. If the system contains clumps of matter, a third ingredient can be in action, dynamical friction.

2.2 Fueling in Bars at Various Scales

1) *Large scales* ($1 \lesssim r \lesssim 10$ kpc): Gravitational torques due to global non-axisymmetric perturbations of the potential such as bars or interactions are the most efficient tool to transfer angular momentum. Fueling is very efficient in bars having x_1 orbits with loops and/or strong curvature (*e.g.* in strong bars) and is proportional to S_{b}, the strength of the bar. 2) *Intermediate scales* ($0.1 \lesssim r \lesssim 1$ kpc): this scale corresponds about to the zone where the ILR can be present. Large scale bars typically compress gas by a factor of ten, and closer to the centre other mechanisms must be invoked: Shlosman *et al.* (1989) have suggested models of embedded bars; self-consistent systems of two bars rotating at two different pattern speeds are indeed feasible and allow gas to be transported deeper into the centre (Friedli and Martinet 1993). In weak bars, Wada and Habe (1992) have invoked a mechanism of gas self-gravitational instability. If a moderate horizontal ILR is present, gas can accumulate on a ring (x_2 orbits) where star formation is expected to occur; the gas inflow is stopped by this. However, if a MBH is present a chaotic region appears (see section 3.2) and the global bar-driven gas fueling is replaced by a local (inside ILR) chaotic gas fueling. 3) *Small scales* ($r \lesssim 0.1$ kpc): MBH dominate the potential ($\Phi \sim 1/r$) up to $r_{\text{lim}}[\text{kpc}] \approx 0.125(M_{\text{bh}}[\text{M}_\odot]/10^9)(200/\sigma[\text{kms}^{-1}])^2$ where σ is the typical velocity dispersion in the nucleus. This scale is not explored here (see *e.g.* Blandford 1990).

2.3 MBH Mass Function

The MBH mass function can be inferred either by direct observations of nuclei kinematics (Kormendy 1992), or by the observed luminosity of AGN where accretion rates are supposed to be close to the Eddington limit. These require $10^6 \lesssim M_{\text{bh}} \lesssim 10^{10}$ M$_\odot$ and the mean relic black hole mass per bright galaxy is $\approx 3 \cdot 10^7$ M$_\odot$. By using quasar luminosity functions, Small and Blandford (1992) constructed evolutionary models to infer the present MBH mass function. They found that a few percent of bright galaxies must harbour MBH with $M_{\text{bh}} \gtrsim 10^9$ M$_\odot$.

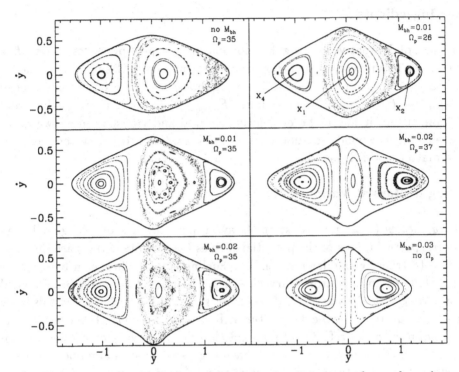

Fig. 1—Surfaces of section at $t = 3,000\,\mathrm{Myr}$ in the plane along the minor axis y. The Jacobi constant is $H = -0.25$. Values of M_{bh} [fraction of M_d] and pattern speed Ω_p [km s^{-1}kpc^{-1}] are indicated in the upper right of each frame. Top left: generic self-consistent barred model with no MBH. Middle and bottom left: MBH added in the generic potential. On the right: self-consistent potentials with initial MBH.

3 SIMULATIONS

3.1 Method and Models

In these simulations, the gravitational forces are estimated with a particle-mesh method using a 3D polar grid (Pfenniger and Friedli 1991, 1993). The hydrodynamical equations are solved with the SPH technique. Pure collisionless models (stars) as well as mixed models with stars and gas have been performed (see Friedli and Benz 1993). The central MBH have been treated as one softened collisionless particle of mass $M_{bh}(t)$. We choose the disc mass $M_d = 2 \cdot 10^{11}\,\mathrm{M}_\odot$ and explore a MBH mass range such that $0.005 M_d \leq M_{bh}(t) \leq 0.03 M_d$. The initial models are axisymmetric with MBH already formed or growing slowly and linearly at a rate of $1\,\mathrm{M}_\odot\mathrm{yr}^{-1}$. We also added MBH in barred models to see changes in the orbital structure and their subsequent evolution. The central softening ϵ_c has been chosen so that $r_{\lim} \lesssim \epsilon_c$.

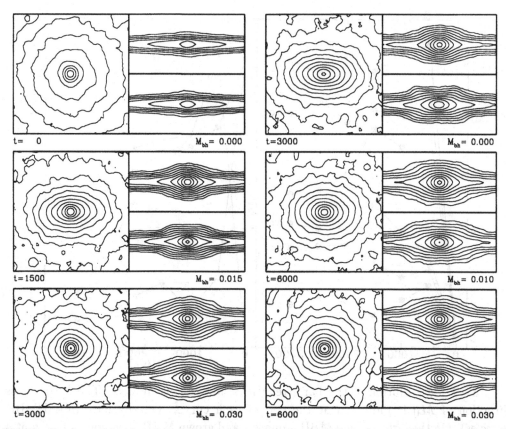

Fig. 2—Projected isodensities in face-on, edge-on and end-on views. The time t [Myr] and $M_{bh}(t)$ [fraction of M_d] are indicated below each frame. Top left and right: generic model with no MBH. Middle and bottom left: model with grown MBH. Middle and bottom right: models with added MBH. Note the bulge appearance.

3.2 MBH as Generators of Chaos

Previous studies in rigid fast rotating barred potentials in 2D (Hasan and Norman 1990) and in 3D (Hasan *et al.* 1993) showed how central chaos is generated by mass concentrations in the nucleus. By adding MBH in self-consistent N-body potentials, the ILR is reinforced and chaos increases with increasing M_{bh} (Fig. 1 left) and decreasing ϵ_c, affecting orbits with larger and larger radius. This phenomenon perturbs most the global dynamics of bars since chaos is "propagated" up to the apocentre of the elongated x_1. The central chaotic zone can be so extended that no stable periodic orbits remain to sustain the bar (x_1 unstable). *The self-consistent response of the potential acts to minimise this generation of chaos by reducing the phase space volume occupied by the x_1.* The retrograde x_4 and the direct x_2 dominate and the central region tends to integrability by becoming nearly spherical (Fig. 1 right).

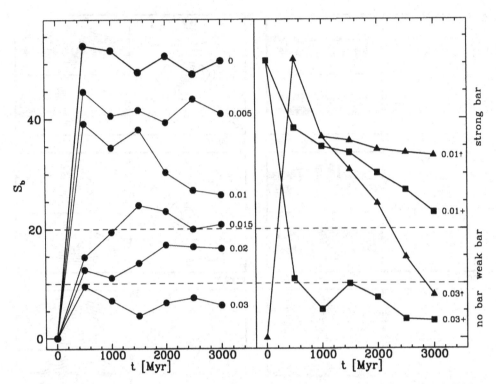

Fig. 3—Time evolution of the bar strength S_b for various models. The MBH mass [fraction of M_d] is indicated at the end of each curve. On the left: initial MBH (circles). On the right: added MBH (squares) and grown MBH (triangles). Horizontal dashed lines indicate a schematic subdivision.

3.3 MBH as Drivers of Large Scale Dynamics

The standard evolution of collisionless disc systems with the so-called box-peanut instability (Fig. 2 top right) induced by the vertical ILR have been extensively described in Combes *et al.* (1990) and Pfenniger and Friedli (1991). The self-consistent response to the addition in the nucleus of MBH strongly modifies the global dynamics (Figs. 2 and 3). If M_{bh}^{lim} is a threshold value, initial MBH with $M_{bh} \lesssim M_{bh}^{lim} \approx 0.02 M_d$ round the central isodensity contours of barred galaxies. The final S_b is progressively decreased with increasing M_{bh}. Initial MBH with $M_{bh} \gtrsim M_{bh}^{lim}$ are sufficient to prevent an otherwise forming bar. If they slowly increase or are added in a barred potential, they produce weaker bars and dissolve them as soon as $M_{bh} \gtrsim M_{bh}^{lim}$, stopping the large-scale gas fueling (see also Friedli and Benz 1993). Barred galaxies are very sensitive to the addition of MBH. In non-strongly interacting galaxies, we thus expect an upper bound to M_{bh} and a decrease of the fueling rate of the largest M_{bh} with time, consistent with models of Small and Blandford (1992). Moreover, some galaxies will change their Hubble type from SB → S (of course, the reverse can also occur).

4 MAIN CONCLUSIONS

1) Significant large scale bar-induced central mass accretion occurs in strongly barred galaxies but it can be much reduced: a) If gas forms a nuclear ring and star formation takes place near a moderate horizontal ILR, for instance as long as $M_{bh} \lesssim M_{bh}^{lim} \approx 0.02 M_d$. b) If the bar is dissolved by a strong ILR, for example as soon as $M_{bh} \gtrsim M_{bh}^{lim}$.

2) Mass accretion to the very centre of large scale strong bars can be either performed by a system of embedded bars (with different pattern speeds) or by chaotic fueling generated by central MBH.

3) MBH candidates for the upper end of the black hole mass function must preferably be searched in SO or Sa galaxies, some of them resulting from dissolved thick bars (see Fig. 2 bottom) and being turned off quasars.

4) Considerably less AGN are observed at $z = 0$ than at high z, although numerous bright galaxies must harbour relics of central active MBH. The existence of these turned off AGN can at least be explained by the lower rate of interactions at the present epoch and/or the dissolution of bars in galaxies having $M_{bh} \gtrsim M_{bh}^{lim}$.

REFERENCES

Blandford, R. D., 1990, in *Active Galactic Nuclei*, Saas-Fee Advanced Course no. 20, eds. T. J.-L. Courvoisier, and M. Mayor (Berlin: Springer), p. 161.

Combes, F., Debbasch, F., Friedli, D., and Pfenniger, D. 1990, *Astr. Ap.*, **233**, 82.

Friedli, D., and Benz, W. 1993, *Astr. Ap.*, **268**, 65.

Friedli, D., and Martinet, L. 1993, *Astr. Ap.*, in press.

Gerhard, O. E. 1986, *Mon. Not. R. Astr. Soc.*, **219**, 373.

Gerhard, O. E., and Binney, J. 1985, *Mon. Not. R. Astr. Soc.*, **216**, 467.

Hasan, H., and Norman, C. 1990, *Ap. J.*, **361**, 69.

Hasan, H., Pfenniger, D., and Norman, C. 1993, *Ap. J.*, **409**, 91.

Kormendy, J. 1992, in *Testing The AGN Paradigm*, AIP Conf. 254, eds. S. S. Holt, S. G. Neff, and C. M. Urry (New-York: AIP), p. 23.

Martinet, L., and Pfenniger, D. 1987, *Astr. Ap.*, **173**, 81.

Norman, C., May, A., and van Albada, T. S. 1985, *Ap. J.*, **296**, 20.

Pfenniger, D., and Friedli, D. 1991, *Astr. Ap.*, **252**, 75.

Pfenniger, D., and Friedli, D. 1993, *Astr. Ap.*, **270**, 561.

Pfenniger, D., and Norman, C. 1990, *Ap. J.*, **363**, 391.

Pfenniger, D., and de Zeeuw, T. 1989, in *Dynamics of Dense Stellar Systems*, ed. D. Merritt (Cambridge: Cambridge Univ. Press), p. 81.

Shlosman, I., Frank, J., and Begelmann, M. C. 1989, *Nature*, **338**, 45.

Small, T. A., and Blandford, R. D. 1992, *Mon. Not. R. Astr. Soc.*, **259**, 725.

Udry, S. 1991, *Astr. Ap.*, **245**, 99.

Wada, K., and Habe, A. 1992, *Mon. Not. R. Astr. Soc.*, **258**, 82.

Self-Gravitating Gas Dynamics: Growing Monsters and Fueling Starbursts in Disk Galaxies

Isaac Shlosman and Clayton H. Heller

Department of Physics and Astronomy, University of Kentucky, U. S. A.

ABSTRACT

The evolution of the gas distribution in a *globally* unstable galactic disk embedded in a 'live' halo is studied numerically on scales ~ 100 pc -10 kpc. The gas and stars are evolved using a 3D hybrid SPH/N–body code and gravitational interactions are fully accounted for. The gas is assumed to obey the isothermal equation of state with $T = 10^4$ K. The effect of a massive object at the disk center is simulated by placing a 'seed' black hole (BH) of 5×10^7 $M_\odot$ with an 'accretion' radius of 20 pc. Modifications introduced by star formation in the disk are discussed elsewhere (Heller and Shlosman, these proceedings).

We find that the global stability of a stellar disk can be heavily affected by the gas, given that the gas mass fraction f_g is high enough and the gas is dissipative. We also find that the rate of radial inflow in disk galaxies is a robust function of global parameters: the inflow is bar-driven for small f_g and dynamical friction-driven for large f_g. Without star formation the radial inflows lead to (1) domination of the central kpc by a few massive clouds that evolve into a single object via a cloud binary system; and (2) sporadic accretion onto the BH.

1 INTRODUCTION

Considerable effort to model the gaseous response to the non-axisymmetric background potential on the galactic scale was made during the last two decades (*e.g.*, Sorensen, Matsuda, and Fujimoto 1976; Huntley, Sanders, and Roberts 1978; Sanders and Tubbs 1980; van Albada and Roberts 1981; Schwarz 1984; Combes and Gerin 1985; Fukunaga and Tosa 1991; Athanassoula 1992; Wada and Habe 1992; *etc.*). These 2D studies have provided important insight into the formation and mainte-nance of spiral structure in disks. They have also established the existence of radial gas flows, but a lack of spatial resolution and limited dynamic range has impaired the ability to follow these flows for more than a decade in radius. In addition, most of these studies have neglected self-gravity and clumpiness in the gas, although these can be viewed as salient properties of the ISM. Dynamically self-consistent treatment was also avoided by *imposing* an oval/spiral distortion and ignoring the backreaction of gas on stars. Thus, the gas was used mainly as a tracer of the stellar gravitational field. Alternatively, some works emphasized the dynamical importance of gas in local and global phenomena in disk galaxies (*e.g.*, Julian and Toomre 1966; Lubow, Balbus, and Cowie 1986; Bertin *et al.* 1989; Norman 1988; Shlosman, Frank, and Begelman 1988, 1989). Fully self-consistent 3D studies of two-component disks have been per-formed by Shlosman and Noguchi (1993) and Friedli and Benz (1993) with the aim of understanding the gaseous response to the large-scale stellar bar instability.

There are indications that the *gas* plays an increasingly important role in the dynamics at progressively smaller radii in disk galaxies. The onset and development of runaway gravitational instabilities on scales $\lesssim$ 500 pc should, therefore, differ dramatically from those in the large-scale disk. In the present work we extend the previous analysis by Shlosman and Noguchi and concentrate on the gas evolution within the central kpc. The global effects of star formation are discussed by Heller and Shlosman (1993, hereafter HS93, and these proceedings).

2 LARGE-SCALE RADIAL FLOWS

To study the fate of dynamically important gas configurations in the central regions of disk galaxies we have constructed a set of globally unstable (*i.e.* bar-forming) models of galactic disks with halo-to-disk mass ratio 1:1 (within 10 kpc). These have been evolved in order to trigger the radial gas inflow. The gas mass fraction in the system f_g was varied and the gas was assumed to obey the isothermal equation of state with 10^4 K. The disk stars and collisionless halo particles have been evolved using an N–body scheme ($N = 26,000$). The gas was modeled using the advanced version of 3D SPH algorithm (HS93). Our modifications include dynamic gravitational softening, multiple timesteps, *etc.* The gravitational interactions between stars and gas have been calculated using the TREE algorithm (Barnes and Hut 1986; Hernquist 1987). For the initial conditions we have selected the Fall and Efstathiou (1980) disk–halo model. To simulate the effect of a massive BH we fix a single SPH particle (in some models) with 5×10^7 M$_\odot$ at the center-of-mass of the disk. This 'seed' BH accretes any *gas* particle within $r_{acc} = 20$ pc. Realistically, we expect magnetic torques to take over somewhere within this radius (Emmering, Blandford, and Shlosman 1992). Further details are given in Heller and Shlosman (1993).

The evolution of the gas distribution in disks depends critically on its ability to form and sustain inhomogeneites. The latter act as scattering centers and redistribute angular momentum. The gas is able to affect the stellar dynamics, if f_g > few percent of the total mass (gas-poor disks). For f_g >0.1–0.15 (gas-rich disks), the gas can suppress the globally unstable stellar mode (Shlosman and Noguchi 1993; Noguchi, these proceedings). We note that in the absence of an energy source (*e.g.,* OB stars) which heats the ISM and limits the lifetime of massive clouds, the effect of clumpiness tends to be overestimated (HS93, and these proceedings). On the other hand, ignoring the tendency of the ISM to clump leads to unrealistic models.

●For the gas-poor disks, gas trapped in the stellar bar forms a prominent and highly eccentric 'ring' with a semi-axis initially only slightly smaller than the bar radius (Fig. 1). The ring's shape is consistent with that of dominant periodic orbits in a strong bar which are aligned with the bar. Growing surface density in the ring results in the Jeans instability, creating nonuniformity in the velocity field, and internal shocks. The ring appears to be robust and the fragments stay within it. The hydrodynamical and gravitational interactions in the ring distort it and bring about

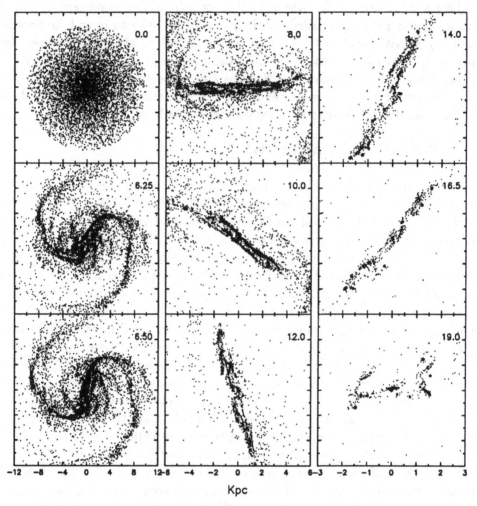

Fig. 1—Evolution of gas ($f_g = 1\%$) in the bar-unstable stellar disk with a 'seed' BH (asterisk), seen face-on. Time is given in units of the dynamical time (upper right corners). The frames are 24, 12 and 6 kpc across. The final BH is $\sim 6 \times 10^8$ M$_\odot$.

a gravitational torque from the stellar bar, depriving the gas of much of its angular momentum. After about two rotations the ring becomes heavily distorted, quickly loses its angular momentum, dumping the clouds in the vicinity of the BH.

•For the gas-rich disks, the stellar bar is damped by the massive gas clumps (in agreement with previous simulations using a 'sticky particle' code [Shlosman and Noguchi 1993]). Without the background bar potential, the gaseous ring does not form as well. Fragmentation in the gas may be related to the formation of GMCs in disk galaxies. Subsequently, these fragments experience dynamical friction with the background disk (due to stellar asymmetric drift) and spiral inwards (Fig. 2).

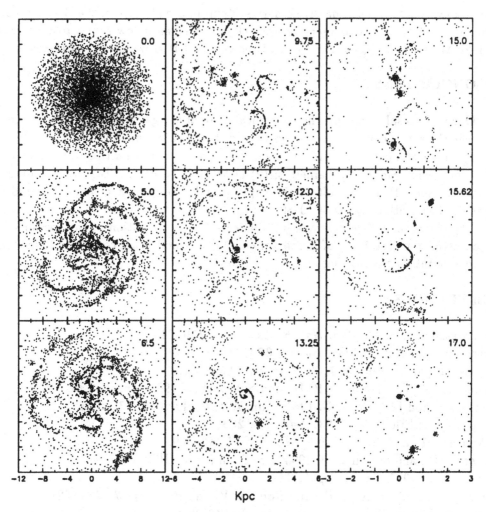

Fig. 2—Same as Fig. 1, but $f_g = 10\%$. The final BH is $\sim 2.5 \times 10^9$ M$_\odot$.

3 BUILDING STRUCTURE WITHIN CENTRAL KPC

Models without energy sources in the gas show a large degree of clumpiness in the central regions. Most of the gas in this area is concentrated in a few clouds which are distributed in the plane and have masses that are a sensitive function of f_g. The rest of the gas is either accreted by the BH or expelled to larger radii by tidal torques from the massive clumps which excite strong density waves within ~ 2 kpc. The clouds populate orbits which occasionally take them close to the BH where they are tidally disrupted and some fragments are captured by the hole, forming disks with radii ~ 60–80 pc (related to the limiting gravitational softening in the gas $\epsilon_g = 75$ pc).

Our results suggest that unless fragmentation is suppressed, the innermost gas in the disk evolves towards a massive binary system (in the absence of the BH) through a hierarchy of successive mergers between clouds. Median inflow rate across 1 kpc is $\sim$ 1–20 $M_\odot$ yr^{-1} (for f_g $\sim$0.1–0.15), and the central BH grows at $\sim$ half this rate.

4 CONCLUSIONS

Our main results concern the dynamical state of the gas within the central kpc, where in the absence of heating sources we find that (1) a few massive clouds dominate the dynamics and evolve towards a single object at the center (with and without the BH), for a wide range of parameters; (2) growth of the central BH has a sporadic character; and (3) as a result of the capture and digestion of clouds by the BH, remnant disks with a radius of 60–80 pc form.

The above models represent a limiting case for the evolution of the gas and its effect on the global stability of the stellar disk. These models are compared to models which account for the heating by massive stars.

REFERENCES

Athanassoula, E. 1992, *Mon. Not. R. Astr. Soc.*, **259**, 345.

Barnes, J., and Hut, P. 1986, *Nature*, **324**, 446.

Combes, F., and Gerin, M. 1985, *Astr. Ap.*, **150**, 327.

Emmering, R. T., Blandford, R. D., and Shlosman, I. 1992, *Ap. J.*, **385**, 460.

Fall, S. M., and Efstathiou, G. 1980, *Mon. Not. R. Astr. Soc.*, **193**, 189.

Friedli, D., and Benz, W. 1993, *Astr. Ap.*, **268**, 65.

Fukunaga, M., and Tosa, M. 1991, *P. A. S. J.*, **43**, 469.

Heller, C. H., and Shlosman, I. 1993, *Ap. J.*, in press. (HS93)

Hernquist, L. 1987, *Ap. J. Suppl.*, **64**, 715.

Huntley, J. M., Sanders, R. H., and Roberts, W. W. 1978, *Ap. J.*, **221**, 521.

Julian, W. H., and Toomre, A. 1966, *Ap. J.*, **146**, 810.

Lubow, S. H., Balbus, A., and Cowie, L. L. 1986, *Ap. J.*, **309**, 496.

Norman, C. A. 1988, in Proc. Caltech Conf. on *Star Formation in Galaxies*, ed. C. J. L. Persson (NASA CP-2466), p. 395.

Schwarz, M. P. 1984, *Mon. Not. R. Astr. Soc.*, **209**, 93.

Shlosman, I., and Noguchi, M. 1993, *Ap. J.*, **414**, 474.

Shlosman, I., Begelman, M. C., and Frank, J. 1990, *Nature*, **345**, 679.

Shlosman, I., Frank, J., and Begelman, M. C. 1988, IAU Coll. No. 134 on *Active Galactic Nuclei*, eds. D. Osterbrock, and J. Miller (Dordrecht: Kluwer), p. 462.

Shlosman, I., Frank, J., and Begelman, M. C. 1989, *Nature*, **338**, 45.

Sorensen, S. A., Matsuda, T., and Fujimoto, M. 1976, *Ap. Sp. Sci.*, **43**, 491.

van Albada, G. H., and Roberts, W. W. 1981, *Ap. J.*, **246**, 740.

Wada, K., and Habe, A. 1992, *Mon. Not. R. Astr. Soc.*, **258**, 82.

Radial Inflows in Disk Galaxies: Effects of Star Formation

Clayton H. Heller and Isaac Shlosman

Department of Physics and Astronomy, University of Kentucky, U. S. A.

ABSTRACT

Results of 3D numerical simulations of a two-component self-gravitating galactic disk embedded in a 'live' halo are presented. The pure stellar disk is chosen to be globally unstable and forms a stellar bar. A 'seed' black hole (BH) of 5×10^7 M$_\odot$ with an 'accretion' radius of 20 pc is placed at the center. The details on the numerical method are given in Shlosman and Heller (these proceedings and references therein). Here we study the effects of star formation (SF) which is introduced when (1) the gas is Jeans unstable; and (2) the gas density exceeds ~ 100 M$_\odot$ pc^{-3}. The formed massive stars account for the gas heating by means of line-driven winds and by supernovae, assuming an efficiency of kinetic-to-thermal energy conversion of a few percent. Models with and without SF are compared and only the robust features are emphasized.

We find that the SF has induced angular momentum loss by the gas and has increased the radial inflow by a factor < 3. We also find that (1) SF is concentrated at the apocenters of the gaseous circulation in the bar and in the nuclear region; (2) the nuclear starburst phase appears to be very luminous (quasar-like) and episodic; and (3) the nuclear starburst phase correlates with the catastrophic growth of the BH. Without the pre-existing BH, the gas at the center (few $\times 100$ pc) becomes dynamically unstable and forms a gaseous bar which drives a further inflow. The gaseous bar may *fission* into a massive cloud binary system.

1 INTRODUCTION

A rapidly mounting body of observational evidence speaks in favor of radial redistribution of the ISM in active galaxies and its accumulation in the nuclear regions where it becomes a significant fraction of the dynamical mass (*e.g.* Henkel, Baan, and Mauersberger 1991; Scoville *et al.*, Turner, these proceedings). Perhaps the most intriguing aspect of high-resolution observations of cold gas in disk galaxies is the prevalence of morphologically disturbed gas in the inner regions. The molecular gas within a kpc of galactic centers is found in elongated structures described as molecular bars or nuclear rings. The former morphology shows clear noncircular motions. In addition, the nuclear gas seems to be highly non-uniform. Gaseous barlike structures have been observed on scales down to 100–300 pc and are typically a few times smaller than the large-scale stellar bars. They also may contain solid-body rotating disks or rings in the innermost regions.

Large concentrations of molecular gas are accompanied by extensive star formation (SF) and/or accretion onto the massive BHs at the centers (*e.g.* Heckman, these proceedings and references therein). There are indications that nuclear starbursts require the initial mass function to be skewed towards more massive stars (*e.g.* Rieke *et al.* 1993). Massive stars recycle the ISM and show a tight correlation with the distribution of GMCs on the galactic scale. They also provide the major energy source for the ISM by means of line-driven winds and supernovae. Both the winds and supernovae inject about equal amounts of energy, though on different timescales (*e.g.* Salpeter 1976). Only a few percent of this energy is typically retained by the ISM, the rest is radiated away. Possible exceptions are nuclear starbursts where the heating efficiency of the ISM can be much higher.

Numerical simulations of gas evolution in a two-component gas+stars galactic disk (Frieli and Benz 1993; Heller and Shlosman 1993; Shlosman and Noguchi 1993) have shown that radial inflow towards the central kpc develops in response to the growing stellar barlike mode. When an isothermal equation of state for the gas has been used together with sufficient dynamic resolution, the gas has exhibited a high degree of clumpiness, *i.e.*, was Jeans unstable. Apart from numerical difficulties, local gravitational instabilities in the gas are expected to lead to SF with the subsequent energy deposition in the ISM. In this work we aim at understanding the physical processes which lead to nuclear starbursts and to the rapid growth of the central BH. Although the detailed physics of SF is highly model dependent, some robust behavior can be inferred and some general statements can be made.

2 RESULTS

The models presented by Shlosman and Heller (these proceedings) have been evolved in the presence of massive SF and a subsequent energy deposition by OB stellar winds and supernovae. A critical density of ~ 100 $M_\odot$ pc^{-3} in a locally gravitationally unstable gas was used as a threshold to activate the OB stars. Such a density is ~ 100 times above the average density (< 5 kpc) in our models and is almost 10^4 times the density of the solar neighborhood. The numerical SF event results in the deposition of energy into the gas and an increase in the gas pressure which is maintained over a period of time (see Heller and Shlosman 1993, for further details). The SF produces a negative feedback on the growth of Jeans instabilities in the gas and limits the lifetime of the Jeans unstable gas. The SF in our models is a self-regulating process and only the rate of energy deposition by a single OB star needs to be prescribed.

On a larger scale, the SF affects the gas circulation within the stellar bar, preventing fragmentation there. At the same time, the SF induces shocks in the flow and mixing of gas with different specific angular momenta. As a result, the radial infall in the stellar bar grows faster and extends to smaller scales than in models without the SF. The SF is concentrated initially at the apocenters of the gaseous circulation (*i.e.*, eccentric 'ring' [see also Shlosman and Heller, these proceedings]) within the bar, and, at a later stage, in the nuclear region. In the gas-poor disks ($f_g < 5$–10%), *i.e.*, those with fully developed stellar bars, almost 90% of the gas within the bar region is channeled towards the center and becomes kinematically decoupled (Fig. 1). In the gas-rich disks more than 2/3 of the gas falls to the center due to dynamical friction

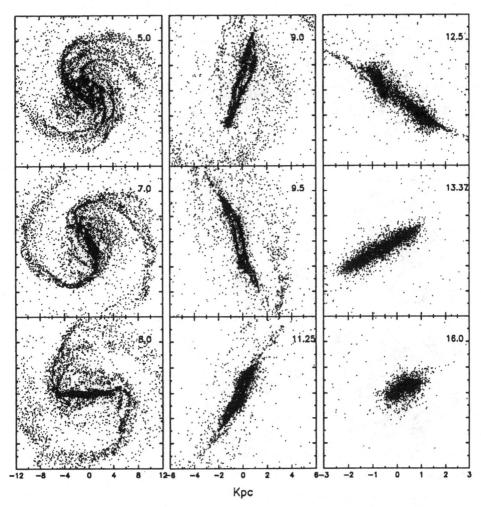

Fig. 1—The evolution of gas ($f_g = 1\%$) in the bar-unstable stellar disk seen face-on with a 'seed' BH (asterisk) and SF. The energy conversion efficiency is 5%. The time is in units of the dynamical time (upper right corners). The frames are 24, 12 and 6 kpc across. The final BH is $\sim 6 \times 10^8$ M$_\odot$.

(Fig. 2). The central (< 2 kpc) starburst phase is luminous $\sim 10^{45} - 10^{46}$ erg s^{-1} and an exhibits episodic character. A typical duration of a burst is 10^7 yrs, in excellent agreement with observations. At the starburst onset, the gas reaches $\sim 10 - 30\%$ of the total mass inside 1 kpc. At the same time, the gas is not in a dynamical equilibrium and $m = 1 - 4$ modes have significant amplitudes in the region. This coincides with the peak of the SF and catastrophic accretion onto the BH which digests most of the gas. The remnant gas cools down and settles in a geometrically thin nuclear disk with a radius of $\sim$0.5–1 kpc and a rotation axis slightly misaligned with the galactic axis.

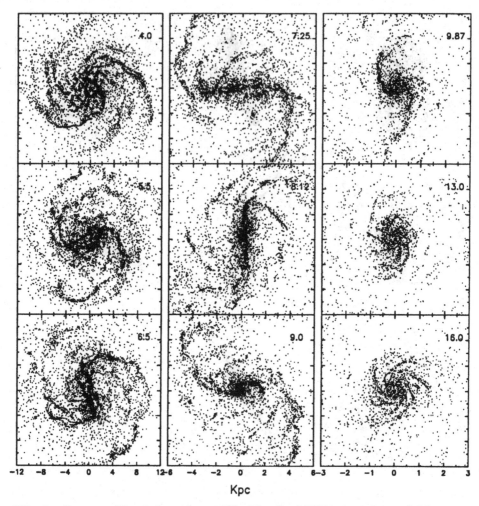

Fig. 2—Same as Fig. 1, but $f_g = 10\%$. The final BH is $\sim 2.8 \times 10^9$ M$_\odot$.

In the *absence* of the 'seed' BH, the gas is not able to cool down quickly as the SF rate monotonically increases with time. Instead, the gas forms a fat disk at the center supported both by gas pressure and rotation. An observational counterpart of such a thick disk may have been recently detected in the form of a molecular torus with a radius of ~500 pc in the starburst galaxy NGC 1808 (Koribalski, Dickey, and Mebold 1993). Dynamically, these nuclear disks are separate entities dominated by self-gravity with the stability parameter $t \lesssim 0.1$ (for definition of t see *e.g.* Binney and Tremaine 1987). We have demonstrated that these disks become globally unstable if and when the SF declines. The barlike mode then dominates and drives a self-similar inflow towards even smaller scales. Our models with high spatial resolution show that the gaseous bar has a tendency to fission followed by the formation of a massive cloud binary. More theoretical work is required to understand the relationship of this inflow to the formation of the central BH itself. Observationally, the difficulties in detection of such dynamically unstable formations of molecular gas in the galactic centers are amplified by their short lifetimes of $< 10^7$ yrs.

3 CONCLUSIONS

Models of bar-unstable two-component galactic disks with SF show suppressed fragmentation in the bar region. In the presence of the central 'seed' BH, these disks are characterized by (1) mixing in the gas confined to the stellar bar, inducing an increase in the angular momentun loss and inflow rate by a factor < 3; (2) SF which is concentrated at the apocenters of the gas circulation in the bar and in the nuclear region; (3) a luminous, episodic, central starburst phase, with a typical burst duration of $\sim 10^7$ yrs; (4) a correlation between the nuclear starburst, the peak accretion rate onto the BH, and the gas-to-dynamical mass ratio reaching ~ 0.1–0.3; and by (5) formation of cold nuclear disks within a few hundred pc.

Nuclear disks also form in models without the 'seed' BH. The starburst phase seems to be prolonged in this case and the disks remain hot and geometrically thick. The subsequent decline in the SF results in the cooling down of the gas and the onset of a bar instability. We find that further evolution leads to a self-similar inflow towards smaller scales, when Jeans instability is suppressed, and otherwise to a fission in the gaseous bar and formation of a massive cloud binary system at the center.

Our modeling has exposed alternatives in the dynamical evolution of self-gravitating gas which may ultimately lead to a luminous starburst and the formation of a supermassive object at the center of an active galaxy. We show numerically that these phenomena may be causally related and constitute a normal evolutionary phase for disk galaxies. We also find that the SF profoundly alters the evolution of the gas distribution in the inner disk.

Acknowledgements: We thank Mitch Begelman, Juhan Frank and Masafumi Noguchi for numerous discussions.

REFERENCES

Binney, J., and Tremaine, S. 1987, in *Galactic Dynamics*, Princeton Univ. Press.
Friedli, D., and Benz, W. 1993, *Astr. Ap.*, **268**, 65.
Heller, C. H., and Shlosman, I. 1993, *Ap. J.*, in press.
Henkel, C., Baan, W. A., and Mauersberger, R. 1991, *Astr. Ap. Rev.*, **3**, 47.
Koribalski, B., Dickey, J. M., and Mebold, U. 1993, *Ap. J. Lett.*, **402**, L41.
Rieke, G. H., Loken, K., Rieke, M. J., and Tamblyn, P. 1993, *Ap. J.*, in press.
Salpeter, E. E. 1976, *Ap. J.*, **206**, 673.
Shlosman, I., and Noguchi, M. 1993, *Ap. J.*, **414**, 474.

Self-Gravitating Gas Dynamics in a Galactic Central Region

Keiichi Wada and Asao Habe

Department of Physics, Hokkaido University, Sapporo, Japan

ABSTRACT

Self-gravity is a key determinant of gas dynamics, especially in a galactic central region. We have investigated self-gravitating gas dynamics with 2-D PM and SPH methods. From simulations of a massive gas disk inside the first ILR, we found a rapid gas fueling accompanied by a forming gas bar which lead the bar potential. The background bar potential and resonances are not important for dynamics of the central self-gravitaing gas in the accreting stage.

1 GAS FUELING PROBLEM

Starburst regions are frequently located in the central regions of barred galaxies or interacting galaxies. A number of studies has been made on triggering mechanism of starbursts, that is, mechanism fueling a large amount of gas into the starburst region. Many people believe that oval distortion of a background potential caused by galactic encounters or a stellar bar can trigger the gas rapid fueling. However, a number of numerical simulations which does not take into account the self-gravity of gas have revealed that the distorted potential itself cannot supply a large amount of gas into a galactic center beyond ILRs, although gas accumulate to form an oval ring near ILRs (*e.g.* Matsuda and Isaka 1980; Schwarz 1985).

2 SELF-GRAVITY OF THE GAS

2.1 Fueling by Collapse of an Elongated Gas Ring

Fukunaga and Tosa (1991), and Wada and Habe (1992) reported that a very elongated gas ring leading a weak background bar potential is formed near ILRs provided that a pattern speed of bar is just below a maximum of $\Omega - \kappa/2$. The gas ring is usually gravitationally unstable; collapse of the ring is triggered by forming dense gas clumps at both sides of the elongated ring. As a result, a large amount of gas ($10^{8-9} M_\odot$) can be supplied to a region far inside the first ILR over a dynamical time scale ($\sim 5 \times 10^7$ yrs, see Fig. 6 in Wada and Habe 1992).

Our video movie of above simulations revealed clearly the collapsing process of the gas ring. We also found that after the gas ring collapses and forms a dense core

($\sim$ 1 kpc size) in the center, the highest density region still has an non-axisymmetric structure, such as an oval ring or a bar. These features, which change on a very short time scale ($\sim 10^{6-7}$ yrs), are traces of the elongated ring formed near ILRs about 10^8 yrs before. We notice that the gas self-gravity dominates dynamics of gas in the fueling stage. That is to say, the background barred potential and the resonances are not important at this stage.

Elongated nuclear rings (they sometimes look like a bar) are often seen in self-consistent gas-star simulations (Barnes and Hernquist 1991; Friedli and Benz 1993; Combes and Elmegreen 1993; Heller and Shlosman 1993 and these proceedings). Is there a common mechanism forming the elongated gas ring in these simulations? We expect that physical reason of the gas ring formation is different in a weak bar and in a strong one. It needs further consideration.

2.2 Gas Motion Inside the First ILR

Observations of starburst galaxies have shown that the molecular masses in the central few hundred parsecs constitute several tens percent of the dynamical mass in the region (see J. Turner, these proceedings). We are interested in the response of a massive gas component inside ILRs to a weak barred potential. In order to analyze dynamics of such self-gravitating gas in the galactic central region by numerical simulations, high numerical resolution is necessary. Since we calculated gas motion from corotation to central (few kpc) region and a fixed Cartesian grid is used for calculating self-gravity, the models in our previous simulations are not advantageous for dealing with massive gas in a galactic center. Therefore we simulate a gas disk only inside the first ILR using a fine grid and a large number of SPH particles in the present paper.

In the following sections our numerical results are shown, and we discuss the origin of the observed asymmetric gaseous structure in galaxies.

3 MODEL AND METHOD

We assume an external potential: $\Phi(R, \theta, t) = \Phi_{\mathrm{axi}}(R)(1 + \epsilon(R) \cos(2\theta - \Omega_b t))$, where Φ_{axi} is the axisymmetic component of the potential, Ω_b is the pattern speed of a barred potential, and $\epsilon(R)$ is the coefficient which represents 'strength of the bar'. We assume a very weak bar, that is, $\epsilon(R_{\mathrm{ILR}_1}) = 0.005$, where R_{ILR_1} is the first ILR radius. For Ω_b in our model, $R_{\mathrm{ILR}_1} = 0.4 R_{\mathrm{CR}}$, $R_{\mathrm{ILR}_2} = 0.6 R_{\mathrm{CR}}$, where R_{CR} is a corotation radius. Gas is uniformly distributed on a disk with $R_{\mathrm{init}} = 0.5 R_{\mathrm{CR}}$ at $t = 0$ and it is treated as 10^4 K isothermal. The initial rotational velocity is given in order to balance the centrifugal force caused by the external potential and its self-gravity. Gas mass fraction to the dynamical mass is 15%.

We use two–dimensional particle–mesh code with FFT to calculate self-gravity of gas, and Smoothed Particle Hydrodynamics method to simulate gas motion. The number of particles is 10^4. Mesh size for calculating self- gravity is 80 pc.

4 RESULTS

Figure 1 shows evolution of gas distribution in the bar rotating frame. The major axis of the bar potential is parallel to the x–axis. After about 10^8 yrs, gas in the central region evolves into a sharp bar-like structure which has about 3 kpc length. Since the gas rotates anti-clockwise (Fig. 1), the gaseous bar leads the potential bar. Gas falls to the central region along a string-like structure from near the 1st ILR in the initial disk. The central high density bar grows denser and steeper ($t = 1.60$). That is to say, the pattern speed of gaseous bar is larger than that of the barred potential. This is caused by the angular momentum brought by the accreting gas. As a result, a spiral-like feature appears after about 10^7 yrs ($t = 1.76$). The whole gas distribution is also highly distorted in the direction leading 45° the bar major axis.

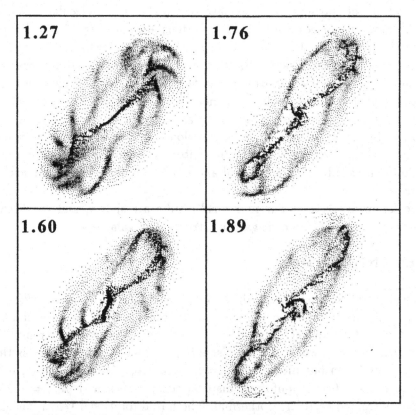

Fig. 1—Evolution of gas disk in rotating frame of the bar. Initial gas disk rotate anti-clockwise. Figures is time (10^8 yrs) from the calculation start. The major axis of bar is horizontally oriented in each frame. Each frame size is 12 kpc and the *first* ILR is located at $R = 4$ kpc at the start.

In Figure 2, velocity vectors of the gas are drawn for the model shown in Figure 1. We can see that the gas moves on oval orbits and two shocks have appeared at the leading sides ($\theta \sim 45°$) of the bar potential. At these shocks the gas loses its kinetic energy and angular momentum, and falls towards the center forming small clumps. The velocity of these clumps is ~ 200 km s^{-1}. These clumps eventually collide with the rotating spiral-like core and shocks are generated again. The gas mass is comparable to the dynamical mass in a region within 500 pc from the center.

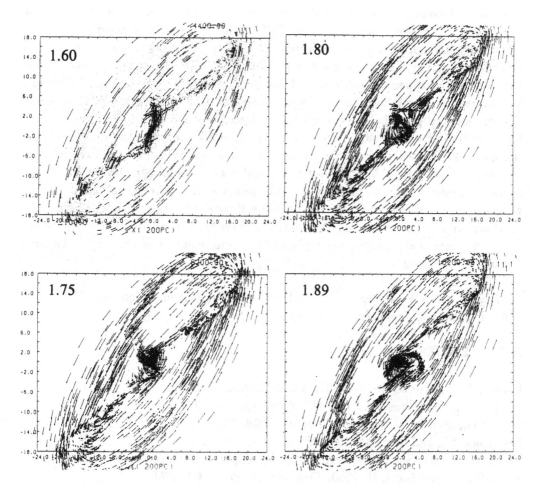

Fig. 2—Gas velocity map for the model shown in Figure 1. Unit length of x and y axes is 200 pc.

5 ORIGIN OF THE OBSERVED NON-AXISYMMETRIC GAS DISTRIBUTION

Recent CO observations with high numerical resolution reveal the structure and kinematics of molecular gas in the starburst or barred galaxies (Ishizuki 1990; see also

J. Turner, J. Kenney, these proceedings). The molecular gas has a large non-circular
motion in such galaxies, and also has various kinds of features such as the bar or
ridges (NGC 2782, IC 342, and M83), and double peaks (NGC 3504, 2782, 6951,
3351, and M101). There is an argument to explain the dynamics of the molecular
gas by the theory of the resonance — closed orbit families (x_1 and x_2) in a bar
potential (Kenney *et al.* 1992). However, we should be careful to apply this argument
to the molecular gas in the central region of the starburst galaxies. Because the
argument based on stellar orbits, not on gaseous orbits. Furthermore, self-gravity was
neglected. However, we observe that it complicates the evolution in the central region
even when a very weak bar is present. Self-gravity is a crucial determinant for gas
dynamics, although resonances are important to produce an initial non-axisymmetric
gas motion. In our model the gas bar leads the potential bar ($t = 1.27$ and 1.60 in
Fig. 1). This happens because of ILR at near outer edge of the initial disk and
the dissipation of gas; oval gas orbits near the first ILR in a weak barred potential
always lead the potential. The central gaseous bar rapidly evolves into a spiral-
like structure ($t = 1.76$ and 1.89 in Fig. 1). It would be expected, however, that
the gaseous bar evolves more slowly due to the gravitational torque exerted by a
stronger bar potential. In this case gas would gradually accrete toward the center
along the gaseous bar, and form two density peaks with trailing spiral arms. In fact,
we observe such structure in our recent 3-D self-consistent simulations (Wada 1993,
in preparation). The result suggests that the observed various features of molecular
gas in starburst galaxies can be explained by time-sequence of the self-gravitating
gaseous bar evolution.

One of us (KW) wishes to thank Hayakawa Yukio Foundation and The Astronomical Society of Japan for travel support.

REFERENCES

Barnes, J., and Hernquist, L. 1991, *Ap. J. Lett.*, **370**, L65.
Combes, F., and Elmegreen, B. 1993, *Astr. Ap.*, **271**, 391.
Friedli, D., and Benz, W. 1993, *Astr. Ap.*, **268**, 65.
Fukunaga, M., and Tosa, M. 1991, *P. A. S. J.*, **43**, 469.
Heller, C. H., and Shlosman, I. 1993, *Ap. J.*, in press; these proceedings.
Matsuda, T., and Isaka, H. 1980, *Prog. Theor. Phys.*, **64**, 1265.
Kenney, J., Wilson, C., and Scoville, N. 1992, *Ap. J. Lett.*, **395**, L79.
Handa, T., Ishizuki, S., and Kawabe, R. 1992, in IAU Colloq. 140 on *Astronomy with
    Millimeter and Submillimeter Wave Interferometry*, ed. S. Ishiguro, in press.
Ishizuki, S., *et al.* 1990, *Nature*, **344**, 224.
Schwarz, M. 1985, *Mon. Not. R. Astr. Soc.*, **212**, 677.
Wada, K., and Habe, A. 1992, *Mon. Not. R. Astr. Soc.*, **258**, 82.

Nuclear Inflow under the Action of Instabilities

Dimitris M. Christodoulou

Department of Astronomy and VITA, University of Virginia

ABSTRACT

During the stage of galaxy formation, or because of subsequent accretion events, cold, rotationally supported matter carrying substantial angular momentum is deposited and organized in disks/rings at the outer parts of many types of galaxies. Examples are warped H I disks in spirals, disks of gas and dust in ellipticals, and polar rings around S0s/ellipticals. The observed nuclear activity in many galaxies has its origin in mass transfer from large radii into the centers of galaxy potential wells. We describe two stages of such a mass transfer process and present related numerical multidimensional hydrodynamical simulations. In the first stage, gas still carrying some angular momentum flows toward the nuclear region as a moderately inclined outer disk is attempting to settle toward an energetically "preferred orientation." In the second stage, a nuclear accretion disk, influenced by the gravitational potential well of a central massive black hole and possibly by a coherent weak magnetic field, suffers one of several known dynamical nonaxisymmetric instabilities that drives gas deeper into the potential well and regulates its accretion onto the black hole.

1 INTRODUCTION

Although gas is predominantly found in spiral galaxies distributed in rotationally supported large-scale disks (~ 10 kpc), recent observations reveal the existence of counter-rotating gaseous disks in the central regions of many elliptical and S0 galaxies (Franx and Illingworth 1988; Bertola *et al.* 1990; Bertola, Buson, and Zeilinger 1992). The accepted interpretation of such observations is that gas in ellipticals and S0s must be of external origin. Cold gas deposited in the gravitational potential wells of these galaxies will generally carry angular momentum and is expected to form rings typically at ~ 10 kpc after only ~ 10 orbits (Rix and Katz 1991). Similarly, warped large-scale disks embedded in nonspherical halos form in cosmological simulations that include both dark matter and gas (Katz and Gunn 1992). Efficient physical mechanisms which can either redistribute or transport away a substantial fraction of the angular momentum in all gaseous disks/rings are not only necessary to explain the nuclear gas content of ellipticals/S0s, they could also explain how gas inflows from scales of ~ 10 kpc down to scales smaller than ~ 1 pc. Such inflow would provide a means of fueling a massive black hole (Gunn 1979; Rees 1980) and of producing the observed nuclear activity characteristic of active galactic nuclei (AGN).

It seems unlikely that a single angular momentum transport mechanism (AMTM) could account for shrinking of a disk by four or more orders of magnitude or that it would operate efficiently at all scales: the outer regions that are dominated by the overall gravitational field of a galaxy; the central region that may be influenced by a massive black hole (Rees 1980; Begelman, Blandford, and Rees 1984) and the presence of magnetic fields (Blandford 1989); and the intermediate transition regions. Consistent with such diversity of influences and length scales, Shlosman, Begelman, and Frank (1990; see also Begelman, Frank, and Shlosman 1989; and Frank, Shlosman, and Begelman 1989) proposed a flexible, unified scenario of AGN fueling that utilizes several AMTMs: the interaction of gas with a pre-existing bar or gas infall at kpc scales, the bar instability due to the self-gravity of the inflowing gas and fragmentation to interacting clouds at pc scales, and the possible formation of an unstable thin accretion disk or torus at 0.01 pc scales.

Each stage of such a complex scenario of gas inflow can be investigated through multidimensional hydrodynamical simulations designed to explore different AMTMs at different length scales. In what follows, we review results from simulations at two length scales. First we discuss the dynamical evolution of gaseous models that inflow down to ~ 1 kpc as they settle toward a preferred orientation of the overall gravitational field. Then we discuss the stability to nonaxisymmetric perturbations of systems composed of a central object and an orbiting thick disk that may be relevant to the structure of AGN to ~ 1 pc scales. Finally, the influence of a coherent weak magnetic field frozen into the gas of the latter models is also discussed.

2 GAS INFLOW TO KILOPARSEC SCALES

Simulations of inclined, precessing gaseous rings inside an external, slightly nonspherical potential indicate that the outcome of each evolution depends critically on the magnitude of differential precession across each ring. Precession varies strongly with ring inclination, approaching zero at polar orientations. Low-inclination models settle toward the equatorial preferred orientation of the potential without inflow (Steiman-Cameron and Durisen 1988, 1990). High-inclination models settle and inflow on time scales longer than a Hubble time, or if they are radially slender and the gas cools efficiently, they may survive close to their original high inclination for more than a Hubble time (Katz and Rix 1992; Christodoulou *et al.* 1992). At moderate inclinations, settling and inflow proceed on comparable time scales within ~ 10 orbits (Christodoulou and Tohline 1993).

Model evolutions become progressively more violent or even catastrophic at all inclinations as the quadrupole distortion of the potential is increased. The influence of self-gravity is also poorly understood (Peletier and Christodoulou 1993). The simulations so far lead us to believe that formation of 1 kpc nuclear disks is favored only for galactic accretion events at intermediate inclinations (~ 40°—60°) relative to the equatorial planes of massive, mildly nonspherical, halo potentials.

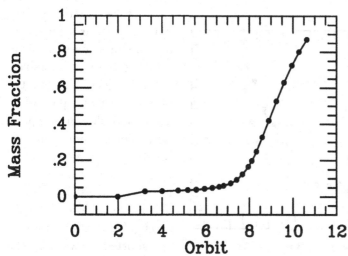

Fig. 1—Fraction of inflowing mass as a function of time for a massless torus inclined by 40° to the equatorial plane of an oblate spheroidal potential with axis ratio 0.95. The initial model shrinks by a factor of 10 in size within 10 orbits and forms a nuclear disk on the equatorial plane of the potential.

Fig. 1 shows the mass inflow as a function of time from a non-self-gravitating ring model initially inclined by 40° to the equatorial plane of an oblate spheroidal external potential with an axis ratio of 0.95. Matter inflow is recorded at about 0.1 of the radius of the initial ring. Gravitational torques applied by the external potential extract angular momentum from the ring which settles toward the equatorial plane and inflows. After only 10 orbits, a nuclear ring has formed on the equatorial plane of the potential. Additional model evolutions leading to nuclear disk formation are discussed in Christodoulou and Tohline (1993).

3 GAS INFLOW FROM PARSEC SCALES

A thick (pressure supported) and possibly self-gravitating accretion torus orbiting around a central mass is subject to nonaxisymmetric instabilities that can redistribute efficiently its angular momentum and drive matter inflow toward the central mass on dynamical time scales (Christodoulou and Narayan 1992, and references therein). Non-self-gravitating and low-mass tori suffer the Papaloizou-Pringle (1984) instability, while self-gravitating tori suffer a fission instability or, at high disk-to-central mass ratios, a Jeans instability (Goodman and Narayan 1988). The Jeans instability does not operate in radially extended, mildly self-gravitating tori. In the same regime of the parameter space, the other two instabilities can only disturb the structure of the torus without causing a complete breakup. They may then drive matter inflow to smaller scales without destroying the orbiting accretion torus (see also Hawley 1991). Related two-dimensional hydrodynamical simulations can be found in Christodoulou (1993).

While the importance of such purely hydrodynamical instabilities to accretion disks and to AGN fueling is still under investigation, a new, more powerful, and potentially more important instability was presented by Balbus and Hawley (1991, 1992). The Balbus-Hawley instability appears in the presence of a weak magnetic

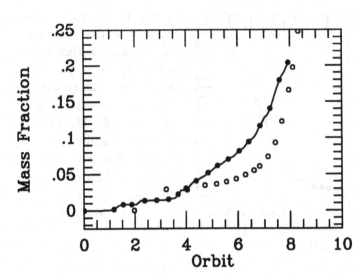

Fig. 2—Fraction of inflowing mass as a function of time for a massless torus orbiting around a massive central object. The points of Fig. 1 are also plotted as open circles for comparison. The system is embedded in a weak vertical magnetic field (plasma parameter $\beta = 10^4$). Inflow occurs within a few orbits as angular momentum is transported outward by the Balbus-Hawley instability.

field and seems capable of initiating efficient matter inflow because it transports angular momentum outward within a disk. (This AMTM is not efficient in strongly magnetized disks where angular momentum is primarily transported away to the surrounding coronal gas by torsional Alfvén waves; *cf.* Shibata and Uchida 1986, 1989; Stone 1990.) Fig. 2 is analogous to Fig. 1 but the torus is now evolving inside a spherically symmetric potential of the form $\Phi(r) = -1/r$ and is embedded in a low-density medium and in a weak vertical magnetic field. As a result of angular momentum transport outward, matter inflows by at least one order of magnitude in scale and much faster than in the model of Fig. 1.

4 SUMMARY

Hydrodynamical simulations indicate that an accreted ring or disk that is moderately inclined ($\sim 40°$—$60°$) relative to the equatorial plane of a mildly nonspherical galaxy potential well loses a substantial fraction of its angular momentum to the external potential. Angular momentum loss and matter inflow are inefficient at low inclinations and proceed on time scales longer than a Hubble time at high inclinations. Matter inflow becomes catastrophic in strongly nonspherical potentials at all inclinations. We conclude that gravitational torques due to a nonspherical potential will cause smooth matter inflow down to ~ 1 kpc at intermediate inclinations (Fig. 1 and Christodoulou and Tohline 1993).

Inflow from ~ 1 pc down to a (possibly existing) central massive black hole is not well understood. Although dynamical nonaxisymmetric instabilities of accretion disks may play a role (Christodoulou and Narayan 1992), the leading inflow mechanism at present appears to be a powerful magnetohydrodynamical instability that depends on the presence of only a weak magnetic field (Balbus and Hawley 1991, 1992). Simulations of weakly magnetized accretion tori (Fig. 2) indicate that this instability causes substantial matter inflow in just a few orbits.

I am grateful to S. Balbus, C. Gammie, J. Hawley, R. Narayan, and J. Tohline for many discussions and suggestions. This work was supported by NASA grant NAGW–1510 and by a grant from the San Diego Supercomputer Center.

REFERENCES

Balbus, S. A., and Hawley, J. F. 1991, *Ap. J.*, **376**, 214.

Balbus, S. A., and Hawley, J. F. 1992, *Ap. J.*, **400**, 610.

Begelman, M. C., Blandford, R. D., and Rees, M. J. 1984, *Rev. Mod. Phys.*, **56**, 255.

Begelman, M. C., Frank, J., and Shlosman, I. 1989, in *Theory of Accretion Disks*, ed. F. Meyer *et al.* (Dordrecht: Kluwer), p. 373.

Bertola, F., Bettoni, D., Buson, L. M., and Zeilinger, W. W. 1990, in *Dynamics and Interactions of Galaxies*, ed. E. R. Wielen (Heidelberg: Springer-Verlag), p. 249.

Bertola, F., Buson, L. M., and Zeilinger, W. W. 1992, *Ap. J. Lett.*, **401**, L79.

Blandford, R. D. 1989, in *Theory of Accretion Disks*, ed. F. Meyer *et al.* (Dordrecht: Kluwer), p. 35.

Christodoulou, D. M. 1993, *Ap. J.*, in press.

Christodoulou, D. M., Katz, N., Rix, H.-W., and Habe, A. 1992, *Ap. J.*, **395**, 113.

Christodoulou, D. M., and Narayan, R. 1992, *Ap. J.*, **388**, 451.

Christodoulou, D. M., and Tohline, J. E. 1993, *Ap. J.*, **403**, 110.

Frank, J., Shlosman, I., and Begelman, M. C. 1989, in *Theory of Accretion Disks*, ed. F. Meyer *et al.* (Dordrecht: Kluwer), p. 387.

Franx, M., and Illingworth, G. D. 1988, *Ap. J. Lett.*, **327**, L55.

Goodman, J., and Narayan, R. 1988, *Mon. Not. R. Astr. Soc.*, **231**, 97.

Gunn, J. E. 1979, in *Active Galactic Nuclei*, ed. C. Hazard and S. Mitton (New York: Cambridge University Press), p. 213.

Hawley, J. F. 1991, *Ap. J.*, **381**, 496.

Katz, N., and Gunn, J. E. 1992, *Ap. J.*, **377**, 365.

Katz, N., and Rix, H.-W. 1992, *Ap. J. Lett.*, **389**, L55.

Papaloizou, J. C. B., and Pringle, J. E. 1984, *Mon. Not. R. Astr. Soc.*, **208**, 721.

Peletier, R. F., and Christodoulou, D. M. 1993, *A. J.*, **105**, 1378.

Rees, M. J. 1980, in *X-ray Astronomy*, ed. R. Giacconi and G. Setti (Dordrecht: Reidel), p. 339.

Rix, H.-W., and Katz, N. 1991, in *Warped Disks and Inclined Rings Around Galaxies*, ed. S. Casertano *et al.* (Cambridge: Cambridge University Press), p. 112.

Shibata, K., and Uchida, Y. 1986, *P. A. S. J.*, **38**, 631.

Shibata, K., and Uchida, Y. 1989, in *Theory of Accretion Disks*, ed. F. Meyer *et al.* (Dordrecht: Kluwer), p. 65.

Shlosman, I., Begelman, M. C., and Frank, J. 1990, *Nature*, **345**, 679.

Steiman-Cameron, T. Y., and Durisen, R. H. 1988, *Ap. J.*, **325**, 26.

Steiman-Cameron, T. Y., and Durisen, R. H. 1990, *Ap. J.*, **357**, 62.

Stone, J. M. 1990, *Ph. D.* Thesis, University of Illinois at Urbana-Champaign.

Mid-IR Imaging of Interacting and Non-Interacting AGNs

R. K. Piña[1], B. Jones[1], R. C. Puetter[1], and W. A. Stein[2]

[1]Center for Astrophysics and Space Science, University of California, San Diego
[2]School of Physics and Astronomy, University of Minnesota

ABSTRACT

We have obtained 10 μm continuum images of a flux-limited sample of bright infrared galaxies with a spatial resolution of 0.8 arcseconds. All observations were made with UCSD's Mid-Infrared Camera on the Mt. Lemmon 1.5 meter telescope, Tucson, AZ. Most of the galaxies imaged display centrally condensed cores of emission. Two galaxies in our sample, NGC 253 and Markarian 171, are well resolved due to their proximity to the Galaxy and show extended emission. In the case of NGC 253, we have also obtained 20 μm continuum images. In this paper we present some results of our observations of NGC 253.

1 INTRODUCTION

Among the class of infrared luminous galaxies established by the *Infrared Astronomical Satellite* (*IRAS*), NGC 253 is a modest example of the "starburst" type. Due to its proximity to the Galaxy ($\sim$ 3 Mpc), it is well resolved at many wavelengths. NGC 253 is an SABc galaxy with an inclination of 78.5°. It displays no peculiarities in morphology. Still, within $R < 500$ pc, the far-infrared luminosity is $\sim 3 \times 10^{10}$ L$_\odot$ (Telesco and Harper 1980).

2 OBSERVATIONS AND RESULTS

The UCSD mid-IR camera, the "Golden Gopher", operates in the spectral region from 5 to 27 μm using a 20 $\times$ 64 element $Si : As$ Impurity Band Conduction (IBC) device manufactured by GenCorp Aerojet Electronics Systems Division. Observations with the camera on the Mt. Lemmon Observing Facility's 1.5 meter telescope yield a noise-equivalent flux density (NEFD) of 23.5 mJy min$^{-1/2}$ arcsec^{-2} at $\lambda = 10$ μm, with $\Delta\lambda = 1$ μm, and a readout frame rate of 366 Hz. Our pixel size is 0.83 arcsec. Details of the instrument have been presented by Piña, Jones, and Puetter (1993).

Panel a of Figure 1 displays a 20 μm contour map of NGC 253 and panel b displays the 10 μm to 20 μm color temperature (assuming a $1/\lambda$ emissivity law). The 10 μm image is very similar in morphology to the 20 μm image (see Piña *et al.* 1992). Our positioning places the mid-infrared peak $\sim$2.5 arcsec to the southwest of the proposed radio nucleus of the galaxy (Turner and Ho 1985). With this positioning, the mid-infrared peak is coincident with an emission peak in the CO bar mapped by

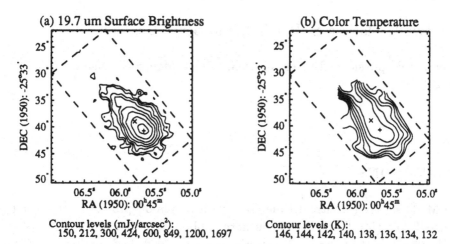

(a) 19.7 um Surface Brightness

(b) Color Temperature

Contour levels (mJy/arcsec²):
150, 212, 300, 424, 600, 849, 1200, 1697

Contour levels (K):
146, 144, 142, 140, 138, 136, 134, 132

Fig. 1—(*a*) 19.7 μm surface brightness contours of NGC 253. The "x" symbol marks the location of the Turner and Ho radio nucleus. (*b*) 11.7 to 19.7 μm color temperature contours. The "+" symbol identifies the mid-infrared peak.

Canzian, Mundy, and Scoville (1988). A peak in CO emission at this location is also indicated by the CO spatial-velocity map of Scoville *et al.* (1985).

The mean color temperature is 139 K and is remarkably constant throughout the nuclear region varying by ±5 K. However, there is a clear trend indicating cooler temperatures along the nuclear ridge and warmer temperatures away from the ridge. This may be consistent with the depletion of small dust grains due to the high UV energy density in the most intense region of the starburst as suggested by Telesco, Decher, and Joy (1989). Another result is that the mid-infrared peak is not discernible in the color temperature map. Since most of the infrared luminosity in starburst galaxies is attributable to dust heated by a large number of co-mingled OB stars, this suggests that the mid-infrared emission peak in NGC 253 is due to a larger quantity of dust (and associated OB stars) rather than an intrinsically different mechanism for producing the infrared luminosity.

REFERENCES

Canzian, B., Mundy, L. G., and Scoville, N. Z. 1988, *Ap. J.*, **333**, 157.

Piña, R. K., Jones, B., and Puetter, R. C. 1993, in *Infrared Detectors and Instrumentation* (SPIE Proceedings), **1946**, p. 66.

Piña, R. K., Jones, B., Puetter, R. C., and Stein, W. A. 1992, *Ap. J. Lett.*, **401**, L75.

Scoville, N. Z., *et al.* 1985, *Ap. J.*, **289**, 129.

Telesco, C. M., Decher, R., and Joy, M. 1989, *Ap. J. Lett.*, **343**, L13.

Telesco, C. M., and Harper, D. A. 1980, *Ap. J.*, **235**, 392.

Turner, J. L., and Ho, P. T. P. 1985, *Ap. J. Lett.*, **299**, L77.

Imaging of MBG Starbursts: Preliminary Results

Roger Coziol[1], Clarissa S. Barth[1,2], and Serge Demers[1]

[1]Département de physique, Université de Montréal, Montréal, Québec, Canada
[2] CNPq Fellow

1 OBSERVATIONS

The MBG survey (Montreal Blue Galaxy) is a spin-off project of the Montreal Cambridge Tololo (MCT) survey of southern subluminous blue stars (Demers *et al.* 1986). Using a subset of the plate material covering 7000 deg^2, with $b \leq -40°$, we pick up all extended UV-bright objects. The analysis of our follow-up spectroscopy has shown that the bulk of our UV-bright candidates consists of H II galaxies or starbursts; the fraction of AGNs being somewhat less than 10% (Coziol *et al.* 1993). We expect to find ~ 500 such galaxies, with magnitude B ≤ 15.5.

Recently, we undertook an imaging follow-up of our candidates to determine their morphology and search for clues to the origin of their activity. Our first sample consist of 11 MBG galaxies with strong H II-region like spectra suggesting an intense phase of star formation. Usually, these galaxies also possess relatively hot *IRAS* color typical of starburst galaxies (Sekiguchi 1987). Using our 1.6m telescope, located on mont Mégantic in Québec, Canada, we obtained CCD images, with BVRI filters. Details of the observations, reductions and analysis will be published elsewhere (Coziol, Barth, and Demers 1993). Following, is a summary of our first results.

We find mostly spiral galaxies with bright nuclear regions. There is a strong tendency to find early type galaxies. Further analysis of the surface photometry reveals that the bursts are located, preferentially, in the circumnuclear region prolonging far into the nucleus. This confirms the earlier spectral classification of our objects, using the excitation diagnostic diagram of [O III]λ5007/Hβ versus [N II]λ6584/Hα, which suggested that MBG objects are mostly starburst nuclei galaxies (see Salzer *et al.* 1989 for a definition). The regions of star formation are associated with different morphological peculiarities like bars, rings, double nuclei or more chaotic structures and could be either the result of interaction with companions galaxies or mergers.

2 DISCUSSION

The UV-bright colors, the H II-region like spectra and the relatively hot *IRAS* colors are all significant indices of an ongoing phase of intense star formation. Using the method proposed by Scoville and Young (1983) to estimate the star formation rate (SFR), we find that MBG's have unusually high SFR for their morphological type (Fig. 1).

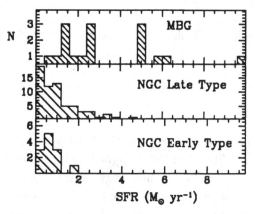

Fig. 1—Comparison of SFR in MBG starbursts and "normal" galaxies. The sample of normal galaxies is from Kennicutt (1983), with SFR recalculated using the method of Scoville and Young. Note that, to reproduce Kennicutt's original results, we have, at least, to multiply our SFR by a factor of three.

From these observations, we conclude that we are observing a recent infall of matter in well formed, pre-existing galaxies. This implies both mass transfer and triggering of star formation. The preferred mechanisms are mild interactions between galaxies or the swallow up of a small galaxy by a bigger one. The redistribution of new matter, and the burst of star formation resulting, alter the morphology of the galaxy host in the form of bars, rings and hotspots. In early type spirals, the presence of a burst in the circumnuclear region enhances the intermediate region between the bulge and the disk creating the effect of a thick lens or false bulge.

Taking into account the well known tendency to find AGNs in early type galaxies, it is interesting that we find many MBG starburst nuclei in early type spirals. Removing the morphological barrier, reinforces the idea of a possible evolution between starbursts and AGNs. In particular, it rules out any interpretation of possible differences, observed between star forming regions in both objects, in terms of different star formation properties (Glass and Morwood 1985; Taniguchi and Mouri 1992).

REFERENCES

Coziol, R., *et al.* 1993, *A. J.*, **105**, 35.

Coziol, R., Barth, C. S., and Demers, S. 1993, in preparation.

Demers, S., *et al.* 1986, *A. J.*, **92**, 878.

Glass, I. S., and Morwood, A. F. M. 1985, *Mon. Not. R. Astr. Soc.*, **214**, 429.

Kennicutt, R. C. 1983, *Ap. J.*, **272**, 54.

Salzer, J. J., MacAlpine, G. M., and Boroson, T. A. 1989, *Ap. J. Suppl.*, **70**, 479.

Scoville, N. Z, and Young, J. S. 1983, *Ap. J.*, **265**, 148.

Sekiguchi, K. 1987, *Ap. J.*, **316**, 145.

Taniguchi, Y., and Mouri, H. 1992, in *Relationships Between Active Galactic Nuclei and Starburst Galaxies*, ed. A. V. Filippenko (San Francisco: ASP-31), p. 365.

Compact Extranuclear Structures of Mkn 298

Piero Rafanelli and Mario Radovich

Department of Astronomy, University of Padova, Vicolo dell'Osservatorio 5, 35122 Padova, Italy

1 RESULTS AND DISCUSSION

Mkn 298 is known as a morphologically peculiar system (Fig. 1), showing a chain (c, d, e) of small compact blue emitting regions (Stockton 1972) aligned on its eastern side up to $\sim$ 80 arcsec from the main body of the galaxy. These regions are aligned, with a tail (b) resembling a spiral arm located on the eastern side of the galaxy (a). On the western side a very faint trace of spiral arm is also visible. d and e are characterized by an emission line spectrum, while spectra of c do not show any trace of either emission or absorption (Stockton 1972; Metik and Pronik 1982).

The contour maps of the extended emission lines Hα, Hβ and [O III]$\lambda\lambda$4959, 5007 have been used to isolate six different emitting regions in a and b. Each strip on the 2D-spectrum has then been mashed into a 1D-spectrum and the diagnostic line ratios proposed by Veilleux and Osterbrock (1987) have been plotted in the diagnostic diagrams giving [O III]λ5007/Hβ versus [S II]$\lambda\lambda$6716+6731/Hα, [N II]λ6583/Hα, [O I]λ6300/Hα. The effect of reddening has been evaluated from the Hα/Hβ ratio using the Whitford (1958) reddening curve as parameterized by Miller and Mathews (1972) and adopting the intrinsic ratio 2.81. At first glance, it appears that the [N II] lines are systematically too faint, log([N II]/Hα) < -0.3, in all six regions to justify the presence of a nonthermal ionizing source as suggested by the intensity ratios [S II]/Hα and [O I]/Hα which are all typical of LINERs, being log([O III]/Hβ) < 0.5, log([S II]/Hα) > -0.4 and log([O I]/Hα) > -1.

We have attempted to reproduce the range of observed line ratios using the photoionization code CLOUDY 84 (Ferland 1993). It has been assumed the presence of a nonthermal power-law source with spectral index $\alpha = -1.4$ and a gas density $n_H = 10^2$ cm^{-3}. Then a grid of models has been calculated for different values of the ionization parameter (log $U = -2.5, -2.8, -3.0, -3.2, -3.5, -3.8$) and of the metallicity ($Z = 1, 0.3, 0.2, 0.1$ Z$_\odot$) and plotted on the diagnostic diagrams. The low [N II]/Hα ratio, which is typical of the H II region-like objects, is well reproduced also by a non thermally ionized model with low metallicity 0.1 Z$_\odot$ $< Z <$ 0.2 Z$_\odot$. This result is confirmed by the other line ratios which are likewise located in a region of comparable low metallicity. There is a clear tendency of the ionization parameter to decrease going from the nucleus towards East and to increase towards West. The observed Hα/Hβ ratio steadily decreasing from West towards East indicates the presence of a larger dust extinction on the western side of the nucleus than on its

Fig. 1—Contour map of the CCD image of Mkn 298 and superimposed the position of the slit (PA=92°).

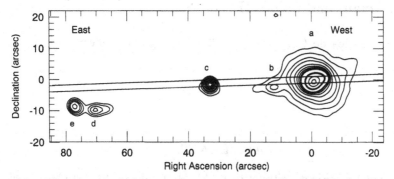

eastern side. The shape of the rotation curve, as derived from the Hα, [N II], [O III] and Hβ extended emission lines, is typical of an early type spiral galaxy and confirms the identification on the CCD image of the eastern and western extensions as spiral arms. The spectrum of the compact *c* source, described as lacking of any line by some authors, results to be the spectrum of a foreground faint late type K star, superposed to a faint extended region fairly emitting in Hα and having the same redshift of Mkn 298. In conclusion, Mkn 298 is likely to be an early type spiral galaxy as suggested by its morphology and rotation curve. Its nucleus shows spectral features typical of a LINER with low metal abundance ($Z \sim 0.1 - 0.2\ Z_{\odot}$). The nonthermal nuclear source ionizes the gas present in the five circumnuclear regions located on its Eastern and Western sides, up to a distance of around 5 kpc. These regions are also characterized by a low metal abundance, comparable with that found in the nucleus. The trend shown by the ionization parameter can be accounted for in terms of different amounts of intervening dust between the nucleus and the circumnuclear regions ionized by its radiation (larger extinction on the West side than on the East side of the nucleus). The presence of close compact blue sources (*c, d, e*) and the membership of Mkn 298 to a crowded field in the Hercules cluster suggest that this galaxy could have captured material of low metal abundance from the outer regions of a close spiral galaxy (Pagel and Edmund 1981; Baum, Heckman, and van Breugel 1992) and as result of the interaction produced the blue knots.

REFERENCES

Baum, S. A, Heckman, T. M., and van Bruegel, W. 1992, *Ap. J.*, **389**, 208.
Ferland, G. J. 1993, *University of Kentucky Internal Report*.
Metik, L. P., and Pronik, I. I. 1982, *Astrofizica*, **17**, 333.
Miller, J. S., and Mathews, W. G. 1972, *Ap. J.*, **172**, 593.
Pagel, B. E. J., and Edmunds, M. G. 1981, *Ann. Rev. Astr. Ap.*, **19**, 77.
Stockton, A. 1972, *Ap. J.*, **173**, 247.
Veilleux, S., and Osterbrock, D. E. 1987, *Ap. J. Suppl.*, **63**, 295.
Whitford, A. E. 1958, *A. J.*, **63**, 201.

NGC 6814: A Very Normal Looking AGN Host Galaxy

John E. Beckman, and Johan H. Knapen

Instituto de Astrofísica de Canarias, E-38200 La Laguna, Tenerife, Spain

ABSTRACT

We have studied the luminosity function (LF) of H II regions in the disk of the Seyfert 1 galaxy NGC 6814. We find that the LF is very similar to LFs of other late-type, not necessarily active, spiral galaxies. Although the Seyfert nucleus shows its character by emitting strongly in Hα, the disk H II regions seem not to be influenced by the active nucleus.

NGC 6814 is an Sbc galaxy with well-defined spiral arms, of type Seyfert 1. Because of the strong X-ray emission, it is considered a key object for understanding nuclear activity.

We have obtained new Hα observations with the 4.2m William Herschel Telescope (WHT) on La Palma, using the TAURUS instrument in imaging mode. The final continuum subtracted Hα image has high sensitivity (H II region detection limit is $L = 10^{36.9}$ erg s^{-1}) and resolution (0.8 arcsec, or about 100 pc at the distance of NGC 6814). From the image, we have measured positions, diameters and fluxes of a total number of 735 H II regions (Knapen *et al.* 1993). We found that the nucleus is a strong Hα emitter, of luminosity $L = 10^{39.9}$ erg s^{-1}.

From our catalog of H II regions we constructed a luminosity function (LF), for all the H II regions in the disk of the galaxy, and for arm and interarm H II regions separately. Figure 1 shows the total LF. The slope of the LF is $a = -2.37 \pm 0.09$, well within the range of slopes measured in the literature for galaxies of similar morphological type. The arm and interarm LF slopes are equal within the fitting uncertainties.

Rand (1992) published a study of the H II regions in the well-known spiral galaxy M51. His study is well comparable to ours in terms of resolution and sensitivity, and the galaxies studied are of very similar morphological type. Rand found that the arm and interarm LF slopes are significantly different in M51, and interprets that difference in terms of a different molecular cloud mass spectrum. The fact that we do not see a difference between arm and interarm LFs in NGC 6814 indicates that such an effect can not be universal. In fact, from the literature we see that some late-type galaxies do have different slopes, and others do not. It is possible that the existence of such different behavior in similar galaxies is influenced by the strength of the spiral arms.

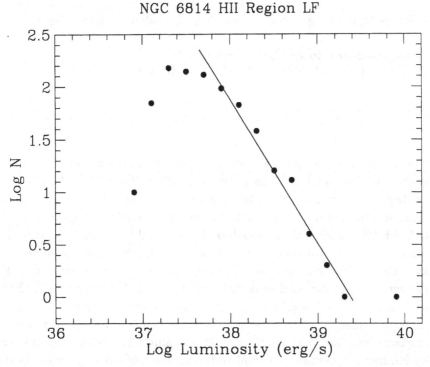

Fig 1—H II region luminosity function of NGC 6814. The single point at $L = 10^{39.9}$ erg s^{-1} corresponds to the Seyfert nucleus. Drawn line indicates best fit.

We do not see any different behavior in the statistical properties of disk H II regions in NGC 6814 when compared to similar galaxies that are not AGNs. The active nucleus apparently does not influence the star formation processes in the disk to a large degree. We do see a zone around the nucleus of some 10″ radius where hardly any H II regions are present. But this effect is not very rare, and in fact is quite common in barred galaxies. Near infrared imaging of this galaxy, which is not obviously barred when seen in optical light, might reveal a bar, which may be stimulating the fueling of the AGN.

REFERENCES

Knapen, J. H., Arnth-Jensen, N., Cepa, J., and Beckman, J. E. 1993, *A. J.*, **106**, 56.
Rand, R. J. 1992, *A. J.*, **103**, 815.

ROSAT PSPC Observations of NGC 3079

G. A. Reichert[1], R. F. Mushotzky[2], and A. V. Filippenko[3]

[1]Universities Space Research Association and NASA Goddard Space Flight Center
[2]Code 668, NASA Goddard Space Flight Center
[3]Department of Astronomy, University of California, Berkeley

NGC 3079 is a remarkable spiral galaxy which exhibits an unusual range of nuclear activity. Viewed edge-on, it harbors a reddened LINER (Low Ionization Nuclear Emission-line Region) that is kinematically complex, with several distinct components of Hα emission. It also contains a compact, flat spectrum nuclear radio source, and shows well-defined, kiloparsec-scale radio lobes of considerable complexity, as well as a smaller loop of Hα+[N II] emission extending approximately along the minor axis of the galaxy. The optical emission lines indicate that gas is outflowing from the nucleus in an energetic, bipolar outflow or "galactic superwind." Armus *et al.* (1990) have suggested that this wind is driven by a powerful, central starburst. A number of different lines of evidence — its infrared brightness (measured by *IRAS*, the extended 10 μm emission, the spatially coincident, circumnuclear molecular gas, and the extremely luminous H_2O maser — point to the presence of ongoing, vigorous star formation within the nucleus, in agreement with this idea. However, other authors (*e.g.,* Irwin and Sofue 1992; Filippenko and Sargent 1992) argue that the nuclear activity originates from an AGN, perhaps supplemented by a powerful starburst.

We have obtained *ROSAT* PSPC observations of NGC 3079 and the adjacent spiral galaxy NGC 3073. Preliminary analysis indicates that X-ray emission from NGC 3079 consists of a point-like source, superposed on lower level emission which is extended by 2.5′ in approximately the same direction as the radio jet, Hα loop, and radio lobes (Figs. 1, 2). The extended emission is roughly coincident with two Hα filaments detected by Heckman *et al.* (1990), and may consist of either a shell-like or a filled structure. The spectrum of the extended emission is quite soft; no extended emission is seen at energies above 0.5 keV. For reasonable spectral assumptions, the extended component contributes a luminosity (0.2–0.5 keV) in the range $(0.2 - 2) \times 10^{39}$ erg s^{-1}, compared to a total luminosity (0.2–2 keV) of about 2.3×10^{40} erg s^{-1}.

The spectrum of the total emission can be well described as the superposition of a simple power law of photon index 1.6 and a Raymond-Smith plasma of temperature 0.4 keV. The best fit spectrum is absorbed by a column of 1.2×10^{20} cm^{-2} ($\lesssim 4 \times 10^{20}$ cm^{-2} at 90% confidence). An equally good fit can be obtained by replacing the power law with a thermal bremsstrahlung component of temperature 3.0 keV. As yet,

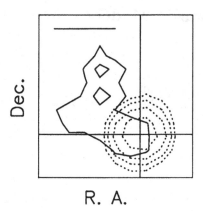

Fig. 1—Contours of low energy ($E \lesssim 0.5$ keV; solid lines) versus higher energy ($0.9 - 2.1$ keV; dashed lines) X-ray emission from NGC 3079. Contours were smoothed using a boxcar 45' on each side. The cross marks the center of the higher energy emission, which coincides with the radio nucleus. A 1' bar is also shown.

Fig. 2—Radial profiles of the soft ($E \lesssim 0.5$ keV) emission about the center of NGC 3079. The solid circles show the profile averaged over the NE quadrant, open circles the profile averaged over the other three quadrants. Solid curve shows the expected PSPC point response for the observed counts distribution plus a constant background. The bottom panel shows the difference.

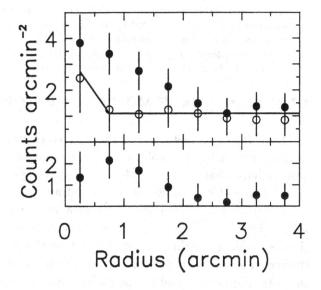

no conclusions can be drawn regarding the nature of the source of the X-ray emission. However, the coincidence of the low energy extended X-ray emission with the radio and optical emission-line structures suggests that this component may be associated with the galactic superwind, as described by Heckman *et al.* (1990).

No X-rays were detected from NGC 3073, to a level of less than 10% the emission from NGC 3079.

REFERENCES

Armus, L., Heckman, T. M., and Miley, G. K. 1990, *Ap. J.*, **364**, 471.
Irwin, J. A., and Sofue, Y. 1992, *Ap. J. Lett.*, **396**, L75.
Filippenko, A. V., and Sargent, W. L. W. 1992, *A. J.*, **103**, 28.
Heckman, T. M., Armus, L., and Miley, G. K. 1990, *Ap. J. Suppl.*, **74**, 833.

MK 231: AGN vs. Starburst? Steps Toward a Model

Wayne A. Stein

School of Physics and Astronomy, University of Minnesota, and
Center for Astrophysics and Space Sciences, University of California, San Diego

ABSTRACT

The infrared luminous galaxy MK 231 appears to exhibit characteristics similar to those of active galactic nuclei — it has been classified as a Seyfert 1 system. However, it has been shown to contain 3×10^{10} M$_\odot$ of gas via CO observations. If the CO is confined to the region shielded by dust as in galactic molecular clouds, the molecular gas occupies a much smaller volume than previously thought. The ultimate questions are which characteristic is primarily responsible for most of the luminosity and whether AGN and starbursts are interdependent or coincidental.

1 THE SUPERGIANT MOLECULAR CLOUD

Scoville *et al.* (1989) and Radford *et al.* (1991) have discussed the physical conditions implied by the molecular observations of MK 231. It is inferred from CO observations, that 3×10^{10} M$_\odot$ of molecular gas (H$_2$) resides within a volume of radius $R_{CO} < 3$ kpc. However, recent infrared observations reveal that the emission from dust at $\lambda \approx 10$ μm arises within a volume of size $R_{IR} < 400$ pc (Keto *et al.* 1992). This is an important complementary result to the CO observations because studies within our galaxy show clearly that the molecular gas is largely confined to the region shielded by dust extinction (Young *et al.* 1982). Thus we must conclude that the volume occupied by the molecular gas is really much smaller than discussed by Scoville *et al.* (1989) and Radford *et al.* (1991) by a factor of about 420 and $R_{gas} \approx 400$ pc. These numbers lead to the conclusion that the average density of gas in the MK 231 molecular cloud is $n_{gas} \gtrsim 5 \times 10^3$ cm^{-3} whereas the result using the much larger volume inferred from the CO observations was only ≈ 10 cm^{-3}.

2 AGN CHARACTERISTICS

Condon *et al.* (1991) have classified UGC galaxies as starburst or "monsters" (AGN) on the basis of the logarithm of the infrared to 4.85 GHz luminosity ratio q. The result is remarkable in that the average $q = 2.62 \pm 0.03$ (standard deviation of the mean) for a sample of 33 starburst systems.

The observed q of MK 231 is $q = 2.0$, much smaller than other starburst systems. This is clearly not because the infrared luminosity is too small (it is one of the most luminous systems) compared to other starburst systems, but rather because the radio core of the object is so luminous. The L_ν of MK 231 at 6 cm is about 200 times larger than the average starburst system and 25 times larger than the next most luminous starburst system . Clearly MK 231 cannot be considered as only a starburst system in spite of all of the interesting results found for the supergiant molecular cloud.

3 FURTHER DISCUSSION

It is inconceivable that we can ignore the potential star forming capacity of the enormous supergiant molecular cloud of MK 231. The 60 μm to 12 μm ratio for MK 231 from *IRAS* data is 1.8, considerably more like star formation systems (*e.g.* M82, 1.9) than AGN (*e.g.* NGC 1068, 1.0). Yet at the same time we cannot ignore the radio luminosity of the core that is indicative of an AGN monster.

The very existence of a mass of $\approx 10^{10}$ $M_\odot$ within a few hundred parsecs suggests that gravitational instability and collapse has led to the formation of a central compact object (Rees 1978) that is now observed via its radio emission. It is probably not necessary to have a supermassive hole from the pre merger existence of the galaxies involved. Whether or not this has any implications for evolution or existence of AGN and QSOs in general is unclear at present.

REFERENCES

Condon, J. J., Frayer, D. T., and Broderick, J. J. 1991, *A. J.*, **101**, 362.

Keto, E., Ball, R., Arens, J., Jernigan, G., and Meixner, M. 1992, *Ap. J. Lett.*, **387**, L17.

Radford, S. J. E., Solomon, P. M., and Downes, D. 1991, *Ap. J. Lett.*, **368**, L15.

Rees, M. J. 1978, *Observatory*, **98**, 210.

Scoville, N. Z., Sanders, D. B., Sargent, A. I., Soifer, B. T., and Tinney, C. G. 1989, *Ap. J. Lett.*, **345**, L25.

Young, J. S., Goldsmith, P. F., Langer, W. D., Wilson, R. W., and Carlson, E. R. 1982, *Ap. J.*, **261**, 513.

Extended Starburst Activity Induced by the Central AGN: A Model for NGC 1068

Smita Mathur[*]

Harvard-Smithsonian Center for Astrophysics
[*]Formerly Smita Shanbhag

ABSTRACT

The effect of an active nucleus on its host galaxy (Shanbhag 1991; Shanbhag and Kembhavi 1988; Begelman 1985) can be large. In particular, heating by the radiation from the nucleus strongly affects the hydrodynamic evolution of gas in the interstellar medium (ISM) of the host galaxy. Enhanced star formation activity on a galaxy-wide scale can be induced. The proximity of NGC 1068 makes it an interesting candidate to study such interaction. The model explains extended X-ray emission and some properties of the diffuse ionized medium (DIM) observed in the galaxy.

1 INTRODUCTION

NGC 1068 is a nearby Sy 1 galaxy disguised as Sy 2 with its "buried" nucleus obscured from direct view (Antonucci and Miller 1985). The unobscured solid angle as seen from the nucleus, is inferred to be $\sim \pi$ (Krolik and Begelman 1986). The emission from the nucleus is modeled by bipolar conical outflows of opening angle $\simeq 82°$ with the symmetry axis inclined to the plane of the galaxy by $\simeq 35°$ (Cecil et al. 1990). Its inferred intrinsic luminosity (Sokolowski et al. 1991) is $\sim 7 \times 10^{43}$ erg s^{-1}. Many observations (Bergeron et al. 1989; Cecil et al. 1990) show existence of activity in the direction of the cone open angle; implying a connection with the active nucleus. The motivation behind this paper is to understand the nature of such interaction.

2 THE MODEL

We consider heating of the gas in the ISM of the galaxy which lies within cone open angle and hence *sees* the nucleus. We model the ISM as the ISM of our Galaxy. The distribution of gas perpendicular to the plane of the galaxy is thus Gaussian plus exponential; while in the radial direction it is exponential. Initial temperature of the gas is assumed to be 10^4 K. Photoionization and Compton heating rates given by Begelman (1985) are used. Hydrodynamic calculations are performed for radii 3–10 kpc with Lagrangean formulation. Details of the technique are given elsewhere (Shanbhag 1991a, b).

Enhanced star formation activity on a galaxy wide scale can be induced by such an interaction. The strength of the starburst is estimated by using the observed surface density of molecular hydrogen in NGC 1068 (Young 1991). Contribution of the burst to the hard X-ray extended emission is estimated by calculating the number of High Mass X-ray Binaries (HMXB) produced in the burst. Number of high mass stars produced in the burst are calculated using Miller and Scalo (1979) Initial Mass Function (IMF) and the flatter IMF given by Garmani *et al.* (1982) for high mass stars ($M \geq 20\ M_\odot$).

3 RESULTS

Heating of the ISM in the galaxy NGC 1068 due to the central AGN severely disrupts its thermal and dynamical equilibrium. Gas starts expanding as it is heated with its scale height increasing with time. The model explains ~ 1 kpc scale height (at $D = 3$ kpc) observed in the DIM (Bland-Hawthorn *et al.* 1991; Bergeron *et al.* 1989). The inferred time scales of $\sim 10^7$ years are similar to those inferred through the observation of CO molecular ring (Myers and Scoville 1987). Temperature of the gas increases to $\sim 4 \times 10^5$ K. The velocity dispersion of ≈ 85 km s^{-1} observed in DIM could thus be thermal. Temperature increases much faster before the velocity gradients develop resulting in a more than an order of magnitude increase in the pressure. This can trigger starburst on a galaxy-wide scale through cloud-cloud collisions and compression. High mass X-ray binaries produced in the burst can account for most/all of the hard X-ray extended (radius ≈ 7 kpc) emission observed in the galaxy (Wilson *et al.* 1992). Induced starburst scenario is also consistent with more than average blue color (Smith *et al.* 1972) and peculiar IR/X-ray color (Wilson *et al.* 1992) of the galaxy.

This work was supported by NASA grant NAGW–2201.

REFERENCES

Antonucci, R. R. J., and Miller, J. S. 1985 *Ap. J.*, **297**,621.
Bland-Hawthorn, J., Sokololowski, J., and Cecil, G. 1991 *Ap. J.*, **375**, 78.
Begelman, M. C. 1985, *Ap. J.*, **297**, 492.
Bergeron, J., Petitjean, P., and Durret, F. 1989, *Astr. Ap.*, **213**, 61.
Cecil, G., Bland, J., and Tully, R. B. *Ap. J.*, **355**, 70.
Garmany, C. D., Conti, P. S., and Chiosi, C. 1982, *Ap. J.*, **263**, 777.
Meyers, S. T., and Scoville, N. Z. 1987, *Ap. J. Lett.*, **312**, L39.
Miller, G. E., and Scalo, J. M. 1979, *Ap. J. Sup.*, **41**, 513.
Shanbhag, S. 1991*a*, *Ap. J.*, **367**, 462.
Shanbhag, S. 1991*b*, *Ph. D.* thesis, Indian Institute of Science, Bangalore.
Shanbhag, S., and Kembhavi, A. K. 1988 *Ap. J.*, **334**, 34.
Sokolowski, J. K., Bland-Hawthorn, J., and Cecil, G. 1991, *Ap. J.*, **375**, 583.
Wilson, A. S., Elvis, M., Lawrence, A., and Bland-Hawthorn, J. 1992, *Ap. J. Lett.*, **391**, L7.

NGC 2782, NGC 4102 and NGC 6764: Evidence for Starburst-Driven Winds

Berto Boer[1,2]

[1]SRON-Leiden, P.O.Box 9504, 2300 RA Leiden, The Netherlands
[2]Astronomisches Institut der Ruhr-Universität Bochum, Germany

ABSTRACT

We used optical spectra to investigate the nuclear regions of the starburst galaxies NGC 2782, NGC 4102 and NGC 6764. In addition to the central starburst, we find evidence for extranuclear shock-ionised gas, with kinematical properties consistent with outflow along the minor axis. The observations are consistent with the presence of dense shock-ionised shells, formed by the starburst-driven winds. The shells, observed in NGC 2782, NGC 4102 and NGC 6764 respectively, appear to be in a different evolutionary phase, which is explained by differences in age and strength of the central starbursts.

1 INTRODUCTION

Evolutionary models of starburst nuclei predict the formation of a thin, dense, shock-ionised shell, surrounding a hot cavity (*e.g.* Tomisaka and Ikeuchi 1988). The shell expands predominantly perpendicular to the galactic plane, gradually elongating until it finally breaks open at the top. The nearby starburst galaxies NGC 253 and M82 are probably examples of the broken-shell phase.

We investigated the starbust galaxies NGC 2782, NGC 4102 and NGC 6764 with the aid of long-slit spectra with high spatial resolution. All three galaxies have been classified previously as starburst galaxies, but the optical line ratios derived from 1-dimensional spectra put the galaxies close to the borderline between starbursts and LINERs in diagnostic diagrams.

2 RESULTS

From the spatial behaviour of the line ratios in our spectra we find, that the high line ratios are due to the presence of an *extranuclear high-ionisation component* in addition to the *central starburst*. The line ratios of the high-ionisation component are in agreement with *shock-ionisation*; its kinematical properties are consistent with *outflow along the minor axis*.

We conclude, that the observations of the galaxies discussed here, are best explained by a nuclear starburst, surrounded by a shock-ionised shell expanding perpendicular to the plane of the galaxy. A detailed comparison of the observations of NGC 2782 with theoretical models by Tomisaka and Ikeuchi (1988), for a shell formed by a 3×10^6 yr old starburst, shows an excellent agreement (Boer and Schulz 1992). The observations of NGC 4102 and NGC 6764 will be discussed in a forthcoming paper (Boer 1993).

TABLE 1
SHELL PROPERTIES

	NGC 2782	NGC 4102	NGC 6764
t_{sh}/yr	3×10^6	1×10^6	4×10^6
$r_{sh}/(\mathrm{h}^{-1}\ \mathrm{pc})$	$\lesssim 300\times\ \sim 720$	~ 250	~ 520
$v_{sh}/(\mathrm{km\ s^{-1}})$	~ 250	~ 250	~ 130
$L(\mathrm{H}\alpha)/(\mathrm{h}^{-2}\ \mathrm{erg\ s^{-1}})$	2×10^{41}	$\gtrsim 6 \times 10^{40}$	1×10^{40}
$\nu_{SN}/(\mathrm{h}^{-2}\ \mathrm{yr}^{-1})$	0.02	$\gtrsim 0.006$	0.001

3 DISCUSSION

Although the three galaxies discussed here are similar, in that they all have a nuclear starburst driving a shockionised shell, they seem to represent different stages in the evolution of the shell. Table 1 lists the relevant parameters of the three galaxies, whereby the supernova rate was calculated with $\nu_{SN} = 3.48\times10^{-10}$ $[L(\mathrm{H}\alpha)/L_\odot]$ (Condon and Yin 1990).

The two parameters, which determine the stage of the shell evolution, are the *supernova rate* and the *time scale*. If the supernova rate is high, the shell will expand fast, elongate and eventually break open at the top. If the supernova rate is low, the expansion is slow and the shell will remain nearly spherical and smaller.

NGC 2782 has a relatively large Hα luminosity, indicating a high supernova rate, and the time scale (3×10^6 yr) is long enough for the shell to have become elongated. As it is a nearby, inclined galaxy, we see the shell spatially separated from the nuclear starburst as an extranuclear maximum in the [O III]λ5007 Å intensity distribution.

NGC 4102 contains a huge amount of dust (Hα/H$\beta \sim 12$), which agrees with the short time scale of the starburst (10^6 yr); the shell has only recently formed, and, although the expansion velocity is similar to NGC 2782, the shell is still compact.

NGC 6764 is an example of a weak, evolved starburst: although it is comparable to NGC 2782 in starburst age, distance and inclination, the shell is by no means as prominent as it is in NGC 2782.

REFERENCES

Boer, B. 1993, in preparation.
Boer, B., and Schulz, H. 1992, *Astr. Ap.*, **222**, 27.
Condon, J. J., and Yin, Q. F. 1990, *Ap. J.*, **357**, 97.
Tomisaka, K., and Ikeuchi, S. 1987, *Ap. J.*, **330**, 695.

AGN Winds and Nuclear Starbursts

Steven J. Smith

High Altitude Observatory and Advanced Study Program, NCAR[1]

This contribution considers the consequences of an active nucleus (AGN) inside a galaxy with a nuclear starburst. General arguments suggest that many AGN generate supersonic winds (Smith 1993a, see also Voit *et al.* 1993) with velocities $v_w \sim 0.1\, v_{0.1}c$, and the interaction of such a wind with a surrounding starburst is considered below. The large number of quasars indicate that some starburst galaxies should contain the remnant black holes of these "dead" AGN. Even if fueled by only a small amount of gas, the resulting AGN wind can have a significant effect on starburst hydrodynamics.

Note that it is unlikely that a black hole (BH) and subsequent AGN could form due to accretion during the lifetime ($10\,\tau_{10}$ Myr) of the starburst. Since Eddington limited accretion has a timescale of ~ 500 Myr, a seed black hole would have to accrete at a rate greater than $\sim 50/\tau_{10}$ times the Eddington rate to grow substantially.

Mass and energy injection by the supernovae and stellar winds of the starburst will form an outflowing wind. A nuclear wind produced by the AGN will evacuate the central region out to the radius where the mass flux injected by the starburst activity is greater than the mass flux in the nuclear wind. At this point the nuclear wind becomes mass loaded, subsonic, and will merge into the developing starburst wind (Smith 1993b). For a uniform starburst the nuclear wind (with kinetic luminosity $L_w = 10^{42} L_{42}$ erg s^{-1}) will become mass loaded at a radius $r_{ml} = [2L_w/(v_w^2 \dot{M}_{sb})]^{1/3} r_{sb}$, or $r_{ml} = 0.15\, L_{42}^{1/3} v_{0.1}^{-2/3} \nu^{-1/3} m^{-1/3} r_{sb}$, where $\dot{M}_{sb} = \nu m\, M_\odot$ yr^{-1} is the starburst mass injection rate, parameterized as the supernova rate times the mass injected per supernova (including stellar winds). The mass injection rate depends on time and the IMF, particularly the upper mass cutoff (Leitherer, Robert, and Drissen 1992).

The central pressure of the starburst wind is $\sim 0.12\,(\dot{E}_{sb}\dot{M}_{sb})^{1/2} r_{sb}^{-2}$ (Chevalier and Clegg 1985). Nuclear wind will come into equilibrium with this pressure at a radius $r_{stag} = 0.1\, L_{42}^{1/2} v_{0.1}^{-1/2} E_{51}^{-1/4} m^{-1/4} \nu^{-1/2} r_{sb}$. This is smaller than r_{ml} unless $L_{42} v_{0.1} m^{1/2} > 12 E_{51} \nu$. So low-luminosity winds stagnate before becoming mass loaded.

The nuclear wind will break out of the starburst region if its luminosity is greater than $L_{42} \sim 280 m\nu v_{0.1}^2$ (where $\dot{M}_w = \dot{M}_{sb}$), or $9\, m v_{0.1}^2$ times the starburst wind luminosity. This wind luminosity (scaling from Heckman, Armus, and Miley 1990), is $0.15\, m v_{0.1}^2$ times the IR luminosity of the starburst. This would require all the mass output within a radius $0.17/\epsilon_{nw}\, r_{sb}$, where ϵ_{nw} is the mass flux to wind luminosity conversion efficiency of the AGN. For (a rather large) $\epsilon_{nw} \sim 5\%$, all the mass loss within $0.5\, r_{sb}$

---

[1] The National Center for Atmospheric Research is sponsored by the NSF.

would have to feed the black hole, which is unrealistic. However a "break out" will occur if the AGN is fueled by more than a fraction $f \sim 10^{-5} \, m v_{0.1}^2 \tau_{10} (\epsilon_{\rm sb}/\epsilon_{\rm nw})[(M/L)/0.01]^{-1}$ of the initial gas from which the starburst formed ($\epsilon_{\rm sb}$ is ratio of initial gas to final starburst stellar mass and M/L the starburst mass to light ratio). This total can be less than the mass of a large molecular cloud.

In summary, the wind from a sufficiently luminous AGN will replace the starburst wind. This level of activity can be achieved if a remnant black hole (or currently active AGN) is fueled by only a small fraction of the initial starburst gas. Dust shrouding may not be as severe as compared to a normal starburst since dust may be destroyed in the mixing process between the high-speed nuclear wind and dusty stellar winds.

There is a range over which the AGN wind still has a greater luminosity than the starburst wind, but does not break out of the starburst region. Observational diagnostics will overestimate the starburst wind luminosity since it has been augmented by the nuclear wind. Emission from the AGN may be visible either of these cases.

For smaller AGN luminosities there will be no observational effects of the presence of a nuclear wind during the active phase of the starburst. Low-luminosity nuclear winds will stagnate at small radii instead of becoming mass loaded. However, since the AGN lifetime could be longer than the starburst lifetime, even low-luminosity winds may aid in clearing away the leftover starburst gas.

1.1 AGN Fueling by Compact Central Clusters

Some authors proposed that central clusters/starbursts fuel AGN. The presence of an AGN wind makes this improbable, even if the wind becomes mass loaded. Wind velocity at the edge of the cluster (of radius $r_{\rm sb} = 10 r_{10}$ pc) is $\sim 0.25 \, v_{\rm w} (r_{\rm ml}/r_{\rm sb})^2$ (Smith 1993b). This is greater than the cluster escape velocity unless wind generation efficiency is $\lesssim 2 \times 10^{-4} f_{\rm edd} v_{0.1}^{1/2} \beta^{3/4} r_{10}^{-3/4} M_8^{-1/4}$. The cluster mass loss rate has been scaled to the Eddington fueling rate as $f_{\rm edd}$ and the total mass of the cluster is β times $10^8 M_8 \, M_\odot$, the mass of the central BH. Even if the cluster is only ~ 1 pc in radius or ten times as massive as the BH this Figure rises only to 10^{-3}. Thus, if AGN converts mass to energy with a 10% efficiency, cluster mass loss will be blown outward if more than 1% of this energy is converted to a wind. This is a certain underestimate since energy input from the cluster and AGN continuum were ignored. Therefore, cluster fueling of an AGN is unlikely unless the cluster mass loss rate is super Eddington.

REFERENCES

Chevalier, R. A., and Clegg, A. W. 1985, *Nature*, **317**, 44.

Heckman, T. M., Armus, L., and Miley, G. K. 1990, *Ap. J. Suppl.*, **74**, 833.

Leitherer, C., Robert, C., and Drissen, L. 1992, *Ap. J.*, **401**, 596.

Smith, S. J. 1993a, *Ap. J.*, **411**, 570.

Smith, S. J. 1993b, *Ap. J.*, submitted.

Voit, G. M., Weymann, R. J., and Korista, K. T. 1993, *Ap. J.*, in press.

Induced Starbursts in Mergers

I. F. Mirabel[1] and P. A. Duc[1]

[1]Service d'Astrophysique, CE-Saclay, 91191 Gif sur Yvette Cedex, France

ABSTRACT

Galaxy-galaxy collisions induce nuclear and extranuclear starbursts. The sudden reduction of angular momentum of the interstellar medium due to the gravitational impact of the encounter leads to the subsequent infall to the central regions of a large fraction of the overall interstellar gas. Starburst galaxies with bolometric luminosities $\geq 10^{11}$ $L_\odot$ have converted most of the H I into H_2 reaching extreme nuclear densities of molecular gas. We also discuss extranuclear starbursts in relation to the formation of dwarf galaxies in mergers. As a consequence of tidal interactions a fraction of the less gravitationally bound atomic hydrogen that populates the outskirsts of the pre-encounter disk galaxies may escape into intergalactic space. We find that the ejected gas may assemble again and collapse, leading to the formation of intergalactic starbursts, namely, tidal dwarf galaxies.

1 "STARBURST GALAXIES"

"Starburst" denotes star formation at higher rates than in normally, self-regulated processes. They are non-equilibrium episodes that last only a small fraction of the total life-time of the host stellar systems. "Starburst galaxies" are stellar systems where the overall energy output is dominated by recently formed stars. In the context of this definition we must distinguish the "extragalactic H II regions" (Searle and Sargent, 1972) from the "nuclear starburst galaxies" (Weedman *et al.* 1981). The first are small, irregular, and dust-poor galaxies where the starburst is encompassing most of the visible galaxy; the second are massive luminous galaxies where the most violent starburst takes place embedded in dust in the central regions. The extragalactic H II regions are identified optically either by the unusual ultraviolet continuum radiated by hot stars, or by the strong narrow emission lines that arise in the interstellar nebulae ionized by massive stars. A large sample of nuclear starburst galaxies were first identified by Balzano (1983) in the Markarian (1967) survey of extragalactic objects with strong ultraviolet continua.

The recent developments in infrared astronomy have permitted a new, less biased way to identify luminous starburst galaxies. Since the most violent nuclear star formation takes place within high optically thick clouds of dust and molecular gas that convert most of the visible and UV light into far-infrared radiation, the far-infrared luminosity has become the best indicator of the starburst bolometric luminosity. The optical absorption along the line of sight to the nuclear starbursts is often so high that a large fraction of the most extreme starburst galaxies had passed unnoticed in the optical surveys.

1.1 Luminous Infrared Galaxies

The Infrared Astronomical Satellite *(IRAS)* discovered a new class of luminous extragalactic sources of infrared radiation. These are galaxies that radiate a large fraction of their total energy in the far-infrared (FIR). The most extreme galaxies of this type, which radiate more than 10^{11} solar luminosities in the *IRAS* broad wavelength bands (12, 25, 60, 100 μm) are named "luminous infrared galaxies". It is striking that the FIR luminosity of these galaxies can be equivalent to the bolometric luminosity of quasars, namely, about 10 times the optical luminosity of the first-ranked cD galaxies in rich clusters.

The study of luminous infrared galaxies has now become a key area in extragalactic research for two reasons: (1) we now realize that among objects with luminosities above 10^{11} $L_\odot$ the *IRAS* luminous galaxies are the dominant population of objects in the Universe; and (2) there is the increasing evidence that luminous infrared galaxies may represent an early phase in the evolution of galaxies. Their study may provide clues for our understanding of the genesis of some elliptical galaxies (*e.g.* Wright *et al.* 1990), quasars (Sanders *et al.* 1988), and radio galaxies (Mirabel 1989).

Most of the FIR emission from galaxies is due to the absorption and re-emission of light by dust. In addition, from the *IRAS* colors we know that a large fraction of the FIR emission in luminous infrared galaxies is produced by warm dust. At present it is debated if the ultimate source of light that heats the dust consists solely of large amounts of recently formed stars, or if, in addition, the formation of massive compact objects with X-ray emitting accretion disks is also required to explain the colossal amounts of thermal energy radiated in the FIR.

1.2 Historic Background

Before the *IRAS* mission some of the ultraluminous Infrared galaxies had already attracted the attention of observational astronomers because of their peculiar properties in the optical, near infrared, and/or radio wavelengths. In fact, some of the prototypes of this class of objects (*e.g.* Arp 220) had enough unusual optical morphologies to warrant their inclusion in Arp's Atlas of Peculiar Galaxies (1966), and had been selected as prototypes of merger galaxies in the work by Joseph and Wright (1985). In addition, near infrared observations by Riecke and Low (1972) had shown that some of the objects that had been classified as quasars and/or active galaxies (*e.g.* Mrk 231, I Zw1) showed abnormally high near infrared luminosities.

Radio astronomers had also identified at centimeter wavelengths some of the ultraluminous infrared galaxies as members of a class of "bright radio-spiral galaxies" (Condon 1980). These appeared as sources with radio-to-optical luminosity ratios more than an order of magnitude above the typical ratio for gas-rich galaxies. The radio spectral properties of these galaxies were somewhat perplexing. At $\lambda 21$ cm they showed very broad atomic hydrogen absorption lines, indicating large amounts of unusually turbulent neutral gas (Mirabel 1982). Another unexpected idiosyncracy of these galaxies was the exceptionally bright hydroxyl maser emission comming from the nuclei (*e.g.* Baan, Wood, and Haschick 1982).

1.3 Optical Morphology: Mergers of Spiral Galaxies

In the last two decades there has been an almost explosive growth of publications reporting evidences that interactions trigger starbursts in galaxies. After the seminal papers by Toomre and Toomre (1972) it was realized that interacting galaxies are more active in the UV (Larson and Tinsley 1978), near infrared (Joseph and Wright 1985), optical emission line strength (Kennicut and Keel 1984), and radio emission (Stocke 1978; Hummel 1981).

However, the optical morphology of ultraluminous infrared galaxies has been a subject of recent controversy. From low resolution images of 10 ultraluminous infrared galaxies of the Bright *IRAS* galaxy sample obtained with the 1.5m Palomar telescope Sanders *et al.* (1988) concluded that nearly all ultraluminous infrared galaxies are strongly interacting merger systems. On the other hand, from images of a sample of luminous *IRAS* galaxies in the North Polar Cap obtained at La Palma, Lawrence *et al.* (1989) concluded that although galaxy interactions may be a common causal factor in luminous IR activity, it may not be an ubiquitous factor as had been suggested by Sanders *et al.* (1988).

In this context, Melnick and Mirabel (1990) obtained New Technology Telescope (NTT) images of the 16 nearest southern ultraluminous galaxies up to $z = 0.13$. The excellent quality of the NTT optics and the good seeing conditions on La Silla were fully exploited to reveal possible faint morphological features (*e.g.* Fig. 1). Melnick and Mirabel (1990) found that all the objects of this nearby sample of southern ultraluminous galaxies show the features that Toomre and Toomre (1972) had identified as resulting from gravity in disk-disk mergers. Tails, wisps and double nuclei are apparent in all galaxies at $cz \leq 25{,}000$ km s^{-1}. However, the faint extended features that are characteristic of tidal interactions become less ostensible at $cz \geq 25{,}000$ km s^{-1}, and only the brighter, distorted main bodies of the galaxies remain easily visible. From this imaging survey Melnick and Mirabel (1990) concluded that the faint extended features provoked by gravity during galaxy-galaxy collisions may become blurred at higher redshifts.

Rowan-Robinson (1990) and Sanders (1992) have in addition reviewed the results of optical imaging of larger samples of luminous galaxies. Sanders (1992) realized that the increase of luminosity is correlated with the degree of interaction as measured

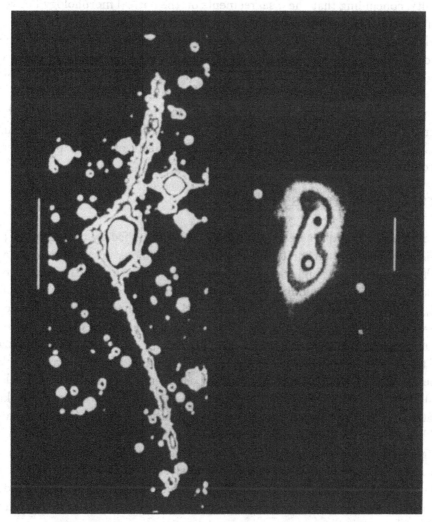

Fig. 1—CCD image of the "Superantennæ", a prototype of ultraluminous infrared galaxy, obtained with the 2.2m telescope of ESO. This remarkable example of ultraluminous infrared galaxy at a distance of 250 Mpc is the product of the collision 10^9 years ago of two giant galaxies. From tip to tip, the tails extend across 350 kpc. The detailed study of this galaxy by Mirabel, Lutz, and Maza (1991) showed that the tails emanate from a merger of giant gas-rich galaxies that harbor two nuclei (one Seyfert, one starburst) separated by 10 kpc. More than 80% of the energy radiated by this powerful infrared system comes from the deeply obscured Seyfert nucleus. On the left, a mosaic image of the whole system is shown, on the right the two nuclei embedded in the central part. The left scale bar corresponds to 1″, the right one to 10″ (= 10 kpc) in the central region.

by the projected separation of galaxy nuclei. Rowan-Robinson (1990) noted that the percentage of interacting or merging galaxies increases steadily with increasing luminosity, concluding that the disagreements on the optical morphology between the different groups are now within the statistical uncertainties.

Melnick and Mirabel (1990) find a critical separation of 10 kpc between the nuclei of the colliding galaxies of their sample of ultraluminous galaxies. In other words, advanced merging seems to be a necessary condition for the greatly enhanced infrared luminosity. We point out that although merging is a necessary condition, it is not sufficient, since there are strongly interacting, and even merging systems that are not ultraluminous in the infrared. This is an indication that in order to bust the luminosity of the mergers, in addition to restricted sets of orbital parameters, the pre-encounter galaxies must be rich in interstellar gas.

2 NUCLEAR STARBURSTS

The diverse forms of nuclear activity are likely to be fed by fresh gas. In particular, the most violent starbursts in the nuclei of luminous infrared galaxies appear to be the consequence of high concentrations of molecular gas in the central regions (Scoville *et al.* 1993). It is found that the most efficient starbursts measured by the large $L_{fir}/M(H_2)$ ratios always have, as expected, the highest space densities of H_2.

The high concentrations of molecular gas in luminous IR galaxies must be a consequence of inward motions of the disk gas during collisions. To concentrate $\geq$ 50% of the overall interstellar gas in the nuclear region, gas from distances $\geq$ 10 kpc must be driven into the center in $\leq 10^9$ years. This implies mean infalling velocities of $\geq$ 10 km s^{-1}. When the systemic velocity of the galaxies are accurately known, observations of H I absorption at $\lambda 21$ cm can be used to probe the kinematics of the cold gas along the line of sight to the nuclear radio continuum source. Since in colliding galaxies the H I absorption has typical half-widths $\geq$ 100 km s^{-1}, and the optical redshifts may have large errors, the analysis must be of statistical nature. Following this idea, Dickey (1986) and Mirabel and Sanders (1988) concluded, with some caveats, that in nuclear starburst galaxies there is infall of H I at a rate of > 1 M$_\odot$ yr^{-1}.

Large scale starbursts are expected to take place in stellar systems that readily form molecular gas out of atomic gas. Such high efficiencies of molecular cloud formation are likely to take place in merging spiral galaxies. When the mean densities become $\geq 10^2$ cm^{-3} most of the H I gas turns into molecular form. To study the overall faction of the interstellar gas that has been converted into molecular form, Mirabel and Sanders (1989) used the results from their H I and CO surveys of galaxies with FIR luminosities in the range of 2×10^{10} L$_\odot$ to 2×10^{12} L$_\odot$. In Figure 2 are shown the diagrams that resulted from this study. Figure 2a shows that the CO(1→0) to H I global luminosity ratios of galaxies are proportional to the overall FIR excess. Of particular interest is the existence of extremely obscured galaxies like Arp 220

(IC 4553) and *IRAS* 10173-0828 where $\leq 15\%$ of the total mass of interstellar gas is in atomic form. Although there may be several factors implicated in the production of high $M(H_2)/M(HI)$ ratios, the relative depletion of H I is likely to be due to an enhancement of molecular cloud formation. Figure 2b shows that the galaxies with enhanced star formation are those that have more efficiently converted H I into H_2.

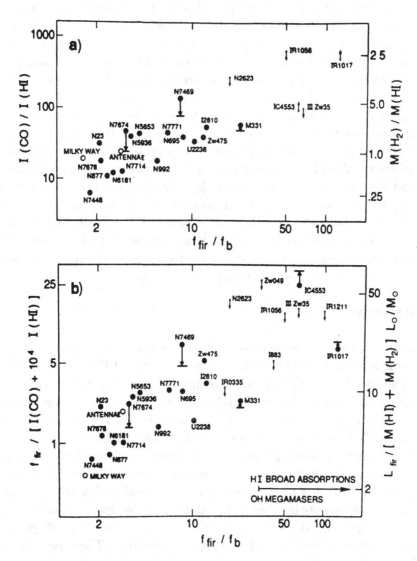

Fig. 2—(*a*) CO(1-0)/H I flux ratio in Jy km s^{-1} measured at $\lambda 2.6$ mm and $\lambda 21$ cm, as a function of the FIR-to-blue flux ratios $f_{\text{fir}}/f_{\text{b}}$. (*b*) Ratio of the FIR luminosity to the sum of the CO and H I spectral fluxes as a function of FIR excesses. These diagrams from Mirabel and Sanders (1989) show that in ultraluminous infrared mergers most of the atomic hydrogen is converted into molecular form.

3 EXTRANUCLEAR STARBURSTS: TIDAL DWARF GALAXIES

The idea that collisions between giant galaxies, in addition of driving matter inward, may eject stars and gas to intergalactic space, out of which star forming dwarf galaxies and star clusters may be formed, was first proposed by Zwicky (1956) and later followed up by Schweizer (1978). These recently formed dwarf galaxies and star clusters may be seen as patches of optically emitting material that usually appear along the tidal tails that emanate from merging disk galaxies. In the particular case of the system shown in Figure 1, the knots along the tails become bluer towards the far-ends. The condensations at the tip have colors $(B - V) \sim 0.4$ and are likely to become detached systems, namely, star forming isolated dwarf galaxies or star clusters.

Kennicut and Chu (1988) have reviewed the question of the formation of young globular clusters and their possible association with giant H II complexes in nearby galaxies. They noted that young populous clusters with massive stars are invariably found in galaxies with widespread star formation bursts such as M82 and NGC 5253. Ashman and Zepf (1992) found the results by Kennicut and Chu (1988) consistent with the hypothesis that young globular clusters form preferentially in interacting galaxies. This hypothesis can provide clues to understand: (*a*) the high frequency of globular clusters around ellipticals; (*b*) the large number of globulars associated to the most massive galaxies at the centers of clusters of galaxies; (*c*) the evidence for higher mean metallicity in clusters around giant galaxies.

3.1 The Antennae

To further explore this hypothesis Mirabel, Dottori, and Lutz (1992) made a spectrophotometric study of the condensation at the tip of the southern tail of the prototype merger NGC 4038/39 (The Antennae) that is shown in Figure 3. They found that this condensation is a dwarf galaxy formed out of the tidal remnants that were ejected to intergalactic space. The tidal dwarf consists of a chain of H II regions ionized by recently formed massive stars, which are embedded in an envelope of H I gas and low surface brightness optical emission. The metallicity of the H II regions is $\sim 1/3$ solar, namely, similar to that of H II regions in the outskirts of galactic disks. The spectrum shown in Figure 3 corresponds to an H II region with a luminosity equivalent to 300 Orion nebulae. Since the material at the tip of the tail was ejected $\sim 7 \times 10^8$ years ago, and the ionizing stars are younger than 2×10^6 years, these massive stars must have been borned well after the ejection. The time elapsed since the detachment of the material from the galactic disks is at least a factor of ten the internal crossing time, so this object will remain bound unless disrupted by stellar winds and supernovae from the massive stars being formed.

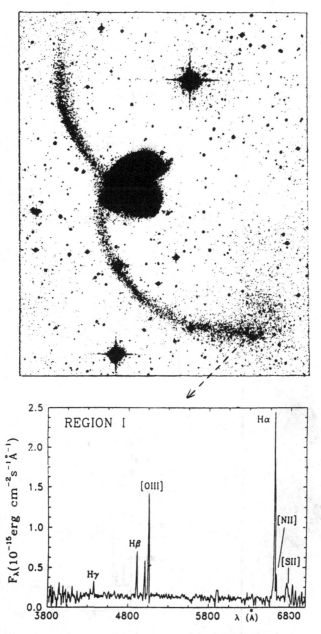

Fig. 3— Optical photograph of the interacting galaxies NGC 4038/39 (the Antennæ). At the tip of the southern tail there is a chain of H II regions embedded in a diffuse low surface brightness envelope. The spectrum is from one of the H II regions at ~ 100 kpc from the merging disks. The H II regions are being ionized by stars formed $\leq 2 \times 10^6$ years ago. Mirabel, Dottori, and Lutz (1992) proposed that the southern tip of the Antennæ is a dwarf irregular galaxy of tidal origin.

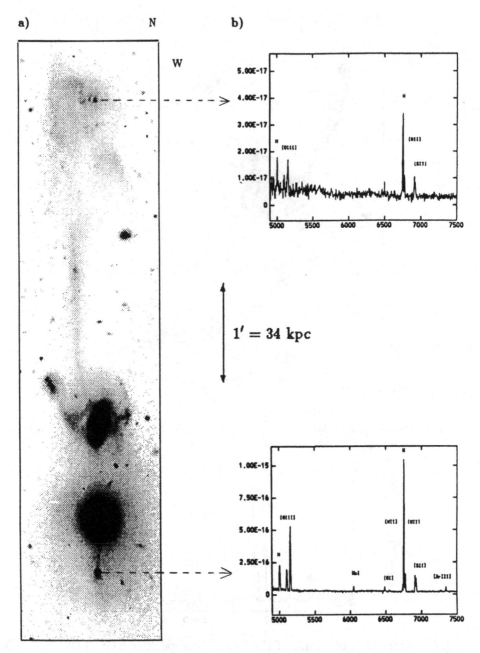

Fig. 4—(*a*) V band image of Arp 105, a multiple collision in the cluster of galaxies Abell 1185. In this remarkable system it is observed the formation of a Magellanic irregular and a blue compact galaxy out of the debris of the collision. (*b*) Spectra of the condensations detected in the forming galaxies. They are typical of star forming regions. From Duc and Mirabel (1993)

3.2 Arp 105: a Multiple Collision in a Cluster

Arp 105 is one of the most remarkable instances of multiple collisions of galaxies. At a distance of $\sim$ 120 Mpc (Figure 4) it consists of several galaxies undergoing encounters near one of the centers of the X-ray cluster A1185. The elliptical is tearing apart a starburst spiral galaxy that radiates 10^{11} $L_\odot$ in the far-infrared. From the spiral emanates a gigantic tail that at a distance of 100 kpc ends in an irregular system that is reminiscent of the Magellanic Clouds. Towards the South of the elliptical emanates a "jet-like" structure that ends in a blue compact object with strong optical emission lines. In addition, the victimed spiral appears to have been penetrated by an object flying by at a speed of 2,600 km s^{-1}.

The most striking result from our radio observations is that the spiral contains 10^{10} $M_\odot$ of molecular gas and $\leq 5 \times 10^8$ $M_\odot$ of HI, namely, more than 95% of the cold interstellar gas in the spiral is in molecular form. On the contrary, the irregulars of Magellanic type at the end of the 90 kpc long tail contain as much as 6×10^8 $M_\odot$ of HI, but less than 10% of the gas is in molecular form. Therefore, Arp 105 is a striking case for the study of the segregation of interstellar gas during the collision of a spiral with an elliptical near the center of a cluster of galaxies. A more detailed description of our research on this system will be published by Duc and Mirabel (1993).

REFERENCES

Arp, H. 1966, *Ap. J. Suppl.*, **14**, 1.

Ashman, K. M., and Zepf, S. E. 1992, *Ap. J.* **384**, 50.

Baan, W. A., Wood, P. A. D., and Haschick, A. D. 1982, *Ap. J. Lett.*, **260**, L49.

Balzano, V. A. 1983, *Ap. J.*, **268**, 602.

Condon, J. J. 1980, *Ap. J.*, **242**, 894.

Dickey, J. M. 1986, *Ap. J.*, **300**, 190.

Duc P. A., and Mirabel I. F. 1993, *Astr. Ap.*, to be submitted.

Hummel, E. 1981, *Astr. Ap.*, **96**, 111.

Joseph, R. D., and Wright, G. S. 1985, *Mon. Not. R. Astr. Soc.*, **214**, 87.

Kennicut, R. C., and Keel, W. C. 1984, *Ap. J. Lett.*, **279**, L5.

Kennicut, R. C., and Chu, Y. 1988, *A. J.*, **95**, 720.

Larson, R. B., and Tinsley, B. M. 1978, *Ap. J.* **219**, 46.

Lawrence, A., Rowan-Robinson, M., Leech, K., Jones, D. H. P., and Wall, J. V. 1989, *Mon. Not. R. Astr. Soc.*, **240**, 329.

Markarian, B. E. 1967, *Astrofizika*, **3**, 55.

Melnick, J., and Mirabel, I. F. 1990, *Astr. Ap.*, **231**, L19.

Mirabel, I. F. 1982, *Ap. J.*, **260**, 75.

Mirabel, I. F., and Sanders, D. B. 1988, *Ap. J.*, **335**, 104.

Mirabel, I. F., and Sanders, D. B. 1989, *Ap. J. Lett.*, **340**, L53.

Mirabel, I. F. 1989, *Ap. J. Lett.*, **340**, L13.

Mirabel, I. F., Lutz, D., and Maza, J. 1991, *Astr. Ap.*, **243**, 367.

Mirabel, I. F., Dottori, H., and Lutz, D. 1992, *Astr. Ap.*, **256**, L19.

Riecke, G. H., and Low, F. J. 1972, *Ap. J. Lett.*, **176**, L95.

Rowan-Robinson, M. 1990, IAU Symp. 146 on *Dynamics of Galaxies and their Molecular Cloud distributions*, eds. F. Combes, and F. Casoli (Dordrecht: Reidel), p. 211.

Toomre, A., and Toomre, J. 1972, *Ap. J.*, **178**, 623.

Sanders, D. B., Soifer, B. T., Elias, J. H., Madore, B. F., Matthews, K., Neugebauer, G., and Scoville, N. Z. 1988a, *Ap. J.*, **325**, 74.

Sanders, D. B. 1992, Proc. Taipei Astrophysics Workshop on *Relationships between Active Galactic Nuclei and Starburst Galaxies*, ed. A. Filippenko (San Francisco: ASP-31), p. 303.

Schweizer, F. 1978, *Structure and Properties of Nearby Galaxies*, ed. E. M. Berkhuijsen, and R. Wielebinski, p. 279.

Scoville, N., Hibbard, J. E., Yun, M. S., and van Gorkom, J. H. 1993, these proceedings.

Searle, L., and Sargent, W. L. W. 1972, *Ap. J.*, **173**, 25.

Stocke, J. T. 1978, *A. J.*, **83**, 348.

Weedman, D. W., Feldman, F. R., Balzano, V. A., Ramsey, L. W., Sramek, R. A., and Chi-Chao Wu. 1981, *Ap. J.*, **248**, 105.

Wright, G. S., James, P. A., Joseph, R. D., and McLean, I. S. 1990, *Nature*, **344**, 417.

Zwicky, F. 1956, *Ergebnisse der Exakten Naturwissenschaften*, **29**, 34.

Dynamics of Gas in Major Mergers

Lars Hernquist[1] and Joshua E. Barnes[2]

[1]Board of Studies in Astronomy and Astrophysics, U.C. Santa Cruz
[2]Institute for Astronomy, University of Hawaii

ABSTRACT

Possible consequences of the dynamics of interstellar gas in merging galaxies are discussed and illustrated with numerical simulations which incorporate both collisionless and hydrodynamical evolution.

1 INTRODUCTION

Over the past 20 years, considerable observational evidence has accumulated implying that "major" mergers of comparable–mass spirals may play an important role in the evolution of galaxies. Based on statistics on well–known merger candidates, Toomre and Toomre (1972) and later Toomre (1977) argued that major mergers are the dominant process by which early–type galaxies form. More recently, infrared and radio surveys have bolstered long–standing suspicions that nuclear starbursts in some peculiar objects may have been triggered by mergers of gas–rich progenitors (*e.g.* Sanders *et al.* 1988a, b; Sanders 1992). Only slightly less compelling are observations implicating galaxy collisions to the onset of activity in bright radio galaxies (*e.g.* Heckman *et al.* 1986) and quasars (*e.g.* Stockton 1990). (For a review, see Barnes and Hernquist 1992a.)

By now, the *stellar–dynamics* of major mergers have been explored in some detail using N-body simulation (*e.g.* Barnes 1988, 1992; Hernquist 1992, 1993a); however studies of the *hydrodynamical* evolution of interstellar gas during these events are less fully developed. Seminal works include simplified calculations, which ignore self–consistency, showing that bars in disks can drive nuclear inflows of gas (Simkin *et al.* 1980), low–resolution models investigating the fate of a dissipative component during major mergers (Negroponte and White 1983), simulations of gas inflow during transient collisions between galaxies (*e.g.* Noguchi 1988, 1991; Combes *et al.* 1990), and possible consequences of star formation and feedback (*e.g.* Mihos *et al.* 1991, 1992).

Here, we review the implications of some of these results in the context of major mergers of gas–rich spirals, employing numerical models developed mainly by us. As discussed below, the simulations demonstrate that mergers of this type are capable of driving large quantities of gas into the nuclear regions of the remnants, modifying their structure and perhaps yielding objects akin to starburst and active galaxies.

2 MODELS

The models we employ consist of several distinct but interacting components. We assume that the stars and dark matter in galaxies are collisionless and their evolution is adequately described by a standard N-body treatment. In particular, gravitational forces are computed using a hierarchical tree algorithm (Barnes and Hut 1986).

For the gas, we adopt a physical description which ignores many of the complexities of a multi-component medium. The gas is also represented by particles, using the method known as smoothed particle hydrodynamics (Lucy 1977; Gingold and Monaghan 1977). In this approach, a fluid is partitioned into elements which are represented computationally as particles. These particles are similar to those used to mimic the collisionless matter in that they evolve according to well-defined equations of motion and respond to gravity. However, they are also subject to local interactions which arise in actual fluids, such as pressure gradients and viscous forces, and they carry with them a thermal energy which evolves in accord with the usual energy equation of compressible fluid dynamics, supplemented by terms accounting for radiative heating and cooling. (For a review, see Barnes and Hernquist 1993a.)

Owing to limited numerical resolution, this description is far from complete and we are forced to make several compromises. For example, radiative cooling is inhibited below 10,000 K to prevent the gas from undergoing a thermal instability which cannot be properly resolved. We have also ignored star formation and feedback. Nevertheless, our calculations do include the effects of large-scale shocks and dissipation and, hence, provide a plausible description of the global evolution of galactic gas.

The galaxy models we employ are composite systems containing disks of stars and gas, dark halos, and bulges. They are built using either a "constructive" approach (Barnes 1988, 1992) or a technique derived from the Jeans' equations (Hernquist 1993b). If these models are evolved in isolation, they develop weak non-axisymmetric structures resembling spirals in actual disks (*e.g.* Hernquist 1991). They are, however, quite stable against bar formation and do not suffer from spontaneous inflows of gas, as might be the case if the models were constructed in a haphazard manner.

In what follows, we describe results from a survey of major mergers. Here, we focus on collisions of equal mass disk galaxies from parabolic orbits, chosen to lead to rapid merging. Among other things, we have varied the relative orientations of the incoming disks, the pericentric separation at first approach, and the parameters associated with the numerical algorithm, such as the particle number, the description of radiative heating and cooling, and the form of the artificial viscosity. Some of these results are discussed at greater length in Barnes and Hernquist (1991, 1992b, 1993b) and Hernquist and Barnes (1991).

3 MERGER DYNAMICS

3.1 An Example

While the various parameters describing a merger of two of our galaxies affect the details of the outcome, many features are common to all the simulations; hence, it is expedient to describe the results by referring to a single, "canonical" model. In this calculation, the merging galaxies each contain a bulge, a disk, and a halo with masses in the proportion B:D:H=1:3:16. The disk consists of stars and gas, with 10% of the disk mass residing in the latter component. The orbit was chosen to give a pericentric separation of 2.4 disk exponential scale-lengths at first close approach for the ideal parabolic trajectory. One of the disks lies in the orbit plane and suffers an exactly prograde encounter, while the other is initially inclined by 70 degrees with respect to the orbit plane.

In the following discussion, results are expressed in a dimensionless system of units in which the gravitational constant $G = 1$, each galaxy has a total mass $M = 5/4$, and the exponential scale–length of each disk is $h = 1/12$. Scaled to the Milky Way, unit length, time, and mass correspond approximately to 40 kpc, 250 Myrs, and 2.2×10^{11} $M_\odot$, respectively.

The global dynamics characterizing a merger of two such galaxies are illustrated in Figure 1, which shows the state of the disk stars at various times during the simulation. Note, in particular, the spirals which develop in each disk prior to their encounter, as in isolated models; the strong tidal response at first pericenter, which occurs at a time $t \approx 1$; the transfer of material from one disk to the other, beginning near $t = 1.5$; the rapid coalescence of the disks following their second close passage; and the relatively smooth appearance of the inner parts of the remnant achieved by $t = 3$.

3.2 Disk Response

On these large scales the response of the disk gas is similar, although some differences are evident. For example, the tails formed from the gas are crisper than their stellar counterparts. More important, the distributions of gas and stars in the inner regions of each disks become increasingly distinct with time. In fact, prior to merging, the disks accumulate large quantities of gas in their central regions. This effect is illustrated in Figure 2, which shows face–on views of the gas and stars in each disk after first pericenter but before the galaxies have coalesced. In response to their mutual tidal interaction, each disk develops non–axisymmetric features of large amplitude resembling bars. These structures arise in both the gas and stellar components; however there are subtle differences between them. In the present example, the gas tends to collect in shocks at the leading edges of the stellar "bars" and is subjected to strong gravitational torques. These torques lead to a secular transfer of

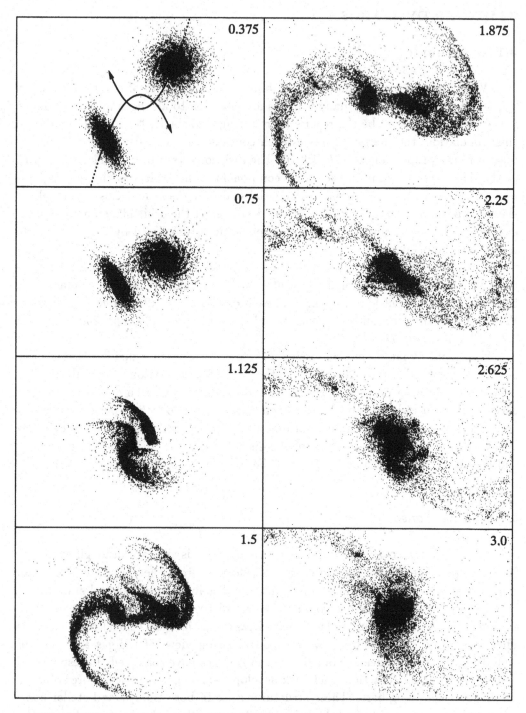

Fig. 1—Time evolution of stellar disks in a merger between two identical model galaxies, projected onto orbit plane. Scaling the progenitors to the Milky Way, the panels measure 100 × 150 kpc and are separated in time by nearly 100 Myrs.

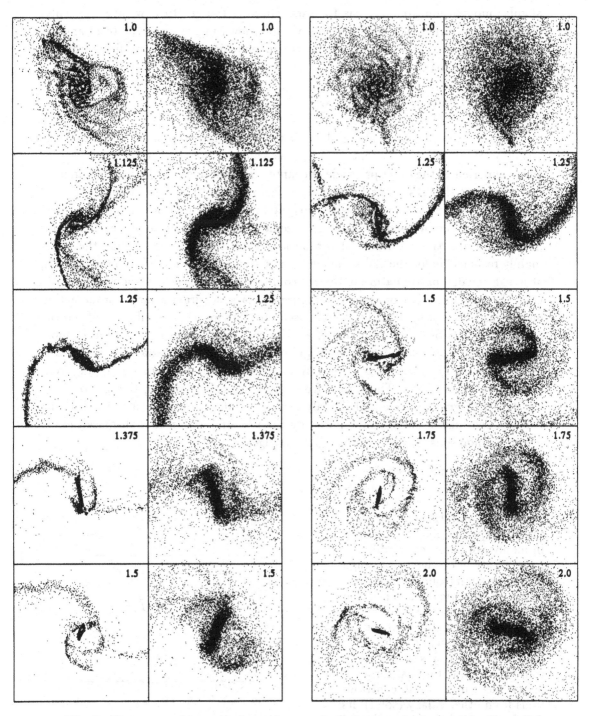

Fig. 2—Response of inner regions of prograde (left two columns) and inclined disk (right two columns) at times shortly after first pericenter. Left and right columns in each sequence show disk gas and stars, respectively.

angular momentum from the gas to stars, allowing the gas to flow into the center of each disk. During the crucial period of time shown in Figure 2, the total angular momentum of the gas which collects in the inner regions of the two disks drops by more than a factor of 10.

3.3 Physical Nature of Gas Inflow

This effect is illustrated in detail in Figure 3, where the top panel shows the time evolution of the angular momentum of the gas which is driven to the center of the disk which experiences the exact prograde encounter. During this time period, the angular momentum of this gas drops by roughly a factor of 30. The rate of change of this angular momentum is compared with the gravitational and hydrodynamic torques on the gas in the middle panel. The close agreement between the rate of loss of angular momentum and the gravitational torque acting on the gas proves that the effect mainly responsible for the radial inflows of gas seen in these models is *gravitational*, although the dissipative nature of the gas component enables it to respond differently from the stars. The bottom panel in Figure 3 shows the gravitational torque acting on the gas in the prograde disk, decomposed into its separate contributions from the other galaxy, its own halo plus bulge, and its own disk. During the period of most rapid angular momentum loss, the torque responsible for the radial inflow of gas is clearly dominated by the galaxy's own disk, with only a modest amount arising from its own bulge and halo. The torque from the other galaxy during this period is of the *opposite* sign. A more detailed analysis shows that the torque contributed by the galaxy's own disk is further dominated by the material within its half-mass radius, where the non–axisymmetric features are strongest.

Fig. 3—Top panel: Evolution of angular momentum of gas which collects near center of prograde disk. Curves show results for two experiments differing only in number of particles. Middle panel: Rate of change of gas angular momentum (solid curve), computed by numerical differentiation of curve in top panel. Solid and open points indicate gravitational and hydrodynamical torques acting on the gas, respectively. Bottom panel: Gravitational torque on gas, decomposed into contributions from other galaxy (solid), its own bulge plus halo (dashed), and its own disk (dotted).

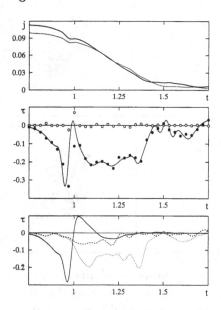

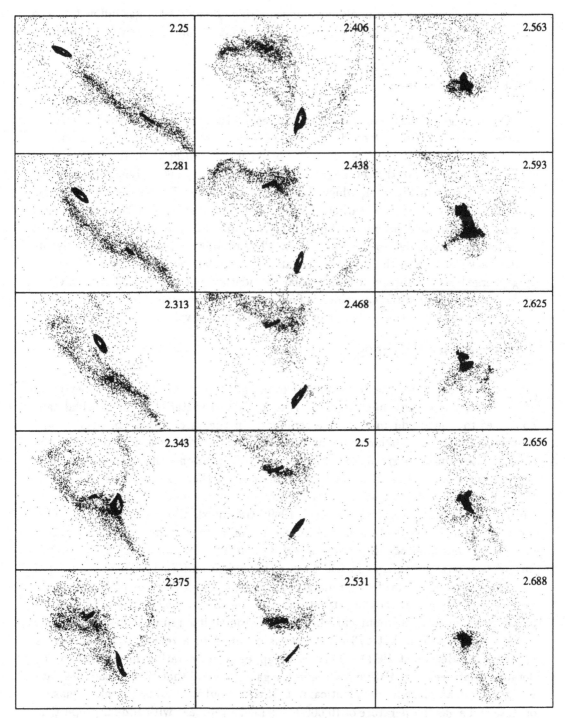

Fig. 4—Final stages of coalescence of gas rings from the two merging galaxies. Scaling the progenitors to the Milky Way, the panels measure 12.6 × 16.8 kpc and are separated in time by roughly 8 Myrs.

Similar effects are seen in all our models, although the rate of gas inflow tends to be greatest in close, prograde encounters. This result is quite insensitive to the form of the artificial viscosity.

3.4 Final Coalescence

Once the gas is driven into the middle of each disk in response to the bar forcing, it typically settles into flattened disks or rings that are supported by rotation. As the galaxies merge, these disks or rings, being much denser than the surrounding stellar material, sink to the center of the merger remnant owing to dynamical friction. This can be seen in Figure 4, which shows the final interaction between the two gas rings in the model depicted earlier. By the final time displayed, 66% of all the gas originally distributed throughout both disks is confined to a region $\sim$ 200 pc across, if the galaxies are scaled to the Milky Way. The gravitational torques exerted by the stellar background and the direct cancellation of angular momentum as the disks or rings merge lead to a further loss of angular momentum by the gas. Consequently, the angular momentum of the gas which is finally located near the center of the remnant has dropped by more than a factor of 100 from its initial value.

4 MERGER REMNANTS

As the most tightly bound material from each galaxy merges, the inner parts of the nascent remnant quickly relax. Loosely bound material from the tidal tails continues to slowly add mass to the remnant. In the case of the canonical model described above, the stellar remnant is nearly oblate. The gas accreted from the tails at lates times forms an extended, warped disk around the remnant which is reminiscent of warped gas disks in some ellipticals. Furthermore, as the stellar debris from the tails oscillates back and forth through the remnant, it produces "shells" through phase–wrapping (*e.g.* Barnes 1989; Hernquist and Spergel 1992). This mechanism for the origin of shells in ellipticals is similar to that involving the accretion of material from small satellite galaxies (*e.g.* Quinn 1984; Hernquist and Quinn 1988), but offers several advantages in select cases (Hernquist and Spergel 1992).

The stellar remnants of major mergers are generally well-fitted by de Vaucouleurs' profiles, indicating that they are structurally similar to elliptical galaxies. In the case of the remnant formed in our canonical model, the stellar remnant has an effective radius of order 4 kpc, but 2/3 of the gas is confined to a region $\sim$ 200 pc across (Barnes and Hernquist 1991). This result appears to be rather insensitive to the detailed treatment of radiative cooling in the gas, provided that the gas can dissipate energy. For example, a model identical to that above but with an isothermal equation of state and a gas temperature of 10,000 K, yielded a remnant with similar properties, again having 2/3 of all the gas confined to a region $\sim$ 200 pc across near its center. However, if radiative cooling is disabled, the gas is simply shock heated to the virial

Fig. 5—Fraction of gas mass at given densities in remnants produced in identical mergers in which only the radiative microphysics was varied: Solid curve is for canonical model which includes radiative cooling and heating; Dashed curve is for an isothermal gas with temperature 10,000 K; Dotted curve is for run with no cooling.

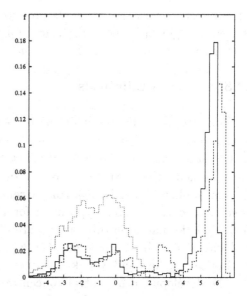

temperature of the remnant and this radial inflow is completely suppressed. This effect is illustrated in Figure 5 where we plot the differential mass distribution of the gas against density. The large accumulation of high density gas in the models with radiative cooling or an isothermal equation of state reflect the radial inflows noted earlier; however, no such high density gas is evident in the simulation where cooling is inhibited. Thus, the build–up of large quantities of gas near the centers of the remnant is mediated by several physical processes: the gravitational torques exerted on the gas by the surrounding stellar material, the fact that the gas can dissipate energy in shocks and lead the stellar response to tidal torques, and the propensity of the gas to collect in dense knots when it can cool radiatively.

All the models in our survey of disk inclinations and pericentric separations yielded results in qualitative agreement with those discussed here. Even perfectly retrograde encounters eventually lead to similar accumulations of gas at the centers of the remnants.

5 IMPLICATIONS

5.1 Relation to Active Galaxies

Simulations like that described above suggest an expanation for the onset of intense star formation in the inner regions of ultra–luminous infrared galaxies. In particular, the concentrations of dense gas may be plausibly identified with those in the most luminous *IRAS* sources in terms of their masses, sizes, and the time–scales over which they are formed. Since our present models do not include the effects of star formation we cannot predict the ultimate fate of the gas in our models. However, a number of models predict that further evolution in gas concentrations of this nature may lead to the production of an AGN (*e.g.* Begelman *et al.* 1984). In this regard, it is interesting that the remnants produced in our simulations are morphologically similar to bright radio galaxies, which typically show tell–tale signs of a past merger event, such as tidal tails (Heckman *et al.* 1986).

5.2 Structure of Ellipticals

Provided that the gas which is driven to the centers of the remnants in our calculations is not dispersed by feedback from star formation, the conversion of this gas into stars will have a significant effect on the structure of the inner regions of the newly formed object. Since gas is not subject to Liouville's theorem, it can boost the central phase space densities of the remnants, perhaps circumventing phase space objections to the merger hypothesis for the origin of elliptical galaxies (Carlberg 1986). Star formation in this gas will also likely establish radial color and metallicity gradients; it remains to be seen if these gradients are in accord with those observed in elliptical galaxies and well–known merger candidates.

The development of central concentrations of gas also affects the shapes of the *stellar* remnants in our models. For example, when compared with pure stellar–dynamical analogues, the models which incorporate gas dynamics produce remnants that are more nearly oblate and possess a larger fraction of tube orbits.

5.3 Kinematically Distinct Cores

An interesting outcome of some of our models has been the finding that the gas driven to the center is kinematically distinct from the surrounding more diffuse gas and stars. In some cases, the dense gas resides in thin disks which *counter–rotate* with respect to outlying material (Hernquist and Barnes 1991). At present, we do not have a sufficient data–base to determine the fraction of mergers that might give rise to kinematic subsystems in the remnants. However, it is interesting to note that NGC 7252 appears to contain a gas disk in its inner regions rotating in a different sense from the surrounding stars (Schweizer 1982). Depending on how the gas in our models is converted into stars, the simulations may also explain the origin of kinematic subsystems of stars in many ellipticals, assuming they originated through major mergers.

5.4 Fragmentation of Tidal Tails

A final curious result of our models is the tendency for the tails drawn from the disks by their tidal interaction to fragment and produce bound substructures. This effect can be seen in Figure 1, where the tail extending to the upper left of the final panel contains a self–gravitating object consisting of both gas and stars. In fact, the two tails in this model include some 23 bound entities, although the one clearly visible in Figure 1 is by far the most massive. Such features also develop in purely stellar–

dynamical mergers (Barnes 1992), although their formation appears to be aided by including gas in the simulations.

Objects like those in the tails of our canonical model have masses and sizes characteristic of dwarf galaxies and the fragmentation of tidal tails apparently provides yet another mechanism by which galaxies can be formed. Less speculatively, observations indicate that the material in tails is often clumped and not smoothly distributed (*e.g.* Mirabel *et al.* 1991, 1992). The models here lend credence to the notion that the irregular distribution of matter in tidal tails is the result of their dynamical evolution following a major merger.

5 ISSUES

While are simulations are suggestive and can, in principle, explain a number of outstanding questions in galactic evolution, it is clear that many important issues remain. Owing to the relatively small number of models we have run (of order 10), we have not fully established the conditions under which radial inflows of gas can be induced in disks by tidal interactions. It is unclear whether to what extent this effect depends in detail on the orbital geometry or the structure of the progenitors. Our calculations have ignored a variety of physical effects that must eventually come to dominate the dynamics of the gas, including star formation and feedback, a description of the multi–phase nature of real interstellar gas, and magnetic fields.

Consequently, our models are, at present, incapable of establishing the ultimate fate of gas driven to the centers of remnants by major mergers and cannot address questions such as the origin of superwinds from galaxies. Nevertheless, the dramatic differences between the evolution of the gas and stellar components of our simulations imply that even modest quantities of gas are capable of significantly modifying our understanding of the essential dynamics of major mergers and that comprehensive models of forming and evolving galaxies must include both collisionless and dissipative components.

This work was supported in part by the Pittsburgh Supercomputing Center, the Alfred P. Sloan Foundation, NASA Grant NAGW–2422, and the NSF under Grant AST 90–18526, and the Presidential Faculty Fellows Program.

REFERENCES

Barnes, J. E. 1988, *Ap. J.*, **331**, 699.
Barnes, J. E. 1989, *Nature*, **338**, 123.
Barnes, J. E. 1992, *Ap. J.*, **393**, 484.
Barnes, J. E., and Hernquist, L. 1991, *Ap. J.*, **370**, L65.
Barnes, J. E., and Hernquist, L. 1992a, *Ann. Rev. Astr. Ap.*, **30**, 705.
Barnes, J. E., and Hernquist, L. 1992b, *Nature*, **360**, 715.
Barnes, J. E., and Hernquist, L. 1993a, *Physics Today*, **46**, No. 3, 54.

Barnes, J. E., and Hernquist, L. 1993*b*, *Ap. J.*, in preparation.

Barnes, J. E., and Hut, P. 1986, *Nature*, **324**, 446.

Begelman, M. C., Blandford, R. D., and Rees, M. C. 1984, *Rev. Mod. Phys.*, **56**, 225.

Carlberg, R. 1986, *Ap. J.*, **310**, 593.

Combes, F., Dupraz, C., and Gerin, M. 1990, in *Dynamics and Interactions of Galaxies*, ed. R. Wielen (Berlin: Springer–Verlag), p. 205.

Gingold, R. A., and Monaghan, J. J. 1977, *Mon. Not. R. Astr. Soc.*, **181**, 375.

Heckman, T. M., Smith, E. P., Baum, S. A., van Breugal, W. J. M., Miley, G. K., Illingworth, G. D., Bothun, G. D., and Balick, B. 1986, *Ap. J.*, **311**, 526.

Hernquist, L. 1991, *Int. J. Supercomput. Appl.*, **5**, 71.

Hernquist, L. 1992, *Ap. J.*, **400**, 460.

Hernquist, L. 1993*a*, *Ap. J.*, **409**, 548.

Hernquist, L. 1993*b*, *Ap. J. Suppl.*, **86**, 389.

Hernquist, L. and Barnes, J. E. 1991, *Nature*, **354**, 210.

Hernquist, L., and Quinn, P. J. 1988, *Ap. J.*, **331**, 682.

Hernquist, L., and Spergel, D. N. 1992, *Ap. J.*, **399**, L117.

Lucy, L. 1977, *Astron. J.*, **82**, 1013.

Mihos, J. C., Richstone, D. O., and Bothun, G. D. 1991, *Ap. J.*, **377**, 72.

Mihos, J. C., Richstone, D. O., and Bothun, G. D. 1992, *Ap. J.*, **400**, 153.

Mirabel, I. F., Dottori, H., and Lutz, D. 1992, *Astr. Ap.*, **256**, L19.

Mirabel, I. F., Lutz, D., and Maza, J. 1992, *Astr. Ap.*, **243**, 367.

Negroponte, J., and White, S. D. M. 1983, *Mon. Not. R. Astr. Soc.*, **205**, 1009.

Noguchi, M. 1988, *Astr. Ap.*, **203**, 259.

Noguchi, M. 1991, *Mon. Not. R. Astr. Soc.*, **251**, 360.

Quinn, P. J. 1984, *Ap. J.*, **279**, 596.

Sanders, D. B., Soifer, B. T., Elias, J. H., Madore, B. F., Matthews, K., Neugebauer, G., and Scoville, N. Z. 1988*a*, *Ap. J.*, **325**, 74.

Sanders, D. B., Scoville, N. Z., Sargent, A. I., and Soifer, B. T. 1988*b*, *Ap. J. Lett.*, **324**, L55.

Sanders, D. B. 1992, in *Relationships Between Active Galactic Nuclei and Starburst Galaxies*, ed. A. V. Fillipenko (San Francisco: ASP-31), p. 303.

Schweizer, F. 1982, *Ap. J.*, **252**, 455.

Simkin, S. M., Su, H. J., and Schwarz, M. P. 1980, *Ap. J.*, **237**, 404.

Stockton, A. 1990, in *Dynamics and Interactions of Galaxies*, ed. R. Wielen (Berlin: Springer–Verlag), p. 440.

Toomre, A. 1977, in *The Evolution of Galaxies and Stellar Populations*, eds. B. Tinsley and R. Larson (New Haven: Yale University Press), p. 401.

Toomre, A., and Toomre, J. 1972, *Ap. J.*, **178**, 623.

Kinematic Instabilities, Interactions, and Fuelling of Seyfert Nuclei

William C. Keel

Department of Physics and Astronomy, University of Alabama

ABSTRACT

I describe studies of the incidence of nuclear activity and star-formation rates for galaxies in two paired samples, as functions of the encounter direction and kinematic properties of the disturbed disk. One sample, designed to test star-formation diagnostics, is a geometrically derived subset of the Karachentsev catalog. A separate sample, of paired Seyferts, is used to search for common kinematic characteristics among interaction-triggered AGN. Both starbursts and Seyfert nuclei occur with about equal frequency in direct and retrograde encounters. Nuclear and disk star formation are correlated with the form of the velocity curve, and with the normalized amplitude of velocity disturbance in the disk. Seyfert nuclei in pairs show a high fraction of galaxies with large solid-body regions in the rotation curves. Such kinematic properties are associated with higher then normal star-formation rates in the K-pair spirals (but not the highest). For both AGN and star-formation processes, the theoretical scheme most nearly accounting for the observations gives a prominent role to a Toomre-style disk instability on large scales, perhaps driving more local processes such as cloud collisions or pressure-induced cloud collapse. Including kinematic information offers a more refined way to identify externally-triggered phenomena than do disturbed morphology or presence of companions alone.

1 INTRODUCTION: TRIGGERING OF STARBURSTS AND ACTIVE GALACTIC NUCLEI

Numerous studies have suggested links between galaxy interactions and either high rates of star formation or enhanced probability of harboring an active nucleus. These results are obtained across the spectrum, from optical and near-UV colors (Larson and Tinsley 1978; Keel and van Soest 1992), through infrared luminosities (Soifer et al. 1984; Lonsdale, Persson, and Matthews 1984; Lawrence et al. 1989; Sulentic 1989), to radio-continuum statistics (Hummel 1981; Condon et al. 1982; Hummel et al. 1987). Perhaps the most spectacular manifestations of both processes are the "ultraluminous" IRAS galaxies, which have bolometric luminosities as high as 10^{11} $L_\odot$ and are mostly (Lawrence et al. 1989) or completely (Sanders et al. 1988) found in strongly interacting or merging systems.

There have been many discussions of the statistics of pair-induced phenomena and their significance (see the review by Heckman 1990), and likewise many proposed physical mechanisms for these enhancements. With a variety of proposed processes, a directly observational test is in order. I present here a study designed to test the predictions of several of these models for both star-forming and active galaxies, and discuss how well these fare in the light of new data.

Statistical studies of interaction effects suffer from the inevitable fraction of false companions projected near galaxies, and more basically from the fact that sample comparisons as usually done must average over a wide range of orbital phases and encounter properties. I will approach the problem from a different direction — looking only at populations of paired galaxies, are there kinematic similarities among the galaxies with active nuclei or strong star formation, which might lead to a more physically refined picture for stimulation during interactions?

2 DIRECT AND RETROGRADE ENCOUNTERS

Many models predict a strong difference in the rate of gas inflow between the cases of direct and retrograde companion orbits (for example, Mihos, Richstone, and Bothun 1992) or between galaxies with and without stellar bars (Noguchi 1988; Shlosman, Begelman, and Frank 1990). I have collected kinematic and imaging data for two samples designed to test these putative connections.

For a statistically complete set of interacting pairs, a subset was identified from the Karachentsev (1972, 1987) catalog which must be dominated by galaxies undergoing a near-planar encounter and viewed with both disk and orbit planes near the line of sight (Keel 1991, 1993). The selection criteria included galaxy axial ratio $b/a < 0.5$, angle between projected major axis and companion direction $\theta < 30°$, and velocity difference in the range 150–400 km s^{-1}. For these galaxies, a rotation-curve measurement plus the sign of the pair velocity difference can tell which sense of encounter is involved. I obtained rotation curves (or for the more pedantically inclined, velocity slices) using the *KPNO* 2.1m and *BTA* 6m telescopes, augmented by a uniform set of R images from the Lowell 1.1m Hall telescope. Star-formation rates were assessed from nuclear and disk-integrated Hα strength, and in many cases *IRAS* fluxes separated between pair members. This sample is useful for testing the response of nuclear and disk star formation to encounter properties; no strong Seyfert nuclei appear among them.

For this K-pair sample, there is no significant difference in star-formation properties between direct and retrograde cases; starbursts appear at comparable frequency for both.

For Seyfert galaxies, the issue is complicated by the fact that internal obscuration hides active nuclei in nearly edge-on galaxies from optical view (Keel 1980), removing just those galaxies that give the cleanest test. Applying the criteria used for the K-pairs gives a nearly null set for catalogued Seyferts. The paired-Seyfert sample was selected from known paired NGC and Markarian Seyferts selected only not to be viewed nearly face-on, augmented by some listed as pair members by Lipovetskii, Neizvestnii, and Neizvetsnaya (1987). Orbit sorting for the Seyferts relied on comparison between direct images and a grid of *N*-body models generated by Howard *et al.* (1993). Tests on the K-pairs show that when classifiable tidal structure is seen, its classification agrees with that from spectroscopy in every case as to encounter direction. Even with such a confidence test, orbit sorting for the Seyferts is sufficiently uncertain that, from available material, it can only be said that Seyfert nuclei do not strongly favor any particular direction of encounter. There are examples among Markarian Seyferts of clearly direct, clearly retrograde, and probably polar encounters.

3 INTERNAL KINEMATICS AND DISK RESPONSE

The motions in a tidally disturbed disk must be directly linked to the processes that govern star formation and AGN; anything that happens to move the gas within the disk, or changes its phase, will be reflected in some way in the observed motions. We might even see disturbed kinematics before the morphological disturbance becomes significant.

3.1 Star Formation

Star-formation indicators among the K-pair spirals show a correlation between overall *form* of the rotation curve and intensity of star formation, and between a normalized indicator of velocity disturbance and star formation. Both Hα and far-infrared excess increase in the same sequence: from galaxies with "normal" rotation, through those with dominant areas of rigid-body rotation, then through those with more general patterns of disturbance, peaking with "slow" rotators that have elongated shapes but very small velocity amplitudes (30 km s^{-1} or less). Nuclear and disk Hα behave in the same way; conditions near the nucleus are connected to disturbances as much as 10 kpc out in the disk. This is further shown by a correlation between Hα equivalent width (nuclear or integrated) and the amplitude of velocity disturbance when normalized to the peak "circular" velocity. The full data sets for the K-pair sample are collected by Keel (1993).

3.2 Nuclear Activity

The paired Seyfert galaxies show a striking number of solid-body rotation curves, 80% of the 39 observed. The size distribution of the solid-body regions is similar to that found in the K-pair sample, and larger than found in published sets of more isolated spirals (such as Mathewson, Ford, and Buchhorn 1992), which contain only 30–40% of spirals with resolved solid-body ranges (depending on how such factors as Hubble type and luminosity are matched between samples). This comparison suggests a connection between this particular kind of disk disturbance and the occurrence of Seyfert activity.

4 PHYSICAL MECHANISMS

The crucial physics in inducing nuclear starbursts and episodes of AGN activity lies in the transport of angular momentum. Various schemes have been proposed to do this via cloud-cloud collisions in the interstellar medium (Olson and Kwan 1990; Mihos, Richstone, and Bothun 1992), via torquing by inducing bars on various scales (Noguchi 1988; Shlosman, Begelman, and Frank 1990), and through an instability of the kind analyzed by Toomre (1964), as applied in this context by Lin, Pringle, and Rees (1988). These processes are all physically plausible and likely to operate to some degree in nature; the question that needs to be addressed observationally is whether one is dominant in producing the observed enhancements in nuclear fireworks.

For driving both nuclear activity and star formation in the inner disk, some variant of the Toomre instability comes closest to fitting the observed relations. It operates with almost the same strength for direct and retrograde encounters, does not require the presence of stellar bars, and is very strong in regions of rigid-body rotation. Given the multiphase nature of the ISM, it seems likely that mechanisms such as cloud-cloud collisions or pressure-driven cloud collapse (Jog and Solomon 1992) will be driven in a cascade toward the nucleus.

For active nuclei, there has remained some statistical question about the extent of triggering by interactions, because the results can depend critically on details of selection of the control sample (Fuentes-Williams and Stocke 1988). The approach I have taken here, looking for common pattersn in the host disks of paired AGN, has shown clearly that there are specific kinds of disturbance that are linked to induced AGN; it is perhaps among members of this sub-population that the clearest study of triggering will be possible. Disk kinematics seem to be a more accurate predictor of the incidence of both AGN and starbursts than either existence of a companion or morphological disturbance.

I acknowledge allocations of time at Kitt Peak National Observatory, Lowell Observatory, and the Spetsialnaya Astrofizicheskaya Observatoriya. This work has been supported in part by EPSCoR grant RII-8610669 and the NSF Large Foreign Telescopes program.

REFERENCES

Condon, J. J., Condon, M. A., Gisler, G., and Puschell, J. J. 1982, *Ap. J.*, **252**, 102.
Fuentes-Williams, T., and Stocke, J. 1988, *A. J.*, **96**, 1235.
Heckman, T. M. 1990, in *Paired and Interacting Galaxies*, ed. J. W. Sulentic, W. C. Keel, and C. M. Telesco (NASA CP-3098), 359.
Howard, S. A., Keel, W. C., Byrd, G. G., and Burkey, J. M. 1993, *Ap. J.*, in press.
Hummel, E. 1981, *Astr. Ap.*, **96**, 111.
Hummel, E., van der Hulst, J. M., Keel, W. C., and Kennicutt, R. C., Jr., 1987, *Astr. Ap. Suppl.*, **70**, 517.

Jog, C., and Solomon, P. M. 1992, *Ap. J.*, **387**, 152.

Karachentsev, I. D. 1972, Catalog of Isolated Pairs of Galaxies, *Soobsch. S. A. O.*, **7**, 3.

Karachentsev, I. D. 1987, *Dvoinye Galaktiki*, (Moscow: Nauka).

Keel, W. C. 1980, *A. J.*, **85**, 98.

Keel, W. C. 1991, *Ap. J. Lett.*, **375**, L7.

Keel, W. C. 1993, *A. J.*, submitted.

Keel, W. C. and van Soest, E. T. M. 1992 *Astr. Ap. Suppl.*, **94**, 553.

Larson, R. B. and Tinsley, B. M. 1978, *Ap. J.*, **219**, 46.

Lawrence, A., Rowan-Robinson, M., Leech, K., Jones, D. H. P., and Wall, J. V. 1989, *Mon. Not. R. Astr. Soc.*, **240**, 329.

Lin, D. N. C., Pringle, J. E., and Rees, M. 1988, *Ap. J.*, **328**, 103.

Lipovetskii, V. A., Neizvestnii, S. I., and Neizvestnaya, O. M. 1987, *Soobsch. S. A. O.*, **55**, 5.

Lonsdale, C. J., Persson, S. E., and Matthews, K. 1984, *Ap. J.*, **287**, 1009.

Mathewson, D. S., Ford, V. L., and Buchhorn, M. 1992, *Ap. J. Suppl.*, **81**, 413.

Mihos, J. C., Richstone, D. O., and Bothun, G. D. 1992, *Ap. J.*, **400**, 153.

Noguchi, M. 1988, *Astr. Ap.*, **203**, 259.

Olson, K. M., and Kwan, J. 1990, *Ap. J.*, **349**, 480.

Sanders, D. B., Soifer, B. T., Elias, J. H., Madore, B. F., Matthews, K., Neugebauer, G., and Scoville, N. Z. 1988, *Ap. J.*, **325**, 74.

Shlosman, I., Begelman, M. C., and Frank, J. 1990, *Nature*, **345**, 679.

Soifer, B. T. *et al.* 1984, *Ap. J. Lett.*, **278**, L71.

Sulentic, J. W. 1989, *A. J.*, **98**, 2066.

Toomre, A. 1964, *Ap. J.*, **139**, 1217.

Stellar Velocity Dispersion in NGC 6240 and Arp 220

René Doyon[1], M. Wells[2], G. S. Wright[3], and R. D. Joseph[4]

[1]Département de Physique, Université de Montréal, Canada
[2]Royal Observatory Edinburgh, Scotland
[3]Joint Astronomy Center, Hilo, Hawaii
[4]Institute for Astronomy, Honolulu, Hawaii

ABSTRACT

We present high resolution spectroscopic measurements of the 2.3 μm CO band of the luminous *IRAS* galaxies NGC 6240 and Arp 220. A convolution analysis yields velocity dispersions σ (FWHM/2.354) of 355 and 150 km s^{-1} for NGC 6240 and Arp 220, respectively. The velocity dispersion found for NGC 6240 is amongst the highest ever found in a galaxy and it is probably due to violent relaxation associated with the merging of two galaxies. The stellar velocity dispersion of Arp 220 is much smaller than that inferred from the *Br*α measurement of DePoy *et al.* (1987). Our result implies that there is no dynamical evidence for an AGN in this galaxy. From the dynamical mass derived from our measurements, we infer an infrared mass-to-light ratio M/L_K of 0.3 and 0.1 M$_\odot$/L$_\odot$ for NGC 6240 and Arp 220, respectively. For comparison, the ratio typically found in bulges of normal spiral galaxies is 0.6. These observations suggest that the bulk of the 2.2 μm luminosity emitted from the nucleus of NGC 6240 is associated mostly with normal giant stars while the 2.2 μm continuum of Arp 220 seem to have a significant contribution from young red supergiants.

1 INTRODUCTION

The velocity dispersion is a fundamental quantity for studying the kinematics and stellar populations of galaxies. In normal galaxies, the velocity dispersion is easily obtained by measuring optical emission or absorption lines. Unfortunately, the situation is more complicated for luminous *IRAS* galaxies. Since a good number of those objects are merging systems showing very high concentration of molecular gas, they usually suffer from relatively high extinction, and hence, infrared measurements are needed to sample deep enough into the nucleus. Also, since merging systems are subject to non-gravitational motions of the gas, it is by far preferable (even though more difficult) to use absorption lines as they sample stars.

The CO band absorption at 2.3 μm is well suited for studying the dynamics of luminous *IRAS* galaxies since this prominent absorption feature arises from the atmospheres of late-type stars (giants and supergiants). Potential extinction problems are also minimized at this wavelength. In this paper, we present velocity-resolved measurements of the 2.3 μm CO band in two luminous *IRAS* galaxies: NGC 6240 and Arp 220.

Fig. 1—Spectra of the 2.3 μm CO band in NGC 6240, Arp 220 and the template star γ Dra. The observed galaxy spectra are represented by points joined by a solid line. The smooth and thick solid line superimposed on each spectrum represents the best CO band fit. The resulting velocity dispersion σ is indicated for each galaxy. The galaxy spectra have been shifted in the restframe.

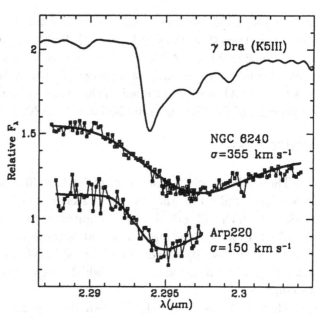

2 OBSERVATIONS

The observations were obtained on the UKIRT on the nights of 5–7th August 1993. The long-slit cool-grating spectrometer CGS4 was used to acquire the observations. The echelle ruling combined with the 150 mm long camera provided a resolving power of 7,000 at 2.33 μm. The slit provided a spatial resolution of 2.2×2.2 arcsec^2 on source. For NGC 6240, the slit was aligned along the line joining the two nuclei at a position angle of 17°. Our spatial resolution was sufficient to separate the spectra of the two nuclei onto two adjacent rows of the detector. For Arp 220, the slit was simply aligned along the north-south direction. Both nuclei found by Graham et al. (1990) were included in the aperture.

3 RESULTS AND ANALYSIS

The CO band spectra of NGC 6240, Arp 220 and the star γ Dra (K5III) are shown in Figure 1. The shallow slope of the CO bandhead seen in the galaxy spectra is clearly the result of velocity dispersion. Velocity dispersions were determined by a convolution analysis. Although a number of late-type star were observed with CGS4 for the purpose of forming a stellar template, they were not used in the final analysis because of their relatively small wavelength coverage which turned out to be insufficient to properly analyze the very broad CO band of NGC 6240. Instead, the high resolution spectrum of γ Dra, taken from the atlas of Kleinmann and Hall (1986), was used as template after taking into account the fact that this spectrum was obtained at a lower resolution (yet very similar) compared with our CGS4 data.

Three parameters were used for fitting the galaxy spectra: the velocity dispersion

σ=FWHM/2.354, the recessional velocity of the galaxy and a dilution factor which varies the strength of the CO band. All three parameters were fitted simultaneously by minimizing the χ^2. A Monte Carlo simulation was used for determining the most probable uncertainty of each parameter. The best fitted spectra are shown in Figure 1 as a thick solid line superimposed on the galaxy spectrum. Our analysis yields velocity dispersions of 355±30 and 150±30 km s^{-1} for NGC 6240 and Arp 220, respectively.

4 DISCUSSION

The stellar velocity dispersion of NGC 6240 is amongst the highest ever found in a galaxy. Indeed, in the catalog of stellar velocity dispersions of Whitmore *et al.* (1985) in which 725 galaxies are compiled, only 14 have velocity dispersions larger than NGC 6240. Interestingly, nearly all of those galaxies are ellipticals. This suggests that the large velocity dispersion observed in NGC 6240 is due to violent relaxation associated with an ongoing merger which might eventually lead to the formation of an elliptical.

The results are strikingly different for Arp 220. Although this galaxy is a well known merger, showing two distinct nuclei at 2.2 μm (Graham *et al.* 1990) and an outer 2.2 μm light profile that obey a $r^{1/4}$ de Vaucouleurs law (Wright *et al.* 1990), its velocity dispersion inferred from our CO band spectrum is relatively modest compared with NGC 6240 and it is certainly not typical of elliptical galaxies. The explanation of this apparent inconsistency is not clear.

The stellar velocity dispersion of Arp 220 corresponds to a line width of about 350 km s^{-1} which is much smaller than the line width of 1,300 km s^{-1} inferred from the Brα profile of DePoy *et al.* (1987). It is difficult to reconcile those two different results. DePoy *et al.* used the broad Brα as a dynamical evidence for the presence of an AGN in Arp 220. Unless it can be shown that the nucleus of Arp 220 is optically thick at 2.2 μm and not at 4 μm, *our observations suggest that there is no dynamical evidence for an AGN in this galaxy.*

4.1 Mass-to-Light Ratio and Stellar Population

If we neglect the rotation of the galaxy and assume no gradient in the velocity dispersion, then the dynamical mass enclosed within a radius r is given by

$$M(<r) = -\frac{r\,\sigma^2}{G}\frac{d\ln\rho(r)}{d\ln r} \tag{1}$$

where σ is the velocity dispersion, G is the gravitational constant and ρ is the space density. For an isothermal core described as $\rho(r) \propto r^\alpha$, $\alpha = 2$, and this is the average value determined for the central 2 kpc of normal galaxies (Devereux *et al.* 1987). One should note that the mass derived from Equation 1 is systematically *overestimated* (by a maximum factor of two) if rotational motions are important.

Although our spectra were obtained with an aperture of 2.2×2.2 arcsec^2, the *effective* aperture was smaller because most of the 2.2 μm continuum is distributed in a much tighter core of about 1.2″ for NGC 6240, as revealed by inspecting high spatial resolution near-infrared images obtained on CFHT (Doyon *et al.*, in preparation). A similar core radius is found for Arp 220 (Neugebauer *et al.* 1987). As an effective aperture we use a nominal radius of 0.6 arcsec for estimating the dynamical mass.

For NGC 6240, we derive a mass enclosed within a radius $r < 290$ pc (at a distance of 98 Mpc) of 2×10^{10} M$_\odot$ which is an order of magnitude higher than the mass of a normal galaxy within a similar region (Devereux *et al.* 1987). Based on the photometry in the literature and adopting an extinction $A_K = 0.6$ mag (Doyon 1991), we derive a 2.2 μm luminosity within the central 1.2 arcsec $L_K = 6 \times 10^{10}$ L$_\odot$, yielding an infrared mass-to-light ratio $M/L_K \sim 0.3$ in solar units. For comparison, the giant-dominated stellar populations of normal galaxies have a typical M/L_K of ~ 0.6. Thus, the observed M/L_K in NGC 6240 seems to suggest that the bulk of the 2.2 μm continuum arises from old giant stars with little contribution from young supergiants. However, we must be cautious with this interpretation given the uncertainties on the dynamical mass and the extinction.

Similarly for Arp 220, we derive a dynamical mass of 5×10^9 M$_\odot$ within a radius of 210 pc and a corresponding 2.2 μm luminosity (corrected for extinction with $A_K = 1.1$ mag; Doyon 1991) of 4×10^{10} L$_\odot$, yielding $M/L_K \sim 0.1$. It is unlikely that we have underestimated this mass-to-light ratio by more than a factor of two, and hence, the relatively small M/L_K of Arp 220 implies that a significant fraction of the K continuum is contributed by young supergiants produced by a starburst.

REFERENCES

Doyon, R. 1991, *Ph. D.* thesis, Imperial College (London).

Devereux, N. A., Becklin, E. E., and Scoville, N. Z. 1987, *Ap. J.*, **312**, 529.

DePoy, D. L., Becklin, E. E., and Geballe, T. R. 1987, *Ap. J. Lett.*, **316**, L63.

Graham, J. R., Carico, D. P., Matthews, K., Neugebauer, G., Soifer, B. T., and Wilson, T. D. 1990, *Ap. J. Lett.*, **354**, L5.

Kleinmann, S. G., and Hall, D. B. 1986, *Ap. J. Suppl.*, **62**, 601.

Neugebauer, G., Elias, J., Matthews, K., McGill, J., Scoville, N. Z., and Soifer, B. T. 1987, *A. J.*, **93**, 1057.

Whitmore, B. C., McElroy, D. B., and Tonry, J. L. 1985, *Ap. J. Suppl.*, **59**, 1.

Wright, G. S., James, P. A., Joseph, R. D., and McLean, I. S. 1990, *Nature*, **344**, 417.

Possible Atomic-to-Molecular Gas Transition in the Center of Merging Galaxies

Zhong Wang[1] and Xiaohui Hui[2]

[1]100–22 IPAC, California Institute of Technology, Pasadena, CA 91125
[2]105–24 Astronomy Department, California Institute of Technology, Pasadena, CA 91125

ABSTRACT

We examine the distribution and kinematics of atomic and molecular gas mapped in a number of galaxies suspected to be in the process of merging. In most cases, the nuclear region of the merger has a high concentration of molecular gas, and a deficiency of atomic gas as compared with larger radii. Thus the total surface mass density of gas often has a minimum at an intermediate radius. In cases where the gas rotation curve is measured, the transition from regions dominated by molecular gas to those of atomic gas corresponds to abrupt changes in rotation characteristics. We propose that the merger is efficiently converting ISM from atomic into molecular form in central region of these galaxies, and that the dense clouds are experiencing radial accretion at a higher rate than diffuse gas.

1 INTRODUCTION

Since the early work of Toomre and Toomre (1972), much progress has been made in understanding dynamical processes of merger galaxies (*cf.* Barnes and Hernquist 1993). One result that appears consistently in theoretical studies and numerical simulations alike is that the end product of the merger is an early type galaxy. Increasingly, this scenario has acquired observational support as well. For example, the K-band light profiles of many Arp galaxies, mostly advanced mergers, show $r^{1/4}$ law typical of ellipticals (*e.g.*, Wright *et al.* 1990; Stanford and Bushouse 1991). Yet, the process in which the merging disks shed their abundant gas mass remains unclear, and numerical simulations are far from adequately resolving this problem given the enormous dynamical range required to mimic the changes in the ISM.

Recent radio, millimeter/submillimeter, and infrared observations have made available high resolution maps of many suspected merger systems. In particular, detailed millimeter CO line and 21-cm H I line mappings can now be combined with surface photometry in optical and near infrared (NIR) to study the distributions of molecular and atomic gas. The velocity information in the spectral line studies further provide kinematic diagnosis of these systems. In this contribution, we present a preliminary analysis of the gas distribution in a sample of possible mergers, and try to examine the role that ISM has to play in the evolution of these very special cosmic events.

2 THE CO AND H I DATA

Over the last few years, coordinated efforts have been made to map the 2.6 mm CO (1–0) emission of many far-infrared (FIR) bright, high luminosity galaxies with the Owens Valley Millimeter Array (*cf.* Scoville *et al.* 1993, these proceedings). Most of these are highly disturbed systems, possibly advanced mergers (Sanders *et al.* 1988; Scoville and Soifer 1991). These include such well-known galaxies as Arp 55, Arp 220, Mrk 231, NGC 3690, and NGC 6240 (Scoville *et al.* 1989, 1991; Sargent and Scoville 1991; Wang, Scoville, and Sanders 1991, 1993). A number of active, clearly interacting systems of somewhat lower luminosity such as NGC 520, NGC 4038/9, and NGC 7252 (Sanders *et al.* 1988; Stanford *et al.* 1990; Wang, Schweizer, and Scoville 1992), are also studied in some detail. Having been mapped in CO emission as well are a subgroup of elliptical and lenticular galaxies which are often considered possible merger products. These are characterized by strong dust lanes, moderately bright FIR emission, and rich molecular gas — all rather unusual among early type galaxies (*cf.* Wiklind and Henkel 1989; Wang, Kenney, and Ishizuki 1992). These include NGC 3928, NGC 4710, NGC 5866, and NGC 7625 (Li *et al.* 1993*a*, *b*; Wrobel and Kenney 1992; Hui and Wang 1993)[3]. In millimeter interferometric observations, one can normally reach a sensitivity of 100 mJy beam^{-1} at an angular resolution of $2 - 3''$, and velocity resolution of ~ 10 km s^{-1}. If the velocity range within the beam is 10^2 km s^{-1}, the corresponding molecular surface mass density is about 100 $M_\odot$pc^{-2}.

In the mean time, many of these galaxies have also been observed in the H I 21 cm line with the *VLA* in its longer (C and B) configurations (*e.g.,* Stanford and Woods 1989; Stanford 1990; Li *et al.* 1993*a*, *b*; Hibbard *et al.* 1993). These observations can reach a resolution of $\sim 5 - 10''$, with a 3σ flux level of approximately 1 mJy beam^{-1} or better. Assuming a velocity width of 50 km s^{-1} within the beam, the above numbers correspond to a detection limit for H I surface mass density on the order of 1 $M_\odot$pc^{-2}. Since the atomic gas component often shows extended structures, the large field of view at centimeter wavelength is also ideally suited for this task.

3 MOLECULAR AND ATOMIC GAS DISTRIBUTIONS

In almost all cases of possible merger galaxies, the CO mappings have resulted in the discovery of a highly compact central molecular gas distribution. The peak surface mass density, assuming a conventional CO surface brightness to H_2 column density conversion factor[4], can be as high as several times 10^4 $M_\odot$pc^{-2}, more than a few orders of magnitude greater than those found in ordinary late-type galaxies. With few exceptions, the spatial extent is only 1–2 kpc in radius, while the ratio of

---

[3] Several nearby galaxies such as the interacting pair NGC 3256 and giant elliptical NGC 5128 (Cen A) have been mapped in CO (2–1) line with the *CSO* and *IRAM* Telescopes (Sargent *et al.* 1989; Casoli *et al.* 1991; Quillen *et al.* 1992).

[4] This conversion factor, usually $N(H_2) = 2$–3×10^{20} I(CO) cm^{-2} (K km s^{-1})$^{-1}$, may be uncertain here based on the ^{13}CO/^{12}CO line ratio measured in merger galaxies (Casoli *et al.* 1992).

molecular-to-total dynamical mass within that radius can be as high as near unity.

In contrast, the atomic gas is distributed mostly at large radii. The region in which molecular gas is found only contains a tiny, often negligible amount of H I. In a few galaxies, such as NGC 7252, there is even a complete lack of atomic gas within the central 6 kpc of the merger (see Fig. 1). Morphologically, the H I gas traces the tidally disrupted spiral arms/tails, with faint, extended structures sometimes reaching out to as far away as 100 kpc from the galactic center while having no molecular emission.

Fig. 1— Radial distribution of gas in two merger galaxies: (upper) NGC 7252, and (lower) NGC 520. The vertical axis is the peak surface mass density. Filled dots are for molecular (H_2) mass derived from CO mapping observations (Wang *et al.* 1992, 1993); and crosses are for neutral hydrogen gas measured from *VLA* maps (Standford 1991; Hibbard *et al.* 1993). Resolution of the CO maps is 2–3″; the H I maps are smoothed to $\sim 25″$ resolution.

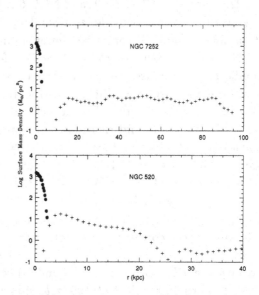

The total gas mass in these galaxies ranges from a few times 10^8 to 10^{10} M$_\odot$, amounting to 5–20% of the total galaxy mass in the optical disk. The fractional mass in molecular (H_2) form is often comparable to that of atomic gas (H I). These are not very much unlike normal spirals, suggesting that the progenitors of a merger are probably regular late-type galaxies. The different distributions between H I and H_2 often create a minimum at an intermediate radius (see Fig. 1)[5].

4 KINEMATIC SIGNATURES

The kinematics of the molecular and atomic gas in mergers can also be drastically different. High resolution observations of CO usually show a very rapidly rising rotation curve in the core regions. Interestingly, we find indications suggesting that the optical rotation curves are often less steep and can be misleading because of the heavy obscuration in the nuclei (Hui and Wang 1993). In a typical case, the molecular rotation curve reaches its maximum (about ± 150 km s^{-1}) within 0.5 to 1.0 kpc, while the optical rotation curve rises slowly to the same peak value at about 3–4 kpc radius.

---

[5] In some of the galaxies where a strong radio continuum source is present at the nucleus, the H I mapping is hampered by the absorption effect and limited resolution.

While H I gas at large radii usually share the kinematics of their stellar counterparts, the molecular gas around the nucleus often appear to be *kinematically decoupled* from the rest of galaxy (Wang *et al.* 1992). For example, the rotation axis of the gas disk may change its orientation abruptly at the radius where the main gas component changes from molecular to atomic. This change may be related to the drastically different dynamical time scales in the inner and outer parts of the merger (Hernquist and Barnes 1991). Hence, within certain intermediate radius where H_2 dominates, the kinematic characteristics tend to be quite different from outer regions where H I takes over and behaves more consistently with the stellar masses.

5 DISCUSSION

Given that the pre-merging galaxies are late-type spirals with normal gas distribution, one obviously has to examine the dynamical interactions between two galaxies in order to understand the observed gas morphology and kinematics.

Giant molecular clouds in the galactic disk are massive objects, thus they tend to fall first into the bottom of the gravitational potential as a result of dynamical friction. Therefore, even without direct conversion of diffuse gas into molecular form, there is net accretion of dense gas toward the center of merger. If the time scale for this infall is sufficiently short, it could account for the curious absence of molecular gas in the off-center regions. This is also likely a main reason for the active star formation in the nuclei of merging galaxies, for the great increase in local gas mass density would certainly create instabilities similar to those in active regions of normal disks (*cf.* Kennicutt 1989). Furthermore, inelastic collisions between clouds are a potential source of gravitational collapse and nuclear starburst (*e.g.*, Harwit and Fuller 1988).

But there must also be efficient conversion of atomic into molecular gas in the central region of mergers. This is because diffuse gas would otherwise accumulate at the galactic center as a result of the merger. Moreover, the enhanced radiation field combined with the increasing amount of molecular gas would create additional atomic gas through photodissociation. It would probably be quite difficult to ionize these gas completely, or to clear them from the central region via dynamical processes.

A possible scenario is a combination of slow radial accretion and fast phase transitions of the diffuse gas. Within certain radius (usually a few kpc), the local kinematics forces the gas to accumulate at the center. Cloud-cloud collisions that occur frequently can compress and effectively cool the gas, which in turn becomes self-shielding and form molecules rapidly. In this picture, the deficiency of atomic gas in the central region is determined by the extent to which the disturbed system can funnel the diffuse gas to the very small central region. The diffuse gas at larger radii would be less affected until much later, when stellar wind (or the superwind of massive nuclear star formation) arrives and pushes them away.

One of the satisfying aspects of such a model is that the coincidence between molecular/atomic gas transition zone and the region where gas kinematics changes abruptly may be naturally explained in terms of bar-driven inflows. Clearly, simulations that account for the multi-phase composition of ISM in merging disks are valuable in helping to determine timescales and relative importance of these processes.

REFERENCES

Barnes, J. E., and Hernquist, L. 1992, *Ann. Rev. Astr. Ap.*, **30**, 705.
Casoli, F., Dupraz, C., Combes, F., and Kazes, I. 1991, *Astr. Ap.*, **251**, 1.
Casoli, F., Dupraz, C., and Combes, F. 1992, *Astr. Ap.*, **264**, 55.
Harwit, M., and Fuller, C. 1988, *Ap. J.*, **328**, 11.
Hernquist, L., & Barnes, J. E. 1991, *Nature*, **354**, 21
Hibbard, J., van Gorkom, J., and Schweizer, F. 1993, in preparation.
Hui, X. and Wang, Z. 1993, in preparation.
Kennicutt, R. C., Jr. 1989, *Ap. J.*, **344**, 685.
Li, J. G., Seaquist, E. R., Wrobel, J. M., Wang, Z., and Sage, L. J. 1993*a*, *Ap. J.*, in press.
Li, J. G., Seaquist, E. R., Wrobel, J. M., and Wang, Z. 1993*b*, *Ap. J.*, submitted.
Quillen, A. C., de Zeeuw, P. T., Phinney, E. S., and Phillips, T. G. 1992, *Ap. J.*, **391**, 121.
Sanders, D. B., Scoville, N. Z., Sargent, A. I., and Soifer, B. T. 1988, *Ap. J. Lett.*, **324**, L55.
Sargent, A. I., and Scoville, N. Z. 1991, *Ap. J. Lett.*, **366**, L1.
Sargent, A. I., Sanders, D B., and Phillips, T. G. 1989, *Ap. J. Lett.*, **346**, L9.
Scoville, N. Z., Sanders, D. B., Sargent, A. I., and Tinney 1989, *Ap. J. Lett.*, **345**, L25.
Scoville, N. Z., and Soifer, B. T. 1991, in *Massive Stars in Starbursts*, eds. C. Leitherer *et al.* (Cambridge: Cambridge Univ. Press), p. 233.
Stanford, S. A. 1990, *Ap. J.*, **358**, 153.
Stanford, S. A., and Bushouse, H. 1991, *Ap. J.*, **371**, 92.
Stanford, S. A., Sargent, A. I., Sanders, D. B., and Scoville, N. Z. 1990, *Ap. J.*, **349**, 492.
Stanford, S. A., and Woods, D. 1989, *Ap. J.*, **346**, 712.
Toomre, A., and Toomre, J. 1972, *Ap. J.*, **178**, 623.
Wang, Z., Kenney, J. D. P., and Ishizuki, S. 1992, *A. J.*, **104**, 2097.
Wang, Z., Schweizer, F., and Scoville, N. Z. 1992, *Ap. J.*, **396**, 510.
Wang, Z., Scoville, N. Z., and Sanders, D. B. 1991, *Ap. J.*, **368**, 112.
_____, 1993, in preparation.
Wiklind, T., and Henkel, C. 1989, *Astr. Ap.*, **225**, 1.
Wright, G. S., James, P. A., Joseph, R. D., and McLean, I. S. 1990, *Nature*, **344**, 417.
Wrobel, J. M., and Kenney, J. D. P. 1992, *Ap. J.*, **399**, 94 .

S0s with Counter-Rotating Gas: NGC 3941 and NGC 7332

David Fisher

Lick Observatory, Board of Studies in Astronomy and Astrophysics, University of California, Santa Cruz, CA 95064

ABSTRACT

Two S0 galaxies with extended counter-rotating [O III] 5007Å gas are presented. The first, NGC 3941, is a barred field S0 inclined 50° to the line-of-sight. The second, NGC 7332, is an edge-on field S0 with a prominent boxy bulge. Multiple stellar components are not obviously present in either object nor do their structures have major distortions. This suggests that the progenitors of the extended emission were low mass dwarf/satellites or tidally stripped material.

1 INTRODUCTION

Early-type galaxies offer a unique laboratory in which to study galaxy interactions and their effects on the stellar and nonstellar components of the participants. The initial low gas and dust content of the host galaxy and the generally smooth morphology of early-type galaxies facilitates the detection of the perturbing influence of interactions and mergers. Whether these signatures of interaction manifest themselves as starbursts, bars, dust lanes, counter-rotating disks, or other phenomena depends on specifics such as the nature of the secondary object, the parameters of the collision, and the resulting evolution of the material.

The origins of the features displayed by the two objects mentioned here are suggestive of galaxy interactions playing an important role in the evolution of some early-type galaxies. These galaxies join the list (see Bertola *et al.* 1992) of S0 galaxies displaying a decoupling of the angular momentum between their gas and stars.

2 NGC 3941

An undistorted disk with an exponential brightness profile (for $r \gtrsim 30''$) surrounds the inner bar of NGC 3941 (see Fig. 1). An ellipse fitting algorithm applied to the galaxy shows that at a radius of $20''$ the apparent angle between the major axis and the bar is 15° ($PA_{disk}=10°$). The ellipticity of the $20''$ isophote is $\epsilon=0.44$ while that of the outer disk measures $\epsilon_{disk}=0.35$.

Spectra taken with the Kast Spectrograph on the Shane 3-meter telescope at Lick Observatory were used to derive the major and minor axis kinematics. The stellar rotations and dispersions were produced by a Fourier fitting routine which

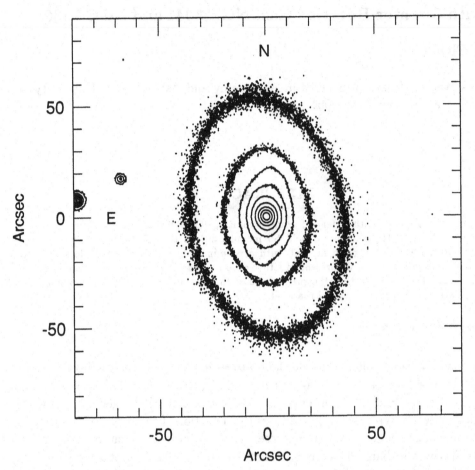

Fig. 1—Contour plot of 300 sec R band CCD image of NGC 3941 taken with 40"
at Lick Observatory.

broadened the spectra of late-type giant templates to the galaxy spectra in the Mg_2/Fe
absorption region. The results of this fit are shown in Figure 2 as are the emission
line velocities as measured from the IRAF Splot task. Typical errors for the stellar
velocities and dispersions are $\varepsilon(V_r) \sim 8$ km s^{-1}, and $\varepsilon(\sigma) \sim 10$ km s^{-1}, while the errors
for the gas measurements are $\varepsilon(V_{5007}) \sim 25$ km s^{-1}.

The stellar rotations show the galaxy to have an observed $V_{max} \sim 130$ km s^{-1} along
the major axis which when corrected for inclination (assuming circular rotation) is
V_{cor}=170 km s^{-1}. The minor axis shows no rotation cosistent with the errors. The
behavior of the stellar velocity dispersion profiles along the major and minor axis
are consistent with those found in other S0 galaxies (Dressler and Sandage 1983). In
particular, the central value $\sigma_0 \sim 145$ km s^{-1}, and its decrease with radius is a clear
signature of a rotationally supported disk galaxy.

The structure of the counter-rotating [O III] 5007Å gas in NGC 3941 along the

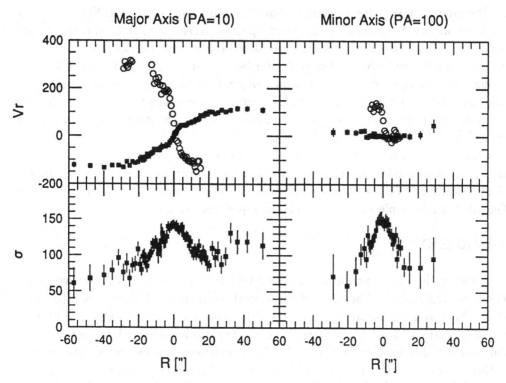

Major Axis (PA=10)   Minor Axis (PA=100)

Fig. 2—Major and minor axis kinematics of NGC 3941. Open circles are [O III] 5007Å velocities and filled squares are stellar measurements.

major axis is in the form of a rapidly rotating disk. At its southern end, the disk extends along the major axis approximately the same distance from the nucleus as the bar (*i.e.* 30″ or 1.8 kpc with H_o=75 km s^{-1} Mpc^{-1}) while along the north it extends 15″. At the south end, the maximum observed rotation velocity of the gas is $V_{max} \sim 300$ km s^{-1} ($V_{cor} \sim$400 km s^{-1}) while at the north $V_{max} \sim 150$ km s^{-1} ($V_{cor} \sim$ 200 km s^{-1}) .

The presence of gas rotating along the minor axis in the inner region is interesting. This indicates that the gas has either not yet reached equilibirum or more likely, that a triaxial potential exists in this bar-dominated region.

3 NGC 7332

Like NGC 3941, NGC 7332 possesses a rapidly counter-rotating extended gas disk as detected from [O III] 5007Å emission (see Fisher *et al.* 1993 for a detailed study of NGC 7332). The gas in NGC 7332 rotates at considerably higher velocities ($V_{5007} \sim 300$ km s^{-1}) than the stars ($V_{max} \sim$140 km s^{-1}) and displays non-circular motions indicating that it has not reached an equilibrium state. These observations are strong evidence in support of an accretion process having occurred in NGC 7332.

Photometric observations clearly show the boxy isophotes that NGC 7332 has long been known to possess while offset spectra taken parallel to the major and minor axes display the rapid cylindrical rotation common to galaxies with box-shaped bulges. The bulge of NGC 7332 is well described by an $r^{1/4}$–law and the outer disk is exponential. There exists, however, a 10″ long region of constant surface brightness along the major axis between the bulge and disk components. Whether this flat component could be associated with a bar is unkown — the detection of an edge-on bar is difficult (Kormendy 1982).

Other than a slight ($\Delta PA \sim 5°$) twisting of the disk, the morphology of NGC 7332 shows no signs of having suffered from a wholesale interaction event. The presence of the Sb spiral NGC 7339 at a projected distance of ~ 40 kpc ($\Delta z \sim 55$ km s^{-1}) from NGC 7332 is a possible source of tidally stripped material.

4 DISCUSSION

Whether the counter-rotating gas in NGC 3941 or NGC 7332 will eventually form stars is intriguing. The mass of gas in both systems is relatively small (of order 10^8–10^9 M$_\odot$) and might not remain in stable disks for a sufficient period of time for star formation to occur. If stars were to form, these galaxies have not accreted enough material to form an S0 like NGC 4550 which possesses two co-spatial counter-rotating stellar disks (Rubin *et al.* 1992; Rix *et al.* 1992). The counter-rotating stellar cores and disks found in elliptical galaxies (Franx and Illingworth 1988; Bender 1988; Jedrzejewski and Schecter 1989) provides a more likely example of a possible end state for NGC 3941 and NGC 7332 if they were to form stars in their counter-rotating components.

REFERENCES

Bender, R. 1988, *Astr. Ap.*, **202**, L5.

Bertola, F., Buson, L. M., and Zeilinger, W. W. 1992, *Ap. J. Lett.*, **401**, L79.

Dressler, A., and Sandage, A. 1983, *Ap. J.*, **265**, 664.

Fisher, D., Illingworth, G. D., and Franx, M. 1993, *A. J.*, submitted.

Franx, M., and Illingworth, G. D. 1988, *Ap. J. Lett.*, **327**, L55.

Jedrzejewski, R., and Schecter, P. 1989, *A. J.*, **98**, 147.

Kormendy, J. 1982, in Saas-Fee Lectures on *Morphology and Dynamics of Galaxies*, eds. L. Martinet, and M. Mayor (Geneva Observatory), p. 115

Rix, H.-W., Franx, M., Fisher, D., and Illingworth, G. D. 1992, *Ap. J. Lett.*, **400**, L5.

Rubin, V., Graham, J., Kenney, J. 1992, *Ap. J. Lett.*, **394**, L9.

Evidence for a Tidal Interaction in the Seyfert Galaxy Markarian 315

John W. MacKenty[1], Susan M. Simkin[2], Richard E. Griffiths[3], and Andrew Wilson[1]

[1]Space Science Telescope Institute
[2]Michigan State University
[3]Johns Hopkins University

ABSTRACT

We have detected a diffuse, continuum knot which may be a reminant nucleus, in the inner regions of Markarian 315. This knot is associated with a complex, ring-like structure in both the continuum and ionized gas emission. The occurance of this feature in combination with an extensive, ionized streamer, or tidal tail, and highly non-circular kinematics in the ionized gas, suggests that this galaxy has suffered a disruptive, tidal interaction whose influence extends well into the inner one kiloparsec region.

1 INTRODUCTION

Markarian 315 (Markarian and Lipovetskii 1971) (also IIZw187), is a moderately luminous Seyfert 1.5 galaxy (Koski 1978). It has a redshift of 11,820 km s^{-1} relative to the galactic center and $M_v = -21.6$ (Sargent 1970). For this discussion, we adopt a scale of 0.57 h^{-1} kpc arcsec^{-1}.

Radio frequency images of Markarian 315 show that it is a steep spectrum source with a diffuse morphology and a total extent of 2.9 h^{-1} kpc (Wilson and Willis 1980; Ulvestad, et al. 1981). This extended structure is the largest in the sample of Seyfert galaxies studied by these authors. (These typically ranged between 0.4 and 1.0 h^{-1} kpc). It is, however, consistent with an extended starburst in the galaxy and the IRAS fluxes are also consistent with this interpretation (MacKenty 1989).

MacKenty (1986) discovered an extraordinary, 80 kpc, streamer of ionized gas emerging from near the nucleus, extending in a straight line for 60 kpc then bending back in a hook. He suggested two possible origins for this feature: a tidal interaction or a dormant radio jet. The observations reported here strongly support the idea of a tidal origin and establish Markarian 315 as an extreme example of a active galaxy with a strong tidal disruption in the gas-rich material close to its nucleus.

2 OBSERVATIONS

2.1 HST Image

A single 230 s exposure was obtained with the Planetary Camera (PC) at a scale of 43 mas on 15 June 1992 UT with the F785LP filter, ($\bar{\lambda} = 887$ nm; half-width, 780

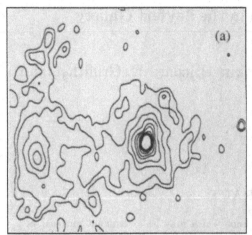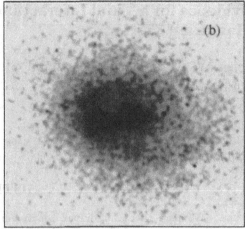

Fig. 1—*HST* F785LP image: *a*) nuclear regions; *b*) display showing the SE arc.

nm to 1,000 nm, Westphal *et al.* 1982). This image was processed with the standard
WF/PC pipeline software (see *e.g.* MacKenty *et al.* 1992) and deconvolved using both
the Lucy-Richardson algorthim and a Maximum Entropy algorithm (MEMSYS) with
a front end developed by N. Weir (1991*a, b*), and a PSF created with the "tinytim"
optical modeling package (Burrows *et al.* 1991; Krist 1993).

The *HST* image shows a second, diffuse peak 2.27″ east (P.A. = 95°) of the
diffraction limited, stellar Seyfert nucleus (Fig. 1*a*). This second peak is clearly
resolved (*i.e.* non-stellar) with a gaussian FWHM of 0.66″ (corresponding to 0.38
h^{-1} kpc at the distance of Markarian 315). A fainter ring, or spiral-like structure is
present 2.5″ – 3″ south of the nucleus (∼1.5 kpc) opening towards the SW (Fig. 1*b*).

2.2 Ground-Based Images

Ground-based images in the *KPNO*/Mould I-band, (720 – 930 nm) and the redshifted bands of [O III] and Hα, were obtained with the University of Hawaii 2.24m
telescope at Mauna Kea Observatory (MKO) in 1983 and 1984. The observational
details are presented in MacKenty (1986).

These suffered from guiding errors of 2″ to 2.5″ which we corrected using the
Weir (1991*a, b*) MEMSYS interface routines and an observed star 30″ SSE of the
galaxy as a template. The corrected images have PSFs with 1.5″ FWHM (*i.e.* they
have not been "superresolved"). The small-scale features in the resulting images were
enhanced by subtracting the large-scale, smooth, galaxy structure. The [O III] image
processed in this way is shown in Figure 2*a*.

The deconvolved [O III], Hα, and I-band images clearly show that the arc-like
feature which is just visible in the *HST* image (Fig. 1*b*) is a ring (Fig. 2*a*). The start
of the external ionized gas streamer is also visible in the [O III] image (Fig. 2*a*) and
appears to have the ring as its endpoint. The diffuse continuum peak (discovered in

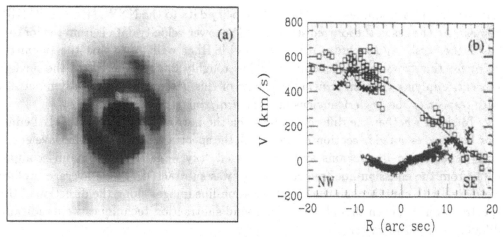

Fig. 2—*a*) [O III] image   *b*) ionized gas velocities (see text).

the *HST* images) is not present in the narrow-band emission line images but does appear in the I-band image. Since neither the F785LP nor the *KPNO*/Mould I-band pass contains any strong emission lines at this redshift, we conclude that this feature is composed of stars.

It is noteworthy that the brightest regions of Hα emission are *not* coincident with the brightest regions of [O III] emission. The [O III] ring-like structure is very close to circular and concentric with the Seyfert nucleus while the Hα emission more closely resembles a spiral arm with both an elongation in the east-west direction and a significant opening angle. In addition, there is a gap in the [O III] structure at the location of the second peak (Figs. 1*b* and 2*a*). These differences are confirmed by long-slit spectroscopy (see below).

2.3 Ground-Based Spectroscopy

Four long-exposure, long-slit spectra were obtained in September 1985 with the University of Hawaii Institute for Astronomy's 2.24m telescope and Faint Object Spectrograph with a TI 3-phase CCD camera. These were positioned along the ionized gas streamer, in P.A. –36° with a wide slit (4.7″) and a scale along the slit of 0.68″ pixel^{-1}. Two of these spectra covered the wavelength region including Hβ and [O III] 5007/4958 (492 to 567 nm), while the other two covered Hα, [N II] and the red [S II] lines (657 to 707 nm). The dispersion was roughly 1.5Å pixel^{-1}. Both nights were non-photometric thus no absolute fluxes were measured.

The emission lines in the reduced spectra were measured for total (relative) integrated flux and redshift (relative to the nucleus). The only emission lines beyond 5 arc seconds from the nucleus with significant flux (3 sigma above the noise in the underlying continuum) were Hα and the [O III] 4958/5007 lines in the streamer.

Figure 2*b* shows the velocities from these spectra. The plotted curves for these data are from least-squares fits. The high velocity data to the NNW, (Hα and [O III]), comes from the gas in the ionized streamer. The lower velocity data is from the ionized gas in the "disk" and includes Hβ, [N II], and [S II] as well. Note that the *symmetry* point for this "low-velocity" component lies roughly 2.5″ to the SE of the Seyfert nucleus, implying that the *kinematic* center of this gaseous system is asymmetrical with respect to the Seyfert nucleus and the arm/ring structure.

To check whether the differences between the narrow-band Hα and [O III] emission line images noted in section 2.2 were real, the spectra were averaged in wavelength over the emission–line regions and an averaged "sky + galaxy continuum" component, from the emission-line free continuum, was subtracted. These average profiles agree well with cuts through the direct emission-line images along the direction of the spectrograph slit, even though the background subtraction techniques were radically different in the two cases.

2.4 *VLA* Observations

Following the discovery of the external ionized gas streamer, an improved 20 cm *VLA* image was obtained on 3 February 1985 (UT) with the array in the "A" configuration. This image shows no evidence for a radio counterpart to the optical emission-line feature. However, it does show two compact features in P.A. ∼ 80° separated by 2.55″. The brightest feature is most likely identified with the Seyfert nucleus but the *HST* to *VLA* astrometry is not accurate enough to verify this identification. The eastern radio source is not coincident in either position angle or radial distance with the second peak seen in the *HST* and *MKO* continuum images nor with the knots in either of the emission line images.

3 CONCLUSIONS

The features seen in the data for Markarian 315 are consistent with an interpretation of the secondary continuum knot being a recently captured galaxy remnant.

The knot is at the leading arc of what appears to be a stellar, spiral density enhancement. The inner ionized gas structures and the *VLA* 20 cm morphology are consistent with recent star formation, and subsequent supernova explosions in gas compressed by tidally induced cloud-cloud collisions. The asymmetrical velocity pattern in the gas kinematics suggests that the center of gravity for the system is displaced from the Seyfert nucleus. All of these different features are symptomatic of the type of gravitational forcing which leads to mass inflow towards the nucleus (*cf*. Osterbrock 1991 for a review), Finally, the external streamer of ionized gas is consistent with type of gaseous remnant from a galaxy-galaxy merger which was discussed by Hernquist (these proceedings).

Altogether, the data provide a strong case for Markarian 315 as a Seyfert galaxy in the midst of a merger which is driving material into its central regions, supporting the hypothesis that galaxy interactions provide a mechanism for stimulating nuclear activity. We believe this object provides a rare example of an AGN "caught in the act" of initial, tidally-induced feeding. As such, it may well prove an excellent laboratory for testing detailed models of AGN activity.

ACKNOWLEDGMENTS

The observations with the NASA/ESA Hubble Space Telescope were obtained at the Space Telescope Science Institute (STScI), and support for this work was provided by NASA through grant number GO05-70600 from STScI which is operated by the Association of Universities for Research in Astronomy, under NASA contract NAS5-26555. The optical data analysis was done with support from NSF AST-89-14567 (to SMS). The 20 cm observations were obtained with the *VLA* telescope of the National Radio Astronomy Observatory which is operated by Associated Universities, Inc., under contract with the National Science Foundation.

REFERENCES

Burrows C. J. *et al.* 1991, *Ap. J. Lett.*, **369**, L21.

Krist J. E., 1993, *Proceedings of the Second Astronomical Data Analysis and Software Systems Conference*, (Boston: PASP), in preparation.

Hernquist, L. 1993, these proceedings.

Mackenty, J. W. 1986, *Ap. J.*, **308**, 571.

MacKenty, J. W. 1989, *Ap. J.*, **343**, 125.

MacKenty, J. W., *et al.* 1992, *Hubble Space Telescope Wide Field/Planetary Camera Instrument Handbook*, (Baltimore: Space Science Telescope Institute).

Osterbrock, D. E. 1991, *Reports of Progress in Physics*, **54**, 579.

Sargent, W. L. W. 1970, *Ap. J.*, **160**, 405. Ulvestad, J. S., Wilson, A. S., and Sramek, R. A. 1981, *Ap. J.*, **247**, 419.

Weir, N. 1991*a*, *ESO Workshop No. 38*, ed. P. J. Grosbol and R. H. Warmels, p. 115.

Weir, N. 1991*b*, *10th International Workshop on Maximum Entropy and Bayseian Methods*, ed. W. T. Grandy and L. H. Schick (Dordrecht: Kluwer), p. 275.

Westphal, *et al.* 1982, *The Space Telescope Observatory*, ed. D. N. B. Hall, (NASA CP-2244), p. 28.

Wilson, A. S., and Willis, A. G. 1980, *Ap. J.*, **240**, 429.

Interaction between the Galaxies IC 2163 and NGC 2207

M. Kaufman[1], D. M. Elmegreen[2], E. Brinks[3], B. G. Elmegreen[4], and M. Sundin[5]

[1]Department of Physics, Ohio State University, Columbus, U. S. A.
[2]Vassar College Observatory, Poughkeepsie, U. S. A.
[3]National Radio Astronomy Observatory, Socorro, U. S. A.
[4]IBM – Watson Research Center, Yorktown Heights, U. S. A.
[5]Department of Astronomy and Astrophysics, Chalmers University, Sweden

ABSTRACT

VLA H I observations of the interacting pair IC 2163/NGC 2207 are presented. The velocity and structural anomalies of IC 2163 agree with predictions of *N*-body galaxy encounter simulations if IC 2163 recently underwent a strong, prograde, in-plane encounter with NGC 2207. The velocity disturbances in NGC 2207 suggest that the main tidal force on NGC 2207 was perpendicular to the disk of NGC 2207.

1 INTRODUCTION

The spiral galaxies IC 2163 and NGC 2207, shown in Figure 1, are involved in a close tidal encounter. IC 2163 has an ocular shape (an eye-shaped central oval with a sharp apex at each end), intense star formation along the eyelid regions, and a double-parallel arm structure on the side opposite its companion, NGC 2207.

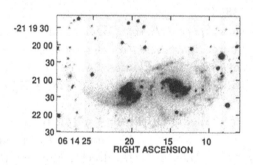

Fig. 1—R-band image taken by H. Deeg of IC 2163 (left) and NGC 2207. The double-parallel arm is on the eastern side of IC 2163.

The optical morphology of IC 2163 is consistent with *N*-body simulations by Elmegreen *et al.* (1991) if IC 2163 underwent a strong, in-plane, prograde encounter with NGC 2207 during the last half-rotation. One component of the double arm is the usual tidal tail; the other component is produced by rapid streaming of tidally perturbed stars and gas from the companion (western) side of the galaxy. The simulations predict a velocity difference of 50 to 100 $km\,s^{-1}$ between the two components of the double arm and large streaming motions along the oval.

358

Using the *VLA*, we made H I observations of this galaxy pair to study the early stages of post-encounter evolution and to look for the velocity anomalies predicted by the numerical simulations.

2 OBSERVED PROPERTIES OF IC 2163

Our most useful *VLA* H I data on this system have an angular resolution of 13″ (2 kpc) and a velocity resolution of 5 km s^{-1}. We were able to separate the H I contributions of the two galaxies kinematically. Figure 2 shows the H I column density N(H I) associated with IC 2163. The central hole in H I coincides with the optical nucleus. In H I, tidal arms are located symmetrically on opposite sides of the nucleus, as predicted by the models (the western tidal arm is obscured optically by NGC 2207).

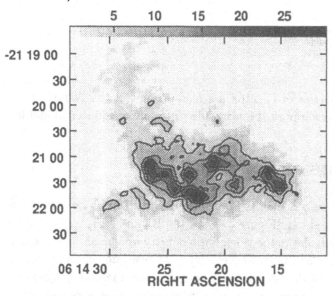

Fig. 2—N(H I) associated with IC 2163. The greyscale wedge is labelled in units of $M_\odot$ pc^{-2}.

Since the two components of the eastern tidal arm in IC 2163 are separated mainly in declination, we made cuts across the double arm at fixed values of right ascension and displayed the H I emission in declination *vs.* velocity diagrams (see example in Fig. 3). We find two main velocity components on the double arm at each value of RA: the component with the higher velocity coincides with the northern component of the optical arm, while the other velocity component coincides with the southern component of the optical arm. The typical difference in the line-of-sight velocity between the two velocity components is 70–100 km s^{-1} and the separation is 20″ – 30″. Thus two orbit streams occur on the tidal arm on the anti-collision side of the galaxy. The magnitude and sense of the velocity differences agree with the simulations in Elmegreen *et al.* (1991).

The kinematic minor axis of IC 2163 is nearly perpendicular to the apparent isophotal (optical and H I) minor axis of the oval. This unusual relative orientation of the axes is reproduced in our new N-body simulations and implies that the intrinsic shape of the disk is extremely oval. Also, from surface photometry in B and R–bands, we conclude that nearly all of the stars have been cleared out of the interarm regions of IC 2163 by the interaction and put into the inner oval and the tidal arms.

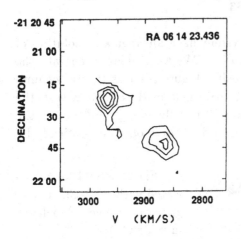

Fig. 3—The two components of the double arm in IC 2163: H I emission in a DEC *vs.* velocity display for a typical cut across the arm at fixed RA.

All of these observed properties of IC 2163 are also present in the N-body simulations. Thus our observations support the prograde, in-plane encounter model for IC 2163.

3 OBSERVED PROPERTIES OF NGC 2207

On short exposure optical images, such as Figure 1, NGC 2207 has a normal spiral pattern and a small, weak, central bar. Figure 4 shows the H I emission attributed to NGC 2207: the display on the left shows the H I column density image, and the display on the right shows the contours of the mean velocity field overlaid on the velocity dispersion image in greyscale. Both galaxies contain unusually massive H I clouds, comparable in mass to dwarf galaxies; the origin of these clouds as a consequence of the interaction is discussed by Elmegreen *et al.* (1993). The main H I emission from NGC 2207 forms a broad, clumpy ring that opens to the south and contains the optical spiral arms. However, on the eastern and western parts of the ring, the H I ridge-line often lies in the interarm region, between the stellar arms. In particular, an H I cloud with a mass of 10^9 M$_\odot$ sits in the interarm region between the two stellar arms on the western part of the ring. The H I ring and the main optical disk are embedded in a large, low density H I puddle that extends nearly 3′ to the south and contains faint optical streamers. While the H I ring probably predates the present encounter, the present interaction has rearranged the distribution of gas in the ring and may have disrupted the southern part of the ring.

The velocity field (see Fig. 4) has a global S-shaped distortion. The gas in NGC 2207 is also very stirred up by the interaction; the H I line profiles in a large part of the main disk are broad and asymmetric. The velocity dispersion image in

Fig. 4—The image on the left shows N(H I) associated with NGC 2207. The optical nucleus is in the H I hole, and the greyscale wedge is labelled in units of $M_\odot \, pc^{-2}$. The image on the right shows the velocity dispersion in NGC 2207 as greyscale with contours of the mean velocity field overlaid. The contour interval for the velocity field is 20 $km \, s^{-1}$. The greyscale wedge for the velocity dispersion is labelled in units of $km \, s^{-1}$.

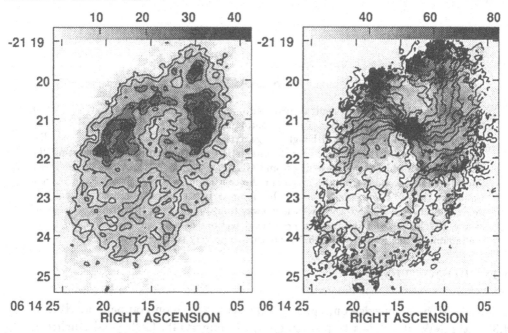

Figure 4 shows that the H I velocity dispersion is high (40 – 60 $km \, s^{-1}$) in a large, butterfly-shaped region that covers much of the main disk and coincides with the S shape of the velocity field. The anomalous velocities suggest that the disk is in the process of being warped by motions perpendicular to the disk, and thus the main tidal force on NGC 2207 was perpendicular to the plane.

The various disturbances in this system suggest that the collision line was nearly in the plane of IC 2163 but not in the plane of NGC 2207.

The National Radio Astronomy Observatory is operated by Associated Universities, Inc., under cooperative agreement with the National Science Foundation. This work was supported in part by NSF Grant AST-8914069 to M. K.

REFERENCES

Elmegreen, B. G., Kaufman, M., and Thomasson, M. 1993, *Ap. J.*, **412**, 90.
Elmegreen, D. M., Sundin, M., Elmegreen, B. G., and Sundelius, B. 1991, *Astr. Ap.*, **244**, 52.

The NGC 5775/4 Interacting System

Judith A. Irwin[1] and Bryan L. Caron[2]

[1]Department of Physics, Queen's University, Kingston, Canada
[2]Department of Physics, University of Alberta, Edmonton, Canada

ABSTRACT

We have made the first neutral hydrogen maps of the galaxy pair, NGC 5775, and NGC 5774, as well as a number of radio continuum maps from several independent observing runs at 6 and 20 cm. Although the galaxies are not strongly distorted, optically, we have discovered clear evidence for an interaction in the form of two connecting H I bridges through which gas is travelling from NGC 5774 to NGC 5775. Along the southern bridge, we have also detected non-thermal radio continuum emission, suggesting (but not requiring) that star formation may also be occuring between the galaxies. In addition, we have discovered H I arcs and extensions away from the plane of the IR-bright, edge-on galaxy, NGC 5775, and can confirm previous detections of arcs and plumes in the radio continuum. In this respect, NGC 5775 appears to be very similar to NGC 891.

1 INTRODUCTION

NGC 5775 is an edge-on, spiral galaxy with a face-on companion, NGC 5774, 4.3' to the NW. It belongs to a small group (Group #148; Geller and Huchra 1983) which, in addition, includes IC 1070, only 3.9' to the SW of NGC 5775. NGC 5775 is the dominant galaxy in the group, it is an IR bright galaxy (Soifer *et al.* 1987), and has a nuclear H II region spectrum (Giuricin *et al.* 1990). NGC 5774 is a barred "low surface brightness" (LSB) galaxy (Romanishin *et al.* 1983); however, its central surface brightness is more than a factor of 5 brighter than the brightest LSB galaxies studied by G. Bothun (these proceedings). Neither galaxy appears strongly tidally disturbed on the POSS prints.

1.1 Observations

We have mapped this system in neutral hydrogen with the *VLA* in the hybrid B/C configuration for a total on-source observing time of 7.4 hours, and with a channel resolution of 42 $km\,s^{-1}$. Both uniformly and naturally weighted maps were made, with spatial resolutions of $\approx 14''$ and $25''$, respectively. The galaxies were also mapped in the radio continuum, as indicated in Table 1, with each entry representing independent observations; spatial resolutions range from $\approx 5''$ to $50''$. Processing and reduction of the continuum data are currently in progress.

TABLE 1

RADIO CONTINUUM OBSERVATIONS

VLA Configuration	On-Source Observing Time (Hours)	
	6 cm	20 cm
D	0.5, 4.9	1.9
C/D		4.7
C		0.5
B		3.2

2 RESULTS

2.1 Neutral Hydrogen

The integrated H I map and mean velocity field are shown in Figures 1*a*, 1*b*, respectively, and clearly illustrate that two H I bridges "connect" the galaxies. The bridge gas is at approximately the systemic velocity of NGC 5774, and there is also an abrupt velocity gradient where the gas from the northern bridge intersects NGC 5775, indicating that H I is travelling from NGC 5774 to NGC 5775. (The southern bridge is actually continuous, and not "broken" as it appears here.) If the transverse velocity in the bridge is equivalent to the radial velocity difference between galaxies, then the timescale for gas to transfer from NGC 5774 to NGC 5775 is $\approx 1 \times 10^8$ yr. Inspection of the POSS (blue) prints also reveals very weak optical emission along both bridges. Therefore, both the gaseous and stellar components are responding, tidally, to NGC 5775, but with the H I relatively more obvious. See also section 2.2, below.

Figure 1*a* also shows the existence of H I loops, or arcs away from the plane, which extend up to ≈ 8 kpc from the major axis, in projection. These appear to be analogous to the Heiles shells which are observed in our own Galaxy (Heiles 1979, 1984) and in several other edge-on systems (*e.g.* NGC 3079, Irwin and Seaquist 1990; NGC 891, Rand 1992).

The global H I properties of the galaxies are listed in Table 2, where we have assumed a single distance to the pair, computed from the mass-weighted mean velocity, and assuming $H_0 = 75$ km s^{-1} Mpc^{-1}. Each of these quantities is within the ranges found for isolated galaxies of the same Hubble type (*cf.* Haynes and Giovanelli 1984; Shostak 1978) and the H I global profiles also show no obvious distortions. Consequently, the interaction is either mild enough or not yet advanced enough to produce any significant global anomalies.

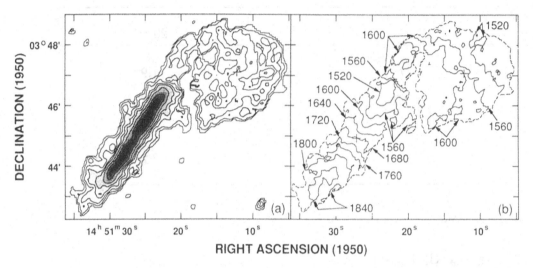

Fig. 1—(*a*) H I column density map (beam size $\approx 13''.5$). Contours are at 1, 5, 10, 17.5, 25, 40, 60, 82.5, and 110 $\times 10^{20}$ cm^{-2}. (*b*) Intensity weighted mean velocity field, with contours as labelled. Crosses mark the optical centers of each galaxy.

TABLE 2

GLOBAL H I PROPERTIES OF THE GALAXIES

Parameter	NGC 5775	NGC 5774
V_{sys} (km s^{-1})	1681 ± 10	1566 ± 10
D_{Hubble} (Mpc)	24.8	24.8
$\Delta V_{20\%}$ (km s^{-1})	463 ± 10	179 ± 10
M_{HI} (10^9 M$_\odot$)	9.1 ± 0.6	5.4 ± 0.4
M_{HI}/L_B (M$_\odot$/ L$_\odot$)	0.32 ± 0.02	0.59 ± 0.04
M_T (10^{11} M$_\odot$)	1.5 ± 0.2	0.5 ± 0.3
M_T/L_B (M$_\odot$/ L$_\odot$)	5.3 ± 0.7	5 ± 3
M_{HI}/M_T	0.06 ± 0.01	0.11 ± 0.07

2.2 Radio Continuum

In Figure 2, we show two independent, 20 cm radio continuum images of the NGC 5775/NGC 5774 system. These are the only two data sets (see Table 1) in which *radio continuum emission* can be seen *between* the galaxies. Given that no intergalactic emission is observed in 6 cm images, we can place an upper limit on the spectral index, α (where $S \propto \nu^{\alpha}$), in this region, of ≈ -1. This is steeper than the measured spectral index ($\alpha = -0.7$) for the integrated emission from NGC 5775, itself, and implies that the emission is predominantly non-thermal in origin — presumably a product (at least in part) of particle acceleration from supernova explosions.

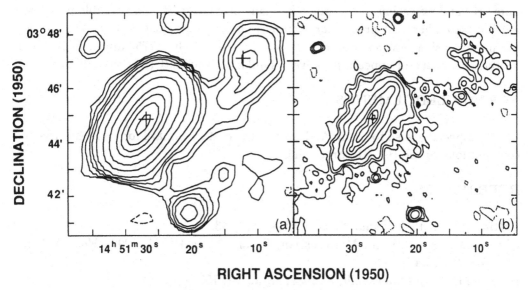

Fig. 2—(*a*) D Array 20 cm image of NGC 5775, NGC 5774 (marked with crosses at their optical centers) and IC 1070 towards the south. Contours are at −0.19, 0.19 ($2.5\,\sigma$), 0.3, 0.5, 0.75, 1.4, 4, 10, 25, 40, 58, and 80 mJy beam^{-1} (beam size = 52″). (*b*) C/D Array 20 cm image, with contours at −0.11, 0.11 ($2.6\,\sigma$), 0.18, 0.28, 0.6, 1.4, 4, 7.5, 12, and 18 mJy beam^{-1} (beamsize = 14″6). Shaded regions indicate decreasing intensity.

The interpretation of this intergalactic emission depends upon whether or not it is produced in situ. If, for example, this emission is produced in the bridge, itself, then massive star formation must also be occurring between the galaxies, since the lifetimes of stars destined to become supernovae (*e.g.* $\tau < 3.4 \times 10^{7}$ yr for $M_{\star} > 8$ M$_{\odot}$; Larson 1974) occur on timescales significantly shorter than the dynamical timescale ($\approx 10^{8}$ yr, section 2.1). On the other hand, particle acceleration may have occurred entirely within NGC 5774 due to star forming activity there, with the magnetic fields (presumably coupled to the gas) and associated relativistic electrons then tidally pulled into a bridge. The lifetimes of these particles, in the absence of further acceleration, can then be computed using the minimum energy assumption. Assuming lower and upper frequency cutoffs, ν_1, ν_2, of 10^7 and 10^{11} Hz, respectively, that the bridge depth equals its width, and that the ratio of heavy particle to electron energy is $k = 100$, we estimate a particle lifetime of $\tau \lesssim 10^8$ yr for $\alpha \lesssim -1$. This timescale increases by a factor of 1.3, if either k decreases by a factor of 2 or ν_2 decreases by a factor of 10. Since the particle lifetime is of the same order as the dynamical timescale, it is therefore also feasible that the radio continuum emission now observed in the bridge may not have originated there. It should be reasonably straightforward to distinguish between these scenarios by searching for other tracers of star forming activity in the bridge, e.g. molecular clouds and H II regions.

We also confirm the previous detection by Hummel *et al.* (1991) of radio continuum arcs and extensions away from the plane of NGC 5775 (also visible in images not illustrated here). A preliminary comparison suggests that the H I and radio continuum extensions are *not* well correlated, spatially. NGC 5775 also displays high latitude Hα arcs and appears to be very similar to NGC 891 (Dettmar 1992).

ACKNOWLEDGEMENTS

We especially wish to thank Andrew Platt and Marc Simard for their enthusiastic contributions to the H I data processing.

REFERENCES

Dettmar, R. -J. 1992, *Fund. Cosmic Phys.*, **15**, 143.

Giuricin, G., Bertotti, G., Mardirossian, F., and Mezzetti, M. 1990, *Mon. Not. R. Astr. Soc.*, **247**, 444.

Haynes, M. P., and Giovanelli, R. 1984, *A. J.*, **89**, 758.

Heiles, C. 1979, *Ap. J.*, **229**, 533.

Heiles, C. 1984, *Ap. J. Suppl.*, **55**, 585.

Hummel, E., Beck, R., and Dettmar, R. -J. 1991, *Astr. Ap. Suppl.*, **87**, 309.

Irwin, J. A., and Seaquist, E. R. 1991, *Ap. J.*, **353**, 469.

Larson, R. B. 1974, *Mon. Not. R. Astr. Soc.*, **166**, 585.

Rand, R. J. 1993, in *Star Forming Galaxies and their Interstellar Medium*, ed. J. Franco, and F. Ferrini (New York: Cambridge University Press), in press.

Romanishin, W., Strom, K. M., and Strom, S. E. 1983, *Ap. J. Suppl.*, **53**, 105.

Shostak, G. S. 1978, *Astr. Ap.*, **68**, 321.

Soifer, B. T. *et al.* 1987, *Ap. J.*, **320**, 238.

High Resolution CO and H I Observations of an Interacting Galaxy NGC 3627

Xiaolei Zhang[1,2], Melvyn Wright[1], and Paul Alexander[3]

[1]Radio Astronomy Laboratory, University of California at Berkeley,
 Berkeley, CA 94720, U. S. A.
[2]Harvard-Smithsonian Center for Astrophysics,
 60 Garden Street, Cambridge, MA 02138, U. S. A.
[3]Mullard Radio Astronomy Observatory, Cavendish Laboratory,
 Madingley Road, Cambridge, CB3 OHE, England

ABSTRACT

A nearby interacting galaxy NGC 3627 was observed in the CO (1-0) transition and in H I using aperture synthesis technique. The combined CO and H I data indicated that, the gravitational torque experienced by NGC 3627 during its close encounter with NGC 3628 triggered a sequence of dynamical processes, including the formation of prominent spiral structures, the central concentration of both the stellar and gaseous mass, the formation of two widely separated and outwardly located Inner Lindblad Resonances, and the formation of a gaseous bar inside the inner resonance. These processes in coordination allow the continuous and efficient radial mass accretion across the entire galactic disk. The observational result in the current work provides a detailed picture of a nearby interacting galaxy which is very likely in the process of evolving into a nuclear active galaxy. It also suggests one of the possible mechanisms for the formation of successive instabilities in the post-interacting galaxies, which facilitates the central channeling of interstellar medium to fuel nuclear activities.

1 INTRODUCTION

In the recent years, it has become increasingly apparent that the interaction among galaxies plays an important role in the evolution of a disk galaxy (Barnes and Hernquist 1992 and the references therein). It is well known that the gravitational tide excited during the galaxy interaction could induce significant inflow of gas to fuel nuclear activities. But since very few interacting galaxies have so far been mapped with high resolution, a detailed picture of the type of instabilities that occur after galaxy interaction, and of the mass accretion process in the inner kiloparsec region of the galaxy is still lacking.

In order to look into the various un-answered questions in the galaxy interaction and evolution process, and especially in order to obtain detailed information for the central region of a post-interacting galaxy, we have mapped a nearby interacting spiral galaxy NGC 3627, which is a member of the interacting galaxy group the Leo Triplet, in the CO (1-0) transition with 7″ resolution using the *BIMA* interferometer, for its central 4.5 kpc, and in the 21 cm emission of neutral hydrogen (H I) with 30″ resolution using the *VLA*, for the entire galaxy. Although classified as a normal spiral, NGC 3627 has certain unusual properties, such as a reasonably strong nuclear Hα emission (Filippenko and Sargent 1985), a high H_2 to H I mass ratio close to that of Seyfert galaxies (Young *et al.* 1983), indicating that it is in fact a borderline galaxy, with properties in between of normal and active galaxies. It is thus an ideal candidate for studying the post-interaction evolution effects. Table 1 lists some of the most important characteristics of galaxy NGC 3627.

TABLE 1
GALAXY PARAMETERS

Parameter	Value
R.A. (1950)[a]	$11^h17^m37.9^s$
Dec. (1950)[a]	$13°16'08''$
RC 2 Type[b]	SAB(s)b
Inclination[b]	60°
Position Angle[b]	173°
V_{LSR}[c]	700 km s^{-1}
Distance[b]	6.7 Mpc
Angular Scale Corresponding to 1 kpc[b]	31''

[a] Dressel and Condon (1976)
[b] de Vaucouleurs, de Vaucouleurs, and Corwin (1976) (RC2)
[c] Young, Tacconi, and Scoville (1983)

2 OBSERVATIONS AND MASS ESTIMATE

The observations of H I was made during 1988 using the *VLA* C and D configurations. The observations of CO was made between 1990 and 1992 using the B and C configurations of the *BIMA* interferometer. Data reduction was done using the MIRIAD package developed at *BIMA*. Figure 1 shows an overlay of the CO (for the central 4.5 kpc) and H I integrated intensities for NGC 3627. A CO bar is seen to reside inside the central H I deficiency. Two CO clumps further out coincide with the H I clumps at the same locations, which trace the beginning of the spiral arms. The H I and CO distributions show evidence of asymmetry and distortions.

Detailed calculations (Zhang *et al.* 1993) gave that the total H I mass obtained by the interferometer observation is 1.2×10^8 M$_\odot$, compare to the single dish value of 4.1×10^8 M$_\odot$ (Haynes *et al.* 1979). The central CO bar indicates a molecular mass of $M_{H_2,bar} = 4 \times 10^8$ M$_\odot$. The interferometer data contains about 50 % of the single dish CO flux (Young *et al.* 1983) near the galaxy center. The interstellar medium (H I + H_2) constitutes about 30% of the galactic dynamical mass near the center of the galaxy, and about 4% for the entire galaxy.

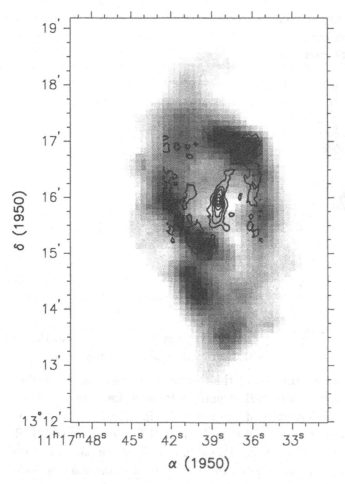

Fig. 1—Overlay of the CO (contours) and H I (halftone) integrated intensity maps. Note that the CO observations cover only the central 140″ by 140″.

3 KINEMATICS AND DYNAMICS OF THE GALAXY

In order to study in more detail the kinematics and dynamics of the galaxy, we used the combined CO and H I position-velocity map along the major axis of the galaxy to derive a rotation curve. Figure 2 shows this composite position-velocity together with a fitted rotation curve. Note that the inclination angle correction for the velocities was not made. Using the derived rotation curve, we can further calculate the angular speed, the pattern speed and the location of the Inner Lindblad Resonances. These are given in Figure 3. We see that the there are two intersections of the $\Omega = \Omega_p$ line with the $\Omega - \kappa/2$ curve. These are designated IILR (Inner Inner Lindblad Resonance, which is the one closer to the galactic center) and OILR (Outer Inner Lindblad Resonance).

Comparing to the CO-H I overlay map of Figure 1, we see that the radius of the IILR (1 kpc) is close to the end of the molecular bar, and the radius of the OILR (3 kpc) is close to the location of the twin peaks. Note that the radii of the twin-peaks from the galactic center are projected through a position angle of $\sim 30°$.

The dynamical basis for the morphology and kinematics of the central gas distri-

Fig. 2—CO–H I composite position-
velocity map and the rotation curve.

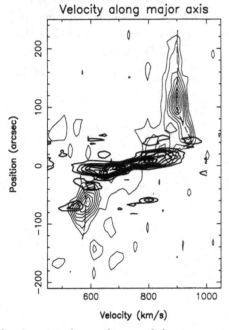

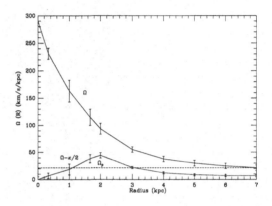

Fig. 3—Angular velocities and the locations of Inner Lindblad Resonances.

bution can be understood from an analysis of the stellar orbits and gas streamlines in a background non-axisymmetric potential. Contopoulos and Mertzanides (1977) showed that in a weakly barred potential, if the mass distribution and the pattern speed is such that two ILRs are present, then outside the OILR and inside the IILR, the dominant periodic stellar orbits are those elongated in a direction parallel to the bar potential (the so-called x_2 orbits); between the two ILRs the dominant periodic orbits are those perpendicular to the bar potential (x_1 orbits). A self-consistent bar model cannot be constructed by orbits mainly of the x_1 type, in which the orbiting mass and the gravitational potential are out of phase. Or, as the numerical models of Schwarz (1984) show, the effect of the appearance of the two ILRs is to clear an annulus between the two ILRs by sweeping the mass towards the center, which may lead to the formation of a bar inside the IILR. On the other hand, near the OILR, the crossing of the two families of periodic orbits, when combined with the pressure of the gas, leads to the crowding of the gas streamlines and thus the formation of the "twin-peaks" gas concentration near the OILR (Kenney *et al.* 1992; Schwarz 1984).

Also notice that the spiral structures near the OILR (the "twin peaks") lag in phase with respect to the background bar potential. This allows the torque exerted by the bar-potential to remove angular momentum from the orbiting mass near the OILR, causing it to accrete inward. Once the inward accreting mass is swept across the space between the OILR and the IILR, it will reach the central bar structure inside the IILR. The linear offset shocks which form at the leading edge of the central bar continue the task of taking away energy and angular momentum, thus facilitate further mass accretion towards the nucleus.

4 IMPLICATIONS ON THE GENERAL PROBLEM OF THE POST-INTERACTION GALACTIC EVOLUTION

The morphology and kinematics of the molecular and atomic gas in the central region of galaxy NGC 3627 suggested that the post-interaction evolution is a well coordinated process in the combined stellar and gaseous disk, down to even the central sub-kiloparsec scale. The spiral structures are usually the first large-scale instabilities formed or enhanced as a result of the tidal encounter. These spiral structures serve to transfer energy and angular momentum outward, thus allow the central concentration of the stellar and ISM mass. The means for the stellar mass distribution to reach higher central concentration was investigated in Zhang (1993). Once the disk mass accretion proceed to the degree such that ILRs form, they could continue the task of removing angular momentum from the disk mass in the inner region of a galaxy, starting right at the location where the power of the spiral structures diminishes.

The instability structure seen in the central region of NGC 3627 is not a separated instability in the self-gravitating gaseous disk (Shlosman *et al.* 1989); nor is the increase in the central mass accretion rate caused chiefly by the increase in cloud-cloud collision (Lin *et al.* 1988). The nuclear structures in NGC 3627 appear as coherence response to the underlying dynamics of the total galactic mass distribution.

Similar nuclear morphology as in NGC 3627 is also observed in NGC 1068 and NGC 4314, as well as in a number of other galaxies. More high resolution images of galaxies are needed in order to determine how common it is for the emergence of the ILRs related instabilities in the nuclear region of the active or nearly active galaxies. This in turn indicates the importance of mass accretion process in the outer disk to the triggering of nuclear activities.

REFERENCES

Barnes, J. E., and Hernquist, L. 1992, *Ann. Rev. Astr. Ap.*, **30**, 705.
Contopoulos, G., and Mertzanides, C. 1977, *Astr. Ap.*, **61**, 477.
de Vaucouleurs, G., de Vaucouleurs, A., and Corwin, H. G. 1976, *Second Reference Catalogue of Bright Galaxies* (Austin: University of Texas Press), RC2.
Dressel, L. L., and Condon, J. J. 1976, *Ap. J.Suppl.*, **31**, 187.
Filippenko, A. V., and Sargent, W. L. W. 1985, *Ap. J. Suppl.*, **57**, 503.
Haynes, M. P., Giovanelli, R., and Morton S. R. 1979, *Ap. J.*, **229**, 83.
Kenney, J. D. P., *et al.* 1992, *Ap. J. Lett.*, **395**, L79.
Lin, D. N. C., Pringle, J. E., and Rees, M. J. 1988, *Ap. J.*, **328**, 103.
Schwarz, M. P. 1984, *Mon. Not. R. Astr. Soc.*, **209**, 93.
Shlosman, I., Frank, J., and Begelman, M. C. 1989, *Nature*, **338**, 45.
Young, J. S., Tacconi, L. J., and Scoville, N. Z. 1983, *Ap. J.*, **269**, 136.
Zhang, X. 1993, submitted to *Ap. J.*.
Zhang, X., Wright, M., and Alexander P. 1993, *Ap. J.* Nov. 1.

Mass–Transfer Induced Starbursts in Interacting Galaxies

J. Christopher Mihos

Board of Studies in Astronomy and Astrophysics,
University of California, Santa Cruz

ABSTRACT

We model a range of flyby galaxy interactions in order to investigate the formation of starbursts in galaxies with induced stellar bars and/or mass loss or accretion. The models indicate that mass transfer is important in triggering radial gas flows and starburst activity if counter-rotating accretion occurs; *i.e.*, accretion from a prograde disk onto a retrograde disk. Such accretion proves effective in shedding rotational angular momentum of the ISM, resulting in the radial gas flow and subsequent nuclear ISM concentration, while leaving behind a relatively unperturbed stellar disk. However, bar formation proves more important under a wider range of interaction scenarios than does mass transfer, and thus bar formation is the dominant process in triggering nuclear activity in interacting systems as a whole.

1 INTRODUCTION

The link between galaxy interactions and elevated star formation rates has been demonstrated through observations of such star formation tracers as optical emission lines (e.g., Kennicutt *et al.* 1987; Bushouse 1987), strong far infrared emission (Lonsdale, Persson, and Matthews 1984), and radio continuum emission (Hummel 1981). However, while this large body of evidence indicates that interactions can cause starbursts, it is not at all clear that they *must* do so. A large fraction of interacting galaxies show little or no increased star formation (Kennicutt *et al.* 1987; Bushouse 1987), suggesting that the triggering mechanism for these starbursts must involve some complicated function of interaction geometry and galaxy properties.

Numerical simulations suggest that the formation of a bar during a close prograde encounter is effective at driving radial inflows of ISM material, perhaps fueling strong starbursts (Noguchi 1988; Barnes and Hernquist 1991; Mihos, Richstone, and Bothun 1992). However, several interacting systems show elevated star formation rates even in the absence of any strong tidal features (Kennicutt *et al.* 1987). In these systems, therefore, bar-driven radial inflows cannot explain these starbursts. However, interactions can also result in significant mass transfer between the two galaxies (e.g., Wallin and Stuart 1992), which may provide a second triggering mechanism for such starbursts. In this work, we survey a range of flyby interactions, examining the effects of both bar formation and mass transfer on the resulting star forming properties of the galaxies.

2 NUMERICAL MODELS

The details of the numerical models used for these experiments have been described in Mihos *et al.* (1992), and will be outlined here. The code is based on L. Hernquist's TREECODE (Hernquist 1987), which employs a tree structure to calculate the gravitational forces, and modifications have been made to the code to model star formation and interactions between ISM gas clouds. Two types of particles are used: "stellar" particles which act as collisionless particles, and "gaseous" particles which can merge with one another or fragment into smaller clouds.

The model galaxies consist of three fully self-gravitating components: a spherical halo, an exponential disk, and a flat distribution of gas particles representing the ISM. The mass ratio of these components is 4:1:0.2, respectively. The particles representing the halo and disk are collisionless, while the gas particles may collide and merge (with typical mean free paths of ~ 2 kpc), as well as fragment into smaller units (on a timescale of ~ 2–3×10^7 yrs). The unit of time is $\sim 5 \times 10^7$ yrs.

Star formation is modeled as a modified Schmidt law (Schmidt 1959). The star formation rate in a cloud given by:

$$SFR_{\text{cloud}} = C \times M_{\text{cloud}} \times \rho_{\text{gas}}, \qquad (1)$$

where M_{cloud} is the cloud mass and ρ_{gas} is the local gas density measured in a 750 pc radius. Volume averaged, this law corresponds closely to a Schmidt law of index $n \sim 1.8$ (Mihos, Richstone, and Bothun 1991). The constant C is set such that the global star formation rate is set at 1 $M_\odot$ yr^{-1}. At each time step, the star formation rate is recalculated and a corresponding mass removed from the cloud, thereby taking into account the gas depletion via induced star formation.

Orbits are chosen to be parabolic, with a closest approach of 8 disk scale lengths at $T = 0$. On such orbits, the galaxies lose only a small amount of orbital energy due to the interaction, and hence do not merge. The orientation of each disk with respect to the orbital plane is defined as: *prograde*, in which the galaxy rotates in the same sense as the orbital motion of its companion; *retrograde*, in which the galaxy rotates in the opposite sense of the orbital motion of its companion; and *orthogonal*, in which each galaxy's disk is inclined 90° to the orbital plane.

3 RESULTS

Figure 1 shows the evolution of global star formation rate for prograde galaxies with companions of differing orientations. Such prograde disks respond strongly to the gravitational interaction, due to the resonance between the rotational angular velocity of the disk and the orbital angular velocity of the companion, and display strong tidal features. These galaxies transfer $\sim 10\%$ of their ISM to the passing companion, and also develop a strong bar in response to the companion's perturbation. This bar allows the gas to shed angular momentum through clouds collisions and gravitational torques

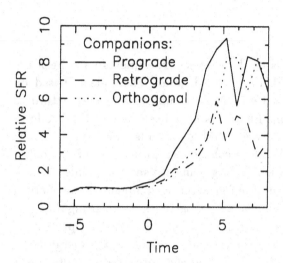

Fig. 1—The evolution of the global star formation rate in prograde disks. Each line corresponds to the response of the star formation rate in the prograde disk to a companion of differing orientation.

(e.g., Noguchi 1988; Mihos *et al.* 1992) and sink to the center of the galaxy. As the gas densities in the central regions grow, the star formation rates rise as well, resulting in a strong, burst of star formation. This burst is very centrally concentrated, with $\sim 70\%$ of the total star formation activity occurring in the central kpc of the disk.

An examination of Figure 1 reveals that these galaxies develop this burst of star formation regardless of the orientation of the companion galaxy. Since the companion galaxy's orientation determines the amount of mass transferred *onto* the prograde disk, this result shows that it is the effects of the bar formation, and not the mass transfer, which provide the dominant trigger for the resulting starburst.

Figure 2 shows the evolution of the global star formation rate in retrograde galaxies with prograde and retrograde companions. Since the retrograde disk rotates opposite the motion of the companion, there is no orbital/rotational resonance; as a result, these galaxies are relatively undisturbed by even these close passages. No strong tidal features develop, nor is there any mass loss or bar formation. As can be seen from Figure 2, when such a disk interacts with another retrograde disk (or, more generally, any galaxy which does not transfer mass to the retrograde disk) the net result is that no burst of star formation develops; rather, the retrograde disk survive the encounter relatively unchanged in both morphology and star forming properties.

However, when a retrograde disk interacts with a prograde companion, the results are quite different. In this case, although the retrograde disk is unperturbed, it now accretes 10% of the ISM from the prograde disk. Furthermore, this material is accreted rotating in the opposite sense to the ISM in the retrograde disk. Collisions between the clouds in the ambient ISM and those making up the counterrotating accreted gas effectively shed the angular momentum of the gas clouds, leading to a radial inflow of gas. As this gas flows inward, the star formation rates in the inner disk rise significantly, again resulting in a strong centrally-concentrated burst of star formation. However, the starburst in this system is neither as strong nor as centrally concentrated as those triggered by the formation of bars in prograde disks. The end result of such an encounter is a morphologically normal disk with a burst of star

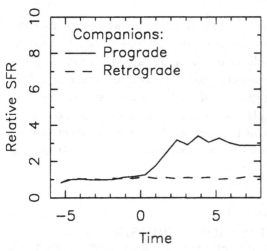

Fig. 2—The evolution of the global star formation rate in retrograde disks. Each line corresponds to the response of the star formation rate in the retrograde disk to a companion of differing orientation.

formation in the inner few kpc.

Finally, Figure 3 shows the evolution of the global star formation rate in orthogonal disks interacting with prograde and orthogonal disks. In this scenario, the orthogonal disks do develop some weak tidal features, but like the retrograde disks, no bar forms, nor does any mass loss occur. Although disturbed in appearance, such galaxies show very little enhanced star formation. Even when gas mass is accreted from a prograde companion, this gas is accreted on polar orbits, such that both the rate and effectivess of cloud collisions is strongly diminished; therefore, no radial gas flows or starbursts occur.

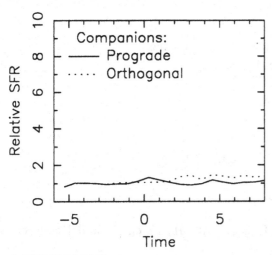

Fig. 3 – The evolution of the global star formation rate in orthogonal disks. Each line corresponds to the response of the star formation rate in the orthogonal disk to a companion of differing orientation.

4 SUMMARY

This small survey of galaxy interactions (summarized in Table 1) has shown that mass transfer plays a relatively small role in the triggering of nuclear starbursts during galaxy encounters. The dominant trigger for such starbursts is the formation of a strong bar associated with prograde interactions. Starbursts were triggered in all prograde galaxies, regardless of the amount of mass transfer from the companion. Mass transfer appears to be important in situations involving counterrotating accretion, such as accretion onto a retrograde disk. In such cases, cloud collisions between the ambient and accreted ISM effectively shed the angular momentum of these clouds, leading to a radial inflow of gas and subsequent central starbursts. This mechanism may be responsible for the triggering of central starbursts in otherwise unperturbed galaxies, which cannot be explained through bar instabilities. However, bar formation occurs under a wider range of interactions than does counterrotating mass transfer, indicating that bar formation is the dominant process in triggering nuclear activity in interacting systems as a whole.

TABLE 1
RELATIVE STAR FORMATION INCREASES

Galaxy Orientation	Prograde Companion	Retrograde Companion	Orthogonal Companion
Prograde (10% mass loss, bar)	8× (10% corotating accretion)	5× (no acc.)	8× (no acc.)
Retrograde (no mass loss, no bar)	3× (10% counterrotating accretion)	1× (no acc.)	... (no acc.)
Orthogonal (no mass loss, no bar)	1× (10% polar accretion)	... (no acc.)	1× (no acc.)

REFERENCES

Barnes, J. E. and Hernquist, L. 1991, *Ap. J. Lett.*, **370**, L65.
Bushouse, H. A. 1987, *Ap. J.*, **320**, 49.
Hernquist, L. 1987, *Ap. J. Supp.*, **64**, 715.
Hummel, E. 1981, *Astr. Ap.*, **96**, 111.
Kennicutt, R. C., Keel, W. C., van der Hulst, J. M., Hummel, E., and Roettiger, K. A. 1987, *A. J.*, **93**, 1011.
Lonsdale, C. J., Persson, S. E., and Matthews, K. 1984, *Ap. J.*, **287**, 95.
Mihos, J. C., Richstone, D. O., and Bothun, G. D. 1991, *Ap. J.*, **377**, 72.
Mihos, J. C., Richstone, D. O., and Bothun, G. D. 1992, *Ap. J.*, **400**, 153.
Noguchi, M. 1988, *Astr. Ap.*, **203**, 259.
Schmidt, M. 1959, *Ap. J.*, **129**, 243.
Wallin, J. F. and Stuart, B. V. 1992, *Ap. J..* **399**, 29.

First HST Images of a Compact Group: Seyfert's Sextet[†]

J. W. Sulentic[1], C. R. Rabaça[1], and H. Arp[2]

[1]Department of Physics and Astronomy, University of Alabama, U. S. A.
[2]Max Planck Institute for Astrophysics, Germany
[†]Based on observations with the NASA/ESA Hubble Space Telescope, obtained at the Space Telescope Science Institute, which is operated by AURA, Inc., under NASA contract NAS5-26555

ABSTRACT

We present details of galaxy morphology in the compact group Seyfert's Sextet obtained with the *HST*-WFC.

1 INTRODUCTION

Seyfert's Sextet (SS, HCG 79) exhibits one of the highest galaxy surface density enhancements ($> 10^3$, Sulentic 1987) outside the core of a rich galaxy cluster. There are convincing signs of interaction between the component galaxies, including: 1) optical evidence of bridges, tails and a common low light level envelope (Sulentic and Lorre 1983), and 2) a distorted distribution of neutral hydrogen (Williams *et al.* 1991).

2 *HST* OBSERVATION AND DECONVOLUTION

SS was observed with the WFC on 12 May 1992. Nine 15-minute exposures were taken with the F439W (B) filter and processed with the standard STScI pipeline. The frames were accurately registered (to within 1 pixel), allowing us to combine them into a single averaged picture. Cosmic rays were removed by ignoring pixels which were significantly (3σ) deviant from the corresponding pixels on other frames.

Deconvolution was performed using both the Richardson-Lucy package bundled in IRAF/STSDAS (refer to the STSDAS User Guide), and the σ-CLEAN algorithm (Keel 1991), optimized to change the point-spread function (PSF) according to position on the chip. PSFs were computed using the Tiny Tim (refer to The Tiny Tim User's Manual) package. A 50 iteration deconvolution was carried out with the Lucy package. After 80,000 iterations with CLEAN the model brightness distribution showed little change. Each of these algorithms presents strong and weak points. Lucy handled very well large structures (low spatial frequencies), such as the halo connecting the galaxies. However, the final brightness distribution seemed to be dependent on the number of iterations used. CLEAN showed its best at the high spatial frequencies, being able to resolve small scale features in the discordant redshift galaxy (e) to nearly the diffraction-limited performance.

3 RESULTS AND SUMMARY

Ground based images provide rather unambiguous evidence that the four accordant redshift galaxies are interacting. The new *HST* data indirectly supports this view with the detection of internal peculiarities in the components (Fig. 1). One very striking aspect of SS is the small size of the galaxies. Dynamical stripping, which is evidenced by the luminous envelope, cannot account for the small sizes. While small, several of the galaxies exhibit well ordered structure more characteristic of higher luminosity galaxies. Galaxy *d* was suspected to be an inclined spiral on previous data. The *HST* data reveals a dust lane on the east side of this object. This can be interpreted as the relatively unperturbed equatorial dust lane of a late type spiral inclined within a few degrees ($\sim 10°$) of edge on. A redshift-implied absolute magnitude of approximately -18 (uncorrected for inclination) reveals a surprisingly well defined disk structure for such a low luminosity spiral, especially in such a dynamically active environment. Galaxy *c* shows clear signs of disk disruption. The disk is much broader and boxier on the side towards its nearest neighbor (component *b).* Galaxies *b* and *c* are connected by a luminous feature imbedded in the larger halo.

The most striking result is found for component *b.* Ground based images suggested that it was a rather boxy SO galaxy. The *HST* images reveal a complex system of dust lanes. A twisted dust lane passes diagonally across the center of the galaxy. In addition, a distinct ring structure is seen extending above and below the disk. The structure is reminiscent of a polar ring (or, perhaps, a failed polar ring galaxy) (see Whitmore *et al.* 1991). The observations strongly suggest that component *b* is a recent merger. It is difficult to interpret its complex internal structure in any other way. The luminous tail that extends to the east of *b* may represent the remnant of the companion acquired by *b.*

SS illustrates well the apparent contradiction between observation and theory for compact groups. Dynamical simulations of such dense systems predict that they will merge rather quickly (~ 1 Gyr) into a single elliptical or cD galaxy. It is surprising that few mergers in progress have been found among the known compact groups (Sulentic 1987; Sulentic and Rabaça 1993). Further, the groups do not show the level of FIR enhancement observed in samples of binary galaxies (Sulentic and de Mello Rabaça 1993). The lack of FIR luminous galaxies in compact groups is a particular surprise if they are merger rich. The *IRAS* beam is too large to permit us to assign unambiguosly the observed FIR flux between the components but if all the flux were assigned to galaxy *b* (unlikely in view of the presence of the spiral component *d*) we would still not single out this galaxy as a likely merger in progress. The *HST* data revealing a strong merger candidate in component *b* is thus significant because it lessens the troubling general deficit of mergers found in the groups by suggesting that "quiet mergers" can occur frequently in these dense aggregates. The "quietness" may be due in part to stripping of nonstellar material from the group components.

REFERENCES

Keel, W. C. 1991, *P. A. S. P.*, **103**, 723.

Sulentic, J. W. 1987, *Ap. J.*, **322**, 605.

Sulentic, J. W., and de Mello Rabaça, D. F. 1993, *Ap. J.*, **410**, 520.

Sulentic, J. W., and Lorre, J. 1983, *Astr. Ap.*, **120**, 36.

Sulentic, J. W., and Rabaça, C. R. 1993, in *Proceedings of the HST Workshop on Groups of Galaxies*, in press.

Williams, B. A., McMahon, P. M., and van Gorkon, J. H. 1991, *A. J.*, **101**, 1957.

Whitmore, B. C., Lucas, R. A., McElroy, D. B., Steiman-Cameron, T. Y., Sackett, P. D., and Olling, R. P. 1990, *A. J.*, **100**, 1489.

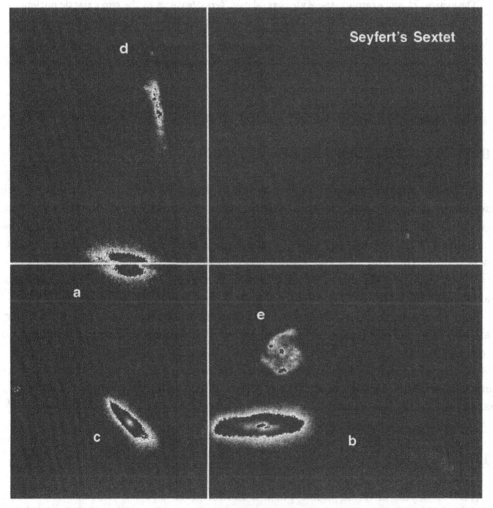

Fig. 1—Deconvolved mosaic of Seyfert's Sextet (SS) using the LUCY procedure.

The X-Ray Structure of Merging Galaxies

Andy Read and Trevor Ponman

School of Physics and Space Research, University of Birmingham, United Kingdom

ABSTRACT

We report the results of *ROSAT* PSPC observations of two merging galaxy systems. The nearest classic example, the *Antennae*, shows X-ray features at the two galactic nuclei, together with features interpreted as large H II regions, all embedded in apparently diffuse emission. Another feature with no radio or optical counterparts may be a hot outflow. NGC 2623, a more infrared-luminous system, is at a later merger stage. Although less X-ray luminous than the *Antennae*, it has a hotter nuclear feature, and a larger outflow. These observations are part of a chronological sequence of merging galaxies, carefully chosen in order to study the evolution of their X-ray properties through the merging process. These early results suggest that, as a merger progresses, the X-ray emission may evolve more rapidly than the infrared, and that massive hot outflows are generated.

1 INTRODUCTION AND OBSERVATIONS

Although merging galaxies have been the subject of intensive multiwavelength studies, particularly over the past decade, studies in the X-ray band have been severely restricted by the limited spatial resolution and sensitivity of instrumentation available. The best study prior to the launch of *ROSAT* was the *Einstein* IPC observation of the *Antennae*, which produced interesting but inconclusive results.

We are studying the X-ray development of merging galaxies at different stages of coalescence. *ROSAT* PSPC results for two such systems are presented here.

The *Antennae*, NGC 4038/9 (Arp 244), has much multi-wavelength information available. It is one of the closest merging systems (hence both large and bright) and its classic two-tailed form makes the presence of an interaction unambiguous. We have 15% of the scheduled 25 ksec observation so far. NGC 2623 (Arp 243) is at a later merger stage than the *Antennae*. One of the brightest infrared and radio galaxies, it has very long tails but the central masses have become nearly indistinguishable. We have a 6.7 ksec observation of this system.

2 THE RESULTS

NGC 4038/9 – Fitting a hot plasma model gives $L(0.1 - 2.0 \text{ keV}) = 8.0 \times 10^{40}$ erg s^{-1} with a galactic hydrogen column of 3.9×10^{20} cm^{-2} and $T \approx 0.7$ keV. Maximum entropy reconstructions show far more structure than could be resolved with the *Einstein* IPC. Several discrete features, with different hard and soft band properties, embedded in spatially extended, apparently diffuse X-ray emission, are seen (Fig. 1). X-ray knots at the two nuclei are observed, together with X-ray features coincident with bright radio and Hα knots. Soft emission, possibly due to the collision interface, and a very soft X-ray feature with no multiwavelength counterpart are also seen.

NGC 2623 — Maximum entropy reconstructions show that NGC 2623's X-ray emission ($L[0.1 - 2.0 \text{ keV}] = 4.1 \times 10^{40} \text{ erg s}^{-1}$) is split into two very distinct components (Fig. 1). The nuclear feature ($L[0.1 - 2.0 \text{ keV}] = 2.0 \times 10^{40} \text{ erg s}^{-1}$) is compact, strong, and hot ($T \gtrsim 1$ keV) and is coincident with published radio, Hα and CO emission. The off-centred source ($L[0.1 - 2.0 \text{ keV}] = 1.0 \times 10^{40} \text{ erg s}^{-1}$) is extended, weaker, and very cool (~ 0.2 keV). It has no multiwavelength counterpart, running along the northernmost edge of the least active (the western) arm.

3 THE SOFT COMPONENTS

We checked the possibility that the very soft X-ray features with no multiwavelength counterparts seen in both systems could be serendipitous background QSOs or foreground white dwarfs. Simulations show that typical QSO spectra observed through these Galactic columns would be too hard to correspond to these soft features. Moreover, the *Antennae's* feature actually lies within the disk of NGC 4038. It is statistically very unlikely that the features are due to white dwarfs as only a few hundred were detected in the entire *ROSAT* all-sky survey.

These very soft features are very reminiscent of the X-ray outflows (or *superwinds*) seen in M82, NGC 253 and the ultraluminous interacting pair, Arp 220. Assuming them to be outflows, then NGC 2623's is far more extended, more massive, and less dense than the *Antennae's*. This suggests that, as a merger progresses, a starburst-driven wind is produced, fed continually by the central starburst(s) until it breaks out from the system and expands.

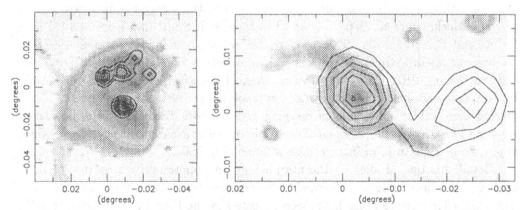

Fig. 1—Maximum entropy reconstructions of NGC 4038/9 (left) and NGC 2623 (right). X-ray contours are superimposed on optical images.

The Galaxy Activity-Interaction Connection in Low Luminosity Radio Galaxies

J. I. González-Serrano[1], R. Carballo[1], and I. Pérez-Fournon[2]

[1]Dpto. Física Moderna, Univ. de Cantabria, Santander, Spain
[2]Instituto de Astrofísica de Canarias, Tenerife, Spain

ABSTRACT

We have performed broad-band CCD imaging of a sample of 24 low luminosity radio galaxies containing radio jets. Their optical properties, photometric parameters, and their local environments have been studied and compared with other samples of radio galaxies.

Radio galaxies can be classified in two groups, powerful radio galaxies (PRG, with total power $P \gtrsim 10^{25}$ W Hz^{-1} at 1.4 GHz) and low luminosity radio galaxies (LLRG, with total power $P \lesssim 10^{25}$ W Hz^{-1} at 1.4 GHz), both from their radio as well as their optical properties. Among the main works that allowed to establish the differences in line emission, photometric structure, and environment between the two classes we should mention those by Hine and Longair (1979), Heckman *et al.* (1985, 1986), Lilly and Prestage (1987), Owen and Laing (1989), and Smith and Heckman (1989a, b). PRG usually show Fanaroff-Riley type II radio morphology (Fanaroff and Riley 1974), strong optical emission lines, and photometric structure similar to normal ellipticals. LLRG generally show FR I morphology, weak emission lines or pure absorption line spectra, and optical luminosity and light profiles similar to bright cluster galaxies.

One feature that powerful and low luminosity radio galaxies have in common is the morphological evidence of ongoing or past interaction/merging processes in these systems (Heckman *et al.* 1986; Lilly and Prestage 1987; Smith and Heckman 1989a, b; Colina and Pérez-Fournon 1990a, b; González-Serrano and Pérez-Fournon 1992 and references therein). However the indications of interaction of each class appear in different form. PRG show sharp morphological peculiarities in the form of tails, bridges, fans and shells, indicating mergers involving at least one disk system. LLRG present more subtle indications of merging, like isophote distortions (*e.g.* boxiness), isophote twists, and the presence of multiple nuclei and extended envelopes, which appear to be signatures of interactions between ellipticals.

Detailed optical studies on the morphology and photometric structure of radio galaxies have been generally devoted to intermediate-to-high power radio galaxies, where data for large samples have been reported in the literature. We have carried out an extensive optical study of 24 elliptical galaxies from the complete sample of

LLRG containing radio jets of Parma *et al.* (1987), aimed at the understanding of the optical structure of these galaxies. The data consist of broad-band V CCD images obtained at the 2.5m (INT) and 4.2m (WHT) telescopes at the Observatorio del Roque de los Muchachos (La Palma, Spain).

Some of the conclusions obtained in this work for LLRG are confirmations of previous results derived in general from intermediate power radio galaxies sharing the LLRG characteristics of being of FR I type or presenting weak emission or pure absorption line spectra. A more extensive account of this study is presented in González-Serrano, Carballo and Pérez-Fournon (1993). We can summarize the main results as follows. A fraction as large as 75% of the galaxies show morphological peculiarities in the optical, indicating gravitational interaction with nearby companions and/or recent merging. The LLRG in our sample have absolute magnitudes similar to bright cluster galaxies, they are more luminous than powerful radio galaxies, and they inhabit regions of higher galaxy density than PRG and radio-quiet ellipticals. Although more observations are inevitably needed, several of the results found for LLRG containing radio jets such us their optical luminosity and local environment, presence of extended envelopes, morphological indications of merging processes, X-ray emission, and presence of dumbbell systems and multiple nuclei, allow us to suggest that they could be the precursors of cD galaxies in clusters or groups of galaxies.

REFERENCES

Colina, L., and Pérez-Fournon, I. 1990*a*, *Ap. J. Suppl.*, **72**, 14.

Colina, L., and Pérez-Fournon, I. 1990*b*, *Ap. J.*, **349**, 45.

Fanaroff, B. L., and Riley, J. M. 1974, *Mon. Not. R. Astr. Soc.*, **167**, 31p.

González-Serrano, J. I., and Pérez-Fournon, I. 1992, *A. J.*, **104**, 535.

González-Serrano, J. I., Carballo, R., and Pérez-Fournon, I. 1993, *A. J.*, **105**, 1710.

Heckman, T. M., Carty, T. J., and Bothun, G. D., 1985, *Ap. J.*, **288**, 122.

Heckman, T. M., *et al.* 1986, *Ap. J.*, **311**, 526.

Hine, R. G., and Longair, M. S. 1979, *Mon. Not. R. Astr. Soc.*, **188**, 111.

Lilly, S. J., and Prestage, R. M. 1987, *Mon. Not. R. Astr. Soc.*, **225**, 531.

Owen, F. N., and Laing, R. A. 1989, *Mon. Not. R. Astr. Soc.*, **238**, 357.

Parma, P., Fanti, C., Fanti, R., Morganti, R., and de Ruiter, H. R. 1987, *Astr. Ap.*, **181**, 244.

Smith, E. P., and Heckman, T. M. 1989*a*, *Ap. J. Suppl.*, **69**, 365.

Smith, E. P., and Heckman, T. M. 1989*b*, *Ap. J.*, **341**, 658.

Multi-Wavelength Observations of "Interactive" Galaxies

Kirk D. Borne[1], and Luis Colina[2]

[1]Space Telescope Science Institute, Baltimore, Maryland, USA
[2]Universidad Autónoma de Madrid, Spain

ABSTRACT

We are studying the optical, radio, and X-ray morphologies of interacting galaxies in which at least one member is also an active galaxy. Deformed gas distributions are seen in galaxies that also show optical evidence of tidal deformation, indicating significant gas redistribution in these interacting systems, thereby providing compelling evidence for a causal connection between tidal and nuclear activity in "interactive" galaxies.

1 THE INTERACTION-ACTIVITY CONNECTION

Stockton (1990) and Heckman (1990) have reviewed the wealth of evidence indicating that galaxy interactions are somehow related to the generation of starburst and AGN activity in galactic nuclei. For example, a large fraction of low-luminosity radio and active galaxies have nearby companions or show evidence for a recent gravitational encounter (*e.g.*, MacKenty 1989). In an optical study of galaxies selected on the basis that they all contain well defined radio jets, it was found that almost half of the sample consists of pairs of elliptical galaxies (Colina and Pérez-Fournon 1990*a*,*b*). Many of these low-luminosity radio galaxies with companions (*e.g.*, 3C31, 3C278, and 3C449) show a well defined distorted radio jet structure at the VLA scale with an *S*– or *C*–shaped morphology. We are currently studying these systems across many wavebands (optical, radio, and X-ray) in an attempt to model the observed morphologies in each case and thereby constrain the various properties of the system (orbital parameters, jet parameters, and hot gas distribution, respectively). These system constraints will hopefully offer some physical insight into the interaction-activity connection.

2 MODELING BENT JETS

We have developed a general numerical simulation algorithm for modeling the propagation and morphology of ballistic radio jets in colliding galaxies. This algorithm has already been used successfully to fit the specific two-sided jet morphology seen in the radio source 3C278 associated with NGC 4782 (Borne and Colina 1993). In our model the morphological evolution of the jets is determined by their response to the simple mechanical forces (*i.e.*, gravity and ram pressure) imposed on them from both the host and the companion galaxies. It is hoped that our studies of bent radio jets in colliding galaxies will ultimately lead to more quantitative determinations, both of (1) the physical conditions required for a galaxy collision to generate nonthermal activity in the nucleus of at least one member of the pair, and of (2) the physical processes associated with the ignition of the active nucleus.

3 THE ROLE OF HOT GAS

From our specific model of the two-sided radio jet structure associated with 3C278, it was determined that ram pressure deflection by the hot interstellar matter (ISM) is the dominant force affecting the morphology of the radio emission. We are using the *ROSAT* HRI to image these systems in order to address the following problems: (1) measuring the quantity of hot circumnuclear gas, one of the fundamental parameters controlling the onset of nonthermal activity in galactic nuclei; (2) mapping the hot gas in these colliding galaxies (thereby revealing the effects of the tidal interaction on this particular galactic component); (3) identifying hot spots in the X-ray surface brightness distribution (perhaps identifying regions of intense jet–ISM interaction); (4) correlating the hot gas distribution with the radio jet morphology (thus testing our ram pressure deflection models in detail); and (5) investigating the connection between the hot X-ray gas and the warm circumnuclear gas detected in our Hα images (thereby investigating the hypothesis of a cooling flow feeding the black hole).

4 SUMMARY

We are studying the optical, radio, and X-ray morphologies of a sample of close pairs of active galaxies. These data give us a general frame of reference from which we can study both the detailed gas dynamical behavior in generic collisions between galaxies and the potential source for the generation of nonthermal activity in AGNs.

This research was supported in part by NASA grant NAG5-1920 (to KB) and by an ESA postdoctoral fellowship (to LC).

REFERENCES

Borne, K. D., and Colina, L. 1993, *Ap. J.*, **416**, in press.
Colina, L., and Pérez-Fournon, I. 1990*a*, *Ap. J. Suppl.*, **72**, 41.
Colina, L., and Pérez-Fournon, I. 1990*b*, *Ap. J.*, **349**, 45.
Heckman, T. M. 1990, in *Paired and Interacting Galaxies*, ed. J. Sulentic, W. Keel, and C. Telesco (Washington: NASA), p. 359.
MacKenty, J. 1989, *Ap. J.*, **343**, 125.
Stockton, J. 1990, in *Dynamics and Interactions of Galaxies*, ed. R. Wielen (Berlin: Springer), p. 440.

Seyfert Nuclei in Interacting/Merging Galaxies

Wolfram Kollatschny and Klaus J. Fricke

Universitätssternwarte Göttingen, Geismarlandstr., D-37083 Göttingen, Germany

243 Seyfert galaxies of type 1, 2, or 3 are listed in the *Catalogue of Quasars and AGNs* of Veron and Veron (1989) having $m_v \leq 15$ and $v_{rad} \leq 20,000$ km s^{-1}. Eight of these Seyfert galaxies show two nuclei (IR 0248-11, Mkn 739, Mkn 266, Mkn 463, Mkn 673, Arp 220, NGC 6240, NGC 7593). The separations of the nuclei are 3 – 10 arcsec corresponding to 2 – 6 kpc.

To confirm the double nuclear structure for these disturbed galaxies we determined the internal velocity field to be sure that they are the result of two galaxies in the late stages of merging (Kollatschny *et al.* 1991).

Making the additional assumption that the Seyfert galaxies Mkn 231 (Kollatschny *et al.* 1992) and Mkn 273 are galaxies in the final stage of merging, having strong tidal arms but unresolved nuclei, one can estimate that 4 percent of all Seyfert galaxies are in the merging process.

The luminosities of the multiple nuclei Seyfert galaxies are extremely high in comparison to morphologically undisturbed Seyfert galaxies. In Table 1, mean values of the visual and blue luminosities and of the far-infrared and radio (6 cm) luminosities as well as the Hα fluxes are listed for both classes. We have separated Seyfert 1 and Seyfert 2 galaxies.

In all cases the luminosities of double nucleus Seyfert galaxies are higher by a factor of more than two with respect to 'undisturbed' Seyfert galaxies — except in the soft X-ray range (*ROSAT*) and the UV. This result might be explained by higher luminosities in the early phases of a Seyfert's life — under the assumption that the nonthermal activity is triggered by tidal interaction — and/or additional starburst phenomena.

On the other hand due to strong nuclear dust absorption, the UV spectra of these merging Seyfert nuclei are unusually weak; this may be valid for the X-ray flux too (Kollatschny *et al.* 1993).

TABLE 1

LUMINOSITY OF MULTIPLE NUCLEUS
AND UNDISTURBED SEYFERT GALAXIES

	multiple nucleus Seyfert	undisturbed nucleus Seyfert	
$\log L_V$ [W]	37.40	36.99	Seyfert 1
	37.14	36.66	Seyfert 2
$\log L_B$ [W]	37.05	36.64	Seyfert 1
	37.09	36.22	Seyfert 2
$\log L_{FIR}$ [W]	37.63	36.70	Seyfert 1
	37.88	36.76	Seyfert 2
$\log L_{radio,6cm}$ [W]	32.12	31.39	Seyfert 1
	32.81	31.71	Seyfert 2
$\log L_{H\alpha}$ [W]	35.96	35.21	Seyfert 1
	34.85	34.21	Seyfert 2
$\log L_{X-ray}$ [W]	(36.0)	37.5	Seyfert 1
	34.5	34.8	Seyfert 2

This work has been supported by DFG grant Ko857/13-1.

REFERENCES

Kollatschny, W. *et al.* 1993, in preparation.

Kollatschny, W., Dietrich, M., Borgeest, U., and Schramm, K. -J. 1991, *Astr. Ap.*, **249**, 57.

Kollatschny, W., Dietrich, M., and Hagen, H. 1992, *Astr. Ap.*, **264**, L5.

Veron-Cetty, M. P., Veron, P. 1989, *Catalogue of Quasars and Active Nuclei*, 4th edition, ESO.

Where Is the Induced Star Formation in Interacting Galaxies?

Sara C. Beck, and Orly Kovo

Tel Aviv University School of Physics and Astronomy of the Raymond and Beverly Sackler Faculty of Exact Sciences and the Wise Observatory

ABSTRACT

It is generally believed that galaxy interactions induce bursts of star formation. We observed a sample of galaxies undergoing different types of interactions in the expectation that the location and nature of the induced star formation could be related to the dynamics of the interaction. We found instead that in almost all galaxies the star formation is concentrated in the nucleus or nuclei, sometimes to a remarkable degree. It appears that extra-nuclear star formation is either difficult to trigger or so short-lived as to be rarely observed. We discuss in detail two galaxies: NGC 5253, site of the most concentrated star formation region yet known, and Arp 30, the exception where star formation is broadly distributed and is seen in both nuclei and in clumps in the bridge connecting them.

1 INTRODUCTION

Studies of the global properties of large samples of galaxies have shown that interactions between galaxies correlate with bursts of star formation and have lead to the generally accepted belief that interactions can trigger such bursts. Attempts to correlate the local star formation properties of individual galaxies and the interactions they have undergone have been much less successful. We therefore undertook a multi-wavelength study of star formation in a sample of interacting or post-interaction galaxies, in the hope of being able to relate the location and type of star formation to the interaction history. The observations include infrared spectra, radio continuum maps, and images in continuum bands, the Wolf-Rayet feature and Hα.

2 OBSERVATIONS AND RESULTS

The galaxies observed are not a complete sample of any one type of interaction but rather include examples of galaxies with the various features usually attributed to interactions such as optical tails, H I tails, multiple nuclei, rings, and bridges. We obtained images in the B, V, R, and I continua and in narrow filters (30 to 50 Angstroms) at Hα and the Wolf-Rayet feature from the Wise Observatory in Mitzpe Ramon, and examined them for faint structures such as plumes, tails and rings and for spectral evidence of star formation. We used multi-wavelength radio continuum maps and infrared spectroscopy, either obtained by ourselves or from the literature, to correct for extinction and to quantify the young stellar population.

We found that for most of the sample galaxies, the enhanced star formation was concentrated towards the nucleus regardless of the interaction and the morphology. Particularly interesting cases were NGC 4438, where the bent ring is bright in the continuum bands but Hα is seen only in the nucleus, and NGC 2782 where we found Hα in the nucleus and in clumps near it, but neither Hα, V, nor R emission in the large H I tail discovered by Smith (1991). The colors of NGC 4438 suggest that stars were formed in the ring in the past but that young stars at present are found only in the center. Similar findings in other galaxies argue that even when an interaction enhances off-nucleus star formation it will be comparatively short lived and thus hard to detect. The only clear exception in our sample is Arp 30, discussed below. In the case of NGC 2783 it appears that the tail is too uniform and of too low density to be the site of star formation clumps such as are seen in other tail galaxies, and we suggest that such quiescent tails, visible only in the H I, may be found in other galaxies that are not usually thought of as starbursts or interacting systems. (M82 is another case where optical and H I structures do not correlate: there is no coincidence between the Hα streamers and envelope we see and the H I Yun and Ho [1993] report, probably because the Hα is seen in the galactic superwind and the H I in the distorted disk). Results for the entire sample of galaxies will appear in Beck and Kovo (1993).

2.1 Exceptional Galaxies: NGC 5253 and Arp 30

Two galaxies in the sample were so unusual that they are the subjects of individual papers. The first is NGC 5253, a small blue amorphous galaxy which is close to M83 but has no obvious signs of interaction. It is the site of the most concentrated starburst and the hottest young stellar population we know. This galaxy and its problematic relation with the equally unusual but very different M83 are discussed in Beck *et al.* (1993). The second exceptional galaxy is Arp 30, an interacting system at $cz = 8,300$ km sec^{-1}. Arp 30 is a strong but not ultra-luminous starburst, about twice as luminous as the M82-M81 system, and the young stars are seen extended over both galaxies and in clumps in the bridge connecting them. This system is described in more detail in Kovo and Beck (1993).

This work was supported in part by the Bi-National Science Foundation grant 89-00070/3.

REFERENCES

Beck, S. C. and Kovo, O. 1993, in preparation.
Beck, S. C., Turner, J. T. and Ho, P. T. P. H. 1993, in preparation.
Kovo, O. and Beck, S. C. 1993, in preparation.
Smith, B. J. 1991, *Ap. J.*, **378**, 39.
Yun, M. S. and Ho, P. T. P. H 1993, *Ap. J. Lett*, submitted.

Interacting Galaxy Pairs and Seyfert Activity

S. M. Simkin, K. E. Haisch, and D. A. Ventimiglia

Michigan State University

ABSTRACT

The idea that Seyfert nuclear activity might be fueled by material inflow induced by a perturbing companion has been fashionable for over a decade. However, the recent literature on the prevalence of Seyfert galaxy companions, is confusing and somewhat contradictory. To clarify this we have constructed a statistical profile of Seyfert galaxy companions in the CfA ZCAT. We use a random sampling technique to estimate the statistical significance of the observed excess of Seyfert companions relative to the rate of pairing found in non-active galaxies. We find the excess of *close* pairs for Seyfert galaxies to be highly significant.

1 RESULTS

More than a decade ago, the morphological similarities between patterns seen in Seyfert galaxies and those produced by gravitational forcing led to the suggestion that Seyfert activity might be fueled by material inflow induced by either a central bar or a perturbing companion (Simkin, Su, and Schwarz 1980). The recent literature on the prevalence of Seyfert galaxy companions, however, is somewhat confusing and, at initial glance, contradictory (*cf.* the extensive review in Osterbrock 1991). Most authors have attributed their disparate conclusions to observational selection effects. (op. cit.). The problems arise because to obtain a valid statistical profile of the non-AGN "control" galaxies requires either an enormous observational effort or a series of *ad hoc* assumptions which differ with each study.

An alternative approach to this statistical problem is to draw the Seyfert galaxy sample from a larger catalog of galaxies all subject to the same measurement errors and selection effects. With this approach, the non-Seyfert galaxies provide the statistical profile necessary for interpretation. We have used this method to analyze a sample of 79 Seyfert galaxies from the CfA redshift catalog (Huchra *et al.* 1982). This sample includes all non-Virgo Seyferts with $cz \leq 4,000$ km s^{-1}, included in the "Catalog of Seyfert Galaxies and Related Objects" by N. Kaneko.

After reviewing the pairing criteria used in earlier studies, we adopted the following selection rules to identify possible physical pairs. (These are independent of H_o, but for purposes of discussion all physical distances are referred to $H_o = 100$):

$$|cz_{sey} - cz_{comp}| \leq 1,000 \text{ km s}^{-1} \text{ , and angular separation} \leq 72(cz)^{-1} \text{ km s}^{-1}$$

With these criteria, 67 of the 79 Seyferts have one or more companions. To assess the significance of this, we randomly chose 1,000 sets of 79 control galaxies from amongst the non-Seyfert galaxies in the catalog (excluding those in the Virgo region). The galaxies in each set were matched one-to-one with the Seyfert galaxies according to the following criteria:

$$|cz_{\text{sey}} - cz_{\text{control}}| \leq 200 \text{ km s}^{-1} \; ; \; |T_{\text{sey}} - T_{\text{control}}| \leq 1 \text{ and } |(b/a)_{\text{sey}} - (b/a)_{\text{control}}| \leq 0.1,$$

where b/a is the axis ratio and T is the de Vaucouleurs T-type.

The resulting statistical profile can be approximated by a Gaussian centered at 63.2 with $\sigma = 3$. Thus, there clearly is no excess of companions for Seyfert galaxies using the broad definition above. However, a plot of the number of companions for the "controls" and the Seyferts as a function of projected separation (Fig. 1) shows that there is a *significant excess of close companions* for the Seyferts. Using the sampling technique outlined above, we derive the probability distribution shown in Figure 2 for close projected companions in a "random" sample of non-Seyfert galaxies from the CfA. The mean of this distribution is 15.10 with $\sigma = 3.19$. Thus the number of Seyfert pairs exceeds that for the controls by more than 4σ and the excess is significant at the level of more than one in 10^4.

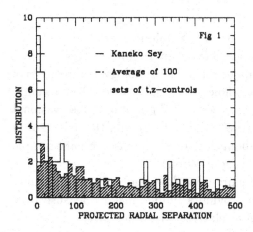
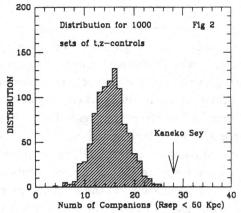

REFERENCES

Huchra, J. P., Wyatt, W. F., and Davis, M. 1982, *A. J.*, **87**, 1628.
Osterbrock, D. E. 1991, *Reports on Progress in Physics*, **54**, 579.
Simkin, S. M., Su, H -J., and Schwarz, M. P. 1980, *Ap. J.*, **237**, 404.

Searching for Mass Transfer in E+S Pairs

D. F. de Mello Rabaça[1,2], J. W. Sulentic[1], R. Rampazzo[3], and
Ph. Prugniel[4]

[1]Department of Physics and Astronomy, University of Alabama, U. S. A.
[2]IAG–USP, São Paulo, Brazil
[3]Osservatorio Astronomico di Brera, Milano, Italy
[4]Observatoire de Haute-Provence, St. Michel L'Observatoire, France

ABSTRACT

We are investigating the properties of gas in the elliptical (E) components of E+S pairs. This is being done both by determining the physical conditions of the gas and by spectral synthesis of the underlying stellar population. A major goal is to determine whether cross fueling occurs in such pairs. We present data for AM 0327-285 which is an E+S pair showing signs of tidal interaction between the E and S components. We present evidence for gas transfer from the spiral to the elliptical component.

1 INTRODUCTION

Paired galaxies as nonequilibrium systems provide a unique chance to study the physics of interaction, galaxy evolution, and the effects of environment on galaxies. If environment at the formation epoch plays a major role on morphology, components of binary galaxies might be expected to show morphological concordance. More recently the formation of elliptical galaxies has been revisited in the light of new observational evidence for fine structure suggesting a "nurture" origin. Therefore, the existence of a large number of true mixed pairs (E+S) (about 25% of an unbiased binary sample are mixed pairs [Sulentic 1990]) raises interesting questions.

We are studying the properties of mixed pairs using low resolution spectroscopy and stellar population synthesis. Such pairs provide a unique insight because they involve one gas rich galaxy in the presence of a relatively clean perturber.

2 OBSERVATIONS AND REDUCTION

We observed a sample of more than 20 mixed pairs using the Boller and Chivens Spectrograph at the 1.52m ESO telescope (La Silla). The spectral range was 4000—9500Å, with a dispersion of 3.7Å pixel^{-1}. The observations were done with the slit aligned along the major axis of the pairs. Spectrophotometric standard stars were observed from the list of Bica and Alloin (1986). Template galaxies were observed in order to verify the compatibility of our data with the results in Bica (1988). We report here on observations of the mixed pair AM 0327-285. It shows signs of tidal interaction and there is evidence for a luminous connection between the two galaxies raising the possibility of cross fueling.

392

3 RESULTS

Preliminary stellar population analysis was done by comparing the spectrum of AM 0327-285 with the ones presented in Bica (1988). Bica has grouped galaxies according to their spectral properties, morphological type and luminosity class. Groups E1 to E4 correspond to early–type galaxies which follow the normal metallicity vs luminosity relation, being dominated by an old stellar population. Groups E5 to E8 mostly show age effects. Late–type galaxies are classified similarly (see Bica 1988 for details).

We used strong absorption features such as CN 4200Å, the G band of CH 4301Å, Mg I+MgH 5175Å and Ca II 8542, 8662Å for comparison. The elliptical galaxy in AM 0327-285 was classified in the group E3 which has no significant contribution from a young population ($T \leq 5 \times 10^8$ yr). This argues against a significant stellar population due to cross fueling in the past few 10^8 years.

In the same spectrum we detect emission from Hα, [N II] $\lambda6548$, 6583, and [S II] $\lambda6717$, 6731. [N II] $\lambda6583$/Hα ratios consistent with an H II region are found in the spiral arm emission knots. The mean redshifts for the spiral and elliptical components derived from the nuclear region in each galaxy are very similar. We see quasi-continuous emission between the two nuclei, especially, in the light of [N II] $\lambda6583$ and the [S II] doublet. The feature is roughly constant in velocity. This is one of the first examples of shocked gas possibly linking two galactic nuclei. Further confirmation is required.

The spectrum of the spiral galaxy was classified as S3 which also suggests little recent star formation. In fact, the spectra of the two galaxies have comparable features suggesting that the star formation histories of the two nuclei are similar. Thus a Holmberg effect usually measured from broad band photometry, may be evidenced in the stellar population analysis.

Our results are still preliminary and more pairs will be analysed using the same criteria in order to verify if other mixed pairs present the same properties as AM 0327-285.

Acknowledgements: We thank Dr. E. Bica for helpful discussions. Extragalactic studies at U. Alabama are supported by NSF/State of Alabama EPSCoR Grant RII-8996152. DFMR thanks financial support by the Brazilian institution CNPq.

REFERENCES

Bica, E., and Alloin, D. 1986, *Astr. Ap.*, **162**, 21.
Bica, E. 1988, *Astr. Ap.*, **195**, 76.
Sulentic, J. W. 1990, *Paired and Interacting Galaxies*, IAU Colloq. 124, ed. J. W. Sulentic, W. C. Keel, and C. M. Telesco (NASA CP-3098), p. 291.

K542, A Hierarchical Pair with Mass Transfer?

Victor Andersen[1], Jack W. Sulentic[1], and Roberto Rampazzo[2]

[1] Department of Physics and Astronomy, University of Alabama, U. S. A.
[2] Osservatorio Astronomico di Brera, Milano, Italy

ABSTRACT

Preliminary results of a search for gas transfer between the galaxies in E+S pairs are presented. In the pair considered K542, ionized gas is detected in the nucleus of the elliptical but no gas in the process of being transferred between the two components has been found.

1 INTRODUCTION

In the past ten years observations of the ionized (Phillips *et al.* 1986) and H I (Knapp *et al.* 1985) gas content of early type galaxies have shown that 50-60% of these galaxies contain a modest amount of interstellar gas. On the basis of the H I data, Knapp and coworkers have suggested that the interstellar gas in ellipticals may be of external origin. At least two external sources for the gas have been suggested: the capture of small gas-rich companions and accretion from nearby gas-rich galaxies. Mixed morphology E+S pairs make an ideal laboratory to test the second hypothesis, since the presumed external source for the gas is relatively unambigous. Two possible strategies exist to test the hypothesis, 1) define a sample of E+S pairs and a control sample of isolated ellipticals and compare the global gas content of the two samples, or 2) examine a sample of E+S pairs that are likely candidates to be currently cross-fueling and attempt to catch them in the act. We have chosen the second strategy, since optical spectra had already been obtained for several E+S pairs.

2 K542: BASIC DATA

K542 is a hierarchical pair from the Catalog of Isolated Pairs (Karachentsev 1972) consisting of a spiral galaxy with a compact elliptical companion. K542 was selected as a promising place to search for mass transfer because of its separation of $8h^{-1}$ kpc ($h = H_0/100$ km sec^{-1} Mpc^{-1}), radial velocity difference of 130 km sec^{-1} and asymmetric H I profile (Haynes and Giovanelli 1991).

Optical spectra were obtained using the twin spectrograph on the 3.5m Calar Alto Telescope. The two wavelength regions covered were approximately 4840Å to 5390Å and 6500Å to 7440Å. Two exposures where taken, one with the slit situated across the S and the E, and a second approximately perpendicular to the first through the E only.

3 THE SEARCH FOR GAS TRANSFER

After reduction, the red spectra were examined to determine if there was any evidence for emission line gas being actively transferred to the elliptical from the spiral. No emission line gas was detected in the area between the spiral and the elliptical, however, we did detect [N II] $\lambda = 6583$Å emission in the elliptical galaxy. The spectra from the nucleus of the elliptical show a relatively broad (FWHM ~ 600 km sec^{-1}) [N II] emission feature, along with a weaker Hα emission feature superimposed upon strong Hα absorption.

With the slit oriented through the elliptical galaxy only, the emission region was unresolved, with the emission detectable at the center of the elliptical only. With the slit oriented across both galaxies, however, the emission lines were traceable to approximately 4″ on either side of the elliptical's nucleus. In addition, the gas with higher recession velocities was found to be preferentially on the side of the elliptical away from the spiral, while the lower velocity gas was found nearer to the spiral.

4 CONCLUSIONS

No evidence for gas which *currently* is being transferred between the two galaxies has been found. Although it is possible that the source of the gas found in the elliptical is the spiral galaxy of the pair, the high frequency of the presence of such emission features in early type galaxies makes this diagnosis highly uncertain.

The kinematics of the gas observed in the elliptical allows for two possible interpretations. The first is that the gas has recently been acquired from the spiral and has not yet had time to settle to the center. The second is simply that the gas lies in a disk which is rotating around an axis parallel to the major axis of the spiral. Again, Phillips *et al.* (1986) found that many of the galaxies detected showed signs of disk-like rotation by the gas in the inner kiloparsec or so of the galaxy. This suggests that the latter may be the preferred interpretation of the data.

REFERENCES

Haynes, M. P., and Giovanelli, R. 1991, *Ap. J. Suppl.*, **77**, 331.
Karachentsev, I. 1972, *Comm. Spec. Astrophys. Obs.*, **7**, 1 (CPG).
Knapp, G. R., Turner, E. L., and Cunniffe, P. E. 1985, *A. J.*, **90**, 454.
Phillips, M. M., Jenkins, C. R., Dopita, M. A., Sadler, E. M., and Binette, L. 1986, *A. J.*, **91**, 1062.

Kar 29: Tidal Effects from a Second or Third Party

P. Marziani[1], W. C. Keel[1], D. Dultzin–Hacyan[2], and J. W. Sulentic[1]

[1]Department of Physics and Astronomy, University of Alabama, Tuscaloosa,
 AL 35487–0324, U. S. A.
[2]Instituto de Astronomia, U. N. A. M., Mexico City, Mexico

ABSTRACT

We describe the results of an imaging and spectrophotometric investigation of the mixed elliptical–spiral pair Kar 29 ($\equiv$ VV 347 $\equiv$ Arp 119). The spiral component ($\equiv$ Mrk 984) shows a strong, extended, LINER-like emission line spectrum. Each line is partly resolved into at least four components, redshifted with respect to the underlying galaxy, and covering $\Delta v_\mathrm{r} \sim 1,300$ km s^{-1}. Line ratios indicate that the dominant ionization mechanism is provided by shocks or a mixture of shocks and photoionization by hot stars. One possible interpretation involves a nearly polar crossing of the spiral disk by the elliptical, with the line emitting gas stripped and accelerated toward the elliptical. Some of the data are however better explained if the large Δv_r between the line components is due to the disk impact of an additional small companion.

1 INTRODUCTION AND OBSERVATIONS

Interaction between galaxies produces a wide variety of observable phenomena; however, most of them are not yet fully understood in physical terms. One important challenge is to explain the response of the gaseous component of a disk galaxy to strong perturbations.

Long slit CCD spectra of the spiral (classified as Sc *pec* or Sdm)/elliptical (E5) pair of giant galaxies Kar 29 were obtained at *KPNO*, San Pedro Martir and ESO. Slits were oriented (1) along the major axis of the pair (P.A. = 8°), (2) along the major axis of the spiral (116°), (3) and at P.A. = 82° (with the slit centered on the knot closest to the nucleus), during several observing runs in 1991/92. The resolution was $\sim 3.5 - 4$Å FWHM for all spectra. CCD images in the B and V bands were collected at *KPNO* by N. Sharp and one of us (JWS) in 1986.

2 RESULTS AND DISCUSSION

The two components of Kar 29 are separated by $\approx 54''$ ($\approx 38\ h^{-1}$ kpc of projected linear distance), and have $\Delta v_\mathrm{r,E-s} \approx 750$ km s^{-1}. Two bright knots are visible to the north of the nucleus of the spiral component ($\equiv$ Mrk 984), approximately along the major axis of the pair; the nucleus itself is visibly off the geometric center of the galaxy. Along the major axis of the pair, the Hα, [N II] $\lambda\lambda$6548, 6583, [S II] $\lambda\lambda$6716, 6731, and [O I] $\lambda\lambda$6300, 6363 lines show structured profiles over a radial velocity range of 1,200 km s^{-1}. The emission extends over 11″ and encompasses the nucleus of Mrk 984 and the knot closest to it. At least four components are contributing to each emission line. The strongest ones are at the most extreme radial velocities, namely $v_r \approx 14,250$ and 15,500 km s^{-1}. All line components are redshifted with respect to the underlying galaxy, whose v_r is $\approx 14,200$ km s^{-1}. Diagnostic diagrams (following to the prescription of Osterbrock and Veilleux 1987) suggest that the dominant ionization mechanism is due to shocks, or to a mixture of shocks and photoionization by hot stars (with the exception of the highest v_r component). No high-ionization gas — suggestive of Seyfert-type activity — has been revealed. All other regions detected across the disk of the galaxy have regular, unresolved profiles and line ratios typical of H II regions. The global star formation rate (SFR) deduced from the FIR luminosity is $\sim 13\ h^{-2}$ M$_\odot$ yr^{-1}. The Hα luminosity on the western side of the galaxy, in the areas covered by our spectra, is as large as that expected for an entire Sd galaxy (SFR ~ 1.5 M$_\odot$ yr^{-1}). These values points toward an enhancement of the SFR.

The multi–peaked appearance of the emission line profiles is reminiscent of super-winds in powerful FIR galaxies, but the lack of blueshifted line components makes this scenario highly implausible. Interaction with a hypothetical third companion could easily explain the existence of several line components. In this view, the intermediate radial velocity components are due to the collision of giant gas clouds or even to part of the disk of the companion. The hypothetical companion should be associated with the highest radial velocity component. This hypothesis is appealing because line ratios for this component are suggestive of H II regions: we may be observing a burst of star formation in the disk of the small intruder. The lack of H I emission at $v_r \gtrsim 14,500$ km s^{-1} (Bothun *et al.* 1984) argues against this view. There is also no clear evidence of a companion in our CCD images. Most of the peculiar features observed on the western side of the spiral are better explained in terms of interaction with the elliptical (E) companion. There are two ring–like structures that are similar to the ones observed in NGC 1144, a perturbed spiral galaxy that collided nearly pole on with an elliptical. A line of faint knots skirting the spiral arms south of the Mrk 984 nucleus indicates that a radial perturbation propagated throught the disk. Viewing Mrk 984 to a nearly edge–on ring galaxy requires that the E-companion has crossed the spiral in a nearly polar encounter. In this case, we must explain why the E-galaxy shows no signs of morphological disturbances: no significant isophotal twisting, no variation of isophotal ellipticity, and only minor deviations with respect to standard $R^{1/4}$ profiles. The relatively high velocity of the encounter $\Delta v \gtrsim 750$ km s^{-1} may have allowed the E to emerge unscathed, without more severe damages to the spiral.

REFERENCES

Bothun G. D., Heckman, T. M., Schommer, R. A., and Balick, B. 1984, *A. J.*, **89**, 1293.

Osterbrock, D. E., and Veilleux, S. 1987, *Ap. J. Suppl.*, **63**, 295.

The Fundamental Plane and Early-Type Galaxies in Binaries

Roberto Rampazzo[1], Pierpaolo Bonfanti[2] and Luca Reduzzi[2]

[1]Osservatorio Astronomico di Brera, Milano, Italy
[2]Dipartimento di Fisica, Univ. di Milano, Italy

ABSTRACT

We use published data for 31 early-type members of binary systems in order to estimate the slope of the (D_n, σ) relation. This is considered to be a representation of the Fundamental Plane (FP) for ellipticals. We find a slope for this relation of $a = 0.92 \pm 0.18$ when computed with a simple model for the distances. A similar slope has been obtained by others for galaxy groups and it is comparable to values obtained for clusters with a low Abell richness. The scatter of pair values around the FP does not correlate with galaxy properties such as ellipticity, isophotal twisting and total color index. The larger deviations from the (D_n, σ) relation tend to involve pairs with smaller projected separations.

1 INTRODUCTION

Concerns about independence of the FP from environmental conditions have been expressed by Djorgovski *et al.* (1988), Lucey *et al.* (1991*a, b*) and De Carvalho and Djorgovski (1992). On the contrary, Burstein *et al.* (1990) suggest that the FP does not depend on environment demonstrating, in particular, that the (D_n, σ) relation does not depend upon cluster properties.

Isolated binaries represent a different environment from that of a cluster center where galaxies have had time to homogenize. Binary galaxy evolution is driven by a combination of three time scales which are roughly of the same order: the orbital period, the rotational period of the individual galaxies and the burst duration of star formation (SF) triggered by the encounters. Theoretical simulations suggest that they merge rapidly (compared to a Hubble time) into E galaxies, because tidal friction is very efficient. General expectations are that pairs will show a larger dispersion around the FP than cluster ellipticals. This scatter would be expected to exist both in σ due to the role of tidal friction and also in D_n due to less homogeneity in the photometric properties. Further, supposing the existence of a difference in slope between cluster and binary members, it is expected that the binary slope will be more similar to values for the field or for ellipticals in groups, since studies of individual objects suggest that the *population of E's* is not very different for these environments.

2 THE FP SLOPE AND SOURCES OF SCATTER IN THE FP

We tried to verify the above expectations about (D_n, σ) plane for early-type in binaries. We determined the FP from data in Faber *et al.* (1989) (FWBDDLT hereafter). Sixteen Karachentsev (1972) isolated binaries (23 early-type galaxies) are present in the above sample. We added to this sample a set of hierarchical pairs from Prugniel *et al.* (1989), five of which (8 galaxies) are present in FWBDDLT. The slope obatained for the FP is $a = 0.92 \pm 0.18$. If we reverse the relation, using now $\log D_n$ as the independent variable, the slope of the FP for pairs is $a = 0.52 \pm 0.10$. The coefficients of the regression line interpolating $\Delta \log D_n = a\mu_e + b$ are $a = -0.039 \pm 0.041$ and $b = 0.829 \pm 0.890$ and indicate that (D_n, σ) relation, we obtained, is an edge-on view of the FP and that the scatter in points cannot be reduced if we tilt it adequately. We searched for possible correlation with the scatter in the FP of binary galaxies. Residuals do not appear to correlate with any of the properties, including ellipticity, maximum twisting or corrected colors. No correlation is observed with pair separation although larger deviations from the FP appear to be present for very small projected separations (less than 20 kpc). The slope of FP we obtained is consistent with the value of $a = 0.98 \pm 0.06$ (Weigelt and Kates 1990) for members of loose groups and with that of Laurikainen (1990) for isolated objects. The slope of the FP computed for E's in pairs is also comparable, within the errors, with clusters like Abell 194 ($a = 0.83 \pm 0.25$), Abell 2199 ($a = 0.88 \pm 0.41$), DC 2345-28 ($a = 0.88 \pm 0.36$) and Eridanus ($a = 1.02 \pm 0.29$) (Burstein *et al.* 1990). All of these clusters have a low Abell richness. The FP slope for E's in pairs appears to be different from values for richer clusters like Coma, Virgo and Fornax where $\langle a \rangle = 1.3$.

REFERENCES

Burstein, D., Faber, S. M., and Dressler, A. 1990, *Ap. J.*, **354**, 18.

Djorgovski, S., De Carvalho, R., and Han, M. -S. 1988, in *Extragalactic Distance Scale*, eds. S. van den Bergh, and C. Pritchet (San-Francisco: ASP Conf. Ser.), 329.

De Carvalho R., and Djorgovski, S. 1992, *Ap. J. Lett.*, **389**, L49.

Faber, S. M., Wegner, G., Burstein, D., Davies, R. L., Dressler, A., Lynden-Bell, D., and Terlevich, R. 1989, *Ap. J. Suppl.*, **69**, 763.

Karachentsev, I. D. 1972, *Catalogue of Isolated Pairs of Galaxies in the Northern Hemisphere, Soobshch. Spets. Astrophiz. Obs.*, **7**, 3.

Laurikainen E. 1990, *Astr. Ap.*, **232**, 323.

Lucey, J. R., Bower, R. G. and Ellis, R. S. 1991a, *Mon. Not. R. Astr. Soc.*, **249**, 755.

Lucey, J. R., Guzmán, R., Carter, D., and Terlevich, R. J. 1991b, *Mon. Not. R. Astr. Soc.*, **253**, 584.

Prugniel, P., Davoust, E., and Nieto, J. -L. 1989, *Astr. Ap.*, **222**, 5.

Weigelt, U., and Kates, R. 1990, *Astr. Ap.*, **240**, 1.

Dumbbell Galaxies and Multiple Nuclei in Rich Clusters: Radio Data

L. Gregorini[1,2], R. D. Ekers[3], H. R. de Ruiter[4], P. Parma[1], E. M. Sadler[5], and G. Vettolani[1]

[1] Istituto di Radioastronomia, Bologna, Italy
[2] Dipartimento di Fisica, Universitá di Bologna, Italy
[3] Australia Telescope National Facility, Epping, Australia
[4] Osservatorio Astronomico, Bologna, Italy
[5] Anglo-Australian Observatory, Epping, Australia

ABSTRACT

We present preliminar results of radio observations of 78 southern rich clusters, whose brightest member is a dumbbell galaxy or a multiple nucleus. We identified 41 radio sources with the cluster brightest member: 23 of the 44 observed have a multiple nucleus, and 18 of the 34 mapped have a dumbbell galaxy.

1 INTRODUCTION

In many galaxy clusters, the first–ranked galaxy is not a single isolated object but has two or more components. Such galaxies show a wide range of morphologies, from *dumbbell systems* (two galaxies of roughly equal brightness ($\Delta m < 1 - 2$) inside a common halo) to galaxies with *multiple nuclei* (two or more condensations visible within the image of a single galactic spheroid, with each secondary nucleus at least two magnitudes fainter than the main system).

It is likely that in some multiple systems the companions are gravitationally bound to the central galaxy (often a cD) and may eventually be cannibalized; while in others we see unbound galaxies whose eccentric orbits in the cluster potential well bring them close to the cluster center where the cD is located (Tonry 1985).

Since many active galaxies are located in dense environments and show signs of interaction, it has often been suggested that gravitational interactions between galaxies may trigger nuclear activity.

To test whether there is a real excess of radio sources in dumbbell galaxies or those with multiple nuclei, we need to observe a complete and unbiased sample of these objects, but most studies of multiple central cluster galaxies have been done only recently and use inhomogeneous samples — since radio galaxies are often associated with dumbbells, for example, many of the known examples come from optical identifications of radio sources.

2 SAMPLES AND RADIO OBSERVATIONS

Because of the lack of existing homogeneous and unbiased samples, Gregorini *et al.* (1992) recently compiled two new, optically–selected samples of dumbbell systems in rich clusters, using the high–quality plate material now available for the southern sky and the Abell *et al.* (1989) catalogue of rich clusters in the southern hemisphere. The first sample is *volume–limited* and contains 17 confirmed and a further 10 possible *dumbbell galaxies* identified in the 171 clusters south of declination $-17°$ and within a comoving distance of 210 h^{-1} Mpc, regardless of Bautz–Morgan class. We observed these clusters at 6 cm with the *Very Large Array or the Australia Telescope*. The second sample contains the 70 *dumbbell systems* found in all 381 clusters of BM type I or I–II south of declination $-27°$, irrespective of distance. Observations at 6 cm were performed with the *Very Large Array* for the clusters with declination north to $-35°$. Other 44 clusters were chosen among all the clusters we classified as a representative sample of clusters in which the central galaxy is not a dumbbell, but may have a *multiple nucleus* or one or more companions. These clusters were observed with *Very Large Array* at 6 cm. Here we present a preliminary analysis of the *VLA* observations, which have a resolution of about 8–20 arcsec and r.m.s. noise of about 0.08 mJy/beam.

3 RESULTS

For each brightest member (dumbbell or multiple nucleus) we listed the nearest radio source within a radius of 9 arcmin, then we calculated the likelihood ratio, *i.e.* the ratio of the probabilities that a radio source is a true identification or a spurious background object (de Ruiter *et al.* 1977). Taking into account also radio sources with complicated and extended structure we found 41 identifications among the 78 observed clusters. The overall reliability of the identifications is 95% (only one object among the 41 proposed identifications is expected to be spurious), while the completeness is at at least 99%. Among these 41 radio sources identified with the cluster brightest member we have *23 multiple nuclei* among the 44 observed, and *18 dumbbell galaxies* among the 34 mapped. The characteristics of their radio emissions are: power in the range 10^{21}–10^{24} W Hz^{-1} and projected dimensions less than 100 kpc.

REFERENCES

Abell, G. O., Corwin, H., and Olowin, R. 1989, *Ap. J. Suppl.*, **70**, 1.
de Ruiter, H. R., Willis, A. G., and Arp, H. C. 1977, *Astr. Ap. Suppl.*, **28**, 211.
Gregorini, L., Vettolani, G., de Ruiter, H. R., and Parma, P. 1992, *Astr. Ap. Suppl.*, **95**, 1.
Tonry, J. L. 1985, *Ap. J.*, **291**, 45.

Tidal Deformation of Galaxies in Binary Systems

John D. Capriotti[1], and Wayne A. Stein[1,2]

[1]Department of Physics, University of Minnesota
[2]Center for Astrophysics and Space Sciences, University of California at San Diego

ABSTRACT

It has been claimed that spherically symmetric, isothermal ($\rho \sim r^{-2}$) halos of invisible material surround certain galaxies and extend to between $\approx$ 40 kpc (Rubin *et al.* 1985) and $\approx$ 1 Mpc (Charlton and Salpeter 1991) from their centers. In this work in progress we consider the tidal effects due to the presence of such halos in binary galaxies. Having investigated the predicted frequency of tidal distortions we then compare with binary galaxy surveys to search for evidence of these effects.

1 TIDAL RADIUS

A tidal radius was calculated based upon a simplified model of a binary galaxy system. Each member of the system is identical and in a circular orbit about the system's center of mass. The dark matter halos extend to the point where they just begin to overlap and the visible disk of each galaxy, despite rotating on its axis, is taken to be approximately spherically symmetric. Furthermore, the rotational angular momentum vector of the material contained in each visible disk is taken to be parallel to the orbital angular momentum vector for the entire system. This information was used to determine the radial equation of motion for a test particle located at the outer edge of either visible disk and from this equation the tidal radius is obtained ($\approx$ 40 kpc). The tidal radius is the separation between members of the binary system at which the test particle just begins to leave the visible disk.

The timescale over which a test particle moves a significant distance from the edge of its galaxy's visible disk when the system separation is at or less than the tidal radius is significantly less ($\approx$ 10%) than one orbital period and thus signs of tidal distortion should arise in a single orbit.

2 BINARY GALAXY SURVEYS

Having considered the tidal radii of binary systems such as the ones described above, observational studies (Schweizer 1987; Schneider and Salpeter 1992) of binary systems were consulted. The galaxy pairs examined in each of these studies were chosen so that there was a high probability that they were bound and isolated from the perturbing influences of other galaxies. After checking for redundancy the galaxies in these two studies were combined and comprise a sample of 142 binary systems. A histogram was constructed and fit with a lognormal function (Schweizer 1987) in order to determine the probability distribution in projected separation for the members of the sample. Under the assumption that the separation vectors of the galaxies in the sample are isotropically distributed in space and with the help of an iterative scheme developed by Lucy (1974), the probability distribution in actual separation was determined.

3 FREQUENCY OF TIDAL DISTORTION

With the knowledge of the probability distributions for projected and actual separations and the tidal radius, one can probabilistically determine the fraction of the galaxies in the sample that should show signs of tidal distortion. This fraction was found to be about twice that indicated by the comments on tidal distortion in the two previously mentioned galaxy surveys. Thus, based upon the above model, the presence of halos of size $\approx$ 20 kpc about the galaxies in the sample implies a frequency of tidal distortion roughly twice as great as that observed. Further matters deserve attention such as angular momentum orientation.

This work constitutes the Ph.D. thesis of J. D. Capriotti at the University of Minnesota.

REFERENCES

Charlton, J., and Salpeter, E. E. 1991, *Ap. J.*, **375**, 517.
Lucy, L. B. 1974, *A. J.*, **79**, 745.
Rubin, V. C., Burstein, D., Ford, W. K., and Thonard, N. 1985, *Ap. J.*, **289**, 81.
Schneider, S. E., and Salpeter, E. E. 1992, *Ap. J.*, **395**, 32.
Schweizer, L. Y. 1987, *Ap. J. Suppl.*, **64**, 427.

Formation of Dwarf Galaxies During Close Tidal Encounters

Michele Kaufman[1], Bruce G. Elmegreen[2], and Magnus Thomasson[3]

[1]Department of Physics, Ohio State University, Columbus, OH, U. S. A.
[2]IBM – Watson Research Center, Yorktown Heights, NY, U. S. A.
[3]Nordita, Copenhagen, Denmark

ABSTRACT

Galaxy interactions that agitate the interstellar medium by increasing the gas velocity dispersion and removing peripheral gas in tidal arms can lead to the formation and possible ejection of self-gravitationally bound cloud complexes with masses in excess of 10^8 $M_\odot$. Some of these complexes may eventually appear as independent dwarf galaxies.

1 MASSIVE CLOUDS IN IC 2163/NGC 2207

VLA H I observations (Elmegreen *et al.* 1993*a*) reveal 10 clouds each with H I mass $> 10^8$ $M_\odot$ in the outer parts and in the main disks of the interacting galaxy pair IC 2163/NGC 2207. Our observations apparently catch this pair in the early stages of massive cloud formation. The clouds, which are comparable in mass to dwarf galaxies, are fundamentally clumps in the gas, not clumps in the stellar component. The H I velocity dispersion in the clouds and in much of the main disk of NGC 2207 is, typically, 40 km s^{-1}, a factor of 4 times higher than in normal disk galaxies. We propose that the high velocity dispersion of the gas is the key to why these clouds are at least 10 times more massive than the largest clouds in normal disk galaxies: the Jeans mass scales as the fourth power of the effective velocity dispersion. Such massive clouds can form by common gravitational instabilities where the Jeans mass is high and where the local value of the instability parameter for the gas is below threshold, *e.g.* in the outer disk and arms. Some of the massive clouds may later become large star formation complexes.

2 *N*-BODY SIMULATION

An *N*-body gas + star simulation of an encounter between a companion and a galaxy with *an extended gas disk* reproduces the elevated velocity dispersions and the large cloud masses. The velocity dispersions and the cloud masses increase with the strength of the perturbation. Our simulations suggest that giant complexes in the tidal tail can escape to form independent dwarf galaxies if the companion mass is comparable to or larger than the galaxy mass. The model (see Fig. 1) also forms an extended 10^9 $M_\odot$ pool of gas at the end of the tidal arm opposite the companion, as most of the outer gas disk moves around to join the tidal tail. This gas pool resembles the massive H I blob observed by van der Hulst (unpublished; see Mirabel 1993) at the end of the tidal tail in the Antennae system.

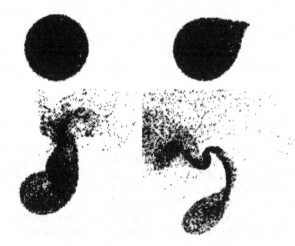

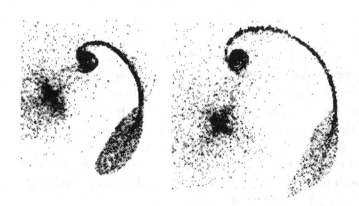

Fig. 1—Six epochs showing the effect of the encounter on the gas particles. The grid extends to 100 kpc. An extended pool of gas develops at the end of the tidal tail. The companion (not shown) has 1.4 times the mass of the galaxy and is located at the swarm of particles left of center in the bottom two frames.

For more details see Elmegreen *et al.* (1993*b*).

This work was supported in part by NSF Grant AST-8914069 to M. K.

REFERENCES

Elmegreen, D. M., Kaufman, M., Brinks, E., Elmegreen, B. G., and Sundin, M. 1993*a*, in preparation.

Elmegreen, B. G., Kaufman, M., and Thomasson, M. 1993*b*, *Ap. J.*, **412**, 90.

Mirabel, I. F. 1993, these proceedings.

Gas Fueling to the Central 10 pc in Merging Galaxies

Kenji Bekki, and Masafumi Noguchi

Astronomical Institute, Tohoku University, Sendai, Japan

ABSTRACT

Some of ultra-luminous galaxy mergers show a sign of quasar-like activity. We have numerically investigated the dynamical evolution of the interstellar gas in the late phase of disk galaxy mergers in order to clarify how the gas dynamics is related to the triggering of quasar-like activity. It is found that in most cases the efficient gas infall to the central 10 pc of the nucleus is realized only after the coalescence of two galaxy cores. This suggests that the quasar-like activity tends to appear only after the merger has been completed.

1 MODELS

The cloud particle scheme has been used to follow the gas response to the time dependent gravitational field made by two galaxy cores spiraling into the mass center of the system. We have neglected the influence of galactic disks because the inner region of a late phase merger is likely to be dynamically dominated by spherical components (*e.g.*, bulges and nuclei). The spherical component (hereafter core) of each galaxy is assumed to be rigid. The self-gravity of the gas has been neglected. The dissipational nature of gas has been included by making gas clouds collide inelastically. Gas clouds are initially distributed in a disk around one of two nuclei. We counted those gas clouds satisfying both of the following conditions, as the clouds have been swallowed by the seed black hole (BH) located at the nucleus:

(1) The cloud is located within 10 pc from the nucleus. This is the radius where $\sim 10^8$ M$_\odot$ BH starts to dominate the gravitational field of the core.

(2) The sum of the kinetic energy and the potential energy due to the BH gravity is negative (*i.e.*, the cloud is gravitationally bound to the BH).

We have examined the dependence of the fueling process on model parameters such as the spin configuration of the gas cloud disk, the scale length of galaxy core, and coalescence rate of two cores.

2 RESULTS

1) Two galaxy cores sinking to the center effectively heat the gas clouds dynam-

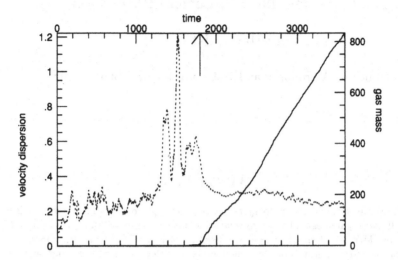

Fig. 1—The mass of the gas which has accumulated within 10 pc from the host core (solid line) for the nearly retrograde merger. Mass of 500 units corresponds to 10^8 M$_\odot$. Time of 500 units corresponds to $\sim 10^8$ yr. The dotted line shows the change of velocity dispersion in unit of 200 km s^{-1} for the gas clouds contained within 200 pc from the core. The arrow indicates the time when two cores coalesce.

ically, increasing their velocity dispersion. This rise-up of velocity dispersion leads to effective (collisional) dissipation in the cloud system, inducing gas infall to the nucleus. In most models these processes remain ineffective when the two cores are far apart so that gas infall to the central 10 pc starts just around the coalescence of two cores (Fig. 1). Thus the quasar-like activity driven by mass accretion onto the nucleus is primarily limited to the post-merger phase.

2) Gas fueling becomes less effective as the inclination angle of the disk increases (*i.e.*, the spin becomes more prograde). This is because an increasingly large number of gas clouds get angular momentum from the cores and are kicked out to the outer regions where they remain for the rest of the time.

3) Gas fueling becomes less effective as the core scale length increases. This is because a less concentrated core is less effective both in heating the cloud system dynamically and in trapping the gas clouds in its potential well.

4) Gas fueling diminishes remarkably in the case the merger proceeds sufficiently slowly. This is because the central part of the gas cloud disk is hardly disturbed.

5) The simulations including gas consumption due to star formation show that star formation event cannot change the epoch when the prominent gas fueling starts but can change the total mass of the gas transferred to the central 10 pc.

In summary, the present numerical study suggests that the galaxy mergers showing a double-nuclei structure will exhibit only starburst activity whereas the quasar-like activity tends to be observed in single-nucleus mergers.

Gas in Shell Galaxies: Non-Spherical Potentials

Melinda L. Weil and Lars Hernquist

Board of Studies in Astronomy and Astrophysics, UC Santa Cruz

ABSTRACT

Simulations which explore mergers like those thought responsible for the shells around many elliptical galaxies find little correlation between the distribution of stars and gas in remnants. Mergers of small companion disks consisting of both gas and stars with non-spherical primary potentials produce shell galaxies with gaseous nuclear rings and clumps.

1 INTRODUCTION

Models which follow the infall of less-massive companion galaxies show that shell galaxies can be formed by accretion. However, it is probable that the sources of material also contain significant amounts of gas. We investigate encounters that produce shells by modeling interactions between non-spherical primary galaxies and companions containing both stars and gas with a three-dimensional code (TREESPH: Hernquist and Katz 1989). Primaries are modeled with rigid elliptical potentials of the form presented by Hernquist (1990) with scale-length $a = 1$. Physical time t' is related to the calculation time unit by $t' \approx 4.3 \times 10^6 t$. The companion is a rotationally supported disk in which particles are distributed according to an exponential surface density profile. Stars have a total mass 1/10 and gas 1/100 that of the primary. In most interactions, the companion potential is disrupted at a small distance from the primary after which the particles evolve in the solitary primary gravitational field.

2 RESULTS

In radial encounters with a primary with axis ratios $a = b = 1.0$ and $c = 0.5$, the stellar component of a companion formed shells, while the gaseous component formed nuclear disks for initial systems with prograde rotation with respect to the orbital plane or a disk rotated 45° out of the orbital plane. For encounters with initially parabolic orbits of impact parameter $p = 1$ and no disk inclination, of $p = 3$ or $p = 5$ and either inclined or flat disks, either disrupted or undisrupted companions, and various mass distributions, gaseous nuclear rings were formed whose sizes depend on impact parameter and method of disruption.

Finer structure arises in some of the gas remnants of the mergers as shown in

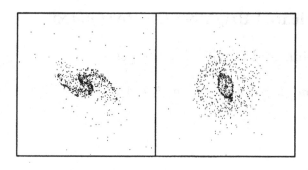

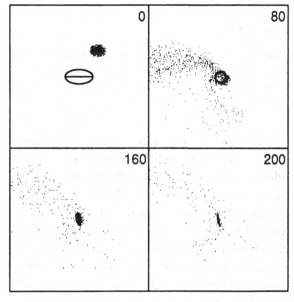

Fig. 1—time 200 $x - y$ and $x - z$ projections of remnant. Companion approached at escape speed from $p = 1$ with $x_0 = 0.0$, $y_0 = 8.0$, $z_0 = 6.0$ and was inclined 45° away from the orbit plane. Primary has axis ratios $a = b = 1.0$ and $c = 0.5$. Frames measure 10 length units per edge.

Fig. 2—time evolution in $y - z$ for companion similar to that of Fig. 1 except with the orbital y and z initial positions and velocities interchanged. The primary has $a = b = 1.0$ and $c = 0.5$; the ellipse shows its geometry. Frames measure 40 length units per edge.

Figure 1. In Figure 2 the gas forms a ring which precesses about the symmetry axis of the primary. Small impact parameter allows the companion to approach the center of the potential where non-sphericity is high enough that the precession time is short.

Overall, these simulations reinforce our earlier conclusions about the distribution of gas in remnants of encounters with spherical potentials. They demonstrate a method for supplying dense clumps of gas to galactic nuclei where star formation may occur. While some of the small impact parameter encounters show interesting warps and other fine features, the practical application of these results to observations is limited. The most interesting result may be that of Fig. 2 in which precessing ring in near-polar alignment is observed. Persistence of such structures may be significant in considerations of polar ring galaxies.

REFERENCES

Hernquist, L. 1990, *Ap. J.*, **356**, 359.
Hernquist, L., and Katz, N. 1989, *Ap. J. Suppl.*, **70**, 419.
Weil, M. L., and Hernquist, L. 1993, *Ap. J.*, **405**, 142.

Merging and Multiply-Nucleated Brightest Cluster Galaxies

Paul W. Bode, Haldan N. Cohn, and Phyllis M. Lugger

Astronomy Department, Indiana University, Bloomington IN 47405

ABSTRACT

We describe a program of N-body simulations of clusters of galaxies. Results concerning merging histories and the kinematic properties of multiple nuclei are presented.

1 THE MODELS

We have completed nine N-body evolutions of models containing 50 galaxies using a total of $N = 4 \times 10^4$ particles. The models are fully self-consistent in that each galaxy is represented as an extended structure containing many particles and the total gravitational potential arises from the particles alone. The evolutions are carried out with a direct N-body code based on the Barnes and Hut (1986) TREE algorithm for computing the gravitational potential; it is the code developed by Lars Hernquist (1987, 1990) with some modifications (Bode, Cohn, and Lugger 1993).

For all the models, 10% of the particles are 'luminous' and the rest represent 'dark' matter; the former are given a smaller softening length than the latter. The dark matter is apportioned between galaxy halos and a smoothly distributed cluster background. The percentage of mass initially in this intra-cluster background, β, is varied from 50% to 90%.

The initial mass distribution of the galaxies follows a Schechter function. For $\beta = 50\%$, the smallest galaxy contains 125 particles and the largest 3500 particles. Galaxies are given a core-halo structure by identifying the most bound particles in each galaxy as luminous. Since the total amount of mass in the cluster is the same for all models, increasing β has the effect of removing mass from the galaxy halos and distributing it through the cluster.

The cluster background particles and the galaxy centers of mass were generated by randomly sampling a King distribution with $W_0 = 6$; such a model provides a good fit to existing poor clusters (Malumuth and Kriss 1986). With gravitational constant $G = 1$ and total mass $M = 1$ the cluster was scaled to have binding energy (ignoring the internal energy of the galaxies) $E = 1/2$. Thus the cluster size, velocity dispersion, and dynamical time are all unity in code units. For the physical scaling $M = 10^{14}$ M$_\odot$ and $R = 1.5$ Mpc, the typical velocity dispersion is 535 km s^{-1} and the core radius of the cluster is 250 kpc.

410

2 RESULTS

In the $\beta = 50\%$ models, merging leads to the development of a dominant central galaxy in about a Hubble time. The merger rate is roughly constant at about one per Gyr; the overall merging history varies little between the runs. The mergers occur within the core radius of the cluster (250 kpc); over half take place within the inner 50 kpc. Early in a typical run, a large merger product settles towards the center of the cluster; most of the subsequent merging involves this dominant central galaxy accreting the others. Larger galaxies (those with mass $M \gtrsim M^*$) are more likely to be involved in merging. By the end of the evolutions there is little scatter in the size of the largest galaxy.

In Abell clusters, about half of all brightest cluster galaxies contain a secondary nucleus within a projected radius of 20 kpc; this is a larger fraction than is expected from chance superpositions (see Blakeslee and Tonry 1992 and references therein). As a dominant central galaxy is formed in our models, the likelihood of seeing secondary nuclei increases. After 10 Gyr, a companion is seen within a 20 kpc projected radius of the dominant galaxy over 30% of the time. Roughly 10% of the time there is a true three-dimensional separation of less than 20 kpc between the dominant galaxy and its nearest neighbor.

The line-of-sight velocity distribution of the secondary nuclei in our models is similar to that of the entire cluster and is well fit by a gaussian. Those galaxies which are in the process of merging have only slightly smaller velocities than the rest. Thus it may be difficult to separate out true secondary nuclei from chance superpositions relying only on velocity data. Interestingly, Lauer (1988) found no dynamical differences between those multiple-nuclei systems which show morphological disturbances and those which do not.

Simulations with 75% or 90% of the mass in the background show a substantially lower merger rate, so that in the latter case a dominant central galaxy may not be created. Increasing β makes the galaxies smaller, lengthening the dynamical friction time scale and decreasing the geometrical cross-section, thereby delaying merging. For $\beta = 90\%$ there are also fewer secondary nuclei and fewer true companions to the brightest cluster galaxy.

REFERENCES

Barnes, J., and Hut, P. 1986, *Nature*, **324**, 446.
Blakeslee, J. P., and Tonry, J. L. 1992, *A. J.*, **103**, 1457.
Bode, P. W., Cohn, H. N., and Lugger, P. M. 1993, *Ap. J.*, **416**, in press.
Hernquist, L. 1987, *Ap. J. Suppl.*, **64**, 715.
Hernqueit, L. 1990, *J. Compt. Phys.*, **87**, 137.
Lauer, T. R. 1988, *Ap. J.*, **325**, 49.
Malumuth, E. M., and Kriss, G. E. 1986, *Ap. J.*, **308**, 10.

Self-Gravitating Simulations of M51 Multiple Encounter History

Heikki Salo[1] and Gene Byrd[2]

[2]University of Alabama, Dept. of Physics and Astronomy, U. S. A.
[1]University of Oulu, Dept. of Astronomy, Finland

ABSTRACT

N-body simulations of M51-system suggest that the companion is currently moving in a highly inclined (75°) but bound orbit ($e = 0.25$) with respect to the main system. In this model the inner spiral structure as well as the extended outer tail follow from excitation during an earlier disk plane crossing about 400 million years ago, while the most recent crossing occurred less than 100 million years ago.

1 INTRODUCTION

The M51 system (NGC 5194/5) is an expectionally well observed spiral galaxy with a clear grand design pattern. Several attempts have been made to explain its structure with N-body modelling, starting with the classic test-particle simulations of Toomre and Toomre (1972). Although these and the later self-gravitating models of Hernquist (1990) have been quite successful in reproducing the tidal bridge and tail structures, they usually predict such a short interval since the principal perturbation at disk plane crossing, that it is hard to explain the existence of the strong inner spiral pattern. Moreover, the arm kinks going from the inner structure to the bridge and tail arms strongly suggest a more complicated process. Finally, the recently observed (Rots *et al.* 1990) large extent hydrogen "far tail" requires serious revisions in existing simulation models. Howard and Byrd (1990) first suggested that all these features could be simultaneously accounted for by assuming a bound companion orbit, with several strong perturbation events at successive disk plane crossing.

2 N–BODY MODEL

In the present study the multiple encounter model is refined and studied with a fully 3-dimensional self-gravitating code, similar to that used in modeling of Arp 86 pair (Salo and Laurikainen 1993). Both stellar and collisional gas cloud components of the disk are included with the morphology of the arms being followed as they propagate into the nuclear regions. By performing an extended survey over the orbital parameter space, a possible orbital history was found, reproducing all the main morphological features of M51 system, as well as accurately matching the constraints set by the projected separation and radial velocity difference between components.

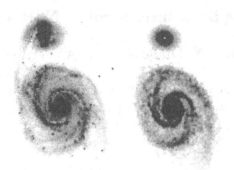

Fig. 1—Simulated M51 multiple encounter arm pattern (right) compared with V-band CCD-image (left). A self-gravitating exponental disk embedded in analytical isothermal halo was perturbed with 0.4 mass companion. Projection to sky-plane assumes 20° inclination and PA 170°. Transitional kinks in the bridge and tail arm are due to most recent crossing less than 100×10^6 yr ago.

Fig. 2—Comparison between observed (Rots *et al.* 1990) and simulated large scale velocity field. Radial velocity range is 380 – 460 km s^{-1} (white to black). Also shown is the relative orbit of the companion, which is now about 15 kpc behind the main disk. The counter rotating tail is due to perturbation at a previous crossing 400×10^6 yr ago.

REFERENCES

Hernquist, L. 1990, in *Dynamics and Interactions of Galaxies*, ed. R. Wielen (Heidelberg: Springer), p. 108.

Howard, S., and Byrd, G. 1990 *A. J*, **99**, 1978.

Rots, A. H., Bosma, A., van der Hulst, J. M., Athanassoula, E., and Crane, P. C. 1990, *A. J.* **100**, 387.

Salo, H., and Laurikainen, E. 1993, *Ap. J.*, **410**, in press.

Toomre, A., and Toomre, J. 1972 *Ap. J.* **178**, 623.

Formation of Ring Structures through N–Body Simulations

A. Curir[1], R. Filippi[2], A. Diaferio[3], and A. Ferrari[1]

[1]Osservatorio Astronomico di Torino, Pino Torinese (Torino), Italy
[2]Istituto di Fisica Generale, Universita' di Torino, Italy
[3]Dipartimento di Fisica, Universita' di Milano, Italy

ABSTRACT

We present numerical N-body simulations of galactic interactions in which a compact body or galaxy penetrates a disklike galaxy. The results are ringlike structures having morphologies in good agreement with what one observes as ring galaxies. The code used is the TREE-code by L. Hernquist.

1 NUMERICAL FEATURES

The disk used as target is imbedded in a massive halo and the whole system is in dynamical equilibrium. It models fairly well the kinematical behaviour of spiral galaxies as far as the velocity distribution is concerned. We performed numerical simulations of collisions between stellar disks (embedded in static halos) and suitable intruders.

The disks have been settled down by solving numerically the Laplace equation in cylindrical coordinates and then immersed in a massive halo structure (King model). The system is evolved several rotation periods to test out the stability. The halo has an important heating effect on the disk during the assessment.

The companions used as intruders are massive points or small king spheres, having different masses and different radii. A series of collisions have been performed varying the direction of the companion's velocity .

The time scale of evolution of the system is around 0.5 Gyr and is consistent with theory (Theys and Spiegel 1977)

2 EVOLUTION OF THE SYSTEM

The passage of the massive point through the disk generates a transient ring-shaped mass distribution. The ring is produced by a single density wave propagating through the disk. The wave has a damped oscillatory behaviour, since after an initial outward propagation, goes backward with decreasing amplitude toward the center of the structure. Correspondingly to the inward propagation, the ring disappears.

The interaction of the disk with a king sphere produces similar behaviour as far as the density in the disk is concerned. The dynamical friction between the two N-

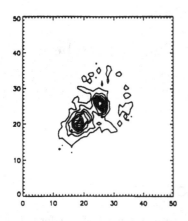

Fig. 1—Isodensity contours of the simulation describing the intrusion of a King sphere on an exponential disk of equal mass at $t/t_{coll} = 0.6$. (The system rotated of 30° around the x-axis and then projected on the plane xy). There is a good agreement with the morphology of the isophotal plots of VV 789 (see the contribution by Mazzei *et al.*, these proceedings).

body systems provides a strong deceleration to the intruder and traps it forming a bounded system: the indruder goes back and crosses the disk again. This situation can simulate fairly well ring galaxies presenting the companion very near to the structure (see Fig. 1).

With suitable initial condition one can observe dense nuclei and knots on the ring (see Fig. 2). By using disks as intruders we succeded in producing ring shaped morphologies also in 3D and not only in the plane of the disk target. These numerical experiments therefore provide some example of the possibility (quoted by Theys and Spiegel 1976) that ring galaxies are shells.

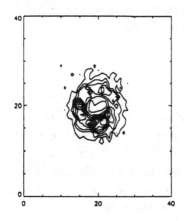

Fig. 2—Isodensity contours (simulations) describing the intrusion of a massive point on an exponential stellar disk having equal mass at $t/t_{coll} = 0.17$. The system is projected on the plane xy. The plot mimics fairly well the isophotal image of VII Zw 466 (see Theys and Spiegel 1977).

REFERENCES

Hernquist, L. 1987, *Ap. J. Suppl.*, **64**, 715.
Theys, J. C., Spiegel, E. A. 1976, *Ap. J.*, **208**, 650.
Theys, J. C., Spiegel, E. A. 1977, *Ap. J.*, **212**, 616.

Self-Consistent Evolution of Ring Galaxies

Paola Mazzei[1], Carlotta Bonoli[1], Anna Curir[2] and Roberta Filippi[2]

[1]Osservatorio Astronomico, Padova, Italy
[2]Osservatorio Astronomico, Torino, Italy

ABSTRACT

Ring galaxies are commonly known as objects where a burst of star formation was triggered by a close encounter with an intruder. Here we present a self-consistent approach to reproduce the observed morphological and photometric behaviour of a sample of ring galaxies using updated N-body simulations and evolutionary synthesis model from UV to far-IR wavelengths.

1 INTRODUCTION

Galaxies were selected from the list of Appleton and Struck-Marcell (1987). BVRI CCD observations of seven ring galaxies were carried out at Padova-Asiago Observatory using a GEC CCD with a pixel size of 22 μm, corresponding to 0.28 arcsec. Fluxes have been translated to the standard BVRI Johnson's system. In the following we briefly outline fundamentals of our models and summarize our results in section 2. Far-IR (FIR) data come from *IRAS* catalogue (Version 2).

1.1 N-Body Simulations

We performed numerical simulations of collisions between stellar disks (embedded in static massive halos) and suitable intruders. The code used is the Hernquist's (1987) TREECODE. The companions used as intruders are massive points or small King spheres, having different masses and different radii. A series of collisions have been performed varying the direction of the companion's velocity. These numerical experiments represent something new with respect to the previous work on the formation of ring structures (Appleton and James 1990) because of the careful and realistic production of the disk target (for details see Curir *et al.*, these proceedings). The passage of the intruder through the disk generates a transient ring-shaped mass distribution. The ring is produced by a single density wave propagating through the disk. After an initial outward propagation, this wave goes backward with decreasing amplitude toward the center of the structure, correspondingly the ring disappears. Finally the disk is more extended than the initial one.

1.2 Evolutionary Synthesis Models

The spectral energy distribution (SED) of ring galaxies has been computed from UV to FIR wavelength range, using the model described in Mazzei *et al.* (1992). They take into account in a self-consistent way chemical evolution, stellar emission, internal dust absorption and re-emission. Dust includes a cold component, heated by the general radiation field and a warm component associated with H II regions. Emission from policyclic aromatic hydrocarbon molecules (PAH) and from circumstellar dust shells are also taken into account.

Following the previous discussion (section 2), we modify the normal disk evolution (for more details see Mazzei *et al.* 1992), at an age of 12 Gyr, with a strong burst of star formation lasting the time of the interaction ($\approx 3 \times 10^8$ yr). It provides very blue colors and higher FIR emission than normal disk galaxies. Moreover the surface density of the diffuse radiation field is decreased by the interaction providing a somewhat larger warm to cold dust ratio (≈ 0.7 instead of ≈ 0.3).

2 CONCLUSIONS

We investigate evolution of ring galaxies in a self-consistent way using N-body simulations and evolutionary synthesis model providing the SED in UV–FIR range. Results are compared with the morphology and the overall SED of 7 ring galaxies selected from the Appleton and Struck-Marcell (1987) list.

The interaction of an unperturbed disk with an intruder object provides a strong burst of star formation lasting $\approx 10^8$ yr. It rapidly disappears leaving a new disk-like configuration. Our N-body simulations predict a flatter surface density distribution owing to the re-arrangement of the galaxy. Computed $L_{\mathrm{FIR}}/L_{\mathrm{bol}}$ ratios of ring galaxies during the burst exceed those of normal disk systems (Mazzei *et al.* 1992) by at least a factor of 2. Moreover, after the burst, the SED in the FIR could be also significantly different from that of a normal disk owing to the lower temperature of the cold dust component. We predict higher relative contribution of the warm component to the overall FIR emission providing a value of f_{60}/f_{100} ratio of about 0.5 instead of 0.3 as for 'standard' disks. Galaxies which have experienced a ring phase would keep warmer FIR colors than normal disk galaxies as a consequence of the past interaction.

REFERENCES

Appleton, P. N., and Struck-Marcell, C. 1987, *Ap. J.* **312**, 566.

Appleton , P. N., and James,R. A. 1990, in *Dynamics and Interactions of Galaxies*, ed. R. Wielen (Berlin: Springer), p. 200.

Curir, A., Filippi, R., Diaferio, A., and Ferrari, A. 1993, these proceedings.

Hernquist, L. 1987, *Ap. J. Suppl.*, **64**, 715.

Mazzei, P., Xu, C., and De Zotti, G. 1992, *Astr. Ap.*, **256**, 45.

Interacting Galaxy Pair Arp 86

Eija Laurikainen[1] and Heikki Salo[2]

[1]Turku University Observatory, Tuorla, Finland
[2]University of Oulu, Department of Astronomy, Finland

ABSTRACT

Arp 86 is studied in terms of 3-dimensional N-body simulations and compared to CCD-observations. Deep, high resolution BVRI images were obtained in order to determine initial parameters for dynamical modelling, and to study star formation properties of these galaxies. The models suggest that the companion galaxy is moving in a low inclination, low eccentricity orbit thus performing several revolutions around the main galaxy, the obtained colors being in agreement with this interpretation. The orbit geometry favors material transfer between the components, which is proposed to be the cause of the anomalously large activity in the companion, and of the ongoing star formation in the bridge. The prolonged perturbation due to dynamically bound companion explains the grand-design structure of the main galaxy.

1 OBSERVATIONS AND N-BODY MODEL

Arp 86 (NGC 7753/54) is a spiral pair resembling M51 system. In order to study its star formation properties deep, high resolution BVRI CCD-photometry has been obtained (Laurikainen *et al.* 1993). Same observations are utilized in determination of initial parameters for dynamical modelling.

Dynamical modelling of the pair is performed with a new fast N-body code (Salo and Laurikainen 1993), capable of following the simultaneous evolution of two or more systems including both stellar and gaseous components. In the code galaxy disks are described in terms of self-gravitating particles, while analytical models are used for the spherical halo components. Potential evaluation is based on multiple, comoving logarithmic spherical potential grids. The code is 3-dimensional and thus allows arbitrary orbital geometry, and retains good spatial resolution simultaneously near the nuclei of both systems as well as in the interaction zone. It also includes routines for iterative fits of the orbit parameters so that the final projected positions of the galaxies and their redshift differences accurately match the observed ones.

The observational initial parameters used for the dynamical modelling are the projected separation, velocity difference, disk orientations and intensity profiles. Using these and some additional initial parameters a systematic search was made through the orbital parameter space (> 100 runs), until a solution (Fig. 1) was found which explains the observed morphology, reproduces the observed velocity field and S-shaped major-axis rotation curve (Marcelin *et al.* 1987).

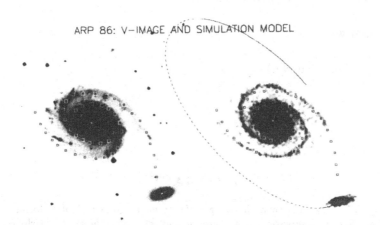

Fig. 1—Comparison between observed and simulated morphology . The simulation system is projected to our observing direction, corresponding to 40° inclination and 53° position angle. In addition to reproducing the main spiral features, the simulation model matches the correct relative separation of the components, as well as the radial velocity difference.

2 RESULTS

(1) Arp 86 was found to be a dynamically bound system, the companion presently moving near to plane of the main galactic disk. The principal perturbation occurred about 600 Myrs ago, and the most recent upward crossing about 50 Myrs ago. Previous downward crossing has excited the strong inner spiral density waves.

(2) The orbit found represents an intermediate phase on the orbital evolution governed by dynamical friction, starting from perhaps only marginally bound orbit and leading to a complete merger. This kind of orbit explains the long lasting grand-design structure of this and probably many similar galaxies.

(3) Optical colors of Arp 86 support the type of orbit found, with long term perturbation being strongest near main galaxy disk plane crossings. This is supported by the very blue B-V color of the bridge as well as by the V-R gradient, suggested to be induced by long-term enhancement of stellar formation.

(4) About 5% of the main galactic gas mass can be transferred to the companion during the interaction. This mass flow is proposed to cause the observed starburst in the companion galaxy.

REFERENCES

Laurikainen, E., Salo, H., and Aparicio, A. 1993, *Ap. J.*, **410**, 574.
Marcelin, M., *et al.* 1987, *Astr. Ap.*, **179**, 101.
Salo, H., and Laurikainen, E. 1993, *Ap. J.*, **410**, 586.

Cold Gas in Early–Type Galaxies

P. T. de Zeeuw

Sterrewacht Leiden, The Netherlands

ABSTRACT

Many elliptical galaxies contain cold gas of external origin, which often forms a warped disk or ring. Recent observations of the atomic and the molecular gas in nearby ellipticals have shed considerable light on the properties of this accreted material. Resulting progress in understanding the details of the settling process are discussed, with attention to questions such as: Are any of the observed gas disks settled? How do the settling times and the observed properties of the gas disks depend on the structure and shape of the galaxy in which it resides? How much mass is delivered to the nucleus?

1 INTRODUCTION

Many elliptical galaxies contain significant amounts of cold gas. This is often detectable by its Hα emission which sometimes extends to 3–5 kpc from the center and appears as a warped disk (Demoulin–Ulrich, Butcher and Boksenberg 1984; Trinchieri and di Serego Alighieri 1991; Shields 1991; Macchetto and Sparks 1992; Buson *et al.* 1993). When seen nearly edge–on, such disks are seen as a dust lane across the image of the galaxy (Bertola and Galletta 1978; Hawarden *et al.* 1981; Ebneter and Balick 1985; Sadler and Gerhard 1985; Bertola 1987). These disks contain 10^3–10^5 M$_\odot$ of ionized gas (Phillips etal 1986), 10^4–10^6 M$_\odot$ of dust (Forbes 1991), and 10^6–10^8 M$_\odot$ of neutral gas (Lees *et al.* 1991; Bregman, Hogg, and Roberts 1992). Large amounts of neutral hydrogen extending to tens of kpc have been found in only a half–dozen elliptical galaxies (van Gorkom 1992).

In many cases the gas is kinematically decoupled from the galaxy, and it is therefore believed that this is accreted material, most likely acquired by capture of a small gas–rich disk galaxy (Athanassoula and Bosma 1985). This is consistent with the observation that a small number of S0 galaxies have polar rings of gas and stars (Whitmore *et al.* 1990; Sackett 1991; Sparke 1991). It is assumed that the infalling gas is cold, and smears out into an inclined ring in a few orbital periods (Tubbs 1980). Differential precession between neighboring orbits then causes dissipative cloud–cloud collisions, which slowly change the orientation and size of the ring. The expected result is that the gas eventually settles on a set of simple non-selfintersecting closed orbits, which occur at most in a few preferred orientations, namely those where the differential precession vanishes (Tohline, Simonson, and Caldwell 1982; Steiman–Cameron and Durisen 1982). Once the gas has reached a preferred plane, its subsequent evolution is very slow.

Many details of this infall scenario are only partially understood. What is the time scale for settling of the accreted material? What are the expected configurations for the gas? Are any of the observed gas disks settled? How do the settling times and the observed properties of the gas disks depend on the structure and shape of the galaxy in which it resides? Can we deduce the intrinsic shape and mass distribution of the host galaxy from measurements of the morphology and kinematics of the disk? How much mass is delivered to the nucleus as food for the central monster (Gunn 1979)?

2 SIMPLE CLOSED ORBITS

Cold gas in an axisymmetric galaxy eventually must settle on the circular orbits in the equatorial plane, unless self–gravity of the gas is important (Sparke 1986). If the galaxy is oblate, the gas disk will appear elongated along the apparent major axis of the galaxy. If the galaxy is prolate, the disk will be extended along the minor axis of the galaxy. Hence in this situation a skewed or warped gas disk would not be settled, and such a configuration would occur infrequently if the settling time were short compared to the age of the Universe. If all elliptical galaxies were axisymmetric, it would be trivial to deduce the flattening of those with settled gas disks.

Elliptical galaxies are slowly tumbling triaxial systems (Binney 1976; de Zeeuw and Franx 1991), and the stable simple closed orbits available to cold gas are not circular, but are approximately elliptic. This results in a number of possible morphologies for a settled gas disk (Steiman–Cameron and Durisen 1982; Merritt and de Zeeuw 1983; Kormendy and Djorgovski 1989).

Infalling gas in a stationary triaxial galaxy will settle on roughly elliptic closed orbits which all lie either in the plane perpendicular to the long axis, or in the plane perpendicular to the short axis of the galaxy (Heiligman and Schwarzschild 1979). When projected on the sky, the kinematic axes of the gas generally will be misaligned from the photometric axes of the galaxy, due to the fact that the projected principal axes of a triaxial light distribution coincide with the major and minor axis of the projected surface brightness distribution only for special directions of observation. The gas disk may also appear warped due to the variation of the surface density caused by the velocity change along an elliptic orbit (de Zeeuw and Franx 1989).

When the figure of the galaxy tumbles, e.g., about its short axis, the gas can settle either on elliptic orbits in the equatorial plane or on the family of so–called anomalous orbits, which do not lie in any of the principal planes, but form a stable configuration which is intrinsically warped (Heisler, Merritt, and Schwarzschild 1982; Magnenat 1982; Mulder and Hooimeyer 1984). The observed properties of such gas disks have not been modeled in any detail (*cf.* van Albada, Kotanyi, and Schwarzschild 1982; Varnas *et al.* 1987), but it is evident that a wide variety of morphologies and velocity fields can be obtained.

The elliptic closed orbits generally become more elongated at smaller radii. This increases the shear between neighboring orbits, and induces turbulence in the gas. This is likely to lead to shocks (van Albada and Sanders 1982), and a steady supply of gas to the nucleus. It follows that the closed orbit description of the gas motions breaks down inside a certain radius. This always occurs in a stationary triaxial galaxy with a homogeneous core, because the elliptic closed orbits in such a system become linear at a finite distance from the center. When the triaxial galaxy rotates slowly and/or has a central cusp (as is frequently observed, Kormendy and Djorgovski 1989; Lauer *et al.* 1991), the simple closed orbits reach a limiting ellipticity in the central regions. The limiting value depends on the intrinsic shape of the galaxy, on the slope of the density profile—with shallower slopes resulting in more elongated orbits—and on the rate of figure rotation. In some cases the limiting ellipticity is sufficiently small for the closed orbit approximation to be valid to very small distances. A settled gas disk in such a system is likely to be long-lived. This suggests that elliptical galaxies with regular gas disks have preferred shapes, and may well be nearly round in the plane of the disk.

3 SETTLING?

The time scale τ_s for settling of infalling gas is a few times the differential precession time scale, which is of the order of 10 or 20 orbital periods. This can be quite long, especially at large radii. Early analytic estimates of τ_s by Tohline, Simonson, and Caldwell (1982) were rediscussed by Steiman–Cameron and Durisen (1988). These authors used a semi-analytic approach in which the cloud–cloud interactions are described by an effective viscosity, and found that τ_s is somewhat larger than thought earlier, and is likely to be of the order of 2–3 times 10^9 years at an effective radius (3–5 kpc). It follows that at any time only part of the gas is settled. The same authors also presented a simple analytic description of the properties of a differentially precessing gas disk in an axisymmetric galaxy with a dark halo (Steiman–Cameron and Durisen 1990).

A number of numerical studies of the fate of infalling gas have been carried out in the past decade (Habe and Ikeuchi 1985, 1988; Varnas 1990; Quinn 1991; Katz and Rix 1992; Christodoulou and Tohline 1993; Dubinski and Christodoulou 1993). These employed a variety of algorithms for treating the gas flow, including sticky particles, smooth particle hydrodynamics (SPH) and Eulerian schemes. The majority of studies considered the evolution of an inclined circular ring of gas in a fixed logarithmic triaxial halo potential, which describes an elliptical galaxy with a massive halo. The actual infall of a satellite and the evolution of an initially toroidal distribution were investigated also. Some studies include the role of radiative cooling and self–gravity of the gas, while others ignore one or both of these effects. Tumbling figures were considered as well. The gas in these simulations usually manages to find the preferred planes, and then stays confined to a ring of simple closed orbits, but there is considerable disagreement on the time scale of this process, and on the amount of mass transferred to the nucleus. For this reason, a number of the main protagonists in this area have compared the results of the different algorithms in detail, and summarized their findings in an excellent review (Christodoulou *et al.* 1992). Their main conclusions are:

1. Settling times of polar rings in slightly flattened oblate galaxies (E1) are long, and agree with the estimates of Steiman–Cameron and Durisen (1988). Polar rings in moderately flattened oblate galaxies (E1–E3) are effectively stable, even without self–gravity (Sparke 1986). The gas quickly finds the long–lived configuration which warps away from the polar orientation, in which the precession is constant with radius so that cloud–cloud collisions are infrequent. Polar rings in strongly flattened oblate models evolve quickly, and disrupt.

2. A polar ring in a prolate galaxy is pulled away from the constant precession configuration by the gravitational field of the galaxy, and as a result disrupts in a few orbital periods, so that the material ends up in the central region.

Katz and Rix (1992) have emphasized the importance of radiative cooling in the simulations. In earlier SPH work the dissipation is due not only to the cloud–cloud collisions, but also to numerically induced supersonic heating. As the gas tries to settle on the elliptic closed orbits, the inclined circular ring deforms, and the velocity varies along the ring. This leads to supersonic compression and associated heating. Not allowing for radiative cooling makes the ring puff up, and hence increases the number of cloud–cloud collisions. This in turn speeds up the settling, and results in rapid mass transfer to the center. Katz and Rix argue that this effect is responsible for the fairly rapid settling, and the large mass transfer rates found by *e.g.*, Habe and Ikeuchi (1985, 1988). Katz and Rix show that with inclusion of realistic radiative cooling the gas finds the long–lived stable warped states, and there is little mass-transfer to the nucleus. Similar results are obtained with a sticky particle code (Quinn 1991).

A lot of work remains to be done in this area, especially for triaxial galaxies with tumbling figures (*cf.* Varnas *et al.* 1987; Arnaboldi and Sparke 1993). Even so, the simulations carried out so far confirm that a substantial fraction of an infalling lump of cold gas can sometimes settle in a preferred plane in a time that is much shorter than a Hubble time, and they show that the closed orbit approximation provides accurate velocity fields for such systems. At the same time, many observed disks may well be only partially settled—if at all— and require careful modeling when used as a diagnostic of the intrinsic shape and mass distribution of the host galaxy.

4 SOME EXAMPLES

We now discuss a few galaxies for which detailed observations of gas and dust shed light on various aspects of the settling process, and on the pitfalls and ambiguities of modeling. Other interesting cases include the peculiar object NGC 3718 (Schwarz 1985; Sparke 1991), the polar ring galaxies NGC 4650A (Sackett and Sparke 1990) and AM 2020–504 (Arnaboldi *et al.* 1993), NGC 4278 (Lees 1993) and NGC 2974 (Amico *et al.* 1993; Cinzano and van der Marel 1993), both of which have an extended H I disk, and the spindle galaxy NGC 2685 (Peletier and Christodoulou 1993). Gerhard (1993) has recently reviewed similar work on the gas motions in the inner part of our own Galaxy.

NGC 4845. The observed velocities of the emission line gas in the bulge of this Sa galaxy can be fitted accurately with a model in which the gas has settled on the elliptic closed orbits in the equatorial plane of a triaxial bulge, which has a shape that is consistent with the observed isophote twist (Bertola, Rubin, and Zeilinger 1989; Gerhard, Vietri, and Kent 1989). The gas orbits are quite elongated, and have axis ratios as low as 1/3. Although in this case there is no need to invoke infalling gas, the fact that the model and the observations match so well provides circumstantial evidence that cold gas can stay for a long time on elongated orbits. It also demonstrates that the two–dimensional velocity field of a gas disk contains useful information on the intrinsic shape of the host galaxy.

NGC 5077. This E3 galaxy has a disk of ionized gas along its apparent minor axis, and gas velocities have been measured along seven position angles out to $\sim 0.5R_e$ (DeMoulin–Ulrich *et al.* 1984; Bertola *et al.* 1991). Constant M/L triaxial mass models with non–rotating figures and the gas in either of the two preferred planes are consistent with the morphology and kinematics of both the gas and the stars. In particular, they reproduce the observed misalignment of 23° between the line of zero velocity of the gas and the apparent major axis of the galaxy as being due to a projection effect. However, a constant M/L oblate model with the gas in a warped polar ring of the kind discussed by Sparke (1986) can be fitted to the data also. The results show that there is no reason to suspect that M/L varies strongly over the luminous part of elliptical galaxies, and furthermore, that it is important to use the proper geometry when deriving density profiles, and hence M/L, from kinematic data. To distinguish between the different models will require high resolution two–dimensional velocity measurements.

IC 2006. This E/S0 galaxy near Fornax is surrounded by a faint counter–rotating ring at 6.5 R_e (18.9 kpc), which contains 4.8×10^8 M$_\odot$ of H I (Schweizer, van Gorkom, and Seitzer 1989). These authors assumed the ring is intrinsically circular, and deduced that two–thirds of the total mass inside the ring is dark. High–resolution *VLA* measurements of the velocity variation along the ring demonstrate that it is indeed nearly circular, with an ellipticity less than 0.03, and confirm the presence of a dark halo (Franx *et al.* 1993). The ring lies close to or in the equatorial plane of IC 2006. The settling time at 6.5 R_e is long, so if the ring is the result of infall—which is not unlikely in view of the counter–rotation—this must have occurred in the distant past, and may be associated with the formation of the entire galaxy. If the gas has indeed settled on a closed orbit, then the near circularity of the ring shows that the dark halo—which dominates the potential—is nearly round in the plane of the ring.

NGC 4753. This is a dusty S0 galaxy with a peculiar—and at first sight chaotic—dust morphology. Steiman–Cameron, Kormendy, and Durisen (1992) have shown that the observed folds and twists of the dust are remarkably well fit by an inclined disk that is twisted by nearly 4π radians over a factor 7 in radius due to differential precession. This model requires the galaxy potential to be dominated by an oblate massive dark halo which is nearly spherical. The disk shows little evidence for settling towards the equatorial plane, and may be less than 10^9 years old. It will be interesting to check and refine this model by kinematic observations of stars and gas.

5 NGC 5128: CENTAURUS A

Centaurus A is one of the most peculiar objects in the sky (Sandage 1961). It is the nearest giant elliptical galaxy, possesses a strong nuclear radio source, has a prominent warped disk of dust and gas, and shows many indications of a recent merger (Ebneter and Balick 1983). It is an excellent laboratory to investigate the infall of gas in detail.

5.1 Dust Lane

The warped dust band contains $\sim 7 \times 10^8$ $M_\odot$ of H I, which extends beyond 5 kpc from the nucleus (van Gorkom *et al.* 1990). The hydrogen in the central 1.5 kpc is predominantly in molecular form, and reveals its presence by various transitions of its tracer molecule CO (Phillips *et al.* 1987; Eckart *et al.* 1990a; Quillen *et al.* 1992: Q92; Rydbeck *et al.* 1993). Hα emission of the disk has been detected out to nearly 3 kpc from the center (Bland, Taylor, and Atherton 1987).

Q92 made detailed dynamical models for the CO(2–1) observations taken with the Caltech Submillimeter Observatory on Mauna Kea*, at 1.3 km s^{-1} resolution. The measured line profiles are very broad, up to 300 km s^{-1}, but have sharp edges and small–scale features. This shows that the emitting gas has a velocity dispersion of less than 10 km s^{-1}. The gas temperature is of the order of 10 K, and the material lies in a disk less than 35 pc thick. The broad swath of dust obscuring the central regions of Cen A is a warped and folded sheet of cold clouds, which have an area filling factor $\lesssim 0.1$.

The broadening of the CO profiles is caused by the convolution of the intrinsic velocity field with the 30″ beam of the CSO. This beam–smearing, well–known in H I studies, must be taken into account in any comparison of the line profiles with dynamical models. Q92 concluded that simple models with the gas on circular or elliptic orbits in a plane indeed reproduce the wide profiles when observed with a 30″ beam, but cannot fit the observed asymmetries in the central region. A warped disk, modeled as a set of concentric circular rings with varying inclination does better. Q92 took the rings initially all at a fixed angle from the symmetry axis of an axisymmetric

---

* Catching Small Waves on the Big Island™

galaxy. Differential precession then results in a twisted disk similar to that used by Steiman–Cameron, Kormendy, and Durisen (1992) for NGC 4753 (section 4). A twisted disk in an *oblate* potential gives a good fit to the CO profiles inside 0.8 kpc but requires 1.2×10^9 yr of differential precession without any settling. This is inconsistent with a settling time scale of about 10^9 yr (section 3). A twisted disk in a *prolate* potential gives an excellent fit to all but the central CO profile (section 5.2), and requires only $\sim 10^8$ yr of differential precession. This is in good agreement with Tubbs' (1980) estimate, and in harmony with the assumption of no settling. The prolate shape of the galaxy is consistent with the stellar absorption line measurements of Wilkinson *et al.* (1986). The integrated emission of the model agrees well with the morphology of the optical dust lane, and with the H I emission in the region of overlap. The peaks in the CO(1-0) emission at 750 pc on either side of the nucleus seen by Eckart *et al.* (1990a) coincide with folds of the disk along the line–of–sight. One of the three absorption features in SN 1986G agrees with the velocity of the disk material at the projected position of the supernova.

The Q92 model agrees remarkably well with the inner kpc of the geometric tilted ring model for the Hα morphology and velocity field, presented by Nicholson, Bland–Hawthorne, and Taylor (1992). Compared with the CO data, the Hα measurements have higher spatial resolution ($\sim 3''$), but lower velocity resolution (36 km s^{-1}). This confirms that the CO and the Hα emission originate in the same cold component.

New JHK data obtained by Quillen *et al.* (1993) show that the inclined circular ring model fits the detailed dust morphology very well inside 1 kpc, but that it can be improved beyond this radius by varying the initial inclination of the precessing rings. The inferred radial variation of the inclination is remarkably well reproduced by a simple merger model in which the gas is tidally stripped from an object that plunges into Cen A. The derived orbit of the infalling galaxy agrees with the location of the stellar shells (Malin, Quinn, and Graham 1983), and with their H I morphology (van Gorkom priv. comm.). It also provides a natural explanation for the inclined rotation axis of the system of planetary nebulae in the halo of Cen A (Hui 1990).

Cen A is a prime candidate for a more detailed simulation. The existence of codes like TREESPH (Hernquist and Katz 1989), which combine accurate SPH with an N–body tree code for following the stellar dynamics, makes this feasible, especially since Quillen *et al.* (1993) have put strong constraints on the orbit of infalling galaxy.

5.2 Circumnuclear Torus

The CO(1-0), (2-1), and (3-2) profiles centered on the nucleus of Cen A have broad wings, which are not reproduced by the Q92 model for the dust lane material. Israel *et al.* (1990, 1991) have proposed that this emission originates in a dense circumnuclear disk or torus which rotates at about 200 km s^{-1}, has a radius of ~ 325 pc, and is responsible for the H$_2$ emission seen in the near infrared. The continuum emission of the associated warm dust (T~ 30 K) has recently been detected at 450 μ (Hawarden *et al.* 1993).

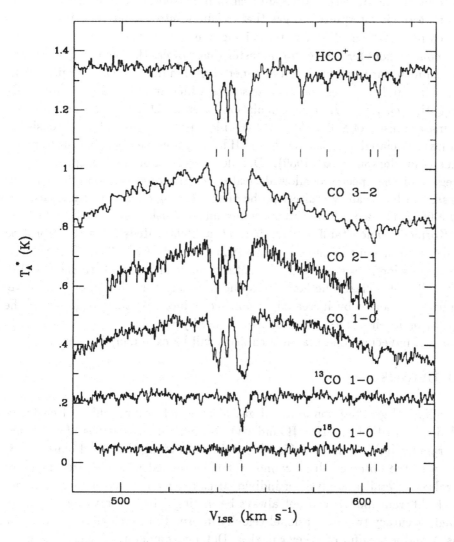

Fig. 1—High resolution spectra of HCO$^+$, ^{12}CO(3–2), (2–1) and (1–0), ^{13}CO(1–0) and ^{18}CO(1–0) toward the nucleus of Cen A, obtained by Israel *et al.* (1991). The spectra have velocity resolution of 0.29, 0.9, 0.23, 0.23, 0.23 and 0.23 km s^{-1}, and have been shifted in T_{A*} by +1.05, +0.4, +0.25, 0.0, –0.05 and –0.20 K, respectively. The vertical dashes indicate the absorption features. The broad absorption at 550 km s^{-1} is associated with the circumnuclear torus.

The nucleus of Cen A is a very strong cm/mm continuum source. Spectra taken towards the nucleus show narrow absorption lines of OH, H_2CO, H I, C_3H_2, NH_3, CO, and HCO^+ (Gardner and Whiteoak 1976; van der Hulst, Golisch and Haschick 1983; Seaquist and Bell 1986, 1990; Israel *et al.* 1990, 1991; Eckart *et al.* 1990*b*). Most lines are redshifted with respect to the systemic velocity of the nucleus (Fig. 1). Israel *et al.* (1990, 1991) interpret the blueshifted material as absorption due to dust lane clouds in front of the nucleus. These are rich in molecules, and have little H I. The redshifted material is interpreted as gas that is falling into the nucleus. It is poor in CO, but rich in HCO^+ and H I. The tantalizing question is whether this material is in fact the nourishment for the central monster that powers the radio source. Nearly edge-on circumnuclear torus may feed an inner accretion disk which is perpendicular to the radio jet, and obscures the nuclear source which may be as powerful as BL Lac (Morganti *et al.* 1991). A crude calculation based on Q92 model suggests a rate of inward mass transfer of $\lesssim 0.03$ $M_\odot$ yr^{-1} (Quillen, priv. comm.). This is consistent with the mass associated with the redshifted H I absorption, and is sufficient to power the nucleus (van Gorkom *et al.* 1989). Detailed observations in the cm, mm, and submm region of the spectrum allow derivation of the physical conditions in the infalling gas, and hence allow a direct probe of the feeding of an active nucleus.

Many elliptical galaxies have similar circumnuclear disks, seen by their dust absorption (Kormendy and Stauffer 1987; Jaffe *et al.* 1993). Redshifted H I absorption has been detected in some of these systems (van Gorkom *et al.* 1989). The velocity width of Hα emission rises steeply in the center of some ellipticals with gas (Phillips *et al.* 1986). It is not clear whether this is due to seeing convolution of a steep velocity gradient, is caused by increased turbulence induced by the elongation of the closed orbits, or is—in some cases—due to a rapidly rotating circumnuclear disk. Observations of molecular material in such disks will be extremely interesting.

6 CONCLUSIONS

Many elliptical galaxies contain cold gas of external origin, which often forms a warped disk detectable in Hα, H I and CO. Numerical simulations of infall are improving rapidly, but it remains difficult to estimate settling times and mass–inflow rates. Many of the observed disks cannot yet have settled in a preferred plane of the host galaxy. Detailed analysis of individual galaxies confirms this expectation, even though different models can not always be distinguished observationally. This requires high–accuracy two–dimensional velocity information, such as is available for Centaurus A and a handful of other galaxies. Detailed modeling of such data gives valuable insight in the settling process, in the intrinsic shape and mass distribution of the host galaxy, and even in the feeding of an active nucleus.

It is a pleasure to thank Joanna Lees, Alice Quillen and Hans–Walter Rix for enlightening conversations.

REFERENCES

Amico, P., Bertin, G., Bertola, F., Buson, L. M., Danziger, I. J., Dejonghe, H. B., Pizzella, A., Sadler, E. M., Stiavelli, M., Saglia, R. P., de Zeeuw, P. T., and Zeilinger, W. W. 1993, in ESO/EIPC Workshop on *Structure, Dynamics and Chemical Evolution of Elliptical Galaxies*, eds. I. J. Danziger, W. W. Zeiliger, and K. Kjär (ESO), p. 225.

Arnaboldi, M., Capaccioli, M., Cappellaro, E., Held, E., and Sparke, L., 1993, *Astr. Ap.*, **267**, 21.

Arnaboldi, M., and Sparke, L. S. 1993, *Astr. J.*, in press.

Athanassoula, E. O., and Bosma, A. 1985, *Ann. Rev. Astr. Ap.*, **23**, 147.

Bertola, F. 1987, in IAU Symposium No. 127 on *Structure and Dynamics of Elliptical Galaxies*, ed. P. T. de Zeeuw (Dordrecht: Reidel), p. 135.

Bertola, F., Bettoni, D., Danziger, I. J., Sadler, E. M., Sparke, L. S., and de Zeeuw, P. T. 1991, *Ap. J.*, **373**, 369.

Bertola, F., and Galletta, G. 1978, *Ap. J. Lett.*, **226**, L115.

Bertola, F., Rubin, V., and Zeilinger, W. 1989, *Ap. J. Lett.*, **345**, L29.

Binney, J. J. 1976, *Mon. Not. R. Astr. Soc.*, **177**, 19.

Bland, J., Taylor, K., and Atherton, P. D. 1987, *Mon. Not. R. Astr. Soc*, **228**, 595.

Bregman, J. N., Hogg, D. E., and Roberts, M. S. 1992, *Ap. J.*, **387**, 484.

Buson, L. M., Bertin, G., Bertola, F., Danziger, I. J., Dejonghe, H., Sadler, E. M., Saglia, R. P., de Zeeuw, P. T., and Zeilinger, W. W. 1993, *Astr. Ap.*, in press.

Christodoulou, D. M., Katz, N., Rix, H. W., and Habe, A. 1992, *Ap. J.*, **395**, 113.

Christodoulou, D. M., and Tohline, J. E. 1993, *Ap. J.*, **357**, 62.

Cinzano, P. A., and van der Marel, R. P. 1993, in ESO/EIPC Workshop on *Structure, Dynamics and Chemical Evolution of Elliptical Galaxies*, eds. I. J. Danziger, W. W. Zeiliger, and K. Kjär (ESO), p. 105.

Demoulin–Ulrich, M. H., Butcher, H. R., and Boksenberg, A. 1984, *Ap. J.*, **285**, 527.

de Zeeuw, P. T., and Franx, M. 1989, *Ap. J.*, **343**, 617.

de Zeeuw, P. T., and Franx, M. 1991, *Ann. Rev. Astr. Ap.*, **29**, 239.

Dubinski, J., and Christodoulou, D. M. 1993, CfA preprint 3556.

Ebneter, K., and Balick, B. 1983, *P. A. S. P.*, **95**, 675.

Ebneter, K., and Balick, B. 1985, *A. J.*, **90**, 183.

Eckart, A., Cameron, M., Rothermel, H., Wild, W., Zinnecker, H., Rydbeck, G., Olberg, M., and Wiklind, T. 1990a, *Ap. J.*, **363**, 451.

Eckart, A., Cameron, M., Genzel, R., Jackson, J. M., Rothermel, H., Stutzki, J., Rydbeck, G. and Wiklind, T., 1990b, *Ap. J.*, **365**, 522.

Forbes, D. A. 1991, *Mon. Not. R. Astr. Soc.*, **249**, 779.

Franx, M., van Gorkom, J. H., Lees, J. F., and de Zeeuw, P. T. 1993, in preparation.

Gardner, F. F., and Whiteoak, J. B. 1976, *Mon. Not. R. Astr. Soc.*, **175**, 1p.

Gerhard, O. E. 1993, in *Panchromatic View of Galaxies*, ed. G. Hensler (Paris: Edition Frontières), in press.

Gerhard, O. E., Vietri, M., and Kent, S. M. 1989, *Ap. J. Lett.*, **345**, L33.

Gunn, J. E. 1979, in *Active Galactic Nuclei*, eds. C. Hazard and S. Mitton (Cambridge: Cambridge University Press), p. 213.

Habe, A., and Ikeuchi, S. 1985, *Ap. J.*, **289**, 540.

Habe, A., and Ikeuchi, S. 1988, *Ap. J.*, **326**, 84.

Hawarden, T. G., Elson, R. A. W., Longmore, A. J., Tritton, S. B., and Corwin, H. G. 1981, *Mon. Not. R. Astr. Soc.*, **196**, 747.

Hawarden, T. G., Sandell, G., Matthews, H. E., Friberg, P., Watt, G. D., and Smith, P. A. 1993, *Mon. Not. R. Astr. Soc.*, **360**, 844.

Heiligman, G., and Schwarzschild, M. 1979, *Ap. J.*, **233**, 872.

Heisler, J., Merritt, D. R., and Schwarzschild, M. 1982, *Ap. J.*, **258**, 490.

Hernquist, L., and Katz, N. 1989, *Ap. J. Suppl.*, **70**, 419.

Hui, X. 1990, Talk presented at Aspen Workshop on *Structure and Dynamics of Galaxies*.

Israel, F. P., van Dishoeck, E. F., Baas, F., Koornneef, J., Black, J. H., and de Graauw, T. 1990, *Astr. Ap.*, **227**, 342.

Israel, F. P., van Dishoeck, E. F., Baas, F., de Graauw, T., and Phillips, T. G. 1991, *Astr. Ap.*, **245**, L13.

Jaffe, W. J., Ford, H. C., Ferrarese, L., van den Bosch, F. C., and O'Connell, R. W. 1993, *Nature*, **364**, 213.

Katz, N., and Rix, H. W. 1992, *Ap. J. Lett.*, **389**, L55.

Kormendy, J., and Djorgovski, S. 1989, *Ann. Rev. Astr. Ap.*, **27**, 235.

Kormendy, J., and Stauffer, J. 1987, in IAU Symp. 127 on *Structure and Dynamics of Elliptical Galaxies*, ed. P. T. de Zeeuw (Dordrecht: Reidel), p. 405.

Lauer, T., *et al.* 1991, *Ap. J. Lett.*, **369**, L41.

Lees, J. F. 1993, *Ap. J.*, in press.

Lees, J. F., Knapp, G. R., Rupen, M. P., and Phillips, T. G. 1991, *Ap. J.*, **379**, 177.

Macchetto, F., and Sparks, W. B. 1992, in *Morphology and Physical Classification of Galaxies*, eds. M. Capaccioli, G. Longo, and G. Busarello (Dordrecht: Kluwer), p. 191.

Magnenat, P. 1982, *Astr. Ap.*, **108**, 89.

Malin, D. F., Quinn, P. J., and Graham, J. A. 1983, *Ap. J.*, **272**, L5.

Merritt, D. R., and de Zeeuw, P. T. 1983, *Ap. J. Lett.*, **267**, L23.

Morganti, R., Robinson, A., Fosbury, R. A. E., di Serego Alighieri, S., Tadhunter, C. N., and Malin, D. F. 1991, *Mon. Not. R. Astr. Soc.*, **249**, 91.

Mulder, W. A., and Hooimeyer, J. R. M. 1984, *Astr. Ap.*, **134**, 158.

Nicholson, R. A., Bland–Hawthorne, J., and Taylor, K. 1992, *Ap. J.*, **387**, 503.

Peletier, R. F., and Christodoulou, D. M. 1993, *A. J.*, **105**, 1378.

Phillips, M. M., Jenkins, C. R., Dopita, M. A., Sadler, E. M., and Binette, L. 1986, *A. J.*, **91**, 1062.

Phillips, T. G., Ellison, B. N., Keene, J. B., Leighton, R. B., Howard, R. J., Masson, C. R., Sanders, D. B., Veidt, B., and Young, K. 1987, *Ap. J. Lett.*, **322**, L77.

Quillen, A. C., de Zeeuw, P. T., Phinney, E. S., and Phillips, T. G. 1992, *Ap. J.*, **391**, 121.

Quillen, A. C., Graham, J. R., and Frogel, J. A. 1993, *Ap. J.*, **412**, 550.

Quinn, T. 1991, In *Warped Disks and Inclined Rings around Galaxies*, eds. S. Casertano, P. Sackett, and F. Briggs (Cambridge: Cambridge Univ. Press), p. 143.

Rydbeck, G., Wiklind, T., Cameron, M., Wild, W., Eckart, A., Genzel, R., and Rothermel, R. 1993, *Astr. Ap.*, **270**, L13.

Sackett, P. 1991, In *Warped Disks and Inclined Rings around Galaxies*, eds. S. Casertano, P. Sackett, and F. Briggs (Cambridge: Cambridge Univ. Press), p. 73.

Sackett, P., and Sparke, L. S. 1990, *Ap. J.*, **361**, 408.

Sadler, E. M., and Gerhard, O. E. 1985, *Mon. Not. R. Astr. Soc.*, **214**, 177.

Sandage, A. 1961, *The Hubble Atlas of Galaxies*, Carnegie Inst. of Washington, Publ. No. 618.

Schwarz, U. J. 1985, *Astr. Ap.*, **142**, 273.

Schweizer, F., van Gorkom, J. H., and Seitzer, P. 1989, *Ap. J.*, **338**, 770.

Seaquist, E. R., and Bell, M. B. 1986, *Ap. J. Lett.*, **303**, L67.

Seaquist, E. R., and Bell, M. B. 1990, *Ap. J.*, **364**, L94.

Shields, J. C. 1991, *A. J.*, **102**, 1314.

Sparke, L. S. 1986, *Mon. Not. R. Astr. Soc.*, **219**, 657.

Sparke, L. S. 1991, in *Warped Disks and Inclined Rings around Galaxies*, eds. S. Casertano, P. Sackett, and F. Briggs (Cambridge: Cambridge Univ. Press), p. 85.

Steiman–Cameron, T. Y., and Durisen, R. H. 1982, *Ap. J. Lett.*, **263**, L63.

Steiman–Cameron, T. Y., and Durisen, R. H. 1988, *Ap. J.*, **325**, 26.

Steiman–Cameron, T. Y., and Durisen, R. H. 1990, *Ap. J.*, **357**, 62.

Steiman–Cameron, T. Y., Kormendy, J., and Durisen, R. H. 1992, *A. J.*, **104**, 1339.

Tohline, J. E., Simonson, G. F., and Caldwell, N. 1982, *Ap. J.*, **252**, 92.

Trinchieri, G., and di Serego Alighieri, S. 1991, *A. J.*, **97**, 363.

Tubbs, A. D. 1980, *Ap. J.*, **241**, 969.

van Albada, T. S., Kotanyi, C. G., and Schwarzschild, M. 1982, *Mon. Not. R. Astr. Soc.*, **198**, 303.

van Albada, T. S., and Sanders, R. H. 1982, *Mon. Not. R. Astr. Soc.*, **201**, 303.

van der Hulst, J. M., Golisch, W. F., and Haschick A. A. 198, *Ap. J. Lett.*, **264**, L37.

van Gorkom, J. H. 1992, in *Morphology and Physical Classification of Galaxies*, ed. M. Capaccioli, G. Longo, and G. Busarello (Dordrecht: Kluwer), p. 233.

van Gorkom, J. H., Knapp, G. R., Ekers, R. D., Ekers, D. D., Laing, R. A., and Polk, K. S. 1989, *A. J.*, **97**, 708.

van Gorkom, J. H., van der Hulst, J. M., Haschick, A. D., and Tubbs, A. D. 1990, *A. J.*, **99**, 1781.

Varnas, S. R. 1990, *Mon. Not. R. Astr. Soc.*, **247**, 674.

Varnas, S. R., Bertola, F., Galletta, G., Freeman, K. C., and Carter, D. 1987, *Ap. J.*, **313**, 69.

Whitmore, B. C., Lucas, R. A., McElroy, D. B., Steiman-Cameron, T. Y., Sackett, P. D., and Olling, R. P. 1990, *A. J.*, **100**, 1489.

Wilkinson, A., Sharples, R. M., Fosbury, R. A. E., and Wallace, P. T. 1986, *Mon. Not. R. Astr. Soc.*, **218**, 297.

Gas Kinematics in the LINER Radio Elliptical Galaxy NGC 4278

Joanna F. Lees

Department of Astronomy and Astrophysics, the University of Chicago, 5640 S. Ellis Ave., Chicago, IL 60637

ABSTRACT

The nearby elliptical galaxy NGC 4278 (D = 8.2 h^{-1} Mpc) has long been known to harbor an active LINER, radio nucleus as well as an extended gas disk. Observations of NGC 4278 in the 21 cm H I line using the *VLA* and recent deep, long-slit optical spectroscopy of the emission line gas are discussed. The atomic gas disk shows regular rotation and extends to over eleven times the half-light radius R_e. Noncircular motions in the central regions can be satisfactorily fit by a triaxial model for the galaxy density distribution in which the gas moves on increasingly elliptic orbits towards the nucleus. These elliptic orbits may help feed gas into the active nuclear source in NGC 4278. The high gas rotational velocities in the outer parts give evidence for a massive, dark halo surrounding this active elliptical galaxy.

1 INTRODUCTION: WHAT CAN WE LEARN FROM GAS KINEMATICS IN ELLIPTICAL GALAXIES?

Contrary to the early definitions of elliptical galaxies as old stellar systems lacking gas and dust, many early-type systems are now routinely detected in all tracers of the cool interstellar medium. About 60% of ellipticals have optical emission lines from ionized gas (Phillips *et al.* 1986), 45% show thermal dust emission at 60–100 μ (Knapp *et al.* 1989), 40% show optical absorption by dust patches (Sadler and Gerhard 1985), and about 20% are detected in H I or CO emission (Knapp 1987; Lees *et al.* 1991). Even though obtaining high signal-to-noise observations of gas in ellipticals is still a challenging undertaking, understanding the gas may be easier than for the typical spiral where dust opacity, gas processes, and the nonlinear feedback from star formation and spiral arms cannot be ignored. Furthermore, the relationship between the interstellar medium, the evolutionary history of the galaxy (perhaps including mergers and interactions), and the fueling of an active galactic nucleus common in early-type systems can be explored.

One of the most important questions in this regard is the origin of this cold gas (see for example T. de Zeeuw 1993, these proceedings). Is it a remnant from the epoch of galaxy formation, accumulated stellar mass loss, or the remains of a recent merger or interaction event? In order to determine whether or not the gas disks in these systems are in equilibrium, it is important to properly model their kinematics, including a consideration of the possibility of a triaxial shape for the main body of the galaxy. Warps and misalignments in gas disks, often used as an argument for an external origin for the gas, can in some cases be modeled successfully as a gas disk in equilibrium in a triaxial potential (for example, IC 2006 [Franx *et al.* 1993] and NGC 5077 [Bertola *et al.* 1991]). However, in other cases such as counter-rotating gaseous and stellar disks (see for example D. Fisher 1993, these proceedings) an external origin appears inevitable. The present paper discusses an attempt to model the extended gas disk in the active elliptical galaxy NGC 4278 as an equilibrium configuration in a triaxial mass distribution.

2 GAS KINEMATIC OBSERVATIONS OF NGC 4278

2.1 Radio Interferometric Observations of the Atomic Gas

Observations of the H I in NGC 4278 were made with the *VLA* in its 1 km (D) configuration with spatial and velocity resolutions of about $60''$ and 41 km s^{-1} respectively (see Lees 1992, 1993*a*). Standard reduction procedures were applied to the data cube, and the subsection of the zeroth- and first-order moment maps showing emission from NGC 4278 and the dwarf S0 NGC 4286 are presented in Figures 1 and 2. The atomic gas in the elliptical NGC 4278 (Fig. 1) appears to be rather smoothly distributed over more than eleven arcminutes (three times the galaxy's optical size) with H I column densities less than 2×10^{20} cm^{-2}; it shows regular rotation (Fig. 2). The l-v diagram and "rotation curve" are displayed in Figure 3. The gas velocities appear to continue rising well outside the optical galaxy.

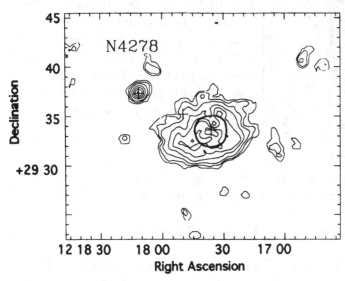

Fig. 1—The total H I surface density images showing emission from NGC 4278 (center) and NGC 4286 (upper left). The position of NGC 4283 between NGC 4278 and NGC 4286 is also marked, although no emission was detected from this galaxy. The contour levels are at 0.7, 1.8, 3.5, 7, 11, 14 and 18×10^{19} cm^{-2}. The optical size of NGC 4278, D$_{25}$, is shown by the thick ellipse.

While the gas disk is on the whole very regular, there are clear non-circular motions in the central parts (which can be seen as a twisting of the velocity contours in Figure 2) and the gas kinematic major axis in the center (at a position angle of

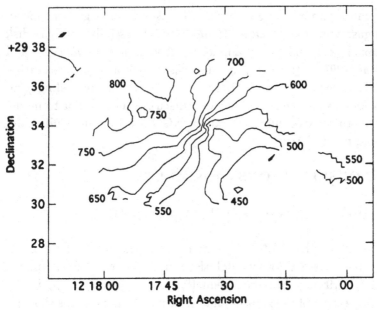

Fig. 2—Observed H I velocity field of NGC 4278. Contour levels in km s⁻² are marked. NGC 4286 appears on the edge of the upper left corner.

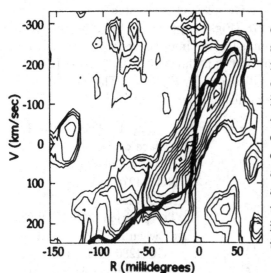

Fig. 3—The l-v plot for the H I emission from NGC 4278 (derived from summing emission along the kinematic minor axis) with the "rotation curve" (a cut along the major axis) superimposed in boldface. The contour levels are at 4, 7, 10, 20, 30,... 80 mJy beam⁻¹. The ordinate is heliocentric velocity in units of km s⁻¹ relative to the central channel at 650 km s⁻¹; the abscissa is distance along the gas kinematic major axis in units of "milli-degrees" (each tick mark is three arcminutes), relative to the central pixel.

about 70°) is strongly misaligned with the central isophotal major axis (at a position angle of 17°). The stellar and gas axes line up at large radii (see Table 1).

2.2 Optical Long-Slit Spectroscopy of Emission Line Gas

Deep (1–2 hour exposures) long-slit spectra of NGC 4278 were obtained in February 1989 at the 60″ telescope at Mount Palomar by the author and J. Gunn. Extended Hα and [N II] emission was observed along the H I kinematic major and minor axes and along slits parallel to the major axis but displaced by ±9″, ±18″, and ±27″. Very diffuse emission, clearly rotating, is detected over more than an arcminute. The

Hα/[N II] ratio appears to remain fairly constant with radius at the low value of the LINER nucleus and the line widths also do not change drastically. The emission line gas kinematics will be compared to those of the H I discussed here in a later paper.

3 TRIAXIAL MASS MODELS FOR NGC 4278

A least-squares fit to the entire H I data cube of NGC 4278 was performed assuming the gas was on equilibrium closed orbits in a principal plane of a triaxial Hernquist potential of the form

$$V_{\rm T}(r, \vartheta, \varphi) = -\frac{GM}{(r + r_0)} + \frac{GMr_1 r}{(r + r_2)^3} \, P_2^0(\vartheta) - \frac{GMr_3 r}{(r + r_4)^3} \, P_2^2(\vartheta, \varphi), \qquad (1)$$

where P_2^0 and P_2^2 are the first two spherical harmonic functions (see Lees 1992, 1993b for details of the fit). The final parameters of the model are shown in Table 1 and the resulting gas velocity field is displayed in Figure 4. This model reproduces the observed twist of the H I velocity contours in the center, as well as the shape and orientation of the optical isophotes. Thus, the observed noncircular motions and kinematic misalignments of the gas disk in NGC 4278 can be fit by an equilibrium triaxial mass model for the galaxy without the need for recent perturbations.

TABLE 1
PARAMETERS FOR NGC 4278

OPTICAL	ATOMIC GAS	MASS MODEL
$L_{\rm B} = 3.8 \times 10^9$ L$_\odot$	$M_{\rm H I} = 2.1 \times 10^8$ M$_\odot$	$M_{\rm tot} = 4.6 \times 10^{11}$ M$_\odot$
$v_{\rm opt} = 643$ km s^{-1}	$v_{\rm H I} = 636$ km s^{-1}	$i = 45°$
Θ_* [a] $= 17°$–$43°$	$\Theta_{\rm H I}$ [a] $= 70°$-$40°$	ϕ [b] $= 44°$
$R_{\rm e} = 35''$	$D_{\rm H I} = 11'$	mass scale $r_0 = 3'.3$
$D_{25} = 3'.5 \times 3'.4$		$b/a = 0.86$–0.94
		$c/a = 0.64$–0.79

[a] Θ_* and $\Theta_{\rm H I}$ are the observed position angles of the kinematic major axes of the stars and atomic gas, respectively, for small and large radii.
[b] ϕ is the azimuthal viewing angle of the galaxy.

The increasing gas velocities with radius even outside the optical galaxy, evident in Figure 3, require a mass-to-light ratio which increases with radius in NGC 4278. The cumulative mass-to-light ratio for NGC 4278 is shown as a function of radius in Figure 5.

This work was supported by NSF grants AST89–21700 and AST89–17744 and by NATO Grant No. RG.217/81.

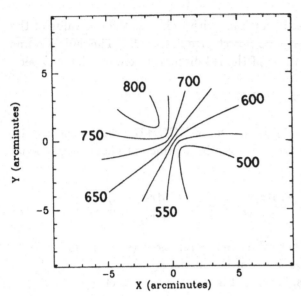

Fig. 4—The velocity field of the best-fit triaxial model for NGC 4278 described in the text. This velocity field was produced from the model data cube in the same way that the observed velocity field of Figure 2 was produced from the *VLA* data cube. Contours are marked in km s^{-1}.

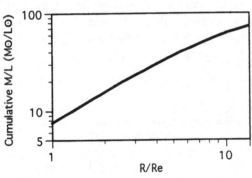

Fig. 5—The cumulative mass-to-light ratio of NGC 4278 as a function of radius (the half-light radius $R_e = 35''$) for the best-fit triaxial model. The light profile is from the R-band imaging of Peletier *et al.* (1990).

REFERENCES

Bertola, F., *et al.* 1991, *Ap. J.*, **373**, 369.

Franx, M., van Gorkom, J. H., Lees, J. F., and de Zeeuw, P. T. 1993, in preparation.

Knapp, G. R. 1987, in IAU Symp. 127 on *Structure and Dynamics of Elliptical Galaxies*, ed. P. T. de Zeeuw (Boston: Reidel), p. 145.

Knapp, G. R., Guhathakurta, P., Kim, D. -W., and Jura, M. 1989, *Ap. J. Suppl.*, **70**, 329.

Lees, J. F. 1992, *Ph. D.* thesis, Princeton University.

Lees, J. F. 1993*a*, *Ap. J. Suppl.*, submitted.

Lees, J. F. 1993*b*, *Ap. J.*, submitted.

Lees, J. F., Knapp, G. R., Rupen, M. P., and Phillips, T. G. 1991, *Ap. J.*, **379**, 177.

Peletier, R. F., Davies, R. L., Illingworth, G. D., Davis, L. E., and Cawson, M. 1990, *A. J.*, **100**, 1091.

Phillips, M. M., Jenkins, C. R., Dopita, M. A., Sadler, E. M., and Binette, L. 1986, *A. J.*, **91**, 1062.

Sadler, E. M., and Gerhard, O. E. 1985, *Mon. Not. R. Astr. Soc.*, **214**, 177.

Environmental Induced Star Formation In Galactic Cooling Flows

Shlomi Pistinner and Giora Shaviv

Department of Physics, Technion Haifa 32000, Israel

ABSTRACT

A 2D hydrodynamic Eulerian code is applied to study the effect of induced star formation in galactic cooling flows. Our main findings are: 1) Star formation activity is induced by ram pressure when the mass loss is dominated by late type giants. 2) Star formation activity is insensitive to the environment when the mass loss is dominated by early type giants. The present calculation supports the claim that the Butcher–Oemler effect is caused by star formation activity and provides a natural explanation for its source.

1 INTRODUCTION

The intra-cluster gas (ICG) is a hot rarefied gas that experiences negligible cooling during a Hubble time. This fact changes as the gas flows into the inner parts of elliptical galaxies and encounters the hot inter stellar medium (ISM). The cooling time in the ISM dominated region may become 6 to 7 orders of magnitude smaller then the Hubble time (*cf.* Portnoy *et al.* 1993, hereafter PPS). Regions in which cooling times are very short are also found in the centers of many cluster of galaxies (for a review *cf.* Sarazin 1988 and references therein). These high density and low temperatures regions are strong sources of thermal X-ray radiation, and are known as cooling flows.

1.1 The ISM Removal Problem

In contrast to cluster cooling flows that were discovered observationally (*cf.* Sarazin 1988), galactic cooling flows (in normal ellipticals in contrast to CDs) were discovered numerically in simulations performed by Shaviv and Salpeter (1982, hereafter ShS). The main motivation for their numerical simulation was the introduction of the cooling process of the gas into the simulations. As a consequence ShS frequently encountered a cooling instability.

1.2 Star Formation In Cooling Flows

The cooling instabilities encountered by ShS led to the introduction of the assumption of star formation. This allowed Gaetz *et al.* (1986, hereafter GSSH) to

437

follow the full evolution of the ISM inside the moving galaxy. The underlying assumption is that *the removed gas will eventually collapse and form stars*. The same assumption and prescription were used by White and Sarazin (1988, hereafter WS) to simulated star formation in cluster cooling flows and spherical galaxies that are in the centers of clusters and are almost stationary. Note, however that since the assumption of spherical symmetry was made, WS could not consider dynamical environmental effects.

We performed a 2D numerical simulation over a range of parameters with the purpose of gaining some insight into environmental effects on star formation within cooling flows in ellipticals ISM. We wish to stress that only a negligible amount of star formation activity was observed in cluster cooling flows and so far no star formation activities were ever observed in ISM galactic cooling flows. Furthermore, recently Radio observation of elliptical galaxies revealed non negligible amounts of cold gas in the form of H I regions and molecular clouds (*cf.* Lees *et al.* 1991 and references therein). At about the same time absorption of X-ray radiation by oxygen in cluster cooling flows was observed (*cf.* White *et al.* 1991). This leads to the conclusion that star formation activity in cooling flows may not be very efficient. However, the results of the study made by us should be regarded as a phase transition from a hot phase to a cold one. Furthermore, we stress that the final fate of the gas, whether it is forming stars, small molecular clouds, or large neutral hydrogen regions does not affect the physical interpretation that shall be given below. The term induced star formation may replaced by induced cold ISM or ICG generation.

2. THE NUMERICAL SIMULATION

The governing equations describing the gas evolution in galaxies interacting with the environment were discussed in great detail by PPS. The numerical technique used to perform the simulation is discussed in some length by GSSH. The cluster parameters for the simulation were the density at infinity, the velocity of the galaxy, and the upstream cluster temperature.

2.1 Galactic Parameters

We follow PPS and assume in the present simulations axial symmetry with no rotational component in the flow. The galaxy is considered to be spherical and the distribution of gas replenishment is assumed to follow the distribution of the spherical stellar mass. We assume the existence of a moderately massive halo, the sole effect of which is gravitational. We assume a generalized King type model for the stellar mass distribution namely:

$$\rho_* = \rho_*(0)\left\{1 + \left(\frac{r}{R_c}\right)^2\right\}^{-\beta}, \quad r < R_{\text{gal}}, \tag{1}$$

where R_{gal} is about ~ 16 kpc, R_c is the core radius $\sim 0.05R_{gal}$ and $\rho_*(0)$ is the central density of stars. The central density is adjusted in such a way, so as to obtain a prescribed stellar galactic mass. The distribution of the mass losing stars is also assumed to follow equation (1). The mass distribution of the halo is given by equation (1) if the $*$ is replaced by *halo*, R_{halo} is the halo's radius, taken to be ~ 32 kpc, with identical R_c, and $\rho_{halo}(0)$ is the central density of the halo material. The central density is adjusted in such a way as to fix the sum of dark and stellar galactic mass to the prescribed value which is $\sim 5.9 \times 10^{11}$ $M_\odot$. The value $\beta = 3/2$ corresponds to the de Vaucouleur law (*cf.* Binney and Tremaine 1987). The mean gas replenishment rate is given by: $\dot{\rho}_{rep} = \alpha_* \rho_*$, where α_*^{-1} is a characteristic time for the rate of replenishment. Typical quoted values of α_* are $\alpha_* = 10^{-11} - 10^{-13}$ yr^{-1} which for our choice of galactic mass correspond to $(5.6 - 0.05)$ $M_\odot$ yr^{-1}. The present calculation does not distinguish between the various sources of gas replenishment in the galaxy and the replenishment is represented by means of two parameters: The rate of gas replenishment $\dot{\rho}_{rep}$ and replenishment temperature T_{rep}. Mathews (1990) and PPS discuss the appropriate value of T_{rep} in detail and following them we assumed $T_{rep} \sim T_{virial} \sim 10^7$ K.

2.2 Star Formation Prescription

The rate of star formation (the sink terms) was taken from WS, with a small variation, namely $\dot{\rho}_{sf} = q\rho/\tau_{cool}$, where τ_{cool} corresponds to the standard definition (*cf.* PPS), ρ is the density of the gas, and q is the star formation "efficiency", and is considered to be of order unity. In our case we assume $q = 1$. Still, note that $q = 2$ reminds the Schmidt law (Schmidt 1959) for star formation in the ISM. The "star formation" algorithm was turned on whenever the number density of the interstellar gas was greater than $n_{gas} = 1.2$ cm^{-3} and $T_{gas} < 3 \times 10^5$ K.

3. RESULTS

In the simulation the temperature of the ICG was $(3.5 - 7) \times 10^7$ K and the velocity of the galaxy was such that $Mach = 1.5$. The rate of replenishment 0.1 $M_\odot$ yr$^{-1} \leq \dot{M}_{rep} \leq 10$ $M_\odot$ yr^{-1}, and the cluster number density at infinity (upstream) was 10^{-3} cm$^{-3} \leq n_\infty \leq 10^{-7}$ cm^{-3}. The simulations were performed with $\beta = 1$ and $3/2$. No star formation was found in cases $\beta = 1$. In the case with $\beta = 3/2$ we find star formation for $\dot{M}_{rep} \geq 1$ $M_\odot$ yr^{-1}. The rate of star formation is independent of n_∞. The values of T_{rep} and $\dot{M}_{rep}$ for which star formation occurs correspond to mass loss from early type giant stars (*cf.* Mathews 1990). For $\dot{M}_{rep} = 0.1$ $M_\odot$ yr^{-1} we find star formation only for 10^{-5} cm$^{-3} < n_\infty < 10^{-3}$ cm^{-3}. The value of $\dot{M}_{rep}$ is such that the main mass contributors are late type giants. The sensitivity of the star formation to the environmental condition is clearly seen. The physical interpretation of the phenomena is simple but counter intuitive. The effect emerges naturally if we write down the kinetic energy equation namely:

$$\frac{\partial}{\partial t}(\rho\frac{v^2}{2}) + \vec{\nabla}\cdot(\rho\vec{v}\frac{v^2}{2}) = -\vec{v}\cdot\nabla P + \rho\vec{v}\cdot\nabla\phi - (\dot{\rho}_{sf} + \dot{\rho}_{rep})\frac{v^2}{2}, \qquad (2)$$

where P is the gas pressure ϕ the gravitational potential and $\vec{v}$ the velocity of the gas. Note that the $\dot{\rho}_{rep}$ term is a dissipative term in the kinetic energy equation. However, it should be stressed that mass replenishment contributes additional energy to the total energy balance. The loss of kinetic energy by the ISM as it interacts with the injected gas impedes its sweeping away. The ICG that succeeds to penetrate the mass injection regions will lose kinetic energy as it flows in the galactic gravitational potential well and mix with the ISM. The nature of the difference between $\beta = 1$ or $3/2$ can be understood if we note that $T_{rep} \sim T_{virial}$. The ISM has the tendency to expand from the region where mass replenishment takes place. This region is deep inside the gravitational potential. For $\beta = 1$ most of the mass is injected in the outskirts of the galaxy. Furthermore, the condition of star formation is rarely met, since the maximum value of the density is not large enough. The gas density at the center of the galaxy in the case $\beta = 3/2$ is considerably larger then in the $\beta = 1$ case, this leads to faster cooling.

High rate of replenishment, Low density and velocity at infinity.
Galaxy in accretion

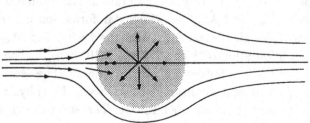

Low rate of replenishment, high density and velocity at infinity
Complete stripping.

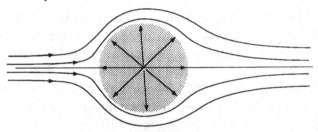

Medium rate of replenishment, density and velocity at infinity.
Ram pressure driven Star Formation in at the center.

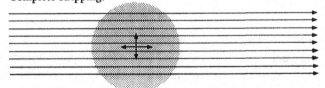

Three different qualitative cases are found, these cases are described in the figure in the previous page. The region in which gas is injected is shaded.

1) A *Stellar Wind* in which the injected ISM does not allow the ICG to penetrate into the injection region. In this case star formation is completely independent of the environment. This case is found for $\dot{M}_{rep} \geq 1$ $M_\odot$ yr^{-1} and $\dot{M}_{rep} = 0.1$ $M_\odot$ yr^{-1} with $n_\infty > 10^{-5}$ cm^{-3}.

2) Full *Stripping* in which the ICG kinetic energy is strong enough to push and sweep the ISM out of the galaxy. Kinetic energy dissipation is still there but, it is not enough to reduce the ICG kinetic energy significantly. This case is found for $\dot{M}_{rep} = 0.1$ $M_\odot$ yr^{-1} and $n_\infty = 10^{-3}$ cm^{-3}.

3) The *Induced Star Formation* case in which the ICG kinetic energy is reduced significantly due to the flow through the dissipative region. The ISM can escape from the galaxy only through a narrow region in the wake, as a result one obtains density enhancement that leads to stronger cooling. The stronger cooling decreases the temperature to a point where gas collapses to form stars.

Its has been postulated in the past that the nature of the Butcher–Oemler effect is a result of environmental induced star formation (*cf.* Sarazin 1988). If the gas removed from the calculation forms stars (and not another form of cold gas), then the present calculation are in support of the last claim.

REFERENCES

Binney, J., and Tremaine, S. 1987, *Galactic Dynamics* (Princeton: Princeton University Press).

Gaetz, T., Salpeter, E. E., and Shaviv, G. 1987, *Ap. J.*, **316**, 530. (GSSH)

Lees, J. F., Knapp, G. R., Rupen, M. P., and Phillips, T. G. 1991, *Ap. J.*, **379**, 177.

Mathews, W. G. 1990, *Ap. J.*, **354**, 468.

Portnoy D., Pistinner S., and Shaviv G. 1993 *Ap. J. Suppl.*, **86**, 95. (PPS)

Sarazin, C. L. 1988, in *X-Ray Emission from Cluster of Galaxies*, (Cambridge: Cambridge University Press).

Schmidt, M. 1959, *Ap. J.*, **129**, 243.

Shaviv, G., and Salpeter, E. E. 1982, *Astr. Ap.*, **110**, 300. (ShS)

White, D. A, Fabian, A. C., Johnstone, R. M., Mushotzky, R. F., and Arnaud, K. A. 1991, *Mon. Not. R. Astr. Soc.*, **252**, 72.

White, III R. E., and Sarazin, C. L. 1987, *Ap. J.*, **318**, 612. (WS)

Photometry of a Complete Sample of Virgo Ellipticals

Frank C. van den Bosch[1], Walter Jaffe[1], Holland C. Ford[2], and
Laura Ferrarese[3]

[1]Leiden Observatory, Leiden, The Netherlands
[2]Space Telescope Science Institute, Baltimore, U. S. A.
[3]Johns Hopkins University, Baltimore, U. S. A.

1 INTRODUCTION

It has become evident in recent years that elliptical galaxies harbor many interesting features. Currently in $\sim$ 40% of all ellipticals dust has been detected (*e.g.*, Sadler and Gerhard 1985). Several galaxies are found to harbor decoupled cores. Although the best evidence for these entities is provided by kinematical data (*i.e.* counter-rotating cores), Bender (1988) found in four cases an interesting correspondence between kinematical decoupling and the ellipticity profile within the region where kinematical decoupling takes place.

Since these features (including the gas and/or dust) are often located in or near the nuclei of these galaxies, many of them may still be undetected due to the atmospheric smearing. We therefore undertook a program of high resolution Hubble Space Telescope *(HST)* imaging of a complete sample of 12 elliptical galaxies in the Virgo cluster. After standard reduction the images were deconvolved by Fourier filtering.

2 RESULTS

The majority of our galaxies show peculiar near-nuclear morphology (Jaffe *et al.* 1993*b*): NGC 4261 (3C 270), one of the two active galaxies in our sample, was found to harbor a small, smooth, dusty disk around a point-like nucleus (Jaffe *et al.* 1993*a*). We found the disk, which we interpret as the outer accretion disk, to be perpendicular to the jet axis. NGC 4476 shows a large circumnuclear ring or disk of dust, whereas the dust is filamentary in NGC 4550 and NGC 4374.

The isophotal analysis (van den Bosch and Ferrarese 1993) showed a similar variety of structures. In three galaxies, NGC 4342, NGC 4570 and NGC 4623 , we found disky isophotes in the inner $1'' - 2''$. Although this might be consistent with a stellar disk embedded in the spheroidal body, we note that without kinematical evidence for a separate stellar component, disky isophotal shapes can be equally well explained otherwise (*e.g.* de Zeeuw *et al.* 1986). All these galaxies also appear to have an outer disk, and therefore resemble S0 galaxies with an inner nuclear disk of $\sim$ 200 pc. For the one galaxy, NGC 4365, where kinematical evidence is available for a decoupled, counter-rotating core (Surma 1993) we find, however, no evidence for such a separate component from our photometry. For the remaining galaxies we often find strange behavior of the isophotes (see Table 1). For the four galaxies with dust in the inner 10″, this is to be expected since dust can easily disturb the isophotes. For the other cases we note that since our passband covers Hα and [N II] small amounts of ionized gas might be present, which also disturbs the isophotal shapes.

Evidently, when studied with *HST* resolution the nucleus of each elliptical seems to have a unique detailed morphology. Table 1 summarizes the results discussed above.

TABLE 1

SAMPLE

Galaxy	RSA	RC2	Characteristic
NGC 4168	E2	E2	photometrically decoupled core
NGC 4261	E3	E2	dust (accretion) disk
NGC 4342	E7	–	indication for stellar disk
NGC 4365	E3	E3	counter-rotating core
NGC 4374	E1	E1	dust lanes
NGC 4473	E5	E5	disky isophotes
NGC 4476	E5pec	S0-	dust ring or disk
NGC 4478	E2	E2	photometrically decoupled core
NGC 4550	E7/S01	SB0	filamentary dust + two counter-streaming disks
NGC 4564	E6	E6	photometrically decoupled core
NGC 4570	S01/E7	S0	indication for stellar disk
NGC 4623	E7	SB0+	indication for stellar disk

REFERENCES

Bender, R. 1988, *Astr. Ap. Lett.*, **202**, L5.

de Zeeuw, P. T., Peletier, R., and Franx, M. 1986, *Mon. Not .R. Astr. Soc.*, **221**, 1001.

Jaffe, W., Ford, H. C., Ferrarese, L., van den Bosch, F. C., and O'Connell, R. W. 1993a, *Nature*, **364**, 213.

Jaffe, W., Ford, H. C., Ferrarese, L., and van den Bosch, F. C. 1993b, in preparation.

Sadler, E. M., and Gerhard, O. E. 1985, *Mon. Not .R. Astr. Soc.*, **214**, 177.

Surma, P. 1993, in *Structure, Dynamics, and Chemical Evolution of Elliptical Galaxies*, ed. I. J. Danziger, W. W. Zeilinger, and K. Kjär (ESO: Garching), p. 669.

van den Bosch, F. C., and Ferrarese, L. 1993, in preparation.

Discovery of Very Red Colors in GPS Radio Galaxies

Chris O'Dea[1], Carlo Stanghellini[2], Eija Laurikainen[3], and Stefi Baum[1]

[1]Space Telescope Science Institute, U. S. A.
[2]Istituto di Radioastronomia, del CNR, Italy
[3]Turku University Observatory, Finland

ABSTRACT

We present $r - i$ colors of GHz–Peaked Spectrum (GPS) Radio Galaxies. We find that most GPS radio galaxies have $r - i$ colors in the range of $0.2 - 0.4$, typical of passively evolving elliptical galaxies. However, several have much redder colors in the range 1–2. We suggest several possible explanations for the very red colors, including (1) large amounts of dust in the galaxy, (2) a significant post-starburst population, and (3) a highly reddened active galactic nucleus. The red colors are consistent with a hypothesis in which significant mass transfer has occurred, producing a dense, clumpy, and possibly dusty ISM in the GPS host galaxy.

1 INTRODUCTION

GPS radio galaxies are characterized by the following properties: (1) a simple, convex, radio spectrum, peaking near 1 GHz, (2) mostly compact (sub kpc) radio structure, (3) low radio and optical polarization, and (4) low variability. We (O'Dea, Baum, and Stanghellini 1991) have suggested that the GPS galaxies have a dense, clumpy ISM (probably acquired externally via cannibalism) that confines and depolarizes the radio source. Our r and i band optical imaging has shown that the GPS galaxies are often in strongly interacting systems (Stanghellini et al. 1993).

2 THE $R - I$ COLORS

We find that most GPS radio galaxies have integrated $r - i$ colors in the range of $0.2 - 0.4$, typical of passively evolving elliptical galaxies. This suggests that most GPS galaxies are not currently undergoing a significant starburst (though they might have in the past), unless the starburst population is highly obscured. However, several galaxies have much redder $r - i$ colors in the range 1–2. We suggest several possible explanations for the very red galaxies, including (1) large amounts of dust in the galaxy, (2) a significant post-starburst population, and (3) a highly reddened active galactic nucleus.

(1) A difference in the $r - i$ of unity require an extra $E(B - V) \simeq 1.4$ which corresponds to an extinction in r of $A(r) \simeq 3.5$ mag or in V of $A(V) \simeq 6$ mag. Extinctions this large would not be consistent with the observed apparent magnitudes of these

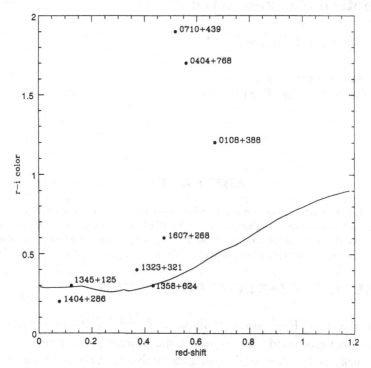

Fig. 1—The integrated $r - i$ colors (Thuan and Gunn system) as a function of redshift for the GPS radio galaxies. The curve is the expected color for a passively evolving elliptical galaxy (courtesy of Stephane Charlot).

objects if the GPS galaxies have hosts which are similar to the powerful 3CR FR II radio galaxies. Better optical spectra are needed to test the dust hypothesis.

(2) S. Charlot has suggested to us that very red integrated colors could be produced by a post-starburst ($\sim 10^8$ yr) population with an IMF truncated below a few $M_\odot$. Better optical spectra are needed to test this hypothesis.

(3) A comparison of multicolor images of 0108+388 (r and i) and 2322–040 (r and K) suggests that in at least these two objects, there is a highly reddened AGN (O'Dea *et al.* 1992). K band images of the GPS galaxies are needed to search for a highly reddened nucleus.

We are grateful to Stephane Charlot for helpful discussions.

REFERENCES

O'Dea, C. P., *et al.* 1992, in *Testing the AGN Paradigm*, AIP Conference Proceedings 254, eds. S. S. Holt, S. G. Neff, and C. M. Urry (New York: AIP), p. 435.

O'Dea, C. P., Baum, S. A., and Stanghellini, C. 1991, *Ap. J.*, **380**, 66.

Stanghellini, C., O'Dea, C. P., Baum, S. A., and Laurikainen, E., 1993, *Ap. J. Suppl.*, in press.

Are All Elliptical Galaxies Active?

Elaine M. Sadler[1] and Bruce Slee[2]

[1]Anglo–Australian Observatory
[2]Australia Telescope National Facility

ABSTRACT

Many bright elliptical galaxies are active in the sense of having compact radio cores of high brightness temperature ('engines') and/or a LINER–like optical emission spectrum. Nuclear activity is very common in the most luminous galaxies (brighter than absolute magnitude $M_B \sim -21$) and essentially absent in those less luminous than $M_B \sim -19$.

1 ACTIVE NUCLEI IN EARLY–TYPE GALAXIES

The presence of LINER emission (Phillips *et al.* 1986) and the fact that many E and S0 galaxies have arcsecond–scale central radio sources suggest that some kind of active nucleus lurks at the centre of most galactic bulges brighter than $M_B \sim -19$ ($H_0 = 100$ km s^{-1} Mpc^{-1}).

Wrobel and Heeschen (1991), however, argue that the central radio emission in many S0 galaxies may be associated with star formation rather than an active nucleus. Furthermore, the emission–line luminosity in E and S0 nuclei correlates more closely with the luminosity of the parent galaxy than with other indicators of activity such as radio emission (Sadler *et al.* 1989), suggesting that the dominant ionization mechanism may be linked to the underlying stellar population rather than to an active nucleus. Optical spectroscopy and arcsecond–scale radio maps, therefore, may not provide an unambiguous test for the presence of low–level 'central engines' in these galaxies.

2 PARSEC–SCALE RADIO CORES

The Parkes–Tidbinbilla Interferometer (*PTI*) is the world's longest real–time interferometer, with a 275 km baseline and a resolution of 0.03 arcsec at 8.4 GHz (corresponding to 3 pc for galaxies at a distance of 20 Mpc). As Norris *et al.* (1990) point out, it is sensitive to parsec–scale quasar cores with brightness temperatures above 10^5 K but not to kiloparsec–scale starburst regions with brightness temperatures around 10^4 K, so provides a powerful way to discriminate between a nuclear starburst and a small 'central engine'.

We recently used the PTI to search for parsec–scale radio cores in early–type

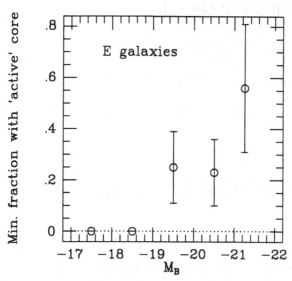

Fig. 1—Detection rate of parsec–scale radio cores ('engines') in nearby elliptical galaxies. The detection rate is a lower limit to the true fraction of 'active' nuclei since some weak cores may fall below the detection limit.

galaxies (Slee *et al.* 1993), and here we use the detection rate in a complete, optically-selected sample of 116 nearby E and S0 galaxies to derive a lower limit to the true fraction of 'active' objects. Figure 1 shows the detection rate (above a 5 mJy limit) as a function of absolute magnitude M_B for elliptical galaxies. The parsec–scale cores in nearby galaxies appear qualitatively similar, in both size and spectral index, to the cores of powerful radio galaxies and quasars, but are several orders of magnitude less powerful ($10^{20} - 10^{23}$ W Hz^{-1} at 5 GHz, compared to $10^{25} - 10^{27}$ W Hz^{-1} for many quasar cores). We conclude that:

• At least 50% of all nearby E and S0 galaxies brighter than $M_B = -21$ appear to have central engines (*i.e.* parsec–scale radio cores with high brightness temperature).

• For galaxies in the range $M_B - 19$ to -21, at least 20% of ellipticals and 5% of S0s have central engines.

• Below $M_B = -19$ central engines appear to be rare, in agreement with the suggestion by Wrobel and Heeschen (1991) that the central radio emission in these galaxies is predominantly driven by star formation rather than an active nucleus.

REFERENCES

Norris, R. P., Allen, D. A., Sramek, R. A., Kesteven, M. J., and Troup, E. R. 1990, *Mon. Not. R. Astr. Soc.*, **359**, 291.

Phillips, M. M., Jenkins, C. R., Dopita, M. A., Sadler, E. M., and Binette, L. 1986, *A. J.*, **91**, 1062.

Sadler, E. M., Jenkins, C. R., and Kotanyi, C. G. 1989, *Mon. Not. R. Astr. Soc.*, **240**, 591.

Slee, O. B., Sadler, E. M., Reynolds, J. E., and Ekers, R. D. 1993, *Mon. Not. R. Astr. Soc.*, submitted.

Wrobel, J. M., and Heeschen, D. S. 1991, *A. J.*, **101**, 148.

Habitat Segregation of Elliptical Galaxies

Yasuhiro Shioya and Yoshiaki Taniguchi

Astronomical Institute, Tohoku University, Japan

ABSTRACT

We have found possible evidence that boxy-type elliptical galaxies favor the environment of clusters of galaxies, while disky-type ellipticals prefer the field environment.

1 MOTIVATION, DATA, AND RESULTS

Recent high-quality imaging and spectroscopic studies have shown that there is a fundamental distinction in elliptical galaxies : the isophotal shapes of elliptical galaxies (*cf.* Bender *et al.* 1989). Elliptical galaxies have three types of isophotes: boxy, elliptic, and disky. It should be stressed that these types are more physically related to the dynamical properties of the elliptical galaxies than are the fine morphologically peculiar features such as shells. The boxy elliptical galaxies owe their shape to the anisotropy of velocity dispersion of stars while the disky type are flattened by rotational motion. This dynamical difference between them suggests that the two types of galaxies have different formation histories. It is therefore natural to ask whether the difference in the isophotal shapes of elliptical galaxies may have some relation to their environments.

Our study is based on the data in Bender *et al.* (1989) who compiled published data as well as their own results of detailed CCD photometry of 109 elliptical galaxies. Among the 109 elliptical galaxies, we assigned the environments for 96 galaxies based on the paper by Faber *et al.* (1989). The elliptical galaxies in Virgo, Coma, Abell 194, and Abell 1367 clusters of galaxies are classified as *cluster ellipticals*. The elliptical galaxies to which no group identification has been given are classified as *field ellipticals*. The remaining ellipticals belong to the group of galaxies. In this way, we obtained a sample of 96 elliptical galaxies.

Figure 1 shows the frequency distributions of elliptical galaxies (boxy, irregular, elliptic, and disky) for environments from rich clusters through groups to the field. It is shown that boxy elliptical galaxies tend to reside within rich clusters of galaxies, while the disky type reside in the field.

448

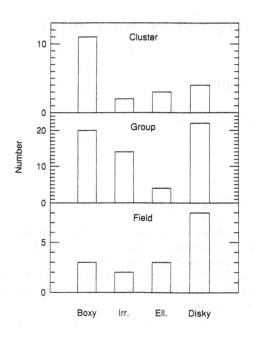

Fig. 1—Frequency distributions of boxy, irregular, elliptic and disky elliptical galaxies among the environments of clusters, groups, and the field.

2 DISCUSSION

We summarize the possible formation mechanisms of boxy- and disky-type elliptical galaxies, while taking account of habitat segregation.

Boxy ellipticals were formed by a dissipative merger of two gas-rich disk galaxies. Since it is expected that merger events occur more frequently in a cluster of galaxies than in the field, the boxy-type ellipticals are a dominant population in the cluster environment.

Disky ellipticals were formed through the classical dissipative collapse of proto galactic gas clouds. During this process a weak disk component was formed. This disk component would be easily destroyed by an interaction or a merging. Since, however, the number density of galaxies in the field is smaller than that in rich clusters of galaxies, merger events would be rarer during the Hubble time. Thus, the disky structure may be longer-lived in the field environment. However, it is likely that field disky ellipticals have experienced some minor merger events with gas-rich smaller galaxies, being responsible for the bluer colors and the presence of morphologically fine structure in field ellipticals (for details see Shioya and Taniguchi 1993).

REFERENCES

Bender, R., Surma, P., Döbereiner, S., Möllenhoff, C., and Madejsky, R. 1989, *Astr. Ap.*, **217**, 35.

Faber, S. M., Wegner, G., Burstein, D., Davies, R., Dressler, A., Lynden-Bell, D., and Terlevich, R. J. 1989, *Ap. J. Suppl.*, **69**, 763.

Shioya, Y., and Taniguchi, Y. 1993, *P. A. S. J.*, **45**, No. 3, in press.

Nebular Properties and the Origin of the Interstellar Medium in Elliptical Galaxies

Joseph C. Shields and Fred Hamann

Department of Astronomy, Ohio State University

ABSTRACT

Previous researchers have suggested that much of the cold interstellar gas in present-day elliptical galaxies is accreted from external sources. The strength of forbidden-line emission in elliptical galaxies provides a constraint on the enrichment history of the gas. Based on photoionization calculations, we conclude that the gas, if accreted, must originate in donor galaxies with metallicities > 0.5 $Z_\odot$. This excludes primordial clouds and Magellanic Cloud-like objects as typical gas donors.

1 INTRODUCTION

Elliptical galaxies often contain modest quantities of interstellar gas that can be generated via normal mass loss by the galaxies' constituent stars, on timescales much shorter than a Hubble time (Faber and Gallagher 1976). Diffuse matter generated by such internal sources may be rapidly removed from the interstellar medium (ISM), however, if this material is heated to X-ray temperatures and expelled in a galactic wind, or compressed to form new stars in a cooling flow. In an alternative scenario, objects which feature significant cold interstellar gas may have acquired this matter by accretion from nearby galaxies or intergalactic clouds. Evidence in support of an external origin for the ISM in ellipticals includes a lack of correlation between interstellar and stellar masses (*e.g.*, Knapp *et al.* 1985), and distinct kinematics for the gaseous and stellar components seen in some objects (*e.g.*, Bertola *et al.* 1990).

2 ABUNDANCES AS A DISCRIMINATOR OF ISM ORIGIN

Stars in large elliptical galaxies are inferred to have average heavy element abundances $\gtrsim 2$ $Z_\odot$, based on observational estimates and predictions from chemical evolution models (*e.g.*, Bica *et al.* 1988). Interstellar matter originating in mass loss from these stars would be expected to have comparable metallicity. In contrast, gas-rich galaxy disks, dwarf irregulars, and intergalactic clouds that comprise likely external ISM donors have metallicities that are likely to be at most $\sim Z_\odot$, and possibly considerably lower. Estimates of heavy element abundances in the ISM of elliptical galaxies thus provide a means of discriminating between internal and external origins for this material.

Nebular emission from elliptical galaxies commonly exhibits relatively strong [N II] and other low-ionization features (*e.g.*, Phillips *et al.* 1986) that may be generated through photoionization by a weak active nucleus, cooling X-ray gas, or hot stars. Nitrogen is particularly attractive as a metallicity diagnostic since it features a Z^2 dependence arising from secondary conversion of C and O into N via CNO burning in stellar envelopes. In order to understand the dependence of [N II] emission on Z, we completed photoionization calculations with G. Ferland's code CLOUDY, with parameters chosen to maximize [N II] emission (*i.e.*, density 10^4 cm^{-3}; abundances of other coolants subject to grain depletion; power-law $f_\nu \propto \nu^{-1}$ ionizing continuum). Predictions of [N II]/Hα as a function of ionization parameter (U; the ratio of ionizing photon to nucleon density) are shown in Figure 1.

Elliptical galaxies commonly have [N II]/H$\alpha > 1$, and this threshold is shown in the Figure. The minimum metallicity at which this ratio can be obtained is ~ 0.5 $Z_\odot$. Very high abundances are not restricted, since emission under less optimal conditions will produce lower ratios than those shown. As a result of this comparison, we conclude that the warm interstellar matter in ellipticals is mostly intrinsic in origin, or else accreted from relatively enriched external donor galaxies. Low-metallicity irregular galaxies (like the Magellanic clouds) and unenriched primordial clouds are unlikely to provide the nebular gas observed in elliptical galaxies.

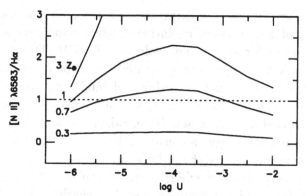

Fig. 1—Predicted [N II]/Hα as a function of ionization parameter (U; ratio of ionizing photon to nucleon density).

REFERENCES

Bertola, F., Bettoni, D., Buson, L. M., and Zeilinger, W. W. 1990, in *Dynamics and Interactions of Galaxies*, ed. R. Wielen (Berlin: Springer), p. 249.
Bica, E., Arimoto, N., and Alloin, D. 1988, *Astr. Ap.*, **202**, 8.
Faber, S. M., and Gallagher, J. S. 1976, *Ap. J.*, **204**, 365.
Knapp, G. R., Turner, E. L., and Cunniffe, P. E. 1985, *A. J.*, **90**, 454.
Phillips, M. M., *et al.* 1986, *A. J.*, **91**, 1062.

Early AGN and Galaxy Formation

S. Djorgovski

Palomar Observatory, 105-24 Caltech, Pasadena, CA 91125, U. S. A.

ABSTRACT

Formation of first AGN probably follows closely the formation of their host galaxies. Both processes may involve similar astrophysical events or processes, with comparable energetics. Young AGN could have had a profound effect on the host galaxies. High-z AGN may be used as markers of galaxy formation sites, or probes of early large-scale structure. Some may well be examples of galaxies in the early stages of formation, although the nature of their dominant energy sources and their evolutionary status remain ambiguous. Future observations at IR through sub-mm wavelengths lead to a discovery of possible obscured protogalaxies and nascent AGN, and probe their formation at very high z.

1 INTRODUCTION: FORMATION OF GALAXIES AND AGN

Probably the two key problems in extragalactic astronomy and cosmology today are the understanding of formation and evolution of galaxies and large-scale structure, and the understanding of formation and evolution of AGN. These problems may be fundamentally connected: it is now generally understood that the same kind of astrophysical processes, dissipative merging and infall, may be central to both formation of galaxies and formation of AGN, and subsequently to be contributing both to the simulation of star forming activity in galaxies and to the feeding of their central engines. AGN may also exert a considerable feedback to their host galaxies, and perhaps even determine some of their global properties.

Evidence for interaction-driven starbursts and nuclear activity in galaxies is both abundant and compelling, and has been discussed at length in other contributions to this volume. Good recent reviews include, *e.g.*, those by Heckman (1990, 1991), Barnes and Hernquist (1992), and many others. It is certainly plausible that dissipative merging can solve the AGN fueling problems (Shlosman *et al.* 1990) by depositing large amounts of gas to the central regions of interacting galaxies, with most of the angular momentum transferred to the stellar and dark matter component (Hernquist 1989; Barnes and Hernquist 1991, *etc.*). This process can equally well lead to the *formation* of central engines, perhaps through a formation of a dense central star cluster, and its subsequent collapse. This does not exclude other possibilities, *e.g.*, an early formation of massive black holes from the initial conditions, etc. Dissipative merging is also the probable mechanism by which elliptical galaxies and bulges were

452

formed; see Kormendy and Djorgovski (1989) for a discussion and references.

It is perfectly possible that most or all ellipticals and luminous bulges today contain remnants of the former AGN, "dead" (or only sleeping?) quasars, and that every one of them has undergone an early AGN phase, at least once (*cf.* Rees 1990, 1992; Small and Blandford 1992; or Haehnelt and Rees 1993, and references therein). There is a kinematical evidence for the quiescent central mass concentrations in the nuclei of several nearby galaxies, including M31, M32, and others (*cf.* Kormendy 1993 for a review). If the ubiquity of dormant engines is confirmed by future observations, it would support the idea that the formation of old, metal-rich stellar systems and nuclear activity in galaxies may have been intimately connected.

1.1 Quasar Pairs at Large Redshifts: Caught in the Act?

Numerous studies have found a preponderance of starbursts or AGN in interacting or post-merger systems at low redshifts. It is less well appreciated that such systems may have been detected at high redshifts as well ($z \sim 2 - 3$), at the epochs where the comoving density of quasars was at its peak. A number of high-redshift radio galaxies shows a morphology suggestive of mergers and interactions, although other effects are probably also at play (Djorgovski 1988*a*; McCarthy 1993, and references therein). But there are also some rather unambiguous pairs of quasars seen.

If quasars represent relatively rare events in galactic nuclei, physically close pairs of quasars seem at first extremely unlikely. Yet, several such pairs at large redshifts, with separations of a few arcsec (tens of kpc in projection), are now known (Djorgovski 1991). These are *not* gravitational lenses: their spectra, redshifts, and other properties are usually sufficiently different, so that the physical pair interpretation seems more likely. Djorgovski (1991) tentatively estimated their frequency of occurrence to be as high as $\sim 10^{-2}$, and probably not less than $\sim 10^{-3}$, but a detailed analysis of the probabilities integrated over the luminosity function, selection effects, *etc.*, still remains to be done. A simple clustering model suggests that the observed numbers of wide quasar pairs (arcmin, or Mpc scale) are about as expected. However, the close pairs, with the ~ 100 kpc separations, appear to be overabundant by about two orders of magnitude. It is intriguing that this happens at the scales corresponding to the size of galactic dark halos, which is roughly the distance where tidal interactions should become important. These systems may represent birth events of quasar pairs, at the redshifts corresponding to the peak of the quasar era.

1.2 Evolution of AGN and the Epoch of Galaxy Formation

It is probably fair to say that we still do not have a complete theory of galaxy and large-scale structure formation. However, in a range of hierarchical models, including the CDM scenario, the peak merging epoch, which may be associated with the "galaxy formation epoch", spans a broad maximum near $z \sim 2 \pm 1$ (White and

Frenk 1991). It is curious that the number density of quasars also shows a broad maximum at the comparable redshifts (Hartwick and Schade 1990, and references therein). Several authors (*e.g.*, Carlberg 1990) attempted to model the evolution of quasar birthrate governed by the merger rate. The observed evolution of AGN could thus be telling us directly about the history of galaxy formation.

This is an appealing concept, but not quite a fully convincing one. There are, after all, quasars and AGN associated with well formed, apparently old galaxies at $z \sim 0$. The epoch of the rise of quasars is at $z > 3$ or 4, and their metallicities suggest already well evolved stellar populations at $z > 4$ (Hamann and Ferland 1993; Matteucci and Padovani 1993; Hamann, these proceedings). Therefore, the appearance of AGN may closely follow, but does not necessarily imply, the formation of the host galaxies. Moreover, several good arguments can be made that galaxy formation at least starts at higher redshifts, $z > 5$ or 10 (Peebles 1989). Searching for young quasars at $z > 5$ is an important observational challenge.

2 ENERGETICS OF AGN AND PROTOGALAXIES

Let us consider the energetics of protogalaxies containing AGN. Similar estimates and arguments have been made by Heckman (these proceedings), and Djorgovski (1992).

There are three basic energy generation mechanisms in protogalaxies: release of binding energy during the protogalaxy collapse and star formation; nuclear burning in primordial starbursts; and energy input from early AGN, if they are present. Consider a typical galaxy at $z \sim 0$. Its binding energy is:

$$|E_{\mathrm{bind,gal}}| \simeq M_{\mathrm{cool}} \langle V_{\mathrm{3d}}^2 \rangle \simeq 1.2 \times 10^{59} \text{ erg} \times \left(\frac{M_{\mathrm{cool}}}{10^{11} \, \mathrm{M_\odot}} \right) \left(\frac{V_{\mathrm{3d}}}{250 \text{ km s}^{-1}} \right)^2, \quad (1)$$

where M_{cool} is the total mass which can cool radiatively (obviously, it could make a major difference if the dark halos are baryonic and have collapsed dissipatively), and V_{3d} is the typical r.m.s. velocity. Binding energy was also released by collapsing protostars, and is of a comparable magnitude:

$$|E_{\mathrm{bind,\star}}| \simeq G \, M_{\Sigma\star} \langle M_\star \rangle / \langle R_\star \rangle \simeq 4 \times 10^{58} \text{ erg} \times \left(\frac{M_{\Sigma\star}}{10^{10} \, \mathrm{M_\odot}} \right) \left(\frac{\langle M_\star \rangle}{\mathrm{M_\odot}} \right) \left(\frac{\mathrm{R_\odot}}{\langle R_\star \rangle} \right), \quad (2)$$

where $M_{\Sigma\star}$ is the total mass converted to stars in the formation phase, $\langle M_\star \rangle$ is the average star mass, and $\langle R_\star \rangle$ is the average star radius, and we ignored small numerical factors due to the exact density distributions. Probably the most important energy source in protogalaxies was the nuclear burning in the initial starbursts:

$$E_{\mathrm{nuc}} \simeq \epsilon \, M_{\Sigma\star} \, c^2 \, \Delta X \simeq 10^{61} \text{ erg} \left(\frac{\epsilon}{0.007} \right) \left(\frac{M_{\Sigma\star}}{10^{10} \, \mathrm{M_\odot}} \right) \left(\frac{\Delta X}{0.05} \right) \quad (3)$$

where $M_{\Sigma\star}$ is the total mass burned in stars during the protogalactic starburst, $\epsilon \simeq 7$ MeV$/m_{\mathrm{p}} c^2 \simeq 0.007$ is the average net efficiency of nuclear reactions in stars, and

$\Delta X \simeq \Delta Z + \Delta Y \simeq 0.05$ is the fraction of the hydrogen converted to helium and metals. High metallicities of the old, metal-rich stellar populations of bulges and ellipticals, which contain about a half of the luminous mass at $z \sim 0$, suggest that primordial starbursts were quite luminous.

Production of the observed metals in stars requires a supernova rate, which depends on the high-mass end of the IMF. For a normal (local) IMF, there is ~ 1 supernova per 100 $M_{\odot}$ of star formation. Thus, protogalaxies may have contained $\sim 10^8$ supernovæ, each of which would release $\sim 10^{51}$ erg, which adds up almost to the binding energies of galaxies. Depending on the efficiency of coupling of the mechanical energy in supernova remnants to the ambient medium, some protogalaxies may have blown themselves apart.

AGN may have been important contributors to the energy budgets of their host galaxies. Assuming, for the sake of this argument, the average luminosity $\langle L_{\text{bol}} \rangle = \epsilon L_{\text{E}}$, where $L_{\text{E}} = 3.4 \times 10^4 (M_{\text{bh}}/M_{\odot}) L_{\odot}$ is the Eddington luminosity, M_{bh} is the mass of the black hole, and ϵ the net efficiency factor, and the duration of the active episode $\Delta t \simeq \tau_{\text{E}} = 4.4 \times 10^8$ yr, the Eddington time, we get:

$$E_{\text{AGN}} \sim \langle L_{\text{bol}} \rangle \, \Delta t \simeq 1.8 \times 10^{61} \text{ erg } \left(\frac{\epsilon}{0.01}\right) \left(\frac{M_{\text{bh}}}{10^8 \, M_{\odot}}\right). \qquad (4)$$

This is comparable to the energy released by the nuclear burning in stars, and is much higher than the binding energy of the host galaxy.

Obviously the fate of a protogalaxy may be severely affected by an early AGN, if one is rapidly formed, while the bulk of its host is still gaseous (*cf.* also Heckman, these proceedings). The feedback from an AGN, or, for that matter, energy released by the primordial starburst and the resulting supernovæ, could have regulated some of the global properties of its host galaxy, *e.g.*, the total mass, the maximum or the average density, *etc.* Ikeuchi and Norman (1991) proposed just such a model.

The energy released in protogalactic starbursts would generate a diffuse background with the energy density today of:

$$u_{\text{nuc}} \sim \epsilon \, \Delta X \, \Omega_{\star} \, \rho_{\text{crit}} \, c^2 \, (1 + z_{\star})^{-1} =$$

$$= 3.5 \times 10^{-15} \text{ erg cm}^{-3} \times h_{75}^2 \left(\frac{\epsilon}{0.007}\right) \left(\frac{\Delta X}{0.05}\right) \left(\frac{\Omega_{\star}}{0.01}\right) \frac{10}{(1 + z_{\star})}, \qquad (5)$$

where ϵ is the average net efficiency of the nuclear reactions in stars, ΔX is the mass fraction of hydrogen converted to stars in the initial starbursts, $\Omega_{\star}$ is the fraction of the critical density in stars generated in these starbursts, and $z_{\star}$ is the typical redshift at which this happened. A comparable amount of energy, times the fraction of protogalaxies which develop them, may have been released from the AGN over the age of the universe. Chokshi and Turner (1992) estimate that the energy density of the integrated quasar radiation is $u_{\text{AGN}} \simeq 1.3 \times 10^{-15}$ erg cm^{-3}. For comparison, the energy density in the CMBR today is $u_{\text{CMBR}} = 4.2 \times 10^{-13}$ erg cm$^{-3} \times (T_{\text{CMBR}}/2.73 \text{ K})^4$.

Thus, protogalaxies may have generated a diffuse background (or backgrounds, depending on the preferred energy release wavelengths) of the order of $\sim 1\%$ of the CMBR, and a comparable diffuse background was generated by the AGN.

2.1 Formation of the Central Engines

Note that we assumed the existence of a compact central engine, *e.g.*, a massive black hole. But how did the engine form? While a formation through a dissipationless gravitational collapse is a theoretical possibility (Rees 1984), the situation in the real world is unlikely to have been so clean. Consider a protogalaxy with a mass of 10^{12} $M_\odot$ within $R = 100$ kpc. Assuming a flat rotation curve, 10^8 $M_\odot$ is initially distributed within $R = 10$ pc, or more, if we allow for a finite core radius. Shrinking that amount of mass to within its Schwarzschild radius of 10^{-5} pc may have involved a substantial dissipation of the binding energy. Depending on the nature and mass fraction of the dark matter involved, and the physical processes affecting the infalling baryons, a non-negligible fraction of the rest mass of the black hole building material may have been radiated away:

$$E_{bh} \sim \epsilon \, M_{bh} \, c^2 \simeq 2 \times 10^{60} \text{ erg} \left(\frac{\epsilon}{0.01}\right) \left(\frac{M_{bh}}{10^8 \ M_\odot}\right), \tag{6}$$

where ϵ is the net efficiency factor.

If massive black holes are ubiquitous among normal galaxies, then their formation may have left us with an observable radiation signature. Models of the evolution of quasar populations, such as those by Small and Blandford (1992), Chokshi and Turner (1992), or Haehnelt and Rees (1993), suggest that the average local density of relict AGN black holes is $\rho_{bh} \sim 10^5$ $M_\odot$ Mpc^{-3}. The resulting radiation density is then:

$$u_{bh} \sim \epsilon \, \rho_{bh} \, c^2 \, (1 + z_{bh})^{-1} \simeq$$

$$\simeq 6 \times 10^{-18} \text{ erg cm}^{-3} \left(\frac{\epsilon}{0.01}\right) \left(\frac{\rho_{bh}}{10^5 \ M_\odot \ \text{Mpc}^{-3}}\right) \frac{10}{(1 + z_{bh})}, \tag{7}$$

where ϵ is the net efficiency factor, and z_{bh} is the average redshift where energy was released. Could this be the origin of the hard X-ray background?

2.2 Ionization of Protogalactic Nebulae

Regardless of the other possible effects quasars may have on their host galaxies, they certainly can photoionize the entire ISM, and even affect the Mpc-scale neighborhood of the protogalaxy, if they were unobscured. An unobscured starburst with a normal IMF and a SFR $\sim$ a few hundred $M_\odot$ yr^{-1} would have a comparable effect.

Consider a typical QSO with $M_B = -25$ ($L \sim 10^{12}$ $L_\odot$), and a power-law spectrum of the form $F_\nu \sim \nu^{-\alpha}$ (typically $\alpha \sim 0.5 - 1$). It will generate $\sim$ a few $\times 10^{56}$ ionizing photons per second. The implied recombination line luminosity is

L (Lyα) $\sim$ $\epsilon\,10^{45}$ erg s^{-1}, where ϵ is the net efficiency factor, depending on the gas clouds filling factor, coverage and beaming geometry, *etc.* The absence of luminous, ubiquitous Lyα nebulosities around high-z quasars implies that for most quasars $\epsilon < 0.01$. Rees (1988) proposed that the recombinant line emission from an extended ($\sim$ 100 kpc, *i.e.*, a few arcsec at $z > 1$) region of cool phase gas ($T \sim 10^4$ K) in the outskirts of a protogalaxy powered by a quasar would generate a line luminosity:

$$L \text{ (Ly}\alpha) \; \sim \; 10^{45}\, f(1-f)^2 \left(\frac{M_{gas}}{10^{11}\,M_\odot}\right)^2 \left(\frac{n_e}{n_e + n_{HI}}\right)^2 \text{ erg s}^{-1}, \qquad (8)$$

where f is the fraction of the condensed gas. Again, the absence of luminous, ubiquitous Lyα nebulosities around high-z quasars implies that $\sim$ 1% or less of the ionizing radiation is reprocessed in this way.

At the typical luminosity distances to high-redshift quasars, $D \sim 10^{29\pm1}$ cm, the expected Lyα line fluxes are $\sim 10^{-16\pm1}$ erg cm^{-2} s^{-1}. This is comparable to the fluxes observed from the Lyα nebulæ associated with powerful radio galaxies and radio-loud quasars at high redshifts (McCarthy *et al.* 1987, 1990; Heckman *et al.* 1991*a,b*; Hu *et al.* 1991). However, these objects are exceptions: most quasars do *not* show such nebulæ, with the typical upper limits one or two orders of magnitude lower (Djorgovski 1988*b*), a fact also noted by Rees (1988). There may be a similar effect with the host galaxies of quasars at low redshifts as well. One possibility is that most quasars are highly beamed or shielded by obscuring disks or tori, so that their host galaxies are exposed to much less ionizing radiation than what these computations imply. Another possibility is that much of the ISM was already quite dusty.

2.3 Are High-Redshift Radio Galaxies and *IRAS* 10214 Protogalaxies?

Protogalactic starbursts and AGN thus have comparable energetics and potentially comparable effects on their host galaxies. Given that the atomic physics was also the same, many observable manifestations of both phenomena (spectra, variability, *etc.*) could also be rather similar. This is indeed the grain of truth behind the Terlevichian Heresy, *viz.*, that high-redshift quasars are nothing but young ellipticals in the process of formation (Terlevich 1992; Terlevich and Boyle 1993, and references therein). Heckman (these proceedings) provides an incisive criticism of this scenario. It would be difficult to fully disentangle the effects of "genuine" AGN, assuming here that such things indeed exist, and spectacular starbursts. Many AGN-powered objects at high redshifts could have been, or resided in young galaxies.

Powerful radio galaxies have often been proposed as candidate protogalaxies (McCarthy *et al.* 1987; Djorgovski 1988*a,b*; Eisenhardt and Dickinson 1992; Eales and Rawlings 1993; *etc.*; see McCarthy 1993 for a review). It is clear, however, that many of their properties are driven by the AGN (the alignment effects, optical polarization, high-ionization emission lines, *etc.*), and so the AGN probably dominate the energetics and the ionization of the gas. We really have no good handle on the star formation rates in them, and even less on the ages of their stellar populations. Whereas they might be young galaxies, the case will be very difficult to prove.

Finally, there is the spectacular object *IRAS* F10214+4724 (Rowan-Robinson *et al.* 1993; Lawrence *et al.* 1993; Clements *et al.* 1993; Soifer *et al.* 1992, and references therein). It is likely that both a massive starburst *and* an obscured AGN are responsible for its great luminosity, and this reviewer suspects that the AGN probably dominates. If all of the luminosity were due to a starburst, the implied SFR would be in excess of $5,000$ M$_\odot$ yr^{-1}, with the resulting supernova rate in excess of 50 per year; not only there is no observed trace of this ostensible supernova activity, but the object would probably blow itself apart in a few million years. Furthermore, there should be about $10^7 - 10^8$ protoellipticals on the sky; there are certainly many fewer equivalents of *IRAS* F10214+4724, maybe ~ 100, or maybe just one. Whatever it is, it must be a very atypical object. Perhaps it can be thought of as an FIR equivalent of the powerful 3CR radio galaxies.

3 QUASARS AS POSSIBLE MARKERS OF PROTOCLUSTERS

It is nevertheless likely that AGN at very high redshifts do reside within young galaxies, simply on account of the timing: the age of the universe is at most 1 or 2 Gyr at $z \sim 4 - 5$, depending on the cosmology. The very existence of luminous quasars at $z > 4$ represents a severe timing problem (Turner 1991). Moreover, they appear to reside within already considerably chemically evolved stellar populations, presumably cores of giant ellipticals (Hamann and Ferland 1993; Matteucci and Padovani 1993; Hamann, these proceedings). A possible answer to this problem may be in the primordial large-scale structure.

Quasars have been used as probes of the large-scale structure and its evolution by several groups (*cf., e.g.,* Hartwick and Schade 1990; Iovino *et al.* 1991; Bahcall and Chokshi 1991; Andreani and Cristiani 1992, and references therein). Wide-separation pairs of quasars have been found by several groups (see Djorgovski 1991 for a list). Particularly interesting are the large groupings discovered by Crampton, Cowley, and Hartwick (1989), and by Clowes and Campusano (1991). These studies presumably probe the growth of the large-scale structure over the last half of the Hubble time, or so. However, quasars at $z > 4$ offer a possibility to probe the *ab initio* large-scale structure, imprinted by the initial conditions of the primordial density field.

The first structures to form via gravitational collapse should be located at the very highest peaks of the initial density field. For just about any reasonable initial density perturbation spectrum, such peaks should be strongly clustered (Kaiser 1984). For a CDM cosmogony, quasars at $z > 4$ should correspond to $\sim 4-5$ σ density peaks (Cole and Kaiser 1989), and should be clustered as strongly as the bright galaxies today (Efstathiou and Rees 1988). If this is true, then practically all $z > 4$ quasars should be pointing towards protoclusters in the early universe, and the optimal sites for galaxy formation. This would also be consistent with their interpretation as being the nuclei of young giant ellipticals.

Clusters have been detected around radio-loud quasars out to $z \sim 1$ (Ellingson *et al.* 1991*a, b*; Hutchings *et al.* 1993; *etc.*). The idea here, however, is that the first objects which formed on galaxian scales were *all* in protoclusters, and that at least some of them have achieved quasarhood by $z \sim 4 - 5$.

3.1 The Case of PC 1643+4631 A+B

A possible example of such a quasar-marked protocluster was found by Schneider, Schmidt and Gunn (1991; SSG). They discovered a pair of quasars at $z = 3.8$, designated PC 1643+4631 A+B. The quasars are separated by $\sim 3.5\ h^{-1}$ Mpc in projection, with the redshift difference $\Delta z = 0.04$, giving a net comoving separation of $\sim 10\ h^{-1}$ Mpc. The Poissonian probability of finding a pair of quasars this close in the entire SSG survey is $P \simeq 10^{-4}$.

We can use this pair to estimate the strength of quasar clustering at $z = 3.8$. Let us assume a standard hierarchical clustering model:

$$\xi(r) = \left(\frac{r}{r_0}\right)^{-\gamma} (1 + z)^{-3-\epsilon} \tag{9}$$

where $\gamma \simeq 1.8$; if $\epsilon = 0$, clustering is fixed in the proper coordinates, and if $\epsilon = \gamma - 3$, clustering is fixed in comoving coordinates. We get a maximum-likelihood estimate of the clustering length r_0: for $\Omega_0 = 0$ and $\epsilon = 0$, $r_0 = 54\ h^{-1}$ Mpc; for $\Omega_0 = 1$ and $\epsilon = 0$, $r_0 = 42\ h^{-1}$ Mpc; for $\Omega_0 = 0$ and $\epsilon = \gamma - 3$, $r_0 = 19\ h^{-1}$ Mpc; and for $\Omega_0 = 1$ and $\epsilon = \gamma - 3$, $r_0 = 15\ h^{-1}$ Mpc. This is quite strong: for a comparison, at $z \sim 0$, normal galaxies have $r_0 \sim 5$ to $8\ h^{-1}$ Mpc, Abell clusters have $r_0 \sim 25$ to $30\ h^{-1}$ Mpc, and quasars and galaxy groups have $r_0 \sim 10$ to $15\ h^{-1}$ Mpc (Hartwick and Schade 1990; Bahcall and Chokshi 1991). For quasars at $z \sim 1 - 2$, one typically finds $r_0 \sim 5$ to $10\ h^{-1}$ Mpc (Hartwick and Schade 1990; Iovino *et al.* 1991; Andreani and Cristiani 1992; *etc.*). This later result is consistent with the growth of large-scale structure from $z \sim 2$ to $z \sim 0$. Strong clustering at $z \sim 4$ indicated by PC 1643+4631 A+B suggests that the two quasars are indeed members of same protocluster.

3.2 A Search for Protoclusters at Palomar

In order to exploit the idea of quasars as markers of protoclusters, we have started a search for faint AGN and star-forming galaxies around the known, $z > 4$ quasars (Djorgovski *et al.* 1991, 1993*a, b*; Smith *et al.* 1993). To date, we have surveyed the fields of some 15 fields using the multicolor technique to select faint quasars. We are typically complete down to $r \sim 24^m$, which corresponds to nearly Seyfert 1 luminosities at these redshifts. We have also obtained narrow-band imaging data on 6 fields so far, in order to search for Lyα emission line galaxies, powered by star formation. We typically reach the flux limits of a few $\times 10^{-17}$ erg cm^{-2}s^{-1}, which corresponds to an unobscured SFR of $5 - 10\ M_\odot$ yr^{-1} at these redshifts.

To date, we have identified several interesting candidate objects, at least one of which is very likely a faint quasar at $z \sim 4.8$ (Smith *et al.*, in preparation), and one or two possible Lyα galaxies. However, these objects are extremely faint, which makes the follow-up spectroscopy extremely difficult. We need better data before we can be sure of their nature and redshifts.

Using one or two tentative detections above $r \simeq 23.5^m$ as an upper limit, we can place a constraint on the cumulative number density of faint quasars, at the luminosity level never probed before at these redshifts. Our magnitude limit translates to the restframe $M_B \simeq -24^m$, and is $\sim 2^m$ deeper that the limits reached in the SSG survey. We find an upper limit to the cumulative number density of $z > 4$ quasars of $\sim 10^{-7}$ per comoving Mpc^3. This is well below the extrapolation using the quasar luminosity function appropriate at $z \sim 2 - 3$, by one or two orders of magnitude (see, *e.g.*, Fig. 4 in Irwin *et al.* 1991). Thus, it appears that the luminosity-dependent turnover in the quasar density at $z > 4$ proposed by SSG (see also Schmidt *et al.* 1991) is real. The data indicate luminosity-dependent density evolution of faint quasars at these redshifts, in the sense that the numbers of faint quasars drop faster with the increasing redshifts than the numbers of brighter objects. This is intriguing, but it is also unfortunate for our survey for protoclusters: we are apparently limited by the number of quasars available at these early epochs!

4 PRIMEVAL GALAXIES AND DOGS IN THE NIGHT

The questions of the origin of galaxies, and the origins of nuclear activity in them are becoming increasingly more intertwined. Apparently, AGN can appear quickly, during the initial stages of galaxy formation, and can affect the outcome of that process. High-redshift AGN can thus be used as signposts of galaxy formation sites. The natural extrapolation of this idea leads to an intriguing possibility.

Searches for Lyα-luminous protogalaxies are now exploring the relevant range of line fluxes and sufficient comoving volumes, yet no obvious population of primeval galaxies has been seen so far (Thompson *et al.* 1991, 1992, 1993; De Propris *et al.* 1993; see Djorgovski and Thompson 1992; Djorgovski 1992; or Djorgovski *et al.* 1993*b* for reviews and summary of the limits). What is the solution to the mystery of silent protogalaxies?

One possible explanation is that all such objects have undergone an AGN phase early on, and have already been detected as high-redshift AGN, as already proposed by Meier (1976) and expanded by Terlevich (1992, and references therein). In other words, *all* young ellipticals or massive bulges may have undergone an early AGN phase immediately, even if a substantial fraction of their luminosities derived from star formation; perhaps there never was a population of "normal" protogalaxies. All objects which we know of at large redshifts through their direct *emission* (of anything, but the Lyα line in particular), are in some ways associated with AGN. Objects identified through their *absorption* at $z > 2$ or so are probably mostly quiescent disks or dwarfs, converting their gas into stars slowly.

This does not exclude another possibility, namely that protogalaxies were at least moderately dusty, which would lower their UV and Lyα luminosities rather dramatically. For discussions of this point see, *e.g.*, Djorgovski (1992); or Charlot and Fall (1993). By the same token, it is possible that there were numerous dusty AGN around as well. The evolutionary scenario of merger $\Rightarrow$ dusty starburst $\Rightarrow$ quasar, proposed for the ultraluminous *IRAS* galaxies at low redshifts (Sanders *et al.* 1988) could have operated at large redshifts just as well. Detections of cold dust around at least some high-redshift quasars (Andreani *et al.* 1993) support this notion. There are even some theoretical arguments which suggest that the first quasars are to be found at very high redshifts, $z > 10$, and that they may have been obscured (Loeb 1993). On the other hand, limits on the distortions of the CMBR spectrum as measured by *COBE* will provide strong constraints for the models of large populations of luminous dusty objects at high redshifts.

In any case, it appears that the observational frontier is moving to the longer wavelengths, from near-IR to sub-mm. The technology is rapidly getting better at these wavelengths, and it may lead us to discoveries of dust-obscured populations of protogalaxies or young AGN at high redshifts, when they first began to form.

Acknowledgements: I wish to thank numerous collaborators, and especially Julia Smith and Dave Thompson, on whose work some of this review is based. Able assistance of the staff of Palomar Observatory and of CTIO are also gratefully acknowledged. This work was supported in part by the NSF PYI award AST-9157412, and a Dudley Award. Finally, thanks to Isaac Shlosman for organizing this excellent meeting, and for his patience while waiting for this manuscript.

REFERENCES

Andreani, P., and Cristiani, S. 1992, *Ap. J. Lett.*, **398**, L13.

Andreani, P., Lafranca, F., and Cristiani, S. 1993, *Mon. Not. R. Astr. Soc.*, **261**, L35.

Bahcall, N., and Chokshi, A. 1992, *Ap. J. Lett.*, **380**, L9.

Barnes, J., and Hernquist, L. 1991, *Ap. J. Lett.*, **370**, L65.

Barnes, J., and Hernquist, L. 1992, *Ann. Rev. Astr. Ap.*, **30**, 705.

Carlberg, R. 1990, *Ap. J.*, **350**, 505.

Charlot, S., and Fall, S. M. 1993, *Ap. J.*, **415**, 580.

Chokshi, A., and Turner, E. 1992, *Mon. Not. R. Astr. Soc.*, **259**, 421.

Clemens, D., van der Werf, P., Krabbe, A., Blietz, M., Genzel, R., and Ward, M. 1993, *Mon. Not. R. Astr. Soc.*, **262**, 23p.

Clowes, R., and Campusano, L. 1991, *Mon. Not. R. Astr. Soc.*, **249**, 218.

Cole, A., and Kaiser, N. 1989, *Mon. Not. R. Astr. Soc.*, **237**, 1127.

Crampton, D., Cowley, A., and Hartwick, F. D. A. 1989, *Ap. J.*, **345**, 59.

De Propris, R., Pritchet, C., Hartwick, F. D. A., and Hickson, P. 1993, *A. J.*, **105**,

1243.

Djorgovski, S. 1988*a*, in *Starbursts and Galaxy Evolution*, eds. T. X. Thuan, *et al.*, (Gif-sur-Yvette: Editions Frontières), p. 401.

Djorgovski, S. 1988*b*, in *Towards Understanding Galaxies at Large Redshift*, eds. R. G. Kron, and A. Renzini (Dordrecht: Kluwer), p. 259.

Djorgovski, S., Smith, J.D., and Thompson, D. J. 1991, in *The Space Distribution of Quasars*, ed. D. Crampton (San Francisco: ASP-21), p. 325.

Djorgovski, S. 1991, in *The Space Distribution of Quasars*, ed. D. Crampton (San Francisco: ASP-21), p. 349.

Djorgovski, S. 1992, in *Cosmology and Large-Scale Structure in the Universe*, ed. R. de Carvalho (San Francisco: ASP-24), p. 73.

Djorgovski, S., and Thompson, D. 1992, in *The Stellar Populations of Galaxies*, IAU Symp. #149, eds. B. Barbuy, and A. Renzini (Dordrecht: Kluwer), p. 337.

Djorgovski, S., Thompson, D., and Smith, J. D. 1993*a*, in *First Light in the Universe*, eds. B. Rocca-Volmerange, *et al.* (Gif-sur-Yvette: Editions Frontières), p. 67.

Djorgovski, S., Thompson, D., and Smith, J. D., 1993*b*, in *Texas/PASCOS'92*, eds. C. Akerlof, and M. Srednicki (Ann. N. Y. Acad. Sci.), **688**, 515.

Eales, S., and Rawlings, S. 1993, *Ap. J.*, **411**, 67.

Efstathiou, G., and Rees, M. 1988, *Mon. Not. R. Astr. Soc.*, **230**, L5.

Eisenhardt, P., and Dickinson, M. 1992, *Ap. J. Lett.*, **399**, L47.

Ellingson, E., Yee, H., and Green, R. 1991*a*, *Ap. J.*, **371**, 49.

Ellingson, E., Green, R., and Yee, H. 1991*b*, *Ap. J.*, **378**, 476.

Haehnelt, M., and Rees, M. J. 1993, *Mon. Not. R. Astr. Soc.*, **263**, 168.

Hamann, F., and Ferland, G. 1993, *Ap. J.*, in press.

Hartwick, F. D. A., and Schade, D. 1990, *Ann. Rev. Astr. Ap.*, **28**, 437.

Heckman, T. 1990, in *Paired and Interacting Galaxies*, eds. J. Sulentic, *et al.* (NASA CP-3098), p. 359.

Heckman, T. 1991, in *Massive Stars in Starbursts*, eds. C. Leitherer, *et al.* (Cambridge: Cambridge Univ. Press), p. 289.

Heckman, T., Lehnert, M., van Breugel, W., and Miley, G. 1991*a*, *Ap. J.*, **370**, 78.

Heckman, T., Lehnert, M., Miley, G., and van Breugel, W. 1991*b*, *Ap. J.*, **381**, 373.

Hernquist, L. 1989, *Nature*, **340**, 687.

Hu, E., Songaila, A., Cowie, L., and Stockton, A. 1991, *Ap. J.*, **368**, 28.

Hutchings, J,. Crampton, D., and Persram, D. 1993, *A. J.*, **106**, 1324.

Ikeuchi, S., and Norman, C. 1991, *Ap. J.*, **375**, 479.

Iovino, A., Shaver, P., and Cristiani, S. 1991, in *The Space Distribution of Quasars*, ed. D. Crampton (San Francisco: ASP-21), p. 202.

Irwin, M., McMahon, R., and Hazard, C. 1991, in *The Space Distribution of Quasars*, ed. D. Crampton (San Francisco: ASP-21), p. 117.

Kaiser, N. 1984, *Ap. J. Lett.*, **284**, L9.

Kormendy, J., and Djorgovski, S. 1989, *Ann. Rev. Astr. Ap.*, **27**, 235.

Kormendy, J. 1993, in *The Nearest Active Galaxies*, eds. J. Beckman, *et al.* (Madrid:

CSIS), in press.

Lawrence, A., *et al.* 1993, *Mon. Not. R. Astr. Soc.*, **260**, 28.

Loeb, A. 1993, *Ap. J. Lett.*, **404**, L37.

Matteucci, F,. and Padovani, P. 1993, *Ap. J.*, in press.

McCarthy, P., Spinrad, H., Djorgovski, S., Strauss, M., van Breugel, W., *Ap. J. Lett.*, and Liebert, J. 1987, **319**, L39.

McCarthy, P., Spinrad, H., van Breugel, W., Liebert, J., Dickinson, M., and Djorgovski, S. 1990, *Ap. J.*, **365**, 487.

McCarthy, P. 1993, *Ann. Rev. Astr. Ap.*, **31**, 639.

Meier, D. 1976, *Ap. J. Lett.*, **203**, L103.

Peebles, P. J. E. 1989, in *The Epoch of Galaxy Formation*, eds. C. Frenk, *et al.* (Dordrecht: Kluwer), p. 1.

Rees, M. J. 1984, *Ann. Rev. Astr. Ap.*, **22**, 471.

Rees, M. J. 1988, *Mon. Not. R. Astr. Soc.*, **231**, 91p.

Rees, M. J. 1990, *Science*, **247**, 817.

Rees, M. J. 1992, in *Physics of Active Galactic Nuclei*, eds. W. J. Duschl, and S. J. Wagner (Berlin: Springer), p. 662.

Rowan-Robinson, M., *et al.* 1993, *Mon. Not. R. Astr. Soc.*, **261**, 513.

Sanders, D., Soifer, B. T., Elias, J., Madore, B., Matthews, K., Neugebauer, G., and Scoville, N. Z. 1988, *Ap. J.*, **325**, 74.

Schmidt, M., Schneider, D., and Gunn, J. E. 1991, in *The Space Distribution of Quasars*, ed. D. Crampton, (San Francisco: ASP-21), 109.

Schneider, D., Schmidt, M., and Gunn, J. E. 1991, *A. J.*, **101**, 2004 (SSG).

Shlosman, I., Begelman, M. C., and Frank, J. 1990, *Nature*, **345**, 679.

Small, T., and Blandford, R. 1992, *Mon. Not. R. Astr. Soc.*, **259**, 725.

Smith, J.D., Thompson, D.J., and Djorgovski, S. 1993, in *Sky Surveys: Protostars to Protogalaxies*, ed. B.T. Soifer, *A.S.P. Conf. Ser.*, **43**, 185.

Soifer, B. T., Neugebauer, G., Matthews, K., Lawrence, C., and Mazzarella, J. 1992, *Ap. J. Lett.*, **399**, L55.

Terlevich, R. 1992, in *The Stellar Populations of Galaxies*, IAU Symp. #149, eds. B. Barbuy, and A. Renzini (Dordrecht: Kluwer), p. 271.

Terlevich, R., and Boyle, B. J. 1993, *Mon. Not. R. Astr. Soc.*, **262**, 491.

Thompson, D. J., Djorgovski, S., and Trauger, J. 1991, in *The Space Distribution of Quasars*, ed. D. Crampton (San Francisco: ASP-21), p. 354.

Thompson, D. J., Djorgovski, S., and Trauger, J. 1992, in *Cosmology and Large-Scale Structure in the Universe*, ed. R. de Carvalho (San Francisco: ASP-24), p. 147.

Thompson, D. J., Djorgovski, S., Trauger, J., and Beckwith, S. 1993, in *Sky Surveys: Protostars to Protogalaxies*, ed. B. T. Soifer, (San Francisco: ASP-43), p. 189.

Turner, E. 1991, *A. J.*, **101**, 5.

White, S. D. M., and Frenk, C. 1991, *Ap. J.*, **379**, 52.

Collisions, Collapses, Quasars, and Cosmogony

Simon D. M. White

Institute of Astronomy, University of Cambridge, United Kingdom

ABSTRACT

I review recent work on galaxy formation and relate it to questions concerning the formation and fuelling of active galactic nuclei. The theory of galaxy formation has developed dramatically in recent years as a result of new analytic methods coupled with substantial programs of direct numerical simulation. Many aspects of how galaxies might form in a universe where structure grows by hierarchical clustering are understood quite well. Others, particularly those that are closely linked to the star formation process, remain highly uncertain. Nevertheless, it is now possible to calculate formation and interaction rates for galaxies with some confidence in a wide variety of cosmogonies. It seems likely that nuclear activity, either starburst or AGN, is an inevitable consequence of the violent, asymmetric, and time-dependent processes which occur during the assembly of galaxies.

1 INTRODUCTION

The idea that quasars might be related to galaxy formation followed quickly after the first measurements of QSO redshifts but was somewhat neglected after the near-universal acceptance of the argument that QSO luminosities are more easily explained by accretion onto a supermassive black hole than by starlight. In such a model black hole formation and fuelling are major issues which must be addressed before QSO's and galaxy formation can be linked. There has always been some dissent from this model, most notably in recent years from R. Terlevich and his collaborators (*e.g.* Terlevich and Boyle 1993), but recent discussions of quasar formation have tended to emphasize how *late* the onset of quasar activity may be in comparison with the initial collapse of a protogalaxy (see, for example, Turner 1991).

Over the last decade numerical simulations of the growth of structure, particularly in the context of the Cold Dark Matter (CDM) cosmogony, have led to substantial changes in our understanding of how and when galaxies may have formed. In this scenario galaxy formation occurs late and is a highly inhomogeneous process closely related to the formation of larger scale structure (*e.g.* White and Frenk 1991; Katz, Hernquist, and Weinberg 1992). This has received some support from observational developments such as the discovery of massive infrared-bright starbursts in the nuclear regions of interacting systems (Sanders *et al.* 1988) and of a faint galaxy population that differs substantially from that in our neighbourhood (Tyson 1988). The former blurs the distinctions between galaxy formation, galaxy interaction, and nuclear activity, while the latter appears to indicate that galaxies have changed much more dramatically in the recent past than had previously been thought likely.

In this contribution I attempt to assess how our increased understanding of galaxy formation in hierarchically clustering universes may be related to the problems of formation and fuelling of AGN. These problems can be posed as a set of questions.

When do galaxies form?

How extended and inhomogeneous is this process?

How is it related to formation of an AGN?

What do forming systems look like – starbursts? QSOs? cooling flows?

Are there real distinctions between galaxy formation, interactions, and merging?

What are the expected rates for formation/interaction/merging?

How do these rates depend on cosmology? on cosmogony?

Do interactions lead to repeated QSO outbursts in a single galaxy?

Does the decrease in QSO abundance since $z = 2$ reflect the end of galaxy formation, or a decrease in the galaxy interaction rate, or the depletion of fuel?

I hope to show that it is now possible to get preliminary answers to a number of these questions within specific hierarchical clustering theories. Further work along the directions I outline below should show whether such theories can account for observed galaxy and AGN evolution.

2 THE SEQUENCE OF COSMOGONY

In the 1970's and the early 1980's there were two major pictures competing to explain the build-up of structure in the Universe. These were often characterised as the "top-down" and "bottom-up" theories, and their major difference lay in their assumptions about galaxy formation. In the first theory, density and velocity fields in the post-recombination universe are assumed to contain no structure below a characteristic scale similar to that of present-day clusters of galaxies. The first nonlinear structures then result from the anisotropic collapse of large "pancakes". Gas is supposed to cool in a thin layer within these pancakes, which fragments to form galaxies. This model is attractive beacause its initial conditions seem the most natural outcome of the processes which might have generated structure in the early universe, and because pancakes offered an explanation for the apparent sheets and voids in the galaxy distribution. However, as advances in numerical techniques allowed details of the theory to be worked out a number of major problems surfaced.

The most obvious difficulty is that the theory predicts galaxy clusters and superclusters to be older than galaxies while observational evidence points towards the opposite ordering. Most galaxies appear to be in equilibrium and to be composed of relatively old stars, whereas galaxy clusters often show substantial irregularities which seem to indicate an unrelaxed and so "youthful" dynamical state. Superclusters are in general still expanding and are far from complete collapse. The ability to simulate cosmological structure formation led to the realization that pancake models predict much of the mass of the universe to be in massive and collapsed lumps very soon after the formation of the first objects. In a neutrino-dominated universe this makes it difficult to produce a sufficient population of old galaxies, or indeed to produce galaxies at all, without overproducing massive galaxy clusters (White *et al.* 1983). However, the most decisive argument against such models comes from the microwave background fluctuation amplitudes measured by *COBE* and recent ground-based experiments. For conventional assumptions about the kinds of density fluctuations which can be produced in the early universe, these are too low to allow the formation of substantial nonlinear structure by the present day and *a fortiori* by redshifts of 1 or more (Holtzman 1989).

The "bottom up" theory has no such timing problems, and its main qualitative uncertainty is whether the apparent coherence of large-scale structure, the sheets and bubbles which are so striking in some recent redshift survey data (*e.g.* Geller and Huchra 1989), can be properly explained in a theory where large objects grow through the aggregation of smaller highly nonlinear lumps. Large-scale numerical simulations suggest this may be possible (*e.g.* White *et al.* 1987; Park 1990) but the question is still open. The Cold Dark Matter model is one specific version of the bottom up theory which has been particularly thoroughly studied in the last decade (*e.g.* Frenk 1991). In its standard form it assumes that the universe has the closure density, that 90% or so of this mass is in some nonbaryonic "cold" form. The initial density fluctuations are those predicted by the simplest models of cosmological inflation. In recent years many variations of this model have been proposed to address a number of relatively small, but nevertheless highly significant discrepancies with observational data. The most important is the fact that the fluctuation amplitude measured by *COBE* implies an overall normalization which for standard CDM is too *large* to be consistent with measures of mass fluctuations on galaxy cluster scales (see, for example, Efstathiou *et al.* 1992).

In bottom-up models galaxies form from gas which cools and condenses at the centre of dark matter halos. As structure builds to larger scales this process naturally produces big galaxies surrounded by satellites (galaxies which condensed in precursor halos), and clusters made of many galaxies. Galaxies form before clusters, as suggested by direct estimates of their relative ages. Also small galaxies form before large ones, a prediction which is less easily reconciled with observation since massive galaxies appear to be made almost entirely of old stars while low mass galaxies often contain gas and young stars in abundance. In such models galaxy cluster masses are set by the current amplitude of density fluctuations, while galaxy masses are set by cooling processes within dark halos. Merging is the primary way in which masses increase, so that cluster mergers, and perhaps also galaxy mergers, are predicted to be frequent. One of the major advances in this field in the last few years has been the development of analytic tools which describe hierarchical clustering in detail and so allow the calculation of merger and interaction rates in different cosmogonies. I will return to this work after describing recent numerical work.

3 SIMULATIONS OF GALAXY FORMATION

Over the last few years advances in computing speed and the adoption of new algorithms have made possible quasi-realistic simulations of galaxy formation. Some studies have focused on the formation of individual systems (Katz and Gunn 1991; Navarro and Benz 1991; Katz 1992; Navarro and White 1993a, b), while others have considered galaxy formation in the context of the formation of clusters or of structure in a "representative" region of the Universe (Carlberg, Couchman, and Thomas 1990; Katz, Hernquist, and Weinberg 1992; Katz and White 1993; Evrard, Summers, and Davis 1993). Of course, there are many aspects of galaxy formation process which cannot really be simulated at all. In particular, simulations of an entire galaxy are unable to resolve structures significantly less massive than a molecular cloud, and as a result the detailed structure of star formation (SF) regions and of the ISM is lost. Thus SF and its effects on gas can only be treated in a crude, *ad hoc*, and highly uncertain fashion. Nevertheless, this work has clarified a number of important issues and has significantly changed our appreciation of how and when galaxies may form.

Two issues related to bottom-up models for galaxy formation were raised by White and Rees (1978) but were not solved by them. The crude simulations available at the time suggested that as larger and larger objects are formed, each new merger product rapidly comes to equilibrium with a single centre and a smooth structure in which its progenitors are no longer discernible (see, for example, Efstathiou *et al.* 1988). Such an object does not resemble a galaxy cluster containing a number of stable and long-lived galaxies. White and Rees therefore argued that dissipational effects during galaxy formation must enhance the binding energy of galaxies and reduce their collision cross-sections so that they survive the merging of their dark halos. Recent simulations show that this process can indeed work (Carlberg, Couchman, and Thomas 1990; Katz and White 1993; Evrard, Summers, and Davis 1993). How well it works is still unclear because of the second issue.

In hierarchical clustering early objects have higher mean densities and lower mean temperatures than later ones. As a result they cool more efficiently. If cooling and collapse to high density are the only requirements for gas make stars, most gas is predicted to turn into small galaxies at early times. Little is left to form large galaxies or to make up the intergalactic medium in galaxy clusters. (See Cole [1991] and White and Frenk [1991] for discussions of this difficulty in the CDM model.) White and Rees avoided this problem by assuming star formation to be inefficient in small galaxies; energy input from the first stars is supposed to prevent most of the gas from collapsing. Dekel and Silk (1986) showed that this could explain many of the observed properties of dwarf galaxies. However, this process is difficult to treat numerically, and where recent simulations have included SF and feedback, the schemes chosen have usually had little effect on the efficiency of galaxy formation, which is determined, in practice, almost entirely by numerical resolution; models with higher resolution allow more gas to cool at early times (see, for example, Evrard,

468 *White*

Summers and Davis 1993). As a result predictions for galaxy masses and for the intergalactic medium should be treated with scepticism. Realistic simulations may never be possible if feedback really is the process limiting early galaxy formation. An alternative is heating of pregalactic gas by a strong UV background (Efstathiou 1992), a process which is much easier to treat numerically because it does not involve small scale structure. It seems likely that progress on these questions will come from observational studies of analogous systems (starbursts, interacting galaxies, cooling flows) rather than from larger and more detailed simulations.

This deficiency of galaxy formation simulations does not, however, detract from the important points they demonstrate.

(1) Objects with the appropriate mass and size to be identified as galaxies *can* indeed form by the cooling of gas within dark matter halos.

(2) Collapse, merging, gas cooling, and galaxy formation are continuous and ongoing processes in hierarchical models. There is no clear epoch of galaxy formation. Present day cooling flows in galaxy clusters should be identified as the tail-end of galaxy formation.

(3) "Galaxies" merge less rapidly than halos as structure grows, thus permitting the development of galaxy clusters.

(4) "Galaxy" merging may nevertheless be too frequent to be consistent with observation in an $\Omega = 1$ universe. Indeed, the whole process of galaxy and cluster formation may be too recent to be tenable in such universe (see below).

(5) If high gas densities are assumed necessary for SF, then cooling often produces centrifugally supported disks which have the correct scale and structure to be identified with galaxy disks.

An example which illustrates several of these points is shown in Figure 1, taken from the work of Navarro and White (1993*b*). This simulation follows the formation of a dark halo with an equivalent circular velocity of ~ 220 km s^{-1} in a standard $\Omega = 1$ CDM universe with a baryon fraction of 10%. Only gravitational, hydrodynamic, and radiative cooling processes were included. As soon as large enough objects form for the Smoothed Particle Hydrodynamics technique to resolve their structure, the gas cools off and sinks to the centre. At the final time almost all the gas is in dense cool clumps. Aggregation of the various lumps into the final system occurs quite late, and indeed the final central object has a moderately massive satellite which would have merged with it quite soon after the simulation was stopped. As the expanded view shows this central object is indeed disk-like and is supported almost entirely by rotational motion. Its surface density profile is quite close to exponential with a scale-length of about 3 kpc. It is, in fact, too compact and too massive compared to observed disks of the same circular velocity. Its small size is largely a result of the transfer of angular momentum from the cold gas to the dark matter during the mergers which produced the final disk-halo system (compare Barnes 1988). This transfer and the overmassive disk would both be reduced if feedback from SF were allowed to keep the gas hot and delay its condensation.

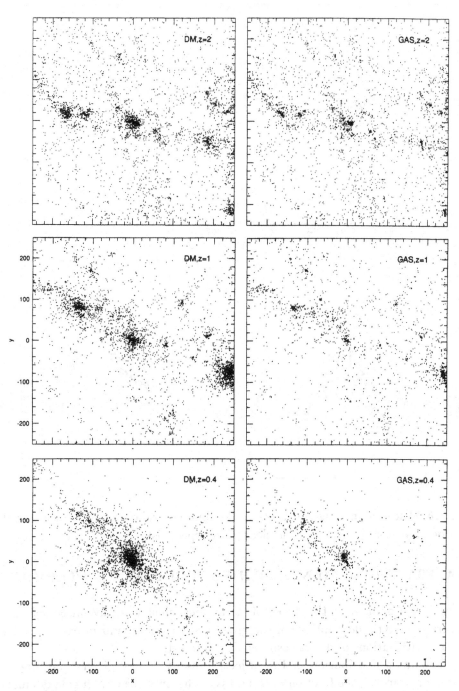

Fig. 1*a*—Formation of the galaxy within a dark halo of present circular velocity 220 km s^{-1}. Left hand panels show the distribution of gas and right hand panels of dark matter at the redshifts indicated. The axes are labelled in kpc. (see also Fig. 1*b*).

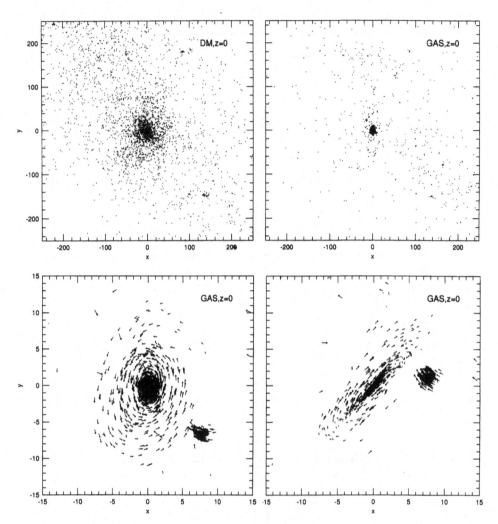

Fig. 1*b*—Face-on and edge-on views of the final central object (from the Fig. 1*a*) with velocity vectors indicated. It is clearly a warped disk.

4 ANALYTIC TREATMENTS OF HIERARCHICAL CLUSTERING

Recent developments in the theory of hierarchical clustering have provided simple but accurate analytic descriptions of the growth of structure, and in particular for a number of distributions of direct cosmogonical interest.

(1) The mass distribution of the objects present at any given time; one can calculate, for example, the fraction of the mass of the universe in rich galaxy clusters or in isolated galaxy halos.

(2) The mass distribution of the progenitors of present-day objects, for example, the fraction of the mass of a rich cluster which was already in a massive enough halos at $z = 3$ to contain a bright elliptical galaxy.

(3) The mass distribution of the objects which currently contain the remnants of high redshift objects, for example the abundance of $z = 3$ quasar remnants in rich clusters, in small groups, or in isolated galaxies.

(4) The distribution of formation times for objects of given mass; for example, one can ask when present-day galaxy clusters or dwarf galaxy halos were formed.

(5) The distribution of survival times; for example, it may be of interest to know how long a galaxy halo lasts before being incorporated into a cluster.

(6) Merger rates as a function of primary and secondary mass and of time. Thus one can estimate how often galaxies acquire new satellites, or how often substructure is regenerated in a galaxy cluster by the accretion of another system.

These results are based on extensions of the theory of Press and Schechter (1974) developed in several recent papers (Bower 1991; Bond *et al.* 1991; Kauffmann and White 1993; Lacey and Cole 1993). The theory begins by assuming a background cosmology specified by the values of H_0, Ω_0, and Λ. Deviations from uniformity at early times are assumed to be small and to be a gaussian random field which can be specified entirely by the power spectrum, $|\delta_\mathbf{k}|^2$, where $\delta_\mathbf{k}$ is the Fourier transform of the relative density fluctuation field. At recent epochs linear fluctuations follow the growing mode for perturbations of a dust universe (amplitude increasing with the expansion factor in an $\Omega = 1$ universe, more slowly in a low density universe). A given mass element is assumed to be part of an object with mass exceeding M if linear theory predicts the overdensity of the surrounding sphere that initially contained mass, M, to exceed a critical value, δ_c. δ_c is usually taken to be the predicted linear overdensity of a spherical perturbation at the instant it collapses; $\delta_c = 1.686$ for an $\Omega = 1$ universe. By manipulating the probability distributions resulting from this *ansatz*, formulae can be obtained for all the distributions listed above. A particularly clear exposition is given by Lacey and Cole (1993).

The current status of this theory is curious. The original derivation by Press and Schechter was far from convincing, but gave mass functions which agreed well with the numerical experiments presented by Efstathiou *et al.* (1988). A more rigorous and convincing derivation was provided by Bond *et al.* (1991) using the theory of excursion sets. However, these authors also showed that the main ansatz is seriously flawed; in numerical simulations there is, in fact, only a weak correlation between the mass of the halo to which any given particle belongs and the halo mass which is predicted for it by the theory. Despite this, all the recent papers show that statistical distributions derived from the theory describe the results of a wide variety of numerical experiments with quite remarkable accuracy. Thus while this extended Press-Schechter theory is incorrect, it nevertheless provides convenient and accurate analytic fitting formulae of great utility.

Applications of the theory allow a number of interesting conclusions to be drawn, all of which apply to the merging of *dark halos* rather than to that of galaxies.

(1) For initial power spectra of the form, $|\delta_\mathbf{k}|^2 \propto k^n$, the halo mass grows more rapidly, and the halo mass distribution is broader, for more negative n.

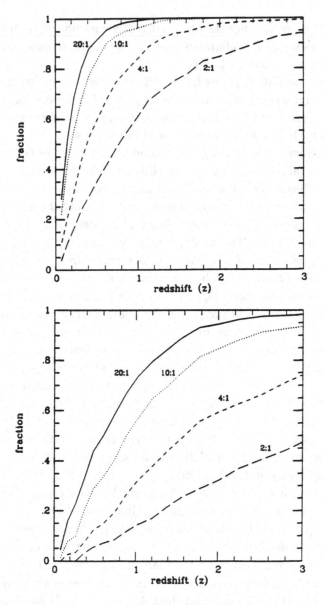

Fig. 2—Merging rates in $\Omega = 1$ (upper panel) and $\Omega = 0.2$ universes. The curves indicate the fraction of present halos with $V_c = 220$ km s^{-1} which have merged with an object of N times smaller mass since the indicated redshift, where $N = 20, 10, 4,$ and 2.

(2) The growth of the characteristic mass of clustering slows substantially once the density of the universe drops significantly below the critical value, but the shape of the mass spectrum is unaffected.

(3) Individual halos grow primarily by merging with relatively massive objects, rather than by accretion. Spherical infall is thus a poor model for the growth of typical objects.

(4) The low mass halos present at any given time were formed earlier and will "die" sooner than the high mass halos.

An example relevant to interaction-induced activity in galaxies is shown in Figure 2 (from Kauffmann and White 1993). These plots give the fraction of present-day halos with a circular velocity of 220 km s^{-1} ("Milky Way look-alikes") which have had a merger with an object one N'th of their mass since redshift, z. Curves are given for $N = 20$, 10, 4, and 2. Both plots are for $H_0 = 50$ km s^{-1} Mpc^{-1} CDM universes; the upper one is for a biased $\Omega = 1$ model with $b = 2.5$ while the lower one is for an unbiased open universe with $\Omega = 0.2$. Thus in a high density universe a Milky Way halo has had a 60% chance of merging with something at least 5% of its mass and a 15% chance of merging with something half its mass since $z = 0.2$. In an open universe one has to go back to $z \sim 1$ before the probabilities rise to similar values. To interpret these plots in terms of interactions of *visible* galaxies clearly requires additional assumptions about how rapidly dynamical friction can bring the two objects together. Nevertheless it is clear that substantially higher interaction rates are predicted in high density universes. Tóth and Ostriker (1992) have used this to argue that the thin stellar disks seen in spiral galaxies could only have survived in a low density universe.

5 GALAXY FORMATION, INTERACTIONS, AND AGN

At low redshift, where quasars are comparatively rare, we see that interactions and mergers often produce nuclear starbursts, and in some cases seem to induce AGN activity. The amounts of material involved range up to about 10^9 M$_\odot$, the timescales are $\sim 10^8$ yrs, and the spatial scales are < 1 kpc. During galaxy formation we expect a much more abundant gas supply than in current spirals, as well as strong nonaxisymmetric dynamical processes (*e.g.* Fig. 1). Thus it seems natural that nuclear activity, both starburst and AGN, should have been brighter and more common at that time than it is today. The decline in quasar activity from its peak at $z = 2 - 3$ may be evidence of a decline in the rate at which large protogalaxies are collapsing. For an $\Omega = 1$ universe, the simulations described in section 3 suggest that large galaxies are indeed being assembled at these redshifts, while analytic models of galaxy formation, based on extensions of the techniques described in section 4 to include gas cooling, SF, and feedback, suggest that efficient conversion of gas into stars also occurs primarily at this time; at earlier times feedback prevents most gas from cooling, while at later times low gas densities lead to inefficient cooling (White and Frenk 1991; Lacey *et al.* 1993; Kauffmann *et al.* 1993). In a low density universe all these processes occur at higher redshift with the difference being much more marked for $\Lambda = 0$ than for a flat, low density universe with $\Lambda \neq 0$ (*cf.* Fig. 2).

Two important questions concern the spatial scale to which material can collapse during galaxy formation and the rapidity with which it can be concentrated into a small region. In the kind of galaxy formation picture discussed here, a substantial fraction of the protogalactic gas can end up in the nuclear regions almost immediately, and the only process which can prevent rapid collapse to very small dimensions appears to be efficient conversion of gas into stars. Thus, despite the late assembly of the model shown in Figure 1, 60% of the gas lost sufficient angular momentum

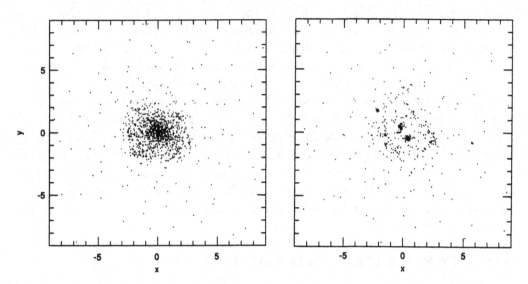

Fig. 3—Two simulations of the collapse of a spherically symmetric rotating perturbation. The simulations are viewed in projection onto the disk plane and differ only in the gravitational softening used.

during formation to settle within 2 kpc of the center. This unresolved central clump is entirely self-gravitating, and its further evolution on short timescales is inhibited only by numerical effects.

This is illustrated by the simplified galaxy formation simulations shown in Figure 3. For these models a spherically symmetric protogalaxy with a shallow initial density profile ($\rho \propto 1/r$) and made of 90% dark matter and 10% gas was given some initial solid body rotation, and then allowed to collapse. The work of Katz and Gunn (1991) shows that with the addition of some small-scale irregularity, such initial conditions collapse to make objects quite similar to real spirals. Our own work (Navarro and White 1993a) comes to the same conclusion for the same initial rotation rate, which is at the high end of the distribution predicted by hierarchical clustering (*e.g.* Barnes and Efstathiou 1987). However, for the experiments shown in Figure 3 the rotation rate was chosen to be on the low side of this same distribution. The gas then contracts twice as much before coming to centrifugal equilibrium. In the left panel of the figure, which shows a time when about 25% of the the system has already collapsed, a small self-gravitating disk (seen face-on here) is clearly forming. Note, however, that the radius of this disk is only about 2 kpc, so it is not properly resolved by this simulation which had a gravitational softening of 1 kpc. The right hand panel shows what happens if the simulation is rerun with a softening of 100 pc. Instead of a disk a central binary is formed in which each component has a scale of about 100 pc!

The reason for this dramatic difference is not difficult to find. In the absence of substantial softening, cold self-gravitating gas disks are well known to be violently unstable to nonaxisymmetric perturbations. Rather than settle to an unstable configuration the gas in the simulation solves its angular momentum "problem" by forming a binary, each component of which continues to radiate binding energy and to contract until it reaches the resolution limit. Several reruns of this simulation with slightly different initial conditions led to similar results, showing that the formation of a few body system, rather than a disk, is generic for rotating cold gas once it has contracted enough to be fully self-gravitating. Models in which the gas was allowed to form stars rapidly and efficiently once it reached molecular cloud densities led to very different results. SF occurred as soon as the gas shocked in the central regions, the stars formed a small "bulge" which dominated the central gravity field; gas falling in later was still able to make a disk.

The conclusion from this simulation seems clear. If the SF efficiency (measured as the fraction of the gas converted into stars per natural dynamical time) is small, then inhomogeneous galaxy formation inevitably leads to the formation of a very compact system with quasar-like luminosity. This is because dissipation and angular momentum loss to the dark matter channel enough gas into the central regions that: (i) the gas can no longer remain hot enough for pressure support to be significant because its high density implies radiative cooling times which are much shorter than the sound crossing time; (ii) the gas cannot settle into a centrifugally supported disk because it dominates the mass budget and self-gravitating cold disks are unstable; and (iii) the gas cannot be prevented from further collapse by hydrodynamic input from supernovae because of the efficiency of radiative losses, and can only be stopped from collapse by radiation pressure if the luminosity of the starburst or central AGN approaches the Eddington luminosity for the relevant mass, $\sim 10^9$ $M_\odot$.

It is, of course, possible that SF during protogalactic collapse is highly efficient so that most of the gas turns into stars before it can settle to the centre. Prompt supernovae might then expel the rest in a wind. This could be a plausible model for the early formation of an elliptical galaxy over a collapse time of a few $\times 10^8$ yrs. If such were the case the observed properties of giant elliptical galaxies then suggest the following scheme for producing a quasar. About 10^8 yrs after the end of the initial star formation phase all the Type II supernovae will have exploded. Immediately thereafter the mass loss rate from the stars remains high, but the energy injection rate drops dramatically. It is therefore plausible that the wind flow reverses and a substantial fraction of the gas lost from stars can cool and flow towards the centre (David, Forman, and Jones 1990). This inflowing gas will have the specific angular momentum of the observed stellar population, and so will accumulate in a centrifugally supported disk with an effective radius about 15% that of the elliptical. This radius contains about 5% of the observed stellar mass, so the disk will become self-gravitating and go dynamically unstable when it grows to a mass of this order. This is expected to take $\sim 10^9$ yrs. When instability occurs the disk will have a mass exceeding 10^9 $M_\odot$, a radius smaller than 1 kpc, and a natural dynamical timescale

$\sim 10^7$ yrs. Thus if it is still made of gas, a very massive starburst should ensue (or a very massive black hole should form). We thus arrive at the model for the QSO luminosity function proposed by Terlevich and Boyle (1992).

The question, of course, is whather the gas can avoid turning into stars as it accumulates in a thin cold disk. Such disks are locally unstable to gravitational fragmentation long before they become globally unstable. In spiral galaxies such instabilities lead to SF. There is no convincing evidence, either theoretical or observational, to suggest that this can be avoided. However, the conditions involved are very different from those in a spiral galaxy so the analogy may be flawed. In the elliptical case the gas settles in an environment where the UV background radiation is intense and may affect cooling and molecule formation. In addition the dust abundance is likely to be low. Since *all* the gas is lost from stars, it is likely to entrain a significant magnetic field. This field may be amplified to equipartition values by the accretion flow and by differential rotation in the disk. Thus it is possible to imagine that the field might suppress fragmentation and SF until the disk is massive enough to become globally unstable. Bar formation would lead to strong shocks, dissipation, and rapid contraction. Although very rudimentary, this model has the attractive feature that many of its characteristic scales can be estimated directly from the observed properties of elliptical galaxies and from standard models for the evolution of their stellar populations.

I thank Julio Navarro for permission to show results from our joint work before publication.

REFERENCES

Barnes, J. E. 1988, *Ap. J.*, **331**, 699.
Barnes, J. E., andf Efstathiou G. 1987, *Ap. J.*, **319**, 575.
Bond, J. R., Cole, S., Efstathiou, G., and Kaiser, N. 1991, *Ap. J.*, **379**, 440.
Bower, R. J. 1991, *Mon. Not. R. Astr. Soc.*, **248**, 332.
Carlberg, R. G., Couchman, H. M. P., and Thomas, P. A. 1990 *Ap. J. Lett.*, **352**, L29.
Cole, S. 1991, *Ap. J.*, **367**, 45.
David, L. P., Forman, W., and Jones, C. 1990, *Ap. J.*, **359**, 29.
Dekel, A. and Silk, J. 1986, *Ap. J.*, **303**, 39.
Efstathiou, G. 1992, *Mon. Not. R. Astr. Soc.*, **256**, 43P.
Efstathiou, G., Frenk, C., White, S. D. M., and Davis, M. 1988, *Mon. Not. R. Astr. Soc.*, **235**, 715.
Efstathiou, G., Bond, J. R., and White, S. D. M. 1992, *Mon. Not. R. Astr. Soc.*, **258**, 1P.
Evrard, A., Summers, F., and Davis, M. 1993, *Ap. J.*, in press.
Frenk, C. S. 1991, *Physica Scripta*, **T36**, 70.
Geller, M. J., and Huchra, J. P. 1989, *Science*, **246**, 897.

Holtzman, J. A. 1989, *Ap. J. Suppl.*, **71**, 1.

Katz, N. 1992, *Ap. J.*, **391**, 502.

Katz, N., Hernquist, L., and Weinberg, D. H. 1992, *Ap. J. Lett.*, **399**, L109.

Katz, N., and White, S. D. M. 1993, *Ap.J.*, **412**. 455.

Kauffmann, G., and White, S. D. M. 1993, *Mon. Not. R. Astr. Soc.*, **261**, 921.

Kauffmann, G., White, S. D. M., and Guiderdoni, B. 1993, *Mon. Not. R. Astr. Soc.*, **264**, 201.

Lacey, C. G., and Cole, S. 1993, *Mon. Not. R. Astr. Soc.*, **262**, 627.

Lacey, C. G., Guiderdoni, B., Rocca-Volmerange, B., and Silk, J. I. 1993, *Ap. J.*, 402, 15.

Navarro, J. F., and White, S. D. M. 1993a, *Mon. Not. R. Astr. Soc.*, in press.

Navarro, J. F., and White, S. D. M. 1993b, *Mon. Not. R. Astr. Soc.*, submitted.

Park, C. 1990, *Mon. Not. R. Astr. Soc.*, **242**, 59P.

Press, W. H., and Schechter, P. L. 1974, *Ap. J.*, 187, 425.

Sanders, D. B., *et al.* 1988, *Ap. J.*, **325**, 74.

Terlevich, R. J., and Boyle, B. J. 1993, *Mon. Not. R. Astr. Soc.*, **262**, 491.

Tóth, G., and Ostriker, J. P. 1992, *Ap. J.*, **389**, 5.

Turner E. L. 1991, in *The Space Distribution of Quasars*, ed. D. Crampton (San Francisco: ASP-21), p. 361.

Tyson, J. A. 1988, *A. J.*, **96**, 1.

White, S. D. M., and Rees, M. J. 1978, *Mon. Not. R. Astr. Soc.*, **183**, 341.

White, S. D. M., and Frenk, C. S. 1991, *Ap. J.*, **379**, 52.

White, S. D. M., Frenk, C. S., and Davis, M. 1983, *Ap. J. Lett.*, **274**, L1.

The Evolution of Quasar Host Galaxies

Fred Hamann

The Ohio State University

ABSTRACT

Analysis of quasar broad emission lines suggests that the emitting gas is substantially enriched, often well above solar at high redshifts. The abundances are like those expected in the cores of massive galaxies early in their evolution, suggesting that observable quasars occur near the end of the epoch when rapid star formation, dominated by high mass stars, has created an enriched interstellar medium. An increase in the derived metallicities with both redshift and luminosity suggests that there is a mass-metallicity relation among quasars analogous (or identical) to the mass-metallicity relation in elliptical galaxies. This relation is consistent with the most massive quasars and/or host galaxies forming only at high redshifts.

1 INTRODUCTION

Observations of strong metallic emission lines in quasars out to redshift $\sim$4.9 imply *some* enrichment at times when the Universe was less than 10% of its present age. If quasars reside in the cores of massive galaxies, their gaseous environments could easily have larger than solar abundances, even at the highest redshifts. Observations and models of giant elliptical galaxies show that metallicities of at least several $Z_\odot$ are attained in less than 1 Gyr (*cf.* Arimoto and Yoshii 1987). In the bulge of our own galaxy the stellar metallicities also reach at least a few $Z_\odot$ (Rich 1988) and the enrichment is again believed to occur in $\lesssim$1 Gyr (Köppen and Arimoto 1990). Since the gas in any evolving star cluster is as chemically enriched as the most recently formed stars, and thus more enriched than the bulk of the stellar population, metallicities above solar may be typical of the gas in massive galactic nuclei. Although this gas may be expelled eventually by a galactic wind, or perhaps consumed by a black hole, it must remain long enough for the evolution to high Z's to occur.

The broad emission lines of quasars offer direct probes of the composition and chemical history of the gas in galaxy cores. We examine quasar abundances and host galaxy evolution by applying spectral synthesis and chemical enrichment models to the emission line observations. We assume the lines form in well mixed interstellar gas. We also discuss the implications of observed trends in the line ratios — as abundance indicators — with the quasar luminosities and redshifts.

2 GALACTIC CHEMICAL EVOLUTION

Galactic chemical enrichment models make specific predictions for the elemental abundances as a stellar population evolves. We constructed one zone models for systems built up over finite times by the accretion of primordial gas (Hamann and Ferland 1992 and 1993, hereafter HF93). The sources of enrichment include stellar winds, planetary nebulae, and types I and II supernovae. Star formation is regulated by power law initial mass functions (IMFs), and the delayed enrichment due to finite stellar lifetimes is included. Our procedure is to first fix the stellar nucleosynthesis with reference to a model of the galactic solar neighborhood, and then vary just the slope of the IMF and the timescales for star formation and primordial infall to model the chemical history of quasar broad line regions (BLRs).

The detailed results for various models are discussed in HF93. The essential point is that the flatter IMFs and shorter timescales expected in galactic nuclei lead to higher metallicities and significantly different abundance mixtures compared to the solar neighborhood. In particular, the models designed to mimic the evolution of massive ellipticals and the bulges of disk galaxies reach $Z \sim 1$ to perhaps ~ 10 $Z_\odot$ in $\lesssim 1$ Gyr. Also, the abundances of N and Fe relative to He, C, O, and the α-process elements exceed solar on about the same timescale. The delayed overabundance of N is due to secondary CNO processing in stellar envelopes, while the late Fe enhancement results from the delayed contribution from type Ia supernovae.

3 ABUNDANCE DIAGNOSTICS IN QUASARS

Our model of the BLR assumes constant density clouds in photoionization equilibrium with a "standard" AGN continuum. The flux in the strong metallic resonance lines in the UV is not sensitive to the global metallicity of the gas because these lines are important coolants (*cf.* Ferland and Hamann 1993). However, the *relative* intensities of some lines *are* sensitive to the selective enrichment of the elements. The prediction that N, in particular, is selectively enriched as a stellar population ages leads us to develop as abundance diagnostics the line ratios of N V 1240Å (hereafter N V) to C IV 1549Å (C IV) and He II 1640Å (He II). These lines are far less sensitive to collisional quenching than the intercombination lines used in previous studies. They are also observed in a much larger sample of quasars. The N V/C IV line ratio is advantageous in low signal-to-noise spectra because C IV is strong and essentially always present. The N V/He II ratio is more difficult to measure because the helium line is often weak, but it is a more robust abundance indicator because the N V emitting region lies within the He II emitting zone. Therefore this ratio is less sensitive to even extreme inhomogeneities in the BLR.

The sensitivity of the line ratios to the physical conditions and the shape of the incident continuum is discussed in detail by HF93, Ferland and Hamann (1993), and Baldwin *et al.* (1993).

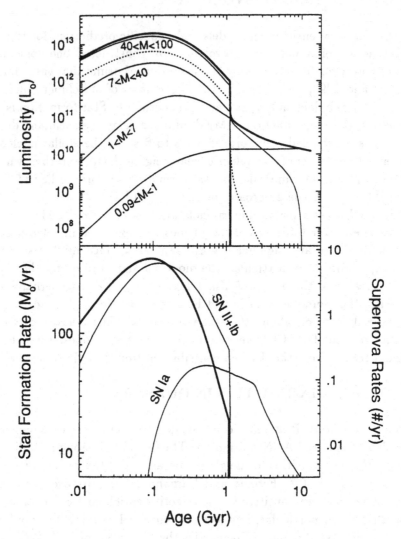

Fig. 1—The evolution of various quantities are shown for a galaxy model with parameters like those developed for the giant ellipticals. This model also provides a good match to the emission lines in many high redshift quasars (see HF93). *Upper Panel.* The thin solid lines show the bolometric luminosities contributed by main sequence stars in the mass ranges shown. The thick solid line shows the total luminosity of the entire (main sequence) population. The dotted line is the total "UV luminosity", *i.e.* νL_ν at 1450Å, derived assuming the stars emit like blackbodies at this wavelength. *Lower Panel.* The thick line and left-hand scale indicate the star formation rate. The thin lines and right-hand scale show the rates of types Ia and II+Ib supernovae. All quantities scale linearly with the evolving galaxy mass, which is set to 10^{11} M$_\odot$ for this plot. The star formation stops at $\sim$1 Gyr when the mass fraction in gas drops below 3%.

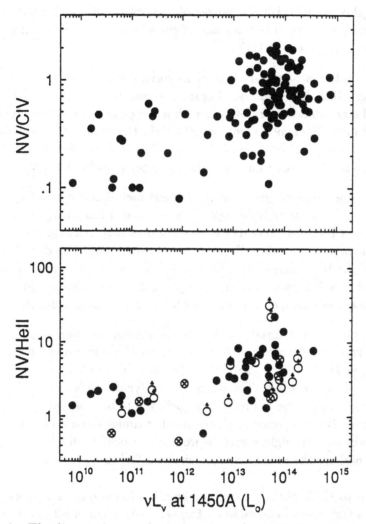

Fig. 2—The line ratios used as abundance/metallicity indicators are shown to correlate roughly with the quasar luminosities.

4 RESULTS AND IMPLICATIONS FOR QUASAR HOST GALAXIES

We use abundances from the enrichment models directly in the spectral synthesis calculations to predict the evolution of the line ratios as a function of the age and metallicity of quasar environments. The detailed results are presented and discussed in HF93. The main points are summarized below.

(1) Quasars are associated with high Z's and vigorous star formation exactly like that expected in massive galactic nuclei. Metallicities from ~ 1 to perhaps $\gtrsim 10\ Z_{\odot}$ are reached before quasars become observable (or "turn on"). At redshifts >3 these metallicities must be reached in $\lesssim 1$ Gyr if $q_{o} = 1/2$ or $\lesssim 2$ Gyr if $q_{o} = 0$. The analogy

with elliptical galaxies and with the bulges of disk galaxies suggests that the IMF favors massive stars more than in the solar neighborhood. The highest Z's can only be reached by such "top-heavy" IMFs.

(2) There is a population of massive galaxies evolving rapidly at high redshifts, i.e at redshifts larger than the quasars. Figure 1 shows the evolution of the stellar luminosities and star formations rates for an elliptical galaxy model that fits much of the high redshift data. For evolving masses $\gtrsim 10^{11}$ $M_\odot$ the total stellar luminosity can exceed 10^{13} $L_\odot$, rivaling that of observed quasars. The integrated stellar spectrum is dominated by the most massive stars and therefore peaks in the far UV.

(3) The observed line ratios suggest that the highest metallicity systems appear only in high luminosity objects at redshifts $\gtrsim 2$. The trend with luminosity in the current data set is shown in Figure 1 (also HF93). We are presently working with expanded samples to deconvolve the trends with redshift and luminosity. The combined L and z trends may be due to a mass-metallicity relation among quasars that is analogous (or identical) to the well known relation in elliptical galaxies. This would imply that the highest mass quasars and/or host galaxies form only at high redshifts.

(4) The models predict a delayed rise in the Fe abundance due to the late time enrichment by type Ia supernovae. The type Ia contribution can be distinguished from that of types II and Ib by examining the ratio of Fe to O, Mg, or other α-process products of type II+Ib's. The delay of ~ 1 Gyr depends only on the (albeit uncertain) lifetimes of the type Ia precursors - believed to be close intermediate mass binaries - and not on the star formation rates or other timescales used in the evolution models. The Fe abundance might therefore provide a "clock" with which to constrain quasar ages and perhaps the cosmology (q_0) when applied to quasars at high redshift.

I am grateful to G. J. Ferland for many helpful discussions. The work of F. H is supported by a Columbus Fellowship at The Ohio State University and by NASA grant NAG 5-1645.

REFERENCES

Arimoto, N., and Yoshii, Y. 1987, *Astr. Ap.*, **173**, 23.

Baldwin, J. A., Ferland, G. J., Hamann, F., Phillips, M., Williams, R., and Wilkes, B. 1993, *Ap. J.*, submitted.

Ferland, G. J., and Hamann, F. W. 1993, in *The Nature of Compact Objects in AGN*, 33rd Herstmonceux Conference, (Cambridge: Cambridge Press), in press.

Hamann, F., and Ferland, G. J. 1992, *Ap. J. Lett.*, **391**, L53.

Hamann, F., and Ferland, G. J. 1993, *Ap. J.*, in press (HF93).

Köppen, J., and Arimoto, N. 1990, *Astr. Ap.*, **240**, 22.

Rich, R. M. 1988, *A. J.*, **95**, 828.

Evolution of the Galaxy Merger Rate: Counting Pairs in HST Fields

Jordan M. Burkey[1], William C. Keel[1], and Rogier A. Windhorst[2]

[1]Department of Physics and Astronomy, University of Alabama
[2]Department of Physics and Astronomy, Arizona State University

ABSTRACT

Using deep *HST*/WFC images, originally taken to study faint radio galaxies, we find 81 clear serendipitous galaxy images, of which 34 are pair members. Based on nearby magnitude-limited samples, this is an excess of more than 4σ above the expected number of pair members. We take this result as strong evidence that the galaxy merger rate was higher in the past, and has declined over time.

1 INTRODUCTION: GALAXY MERGERS AND EVOLUTION

Galaxy interactions and mergers have been implicated as driving galaxy evolution in several ways:

Triggering starbursts, thus making the star-forming history episodic
Driving global winds from starbursts, sweeping merger remnants free of gas and dust
Transforming galaxy morphology through mergers and tidal impulses
Triggering nuclear activity

Counts of local pairs and mergers, plus N-body modelling of orbital decays, suggest that many (perhaps most) present galaxies underwent mergers during cosmic history. This means that the merger rate was probably higher in the past. We are using galaxy and pair counts from deep *HST* serendipitous fields to constrain the merger rate.

We cannot uniformly trace mergers themselves to large redshifts, because (1) cosmological $(1 + z)^4$ surface-brightness dimming makes the characteristic tidal features too faint for detection and (2) at large redshifts, the disturbed structures can be too small for detection given surface-brightness constraints. We therefore trace the merger rate by studying the evolution of *galaxy pairs* some of which are the immediate precursors of mergers. If we see many pairs disappear as we approach the present epoch, the vast majority of the disappearances will have gone into mergers (except for any rate at which orbital decay or capture may increase the number of isolated pairs, and neither process is likely to be important).

2 HST OBSERVATIONS AND DATA PROCESSING

We have used WFC images originally taken for studies of faint LBDS radio galaxies, and are taking the surrounding regions as serendipitous fields. There are currently 7 fields from programs in cycles 1 and 2; we present the first four fields here. In each case, images in F555W ("V") and F785LP ("I") passbands were obtained, with total exposure times per filter 1.5–4 hours. The images were stacked using an iterative cosmic-ray removal scheme.

The structure of individual objects in these fields has been studied both by using deconvolution (Lucy-Richardson or σ-CLEAN) and by model fitting, taking model galaxy components convolved with (real or computed) PSFs and fitted directly to the aberrated data to measure scale lengths and bulge-to-disk ratios. For objects brighter than about V=20, both techniques yield virtually identical structural parameters for symmetric galaxies.

3 OBJECT IDENTIFICATION AND PAIR COUNTS

We attempt to select pairs using criteria similar to those used for nearby samples, to allow a direct comparison of the pairing fraction in magnitude-limited samples (and with luck, eventually as a direct function of redshift). The pairing fraction for several different criteria among nearby galaxies ranges from 8% for UGC spirals (Keel and van Soest 1992) to 12% for ellipticals in the Karachentsev (1972) catalog.

In our fields to date (with 4 sets of WFC frames analyzed), we find 81 clear serendipitous galaxy images, of which 34 are pair members. The expectation number of pair members from nearby magnitude-limited samples is 8 — our result is more than 4σ above the local samples. We have detected an excess of pairs in distant, probably intermediate-redshift ($z = 0.3 - 0.7$) galaxies. A very similar result has been found for galaxies in a cluster environment (McClure, Pierce, and Lavery 1993). Together with a model of the merging population, this will give an estimate of the merger rate; when redshifts are available, we can derive a direct measure of the merger rate over time. These results are strong evidence that *the galaxy merger rate has indeed declined over time.*

This work incorporates observations using the NASA/ESA *Hubble Space Telescope* obtained at the Space Telescope Science Institute, which is operated by AURA for NASA. These results come from programs 2405 and 3545. JMB and WCK acknowledge partial support from EPSCoR grant EHR-9108761.

REFERENCES

Karachentsev, I. D. 1972, *Soobsch. S. A. O.*, **7**, 3.
Keel, W. C., and van Soest, E. T. M. 1992, *Astr. Ap. Suppl.*, **94**, 553.
McClure, R. D., Pierce, M. J., and Lavery, R. J. 1993, *Bull. Am. Astr. Soc.*, **25**, 838.

Stellar Mass Loss in Ellipticals and the Fueling of AGN

Paolo Padovani[1] and Francesca Matteucci[2]

[1]Dip. di Fisica, II Università di Roma "Tor Vergata", I-00133 Roma, Italy
[2]ESO, Karl-Schwarzschild-Straße 2, D-8046 Garching bei München, Germany

ABSTRACT

The connection between some observational properties of active galactic nuclei and their host galaxies is studied by means of a self-consistent model of galactic evolution which reproduces the main features of elliptical galaxies.

1 THE PROBLEM

Thirty years after the discovery of quasars, the problem of their evolution is still open. The optical evolution of the quasar luminosity function can be modelled with a constant comoving space density and a (pure) luminosity evolution, up to $z \sim 2$, of the type $L_o(z) = L_o(0)(1 + z)^k$, with $k \simeq 3.5$ for both $q_0 = 0.1$ and 0.5 (*i.e.* $L(t) \approx t^{-3.5}$ and $\propto t^{-2.3}$ respectively: Boyle *et al.* 1991). The X-ray and radio evolutions are weaker than the optical one.

If the central energy source is primarily gravitational in the form of a massive black hole then $L \simeq \eta \dot{M} c^2$, with η efficiency of the mass-energy conversion (usually taken equal to 0.1) and $\dot{M}$ the mass accretion rate. In this framework, accretion rates in the range $10^{-3} - 10^2$ M$_\odot$ yr^{-1} are needed to explain the observed range of AGN luminosities ($10^{43} - 10^{48}$ erg s^{-1}).

The origin of the required mass supply is not clear. Several papers have studied the evolution of very dense star clusters (with central densities up to $10^8 - 10^9$ M$_\odot$ pc^{-3}) and the resulting fueling of the central black hole through stellar mass loss (*e.g.* Norman and Scoville 1988). These studies fail to explain the bolometric luminosities observed in local AGN by at least two orders of magnitude, while the predicted luminosity evolution is weaker than the one inferred for quasars. Moreover, such high stellar densities have never been observed. More fuel seems in any case to be needed and therefore a larger part of the host galaxy has to be involved.

In this paper we address briefly the following questions: a) can mass loss in elliptical galaxies explain the observed quasar bolometric luminosities? b) how do the predicted ratios between the quasar and host galaxy luminosities compare to the ones inferred from observations? c) can the time dependence of the fuel supply be responsible for the observed evolution of AGN? A full presentation of these results will be given elsewhere (Padovani and Matteucci 1993).

2 OUR RESULTS

We employ a supernova (SN) driven galactic wind model, with quite efficient star formation leading to the development of a galactic wind when the energy input from SNe in the interstellar medium equals the binding energy of the gas. Star formation is assumed to stop at the time of the galactic wind, after which there is only passive evolution. The model includes the most recent ideas on SN progenitors and nucleosynthesis and reproduces some of the main features of normal ellipticals: stellar metal content, luminosity, M/L, SN Ia rate (Matteucci 1992 and references therein). The initial luminous mass range is taken to be $10^{11} - 5 \times 10^{12}$ M$_\odot$. Our standard model assumes a Salpeter initial mass function over the mass range $0.1 \leq M/M_\odot \leq 100$, but we explored the effects of changes in these parameters on our results.

Our main conclusions are the following:

The bolometric luminosities predicted for local AGN residing in ellipticals (i.e. radio-loud quasars) for a typical efficiency $\eta \simeq 0.1$ ($L_{\rm bol} = \eta \dot{M} c^2$), are in good agreement with the observational estimates ($10^{45} - 5 \times 10^{46}$ erg s^{-1}).

The observed ratio of AGN to host galaxy visual luminosity is centered, within a factor of two, around the value predicted by our model. The observed large dispersion might be consistent with the expected model range plus observational errors.

The time dependence of the stellar mass-loss rate in ellipticals ($\propto t^{-0.9} - t^{-1.3}$) is too small to explain the luminosity evolution of optically selected radio-loud quasars, probably similar to, or even stronger than, the one of the more numerous radio-quiet quasars ($\gtrsim t^{-3.5}$ and $t^{-2.3}$ for $q_0 = 0.1$ and 0.5 respectively). Very flat initial mass functions (*e.g.* $\phi(m) \propto m^{1.5}$) would give a much stronger time dependence ($\gtrsim t^{-2}$) but the resulting elliptical galaxies would have mass-to-light ratios orders of magnitude larger than the observed ones, while the resulting AGN bolometric luminosities would be too small to agree with observations. We therefore suggest that: either quasar luminosity evolution is much weaker than currently estimated; or some density evolution has to be present, in which case quasar evolution is not driven only by the declining fueling rate of the host galaxy but also, for example, by galaxy interactions. Recent evidence that the shape of the quasar luminosity function is redshift dependent (*e.g.* Hewett *et al.* 1993) seems to support the latter hypothesis.

REFERENCES

Boyle, B. J., Jones, L. R., Shanks, T., Marano, B., Zitelli, V., and Zamorani, G. 1991, in *The Space Distribution of Quasars*, ed. D. Crampton (San Francisco: ASP-21), p. 191.

Hewett, P. C., Foltz, C. B., and Chaffee, F. H. 1993, *Ap. J. Lett.*, **406**, L43.

Matteucci, F. 1992, *Ap. J.*, **397**, 32.

Norman, C., and Scoville, N. Z. 1988, *Ap. J.*, **332**, 124.

Padovani, P., and Matteucci, F. 1993, *Ap. J.*, **416**, 26.

Star–Forming Galaxies and Large–Scale Structure

G. Vettolani[1], J. Alimi[2], C. Balkowski[2], A. Blanchard[2], A. Cappi[3],
V. Cayatte[2], G. Chincarini[4], C. Collins[5], P. Felenbok[2], G. Guzzo[4],
D. Maccagni[6], H. T. MacGillivray[5], R. Merighi[3], M. Mignoli[3],
S. Maurogordato[2], D. Proust[2], M. Ramella[7], R. Scaramella[8],
G. Stirpe[3], G. Zamorani[3], and E. Zucca[1]

[1] Istituto di Radioastronomia, Bologna, Italy
[2] DAEC, Observatoire de Meudon, France
[3] Osservatorio Astronomico, Bologna, Italy
[4] Osservatorio di Brera, Milano, Italy
[5] Royal Observatory Edinburgh, United Kingdom
[6] IFTCR, Milano, Italy
[7] Osservatorio di Trieste, Italy
[8] Osservatorio di Roma, Monteporzio Catone, Italy

ABSTRACT

From the preliminary analysis of a sample of $\simeq 600$ galaxies with $b_j \leq 19.4$, and spanning a redshift interval up to $z \simeq 0.3$, we deduce that star formation per unit luminosity, as indicated by the O II 3727 line equivalent width, is a function of galaxy luminosity (decreasing at increasing luminosities), redshift (increasing at increasing redshift, or, physically, look–back time) and environment.

1 INTRODUCTION

At the ESO 3.6m telescope at La Silla, we are currently performing a redshift survey of galaxies with $b_j \leq 19.4$, in a rectangular area $\simeq 22° \times 1°$ (plus a nearby area of $\simeq 5° \times 1°$) in a region around the South Galactic Pole. Up to now we have acccumulated spectra for $\sim 2,000$ galaxies over $\sim 70\%$ of the area.

The distribution in distance of the survey galaxies exibits significative peaks above the expectation at $D \simeq 180 h^{-1}$ Mpc and $D \simeq 300 h^{-1}$ Mpc. These peaks correspond to large scale structures extending over a significant fraction of the strip.

A large fraction of galaxies ($\sim 40\%$) shows the presence of one or more emission lines (O II $\lambda 3727$, Hβ, O III $\lambda\lambda 4959, 5007$). These objects can be either spiral galaxies, where lines originate mostly from H II regions in the disks, or galaxies undergoing a significant burst of star formation. The observed peaks in the galaxy distribution are much less pronounced when only emission line galaxies are considered. This suggests that either spiral galaxies are less frequent in the densest regions, thus confirming a large scale validity of the well known morphology–density relation, or starburst phenomena in galaxies occur preferentially in low density environments, or both.

2 STAR FORMATION, LUMINOSITY AND REDSHIFT

The O II doublet, $\lambda 3727$, is the most useful star formation tracer in the blue and can be used as a substitute to Hα in distant galaxies. Kennicut (1992) has shown that if R is the number of solar masses per year in new stars, EW the equivalent width of the O II line, and L the galaxy luminosity in solar units, then R can be roughly estimated, within a factor of a few, through the relation $R \sim 7 \times 10^{-12} L \times EW \times h^{-2}$ where $h = H/100$.

In a subsample of ~ 600 galaxies, we have measured the O II EW for all detections and evaluated, for non–detections, its upper limit as a function of the continuum S/N as $(EW)_{\text{obs}} \leq 2.5 \times (S/N)_{\text{line}}/(S/N)_{\text{c}}$, with $S/N_{\text{line}} = 10$, as appropriate to our O II detections.

We have, therefore, O II $\lambda 3727$ EW estimates, both detections and upper limits, for a large complete sample of galaxies, allowing detailed statistical analysis up to $z \simeq 0.3$. The first relevant question we asked is: *is there any systematic dependence of O II equivalent width* (which is proportional to the number of new stars per unit luminosity) *with galaxy luminosity and/or redshift?*

Application of a bivariate regression analysis to our sample data shows that both correlations are present but with different levels of significance. The correlation of EW with luminosity is significant at 4.5σ and is in the sense that *faint galaxies have larger O II equivalent widths than bright ones.* Part of this effect might be induced by a different mixture of morphological types as a function of magnitude. However, the fact that this correlation is significant also for galaxies brighter than -18.8, where the ratio of spirals to ellipticals is approximately constant, make us to believe that the correlation is true also within a given morphological type.

The correlation with z is significant at 2.5σ only. However, the dependence of EW versus redshift is more complex to study since emission line galaxies, as we mentioned above, do not smoothly follow the large scale distribution. Moreover, the present sample, being magnitude limited, does not sample well enough the faint part of the galaxy luminosity function at high redshift. Adding to ours other data from fainter surveys we have significative evidence that EW is increasing with redshift for galaxies in a given luminosity interval.

Acknowledgement: This work has been partially supported by NATO Grant CRG920150.

REFERENCES

Kennicut, R. C. 1992, *Ap. J.*, **388**, 310.

Summary: Inflows and Outbursts in Galaxies

Richard B. Larson

Astronomy Department, Yale University

1 INTRODUCTION

A wealth of information has been presented at this conference illustrating the basic theme that large-scale gas inflows can result in the release of large amounts of energy in galactic nuclei by starbursts and by non-thermal processes associated with black hole accretion. Particular attention has been paid to two situations in which such phenomena frequently occur: (1) bar-driven inflows in barred spiral galaxies can lead to gas accumulation and star formation in nuclear rings of 'hot spots', and (2) tidal interactions and mergers between galaxies can cause large amounts of gas to fall rapidly into their nuclei and produce luminous starburst and non-thermal activity. Although it is difficult to observe the gas inflows themselves directly, partly because they may be masked by more conspicuous outflows, inflows are commonly predicted by numerical simulations of barred and interacting galaxies, and the success with which these simulations match many of the observed properties of such systems leaves little doubt that inflows occur quite generally whenever a non-axisymmetric gravitational potential is present. Evidence that gas inflows from regions of galactic size are responsible for triggering the most energetic outbursts of activity in galactic nuclei is provided by the fact that in many of the most luminous systems, an amount of molecular gas comparable to the gas content of an entire large spiral galaxy is observed to be concentrated into a small nuclear region only a few hundred parsecs across.

Starbursts and non-thermal activity in galactic nuclei have two major consequences for galactic evolution: star formation and the growth of black holes can build up the central regions of galaxies, and the energy so released can heat and expel residual gas into an intergalactic medium. Thus an understanding of how gas condenses into galactic nuclei is an important part of understanding the evolution of galaxies and of the universe as a whole. Much remains to be learned about the many processes involved, especially on the small scales relevant to black hole accretion, but considerable progress has been made in recent years in understanding the mechanisms that operate on galactic scales, thanks to a fruitful combination of observational and theoretical studies, as has been evident at this meeting. As a result, a good understanding is beginning to emerge of how bars and tidal interactions can cause gas to accumulate in nuclear regions less than a kiloparsec in radius, and the possibility that similar processes operate on even smaller scales and play a role in black hole accretion is currently a subject of active study.

In several respects, there is a close analogy between the condensation of gas in galactic nuclei to form central black holes and the condensation of gas in protostellar clouds to form stars. In both cases, gas that was initially widely distributed condenses under its self-gravity into a region many orders of magnitude smaller in size, and the energy released creates not only a high luminosity but also vigorous outflows that can dominate the observed properties of the system. There are also similar requirements in the two situations that the gas must lose nearly all of its angular momentum if it is to become highly condensed, and the mechanisms involved in the transport of angular momentum may even be similar. The mechanisms that act to redistribute angular momentum on galactic scales and drive gas inflows into galactic nuclei are summarized in the next section; most of the effects discussed also operate in other contexts, such as star forming clouds, and therefore are of more general interest.

2 GRAVITATIONAL MECHANISMS

Most presentations at this conference have dealt with phenomena that occur on scales larger than a hundred parsecs, where gravity is almost certainly the dominant force. The basic theoretical problem, in the galactic as in the stellar case, has usually been taken to be the need to understand how angular momentum is removed from the condensing gas. Angular momentum has played a key role in cosmogonical discussions because it has traditionally been assumed for simplicity that the initial configuration is spherical and has a central force field; angular momentum is then a conserved quantity, and gas that initially has any significant non-radial motion must overcome an 'angular momentum barrier' if it is to become highly condensed. In reality, of course, astronomical systems need not begin with spherical or even axial symmetry, and angular momentum is then not in general a conserved quantity and there need be no 'angular momentum problem'. Nevertheless, since most observed galaxies do show approximate symmetries, it is still a useful approximation to discuss their structure in terms of departures from axial symmetry and their dynamics in terms of torques that can redistribute angular momentum. Several mechanisms for redistributing angular momentum can be distinguished that are associated with different types of non-axisymmetric disturbances in galaxies, and they are summarized below; in reality, these mechanisms are not entirely distinct because they all involve related gravitational effects, and they may even work together in many cases.

2.1 Spiral Patterns

Spiral structure has long been recognized to be capable of transporting angular momentum in galaxies, and it plays an important role in some of the simulations that have been reported here. Two distinct physical effects are involved: (1) the gravitational torque acting between the inner and the outer part of any spiral arm transports angular momentum outward if the arm is trailing, and (2) a spiral density wave has an associated energy and angular momentum that are also transported outward by wave propagation if the spiral pattern is trailing. The energy and angular momentum of a wave in a disk are negative inside its corotation radius and positive outside it, so that if a wave is somehow generated in a disk and then dissipated, for example by shocks, the net effect will be that inner part of the disk loses energy and angular momentum while the outer part gains. The rate at which gas flows inward as a result of these effects depends on the rate at which wave energy is dissipated, and the inflow timescale is essentially the orbital period divided by the fractional energy of the gas dissipated per orbit. In normal spiral galaxies the resulting gas inflow is quite slow, and has a timescale of the order of the Hubble time; thus, while inflows driven by spiral structure may well be important for normal galactic evolution, very strong spiral wave activity is required for this effect to play an important role in the fueling of nuclear activity.

2.2 Bars

Closely related to the effect of spiral patterns is the effect of bar-like distortions in driving gas inflows in galaxies. For present purposes, a bar can be regarded as a particularly strong and open type of spiral pattern, and its effect on the dynamics of the gas in a galactic disk is similar to that of a spiral pattern but more extreme. In a barred galaxy the departure of the gas motions from circular orbits can become very large, and can cause strong shocks to form along the leading edge of the bar; this in turn can result in rapid energy dissipation and hence a strong inflow. Many presentations at this meeting have demonstrated the existence and the importance of such effects, both in numerical simulations and in observations, and have shown that the gas tends as a result to pile up in a ring with a double-peaked density distribution near the Inner Lindblad Resonance (if an ILR is present, as seems usually to be the case). This gas accumulation can then fuel star formation in a nuclear ring of 'hot spots', which is a common observed phenomenon in barred galaxies. The impressive agreement that has been achieved between simulations and observations of barred galaxies shows that a good understanding of their dynamics and evolution is finally being approached, at least with regard to the basic features. The gas inflow rate can be quite high when a bar first forms, but in a steady state it becomes considerably smaller and careful modeling is then required to determine it. It is clear in any case that gas inflows are much more important in barred than in unbarred galaxies, since bars are often observed to be strongly depleted of gas and star formation, both of which appear to have have been moved inward to the nuclear regions of these galaxies.

While it is clear that a bar can funnel gas efficiently from the disk of a galaxy into a central region only several hundred parsecs in radius, the presence of an ILR causes the gas to pile up at this radius and thus appears to present an obstacle to further inflow. Since the fueling of nuclear activity requires the gas to flow inward to much smaller radii, much interest has been expressed at this meeting in the question of whether similar mechanisms can also operate on smaller scales; for example, there has been much interest in the possibility that 'bars within bars' may form and continue to drive inflows to smaller radii. Observations of the innermost parts of some galaxies have provided evidence for the presence of separate nuclear bars or spirals, but there is as yet no clear understanding of the importance of such phenomena or of the mechanisms that might be responsible for producing them. One possibility is that mode coupling might play a role if the corotation radius of an inner bar coincides with the ILR of an outer bar. It is also possible that the self-gravity of the gas becomes very important in many galactic nuclei and contributes to the formation of nuclear bars or spirals; however, the self-gravity of the gas may also lead to other more complex phenomena, as are discussed next.

2.3 Dynamical Friction

A third possible mechanism for driving at least sporadic inflows in galaxies involves dynamical friction, which can cause massive clumps of matter to lose energy and angular momentum and sink rapidly toward the center. The timescale for the sinking of a clump toward the center is approximately its orbital period divided by the ratio of the clump mass to the total mass in the region; thus if sufficiently massive clumps are present in the inner part of a galaxy, either because they fall in from the outside or because they form there by instabilities, the evolution of the central region can become quite rapid and violent. A particularly dramatic example is provided by a merger of two galaxies, in which dynamical friction can cause the nuclei of the two galaxies to spiral rapidly together and merge. Massive clumps might also form as a result of gravitational instability in a nuclear disk or ring of gas if the self-gravity of the gas becomes sufficiently important, as might occur for example if gas is accumulated by bar-driven inflows. Several simulations of the dynamics of gas in galactic nuclei were reported at this meeting, and they show that such phenomena can have major effects, including not only the rapid sinking of massive clumps toward the center but also the occurrence of violent interactions between the clumps; massive clumps also tend to scatter stellar orbits and thus inhibit bar formation or disrupt existing bars.

The actual effects of gravitational instability of the gas in galactic nuclei remain uncertain, however, because the gas clumps that form may soon be disrupted by star formation, thus reducing the importance of dynamical friction and allowing more organized phenomena such as spiral waves or bars to play a more important role. In any event, the results of the various numerical simulations that have been reported here suggest that one type of non-axisymmetric disturbance or another may always occur and cause gas to flow inward, as long as enough gas is present; thus, given the variety of effects that can operate, nature may always manage to find one way or another to make the gas in a galactic nucleus more centrally concentrated.

2.4 Tidal Interactions

A final mechanism that can drive rapid gas inflows in galaxies is the effect of tidal interactions, which can be especially effective because in addition to producing a perturbation in the gravitational potential similar to that of a bar, they can induce strong responses such as spiral patterns and bars in galactic disks that contribute to the outward transport of angular momentum. In the simulations of tidal encounters that were reviewed at this meeting, the most important effect appears to be the formation of strong bars in both the stellar and the gas components of a perturbed galaxy; since the gas bar tends to lead the stellar bar, it experiences a decelerating torque that causes the gas to lose angular momentum and fall rapidly inward. A substantial fraction of the gas in a galactic disk can as a result fall into the nuclear region within a short period of time. Enormous central concentrations of gas comparable to those predicted by the simulations are in fact observed in many interacting galaxies, so it seems almost certain that the energetic activity in their nuclei has been caused by tidally driven inflows from regions of galactic size. In all, the impressive degree to which current numerical simulations of interactions and mergers reproduce many of the observed properties of active galaxies provides compelling evidence for the importance of tidal effects.

3 BLACK HOLE FEEDING

All of the dynamical phenomena that have been discussed so far occur on scales of the order of a hundred parsecs or larger, and can be understood on the basis of relatively simple physics involving only Newtonian gravity and gaseous dissipation. The fueling of nuclear starbursts, which typically occur in regions a few hundred parsecs across, may thus be explainable largely on the basis of well-understood gravitational physics. It is also possible, but presently far from clear, that similar mechanisms operate on much smaller scales and play a role in the fueling of central black holes. This possibility is supported by some of the simulations presented at this meeting which have included the self-gravity of the gas in galactic nuclei and have shown that several different effects can lead to continuing gas inflows, as discussed above. However, the results of these interesting numerical experiments cannot yet be compared directly with any observations, so that their applicability to real galaxies remains unclear. Given the likely complexity of the dynamics of the gas in galactic nuclei, further progress in understanding the many processes involved will require increasingly detailed observational studies, coupled with parallel numerical work directed toward modeling the observations and clarifying the processes involved.

The mechanisms that might be responsible for driving inflows on much smaller scales than have been discussed so far, especially scales smaller than a parsec, remain at present a matter for speculation. Accretion disk models have been considered, but they encounter some difficulties, among which is the fact that such disks are likely to be violently gravitationally unstable. A possible outcome of gravitational instability in such central gas disks is that it might lead to the formation of clumps that continue to spiral inward through dynamical friction, as discussed above. It is unclear whether this or any of the other effects that have been discussed so far could drive accretion all the way into a central black hole, although it is perhaps worth keeping in mind that it is very easy in numerical simulations of self-gravitating systems containing gas, including collapsing protogalaxies and gas-rich galactic nuclei, to end up with configurations that are as centrally condensed as is allowed by the numerical technique used.

Perhaps a more likely outcome of gravitational instability in a disk around a central black hole, as was also suggested at this meeting, is that it leads to rapid star formation, and that various feedback effects of this star formation then mediate the continuing inflow of gas into the black hole; relevant feedback effects might include the generation of turbulence and the generation of strong magnetic fields. Given the likely ionized and turbulent state of the gas near a central black hole, the generation of strong magnetic fields is perhaps inevitable, and these magnetic fields might then dominate the dynamics of the gas in the innermost region and play a major role in black hole accretion. There is evidence that strong magnetic fields are present near the centers of active galaxies (and even of our own relatively inactive galaxy), and that they play an important role in producing or collimating jet-like outflows. Again, there is a possible analogy with star formation: while protostellar accretion disks are believed to play an important role, there is evidence in many cases that these disks are truncated near the central star, and that the final step of the accretion process involves the infall of gas onto the star along magnetic field lines.

If black hole accretion is indeed driven or mediated by the feedback effects of star formation, then clearly star formation is not just an incidental accompaniment to non-thermal activity in galactic nuclei but is a necessary part of the accretion process itself. Such a causal connection between star formation and black hole accretion might provide a physical basis for the often hypothesized evolutionary connection between starburst activity and non-thermal activity in galactic nuclei, i.e. for the possible evolution of starburst nuclei into Seyfert nuclei or quasars.

4 COSMOGONICAL IMPLICATIONS

As has been noted, the processes that have been discussed play an essential role in the formation of the innermost parts of galaxies and their central black holes. Given the very general nature of the transport processes on galactic scales, which require only the presence of non-axisymmetric disturbances and gaseous dissipation, the formation of highly centrally condensed structures in galaxies seems almost inevitable. Even the formation of massive black holes may be an almost inescapable outcome of the tendency of the gas in galactic nuclei to become ever more centrally condensed. In fact, as long as dissipation and cooling can occur, the gas in a galaxy must all eventually condense into stars or into a central black hole, or it must be expelled from the system by the energy released by these processes. All three of these possible fates for the gas apparently do occur in many galaxies, and all can play important roles in galaxy formation and evolution.

The total amount of energy released by star formation in galactic nuclei is of the same order as the amount released by black hole accretion, and this makes it difficult in most cases to argue that one or the other of these sources is dominant. However, black hole accretion is about two orders of magnitude more efficient than star formation in converting mass into energy, and this means that the most luminous active galaxies are almost certainly powered mainly by black hole accretion, since star formation cannot generate enough energy to account for their luminosities from the mass present in the small region from which the luminosity is emitted. The fact that black holes only need to accrete about one percent as much mass as goes into stars in order to liberate comparable amounts of energy may help to explain why black holes account for roughly only one percent of the mass in the inner regions of galaxies: such a limit on their masses might result if black hole accretion is driven by the feedback effects of star formation, but its rate is limited by the condition that the energy released cannot greatly exceed the energy released by star formation, since otherwise the remaining gas might be blown away rather than accreted.

From a cosmic perspective, most of the phenomena that have been discussed at this meeting can be viewed as various aspects or consequences of the inexorable tendency of some part of the gas in the universe to become ever more condensed, while the energy released causes the remaining gas to become more dispersed. Thus, the processes of galaxy formation, galaxy interactions and mergers, bar and spiral dynamics, gas inflows, gravitational instabilities, star formation, magnetic field generation, black hole accretion, jet formation, etc., that have been discussed here can all be seen as playing significant roles in the evolution of the matter distribution in the universe into both more condensed and more diffuse forms, with the consequent release of energy in the densest regions that is responsible for the most spectacular observed cosmic phenomena.

SUBJECT INDEX

OBJECT INDEX

AUTHOR INDEX

Printed in the United States
By Bookmasters